**Transcription Factors
in the Nervous System**

*Edited by
Gerald Thiel*

Related Titles

Koslow, S. H., Subramaniam, S. (Eds.)
Databasing the Brain
From Data to Knowledge (Neuroinformatics)
2005
ISBN 0-471-30921-4

Hefti, F.F.
Drug Discovery for Nervous System Diseases
2004
ISBN 0-471-46563-1

Bähr, M. (Ed.)
Neuroprotection
Models, Mechanisms and Therapies
2004
ISBN 3-527-30816-4

Frings, S., Bradley, J. (Eds.)
Transduction Channels in Sensory Cells
2004
ISBN 3-527-30836-9

Joseph, J. T., Cardozo, D. L.
Functional Neuroanatomy
An Interactive Text and Manual
2004
ISBN 0-471-44437-5

Tofts, P. (Ed.)
Quantitative MRI of the Brain
Measuring Changes Caused by Disease
2003
ISBN 0-470-84721-2

Woolsey, T. A., Hanaway, J., Gado, M. H.
The Brain Atlas
A Visual Guide to the Human Central Nervous System
2002
ISBN 0-471-43058-7

Transcription Factors in the Nervous System

Development, Brain Function, and Diseases

Edited by
Gerald Thiel

WILEY-
VCH

WILEY-VCH Verlag GmbH & Co. KGaA

Editor

Prof. Dr. Gerald Thiel
Department of Medical Biochemistry
and Molecular Biology
University of Saarland Medical Center
Building 44
66421 Homburg
Germany

Cover
Neuron of a hippocampal cell culture ten days after plating. (Courtesy of Thomas Dresbach and Nils Brose.)

■ All books published by Wiley-VCH are carefully produced. Nevertheless, authors, editor and publisher do not warrant the information contained therein to be free of errors. Readers are advised to keep in mind that statements, data, illustrations, procedural details or other items may inadvertently be inaccurate.

**Library of Congress Card No.: applied for
British Library Cataloguing-in-Publication Data:**
A catalogue record for this book is available from the British Library

**Bibliographic information published by
Die Deutsche Bibliothek**

Die Deutsche Bibliothek lists this publication in the Deutsche Nationalbibliografie; detailed bibliographic data is available in the Internet at http://dnb.dbb.de.

© 2006 WILEY-VCH Verlag GmbH & Co. KGaA, Weinheim
All rights reserved (including those of translation in other languages). No part of this book may be reproduced in any form – by photoprinting, microfilm, or any other means – nor transmitted or translated into machine language without written permission from the publishers. Registered names, trademarks, etc. used in this book, even when not specifically marked as such, are not to be considered unprotected by law.

Printed in the Federal Republic of Germany.
Printed on acid-free paper.

Typesetting pagina GmbH, Tübingen
Printing   betz-druck GmbH, Darmstadt
Bookbinding Litges & Dopf Buchbinderei
            GmbH, Heppenheim
ISBN-13: 978-3-527-31285-6
ISBN-10: 3-527-31285-4

Contents

Preface XVII

List of Contributors XXI

Color Plates XXV

Part I Transcription Factors in Neural Development

1 **Roles of Hes bHLH Factors in Neural Development**
 Ryoichiro Kageyama, Jun Hatakeyama, and Toshiyuki Ohtsuka
 Abstract 3
1.1 Introduction 3
1.2 Structure and Transcriptional Activities of Hes Factors 4
1.3 Regulation of *Hes* Gene Expression 8
1.4 Expression of *Hes* Genes in the Developing Nervous System 9
1.5 Maintenance of Neural Stem Cells by *Hes* Genes 11
1.6 Promotion of Gliogenesis by *Hes* Genes 15
1.7 Maintenance of the Isthmic Organizer by *Hes* Genes 16
1.7 Perspective 16
 Acknowledgments 18
 Abbreviations 18

2 **The Role of Pax6 in the Nervous System during Development and in Adulthood: Master Control Regulator or Modular Function?**
 Nicole Haubst, Jack Favor, and Magdalena Götz
 Abstract 23
2.1 Introduction 23
2.2 Molecular Features of Pax6 26

Transcription Factors. Edited by Gerald Thiel
Copyright © 2006 WILEY-VCH Verlag GmbH & Co. KGaA, Weinheim
ISBN 3-527-31285-4

2.2.1	The Paired Domain	26
2.2.2	The Paired-Type Homeodomain	26
2.2.3	Different Pax6 Isoforms	27
2.2.4	Protein-Protein Interactions	28
2.2.5	Post-Translational Modifications of Pax6	28
2.3	Function of *Pax6* in Development	29
2.3.1	Function of *Pax6* in the Developing Eye	29
2.3.2	Function of Pax6 in the Developing Brain	31
2.3.2.1	Telencephalon	33
2.3.2.2	Diencephalon	36
2.3.2.3	Cerebellum	37
2.3.2.4	Spinal Cord	37
2.4	Function of Pax6 in the Adult Brain	38
2.5	Mechanisms of Pax6 Function	39
2.6	Conclusions and Outlook	40
	Abbreviations	41

3 Phox2a and Phox2b: Essential Transcription Factors for Neuron Specification and Differentiation
Uwe Ernsberger and Hermann Rohrer

	Abstract	53
3.1	Introduction	53
3.2	Molecular Characteristics of Phox2 Genes and Proteins	54
3.2.1	Sequence and Gene Structure Conservation in the Animal Kingdom	54
3.2.2	Transcriptional Activation by Phox2 Proteins	54
3.3	Physiological Relevance of Phox2 Transcription Factors	56
3.3.1	Expression Pattern	57
3.3.2	Effects of Phox2 Gene Mutations	58
3.3.2.1	Autonomic Neural Crest Derivatives and Visceral Sensory Ganglia	58
3.3.2.2	Central Noradrenergic Neurons	58
3.3.2.3	Autonomic Centers in the Hindbrain	59
3.3.3	Human Mutations	59
3.4	Molecular Mechanism of Action in Different Lineages	60
3.4.1	Sympathetic Neurons	60
3.4.2	Parasympathetic Neurons	61
3.4.3	Enteric Neurons	62
3.4.4	Visceral Sensory Neurons of the Geniculate, Petrosal and Nodose Ganglia	62
3.4.5	Central Noradrenergic Neurons	63
3.4.6	Autonomic Centers in the Hindbrain	64
3.4.6.1	Afferent Visceral Centers	64

3.4.6.2	Efferent Visceral Centers 64	
3.4.7	Oculomotor (nIII) and Trochlear (nIV) Centers 65	
3.5	Conclusions and Outlook 66	
3.5.1	Distinct or Identical Functions for Phox2a and Phox2b? 66	
3.5.2	Master Control Genes for Noradrenergic Differentiation 67	
3.5.3	Master Control Genes for Autonomic Reflex Circuit Generation 68	
	Acknowledgments 68	
	Abbreviations 69	

4	**Functions of LIM-Homeodomain Proteins in the Development of the Nervous System**	
	Yangu Zhao, Nasir Malik, and Heiner Westphal	
	Abstract 75	
4.1	Introduction 75	
4.2	Common Structural Features and Classification of LIM-HD Proteins 75	
4.3	LIM-HD Proteins and the Development of Invertebrate Nervous Systems 76	
4.3.1	*C. elegans* 77	
4.3.2	*Drosophila* 78	
4.4	Functions of LIM-HD Proteins in the Development of Vertebrate Nervous Systems 79	
4.4.1	The Vertebrate Spinal Cord 80	
4.4.2	The Vertebrate Brain 82	
4.4.3	The Olfactory and Visual Sensory Systems 83	
4.4.4	LIM-HD Genes and Early Patterning Events in the Developing CNS 83	
4.5	Factors that Interact with LIM-HD Proteins 85	
4.6	Downstream Targets of LIM-HD Proteins 87	
4.7	Conclusion and Future Directions 88	
	Abbreviations 88	

5	**The Roles of Serum Response Factor in Brain Development and Function**	
	Bernd Knöll and Alfred Nordheim	
	Abstract 95	
5.1	Serum Response Factor as a Transcription Factor 95	
5.2	Neuronal Expression Patterns of SRF and Partner Proteins 96	
5.3	SRF Target Genes with Brain Functions 98	
5.4	Essential Requirement for SRF in Neuronal Migration 98	
5.5	SRF and Partner Proteins in Neurite Outgrowth and Axonal Guidance 100	

5.6	SRF-Mediated Gene Expression in Learning and Memory	103
5.7	SRF in Neurological Disorders	105
5.8	Perspectives	106
	Acknowledgments	106
	Abbreviations	107

6 RE–1 Silencing Transcription Factor (REST): Regulation of Neuronal Gene Expression via Modification of the Chromatin Structure

Gerald Thiel and Mathias Hohl

Abstract   113

6.1	Tissue-Specific Gene Expression: The Molecular Basis for the Function of a Multicellular Organism	113
6.2	Modular Structure of REST	114
6.3	Biological Activity of REST	114
6.4	Mechanism of Transcriptional Repression by REST: Modulation of the Chromatin Structure	117
6.5	Lessons from the REST Knockout Mouse	120
6.6	Cell Type-Specific Regulation of REST Target Genes	122
6.7	The Role of REST in the Differentiation of Neural Stem Cells	123
6.8	Involvement of REST in Brain Dysfunction and Disease	124
6.9	Conclusion and Prospects	124
	Acknowledgments	125
	Abbreviations	125

7 Roles of Tlx1 and Tlx3 and Neuronal Activity in Controlling Glutamatergic over GABAergic Cell Fates

Qiufu Ma and Le-ping Cheng

Abstract   129

7.1	Introduction	129
7.2	The Dorsal Horn of the Spinal Cord	130
7.3	Neurogenesis in the Dorsal Spinal Cord	131
7.4	The Tlx Family of Homeobox Proteins	131
7.5	*Tlx* Gene Expression Marks Sensory Circuits	134
7.6	*Tlx* Genes Serve as Binary Switches between Glutamatergic and GABAergic Transmitter Phenotypes	134
7.7	Binary Decision between GABAergic and Glutamatergic Cell Fates is a Common Theme	136
7.8	Coupling of Generic Transmitter Phenotypes and Region-Specific Neuronal Identities	136
7.9	The Plasticity of Neurotransmitter Phenotypes	137
7.10	Summary and Unsolved Problems	138
	Abbreviations	139

8	**Transcriptional Control of the Development of Central Serotonergic Neurons**
	Zhou-Feng Chen and Yu-Qiang Ding
	Abstract 143
8.1	Introduction 143
8.2	Transcription Factors in the Development of 5–HT Neurons 146
8.3	Transcription Factors Expressed in 5–HT Progenitor Cells 146
8.3.1	*Nkx2.2* 146
8.3.2	*Mash1* 148
8.4	Transcription Factors Expressed in the Ventricular Zone and Postmitotic 5–HT Neurons 149
8.4.1	*Gata2* and *Gata3* 149
8.5	Transcription Factors Expressed in Postmitotic 5–HT Neurons 150
8.5.1	*Lmx1b* 150
8.5.2	*Pet1* 153
8.6	The Relationship between *Lmx1b* and *Pet1* 155
8.7	Conclusions 156
	Abbreviations 156

9	**Role of *Nkx* Homeodomain Factors in the Specification and Differentiation of Motor Neurons and Oligodendrocytes**
	Jun Cai and Mengsheng Qiu
	Abstract 163
9.1	Introduction 163
9.2	Structural Features of *Nkx* Homeobox Genes Involved in Ventral Neural Patterning 164
9.3	Selective Expression of *Nkx* Homeobox Genes in the Ventral Neural Tube 166
9.4	*Nkx* Genes are Class II Components of the Homeodomain Protein Code for Ventral Neural Patterning and Cell Fate Specification 168
9.5	*Nkx* Genes Control the Fate Specification and Differentiation of Motor Neurons 170
9.5.1	*Nkx6.1* and *Nkx6.2* have Redundant Activities in Promoting Somatic Motor Neuron Fate Specification 170
9.5.2	*Nkx2.2* Represses Somatic Motor Neuron Fate but Promotes Visceral Motor Neuron Fate 171
9.5.3	Nkx6 Proteins Control the Migration and Axonal Projection of Hindbrain vMN 172
9.6	The Role of *Nkx* Genes in Oligodendrocyte Development 172
9.6.1	Nkx6 Proteins Promote *Olig2* Expression and Ventral Oligodendrogenesis in the Spinal Cord 173
9.6.2	Nkx6 Proteins Suppress *Olig2* Expression and Ventral Oligodendrogenesis in the Rostral Hindbrain 173

9.6.3	*Nkx2.2* Controls the Terminal Differentiation of Oligodendrocytes *174*
9.6.4	*Nkx6.2* Homeobox Gene Regulates the Oligodendrocyte Myelination Process *175*
	Acknowledgments *175*
	Abbreviations *177*

10	**Sox Transcription Factors in Neural Development**
	Michael Wegner and C. Claus Stolt
	Abstract *181*
10.1	The Sox Family of Transcription Factors *181*
10.2	Sox Proteins and Neural Competence *182*
10.3	Sox Proteins and the Neuroepithelial Stem Cell *183*
10.4	Sox Proteins and the Neural Crest Stem Cell *185*
10.5	Sox Proteins in Neural Determination and Lineage Decisions *189*
10.6	Sox Proteins in Glial Differentiation *190*
10.7	Sox Proteins in Neuronal Differentiation *191*
10.8	Sox Proteins and their Molecular Mode of Action *193*
10.9	Conservation of Sox Protein Function in Nervous System Development *195*
	Acknowledgments *197*
	Abbreviations *197*

Part II   Transcription Factors in Brain Function

11	**The Role of CREB and CBP in Brain Function**
	Angel Barco and Eric R. Kandel
	Abstract *207*
11.1	Introduction *207*
11.2	The CREB Family of Transcription Factors *208*
11.2.1	CREB Family Members and Close Friends *208*
11.2.2	Structural Features of the CREB Family of Transcription Factors *209*
11.2.3	Gene Structure and the Regulation of Expression of CREB Family Members *211*
11.3	The CREB Binding Protein *211*
11.3.1	Structure and Multifunction *212*
11.4	The CREB Activation Pathway *212*
11.4.1	Post-Translational Regulation of CREB Activity *214*
11.4.2	Regulation of CBP Function *214*
11.4.3	Other Modulators of the CREB Pathway *216*
11.4.4	CRE-Binding Activity and CREB Downstream Genes *216*

11.5	Functions of the CREB Activation Pathway in the Nervous System	220
11.5.1	Regulation of Cellular Responses by the CREB Pathway	220
11.5.1.1	CREB is Important for Neuronal Survival and Neuroprotection	221
11.5.1.2	CREB is Required for Axonal Outgrowth and Regeneration	223
11.5.1.3	CREB has a Role in Neurogenesis and Neuronal Differentiation	223
11.5.1.5	CBP, Epigenetics and Long-Term Changes in Neuronal Function	228
11.5.2	Regulation of Systemic Responses by the CREB Pathway	229
11.5.2.1	CREB and Memory	229
11.5.2.2	CREB and Circadian Rhythms	231
11.5.2.3	CREB Function and Development	232
11.6	Dysregulation of CREB Function and Disease in the Nervous System	232
11.6.1	CREB and Addiction	233
11.6.2	Mental Retardation	233
11.6.3	CREB and Age-Related Memory Impairment	234
11.6.4	CREB and Neurodegenerative Diseases	234
11.6.4.1	Huntington Disease	234
11.6.4.2	Alzheimer's Disease	235
11.6.5	CREB and Mental Disorders: Depression and other Disorders of Mood	236
11.7	Conclusions	236
	Abbreviations	236

12	**CCAAT Enhancer Binding Proteins in the Nervous System: Their Role in Development, Differentiation, Long-Term Synaptic Plasticity, and Memory**	
	Cristina M. Alberini	
	Abstract	243
12.1	The CCAAT Enhancer Binding Proteins (C/EBPs)	243
12.2	The Role of C/EBPs in Development and Differentiation	246
12.2.1	C/EBPs Play a Critical Role in Neurogenesis	247
12.2.2	C/EBPs Play a Critical Role in Neuronal Cell Death	248
12.2.3	C/EBP Expression in Glia	249
12.3	The Role of C/EBPs in Synaptic Plasticity and Memory	250
	Abbreviations	255

13	**The Role of c-Jun in Brain Function**	
	Gennadij Raivich and Axel Behrens	
	Abstract	259
13.1	Introduction	259
13.2	C-Jun Phosphorylation and Upstream Signaling	260

13.2.1	Mitogen-Activated/Stress-Activated Protein Kinase (MAPK/SAPK) Level	260
13.2.2	MAP Kinase Kinase (MEK/MKK) and MAP Kinase Kinase Kinase (MEKK) Level	261
13.2.3	Scaffolding Proteins	261
13.2.3.1	Multimodal Effects of Deletion	262
13.3	Development	263
13.4	Novelty, Learning and Memory, and Addiction	265
13.4.1	Novelty and Pain	265
13.4.2	Learning	266
13.4.3	Addiction	266
13.5	Seizures and Excitotoxicity	267
13.6	Ischemia, Stroke, and Brain Trauma	268
13.6.1	Biochemical Regulation	268
13.6.2	Role of Jun	268
13.6.3	Functional Role of JNK Cascade	269
13.6.4	Direct Evidence	270
13.7	Axotomy	270
13.7.1	Regulation	271
13.7.2	Functional Role: Only Partial Overlap with Jun and JNK	272
13.8	Conclusions	273
	Abbreviations	273

14 Expression, Function, and Regulation of Transcription Factor MEF2 in Neurons

Zixu Mao and Xuemin Wang

	Abstract	285
14.1	Introduction	285
14.2	The MEF2 Family of Transcription Factors	285
14.2.1	MEF2 Genes and Transcripts	286
14.2.2	Structure of MEF2 Proteins	286
14.2.3	Specific Interaction Between MEF2 and DNA	288
14.3	Expression of Mef2 in Neurons	289
14.3.1	Expression of mef2 Transcripts in the Central Nervous System	289
14.3.2	Expression of MEF2 Proteins in Neurons	289
14.4	Function of Mef2 in Neurons	291
14.4.1	The Role of MEF2 in Neuronal Differentiation	291
14.4.2	The Role of MEF2 in Neuronal Survival	291
14.4.3	Regulatory Targets of MEF2 in Neurons	293
14.5	Regulation of MEF2 in Neurons	294
14.5.1	Regulation of MEF2 Transactivation Potential	294
14.5.2	Regulation of MEF2 DNA Binding	296
14.5.3	Regulation of MEF2 Stability	297

14.5.4	Regulation of MEF2 Subcellular Localization	298
14.5.5	Regulation of MEF2 by Alternative Splicing	299
14.5.6	Regulation of MEF2 by Interaction with Co-Regulators	299
14.5.7	Regulation of MEF2 by Calcium Signaling	300
14.6	Future Studies	301
	Acknowledgments	301
	Abbreviations	302

15	**RORa: An Orphan that Staggers the Mind**	
	Peter M. Gent and Bruce A. Hamilton	
	Abstract	307
15.1	Introduction	307
15.2	Identification and Biochemical Properties of RORa	308
15.2.1	Identification	308
15.2.2	Isoforms	308
15.2.3	RORa Binding and Response Elements	308
15.2.4	Crosstalk Between Factors	310
15.2.5	Ligands or Cofactors?	310
15.2.6	Co-activators	311
15.2.7	Co-repressors	311
15.2.8	Activation and Regulation of RORa Expression	312
15.2.9	RORa Expression in the Nervous System	313
15.3	Role of RORa in the Developing Cerebellum	313
15.4	Roles of RORa in Other Tissues	316
15.4.1	Suprachiasmatic Nuclei	316
15.4.2	Peripheral Tissues	317
15.5	*In-Vivo* Identification of RORa Targets	317
15.5.1	Genetic Program Controlled by RORa in the Cerebellum	317
15.5.2	Direct or Indirect Targets?	318
15.5.3	Developmental Signaling Genes	318
15.5.4	Calcium Signaling and Synaptic Function Genes	320
15.6	Implication of RORa in SCA1 Disorder	321
15.7	Summary	321
	Abbreviations	322

16	**The Role of NF-kB in Brain Function**	
	Barbara Kaltschmidt, Ilja Mikenberg, Darius Widera, and Christian Kaltschmidt	
	Abstract	327
16.1	Introduction	327
16.2	The NF-kB/Rel Family of Transcription Factors	327
16.2.1	The IkB Proteins: Inhibitors of NF-kB	328

16.3	Canonical NF-kB Activation	330
16.3.1	Activators of NF-kB	335
16.3.2	Repressors of NF-kB	337
16.3.3	Synaptic NF-kB	338
	Acknowledgments	343
	Abbreviations	343

17	**Calcineurin/NFAT Signaling in Development and Function of the Nervous System**	
	Isabella A. Graef, Gerald R. Crabtree, and Fan Wang	
	Abstract 353	
17.1	Biochemistry of NFAT Signaling 353	
17.1.1	Biochemical Basis of Coincidence Detection and Signal Integration by NFAT Transcription Complexes 353	
17.1.2	The Mechanism of Nuclear Entry of NFATc Proteins. 354	
17.1.3	Discrimination of Calcium Signals and the Nuclear Exit of NFATc Proteins 356	
17.1.4	Combinatorial Assembly of NFAT Transcription Complexes Determines Specificity of Ca^{2+} Responses. 359	
17.1.5	Dedication of CaN to NFATc Family Members 360	
17.1.6	Evolution of the Genes that Encode the Cytosolic Components, the NFATc Family 360	
17.2	Roles of NFAT Signaling in Axonal Outgrowth and Synaptogenesis 361	
17.3	A Possible Role for NFAT Signaling in Defining Pathways for Both Vessels and Peripheral Nerves 362	
17.4	Roles of NFAT Signaling in Later Development: Responses to Spontaneous Activity 366	
17.5	The Role of NFAT in Neuronal Survival 370	
17.6	Small Molecule Inhibitors of CaN are Powerful Probes of Neuronal Development 371	
17.6.1	The Mechanism of Action of FK506 and Cyclosporine A 371	
17.6.2	Use of CsA and FK506 in Studies of Neural Development and Function 372	
17.6.3	Assessing CaN Activity 372	
17.7	NFAT Signaling and Transcriptional Control in Human Disease 373	
17.7.1	Possible Defects in NFAT Signaling in Human Schizophrenia 373	
17.7.2	Down Syndrome and NFAT Signaling 373	
17.8	Conclusion 374	
	Abbreviations 374	

18	**Stimulus-Transcription Coupling in the Nervous System: The Zinc Finger Protein Egr–1**
	Oliver G. Rössler, Luisa Stefano, Inge Bauer, and Gerald Thiel
	Abstract *379*
18.1	Introduction *379*
18.2	Modular Structure of Egr–1 *379*
18.3	Intracellular Signaling Cascades Converging at the *Egr–1* Gene *381*
18.4	The Egr–1 Promoter *383*
18.5	Lessons from *Egr–1*–Deficient Mice *385*
18.6	Egr–1 Regulates Synaptic Plasticity in the Nervous System *386*
18.7	Correlation Between Proliferation of Astrocytes and Egr–1 Biosynthesis *387*
18.8	Egr–1: A "Pro-apoptotic Protein" for Neurons? *387*
18.9	Conclusions and Future Prospects *390*
	Acknowledgments *391*
	Abbreviations *391*

Part III Transcription Factors in Neuronal Diseases

19	**The Presenilin/g-Secretase Complex Regulates Production of Transcriptional Factors: Effects of FAD Mutations**
	Nikolaos K. Robakis and Philippe Marambaud
	Abstract *399*
19.1	Introduction *399*
19.2	Processing of APP and FAD *400*
19.3	The Presenilins *402*
19.4	The Notch1 ICD (NICD) Mediates Transcriptional and Developmental Functions Associated with Notch1 Receptor *403*
19.5	Transcriptional Function of the APP ICD (AICD) *404*
19.6	PS1 and b-Catenin-Mediated Transcription *405*
19.7	PS1 is a Critical Regulator of Cadherin-Dependent Cell-Cell Adhesion and Signal Transduction *406*
19.8	Conclusions *408*
	Abbreviations *408*

20	**Transcriptional Abnormalities in Huntington's Disease**
	Dimitri Krainc
	Abstract *417*
20.1	Introduction *417*
20.2	Mutant Huntingtin Interferes with Specific Components of General Transcriptional Machinery *418*

20.3 Mutant Huntingtin Disrupts Sp1–TAF4 Transcriptional Pathway 422
20.4 Deregulation of CRE-Dependent Transcription in HD 424
20.5 Summary 435
Abbreviations 435
Acknowledgments 436

Index *441*

Preface

Sequencing of the human genome, "the blueprint for life" has revealed that 5% of our genes encode transcription factors [Tupler et al., 2001], demonstrating the importance of gene regulatory proteins in the organization of life. This book provides a comprehensive overview of how transcription factors operate as key regulators for the development and function of the brain. The knowledge of the molecular structure and function of these proteins are essential for understanding how the nervous system develops and how the brain works.

The phenotype of every cell, including the cells of the nervous system, is defined by the set of active genes. Cellular diversity is a remarkable feature of the nervous system structure. There are thousands of distinct neuronal and glial cell types. This complexity excludes the existence of a single "master gene" responsible for the entire gene expression program leading to the many differentiated phenotypes. Rather, the combinatorial action of numerous transcription factors is required for the development and function of the nervous system. Research in the last years in the field of molecular neurogenetics has aimed to decipher these transcription factor codes, and this book tells some of those exciting stories.

The development of the nervous system requires tightly controlled expression of transcription factors and their target genes. The identification of transcription factors that regulate this process offers a mechanism in answer to a key question of neurobiologists, how neuronal and glial fates are determined. Along with control of the formation of neurons and glia cells from uncommitted progenitor cells, transcription factors also determine the subtype of neurons, are involved in the glial subtype determination, and play a pivotal role in neuronal migration.

In the adult nervous system, synaptic activity is a major stimulus for induction of neuronal gene transcription. The induction is mediated by transcription factors that respond to synaptic activity. These proteins have been shown to be essential for long-lasting neuronal plasticity, but are also involved in neuronal survival and differentiation. Naturally, a dysfunction of transcription factors in the nervous system has severe effects, as demonstrated by transcriptional defects in Alzheimers' and Huntington's disease. Moreover, a molecular explanation for spinocerebellar ataxia type 1 has recently been offered, involving a complex of the mutated ataxin–1 protein with the transcription factor Gfi–1 [Tsuda et al., 2005].

Transcription Factors. Edited by Gerald Thiel
Copyright © 2006 WILEY-VCH Verlag GmbH & Co. KGaA, Weinheim
ISBN 3-527-31285-4

The stimulating results of gene targeting experiments, in combination with new imaging techniques, should not let us forget that understanding of the functions of transcription factors always involves the identification of transcription factor target genes that are activated or repressed and are responsible for the phenotypic changes. In this context, the chromatin immunoprecipitation technique, a state-of-the-art method to examine transcription factor interaction *in vivo* with chromatin-packed genes, has increased our knowledge about the interaction of transcription factors with DNA in its natural chromosomal context. Moreover, the "ChIP on chip" technique, the combination of chromatin immunoprecipitation with microarray analysis [Kirmizis and Farnham, 2004], will certainly help to identify additional transcription factor targets in the genome.

A recent genome-scale transcription factor expression analysis identified over 300 transcription factors expressed in the brain of developing mice [Gray at al., 2004]. This book covers many but not all transcription factors involved in the development and function of the nervous system. The balance of death or survival of neurons, for instance, is regulated by p53 and the forkhead transcription factors. A very important issue in brain function are the Ca^{2+}-regulated signaling pathways that are initiated by the influx of Ca^{2+}-ions through L-type voltage-sensitive Ca^{2+}-channels and the NMDA receptors as a result of neuronal activity and depolarization. In addition to the here described transcription factors NFAT, CREB, NF-$\varkappa$B, MEF2 and Egr–1 the transcription factors DREAM (downstream response element antagonist modulator) and CaRF (Calcium-response factor) have to be added to the list of Ca^{2+}-responsive transcription factors, indicating that neurons have many ways to connect an increase in the intracellular calcium concentration with enhanced gene transcription. The proteins described in this book normally bind to DNA via a distinct DNA-binding domain. They can recruit other proteins, termed co-activators or co-repressors that bind via protein-protein interaction and often function via altering the chromatin structure. The discovery that transcription factors are able to recruit chromatin-modifying enzymes has revolutionized our thinking about the regulation of gene transcription [Orphanides and Reinberg, 2002] and has focussed attention on those proteins that modify the chromatin.

Many new and exciting discoveries are anticipated in the near future in the investigation of transcription factors in the nervous system. This book provides a snapshot of the current knowledge of key transcription factors that are essential for proper brain development and function. I thank my editorial partner, Dr. Andreas Sendtko at Wiley-VCH, for the initial suggestion and promotion of the project.

References

Gray PA, Fu H, Luo P, Zhao Q, Yu J, Ferrari A, Tenzen T, Yuk DI, Tsung EF, Cai Z, Alberta JA, Cheng LP, Liu Y, Stenman JM, Valerius MT, Billings N, Kim HA, Greenberg ME, McMahon AP, Rowitch DH, Stiles CD, Ma Q. 2004. Mouse brain organization revealed through direct genome-scale TF expression analysis. Science 306: 2255–2257.

Kirmizis A, Farnham PJ. 2004. Genomic approaches that aid in the identification of transcription factor target genes. Exp Biol Med 229: 705–721.

Orphanides G, Reinberg D. 2002. A unified theory of gene expression. Cell 108: 439–451.

Tsuda H, Jafar-Nejad H, Patel AJ, Sun Y, Chen H-K, Rose MF, Venken KJT, Botas J, Orr HT, Bellen HJ, Zoghbi HY. 2005. The AXH domain of ataxin–1 mediates neurodegeneration through its interaction with Gfi–1/senseless proteins. Cell 122: 633–644.

Tupler R, Perini G, Green MR. 2001. Expressing the human genome. Nature 409: 832–833.

List of Contributors

Cristina M. Alberini
Department of Neuroscience
Mount Sinai School of Medicine
New York, New York 10029
USA

Angel Barco
Instituto de Neurociencias de Alicante
(UMH-CSIC)
Campus de Sant Joan, Apt. 18
Sant Joan d'Alicant 03550
Spain

Inge Bauer
Department of Medical Biochemistry and
Molecular Biology
University of the Saarland Medical Center
Building 44
66421 Homburg
Germany

Axel Behrens
Mammalian Genetics Laboratory
Cancer Research UK
London Research Institute
Lincoln's Inn Fields Laboratories
44, Lincoln's Inn Fields
London WC2A 3PX
UK

Jun Cai
Department of Anatomical Sciences and
Neurobiology
School of Medicine
University of Louisville
Louisville, Kentucky 40292
USA

Zhou-Feng Chen
Departments of Anesthesiology, Psychiatry,
Molecular Biology, and Pharmacology,
School of Medicine
Washington University Pain Center
St. Louis, Missouri 63110
USA

Le-ping Cheng
Institute of Biochemistry and Cell Biology
Shanghai Institute for Biological Sciences
Chinese Academy of Sciences
320 Yueyang Road
Shanghai 200031
China

Gerald R. Crabtree
Howard Hughes Medical Institute
Department of Developmental Biology and
Pathology
Stanford School of Medicine
Beckman Center B211
279 Campus Drive
Stanford, California 94305–5323
USA

Yu-Qiang Ding
Laboratory of Neural Development
Institute of Neuroscience
Chinese Academy of Sciences
320 Yue Yang Road
Shanghai 200031
China

Uwe Ernsberger
ICN
Department of Neuroanatomy
University of Heidelberg
INF 307
69120 Heidelberg
Germany

Jack Favor
Institute of Human Genetics
GSF National Research Center for
Environment and Health
Ingolstädter Landstr. 1
85764 Neuherberg/Munich
Germany

Peter M. Gent
Biomedical Sciences Graduate Program
University of California
San Diego School of Medicine
9500 Gilman Drive
La Jolla, California 92093–0644
USA

Magdalena Götz
Institute of Stem Cell Research
GSF National Research Center for
Environment and Health
Ingolstädter Landstr.1
85764 Neuherberg/Munich
Germany

Isabella A. Graef
Department of Pathology and Developmental
Biology
Stanford School of Medicine
Beckman Center B209
279 Campus Drive
Stanford, California 94305–5323
USA

Bruce A. Hamilton
Department of Medicine
Division of Genetics
University of California
San Diego School of Medicine
9500 Gilman Drive
La Jolla, California 92093–0644
USA

Jun Hatakeyama
Institute of Molecular Embryology and
Genetics
Kumamoto University
Kumamoto 860–0811
Japan

Nicole Haubst
Institute of Stem Cell Research
GSF National Research Center for
Environment and Health
Ingolstädter Landstr.1
85764 Neuherberg/Munich
Germany

Mathias Hohl
Department of Medical Biochemistry and
Molecular Biology
University of the Saarland Medical Center
Building 44
66421 Homburg
Germany

Ryoichiro Kageyama
Institute for Virus Research
Kyoto University
Shogoin-Kawahara, Sakyo-ku
Kyoto 606–8507
Japan

Barbara Kaltschmidt
Institute of Neurobiochemistry
University Witten/Herdecke
Stockumer Strasse 10
58448 Witten
Germany

Christian Kaltschmidt
Institute of Neurobiochemistry
University Witten/Herdecke
Stockumer Strasse 10
58448 Witten
Germany

Eric R. Kandel
Howard Hughes Medical Institute and the
Kavli Institute
Center for Neurobiology and Behavior
College of Physicians and Surgeons of
Columbia University
1051 Riverside Drive
New York, New York 10032
USA

List of Contributors

Bernd Knöll
Department of Molecular Biology,
Interfaculty Institute for Cell Biology
University of Tübingen
Auf der Morgenstelle 15
72076 Tübingen
Germany

Dimitri Krainc
Department of Neurology
Massachusetts General Hospital
Harvard Medical School
MassGeneral Institute for
Neurodegeneration
114 16th Street
Charlestown, Massachusetts 02129
USA

Qiufu Ma
Dana-Farber Cancer Institute and
Department of Neurobiology
Harvard Medical School
1 Jimmy Fund Way
Boston, Massachusetts 02115
USA

Nasir Malik
Laboratory of Mammalian Genes and
Development
National Institute of Child Health and
Human Development
6 Center Drive
Bethesda, Maryland 20892–2790
USA

Zixu Mao
Department of Pharmacology
Center for Neurodegenerative Disease
Emory University
615 Michael Street
Atlanta, Georgia 30329
USA

Philippe Marambaud
Department of Pathology
Forchheimer Building, Room 526
Albert Einstein College of Medicine
1300 Morris Park Avenue Bronx
New York, New York 10461
USA

Ilja Mikenberg
Institute of Neurobiochemistry
University Witten/Herdecke
Stockumer Strasse 10
58448 Witten
Germany

Alfred Nordheim
Department of Molecular Biology
Interfaculty Institute for Cell Biology
University of Tübingen
Auf der Morgenstelle 15
72076 Tübingen
Germany

Toshiyuki Ohtsuka
Institute for Virus Research
Kyoto University
Shogoin-Kawahara, Sakyo-ku
Kyoto 606–8507
Japan

Mengsheng Qiu
Department of Anatomical Sciences and
Neurobiology
School of Medicine
University of Louisville
Louisville, Kentucky 40292
USA

Gennadij Raivich
Perinatal Brain Repair Group
Department of Obstetrics and Gynaecology
Department of Anatomy
University College London
Gower Street Campus
86–96 Chenies Mews
London WC1E 6HX
UK

Nikolaos K. Robakis
Mount Sinai School of Medicine
New York University
One Gustave Levy Place, Box 1229
New York, New York 10029
USA

Hermann Rohrer
Max-Planck-Institut for Brain Research
Research Group Developmental
Neurobiology
Deutschordenstr. 46
60528 Frankfurt/Mainz
Germany

Oliver G. Rössler
Department of Medical Biochemistry and
Molecular Biology
University of the Saarland Medical Center
Building 44
66421 Homburg
Germany

Luisa Stefano
Department of Medical Biochemistry and
Molecular Biology
University of the Saarland Medical Center
Building 44
66421 Homburg
Germany

C. Claus Stolt
Department of Biochemistry
University Erlangen-Nuremberg
Fahrstrasse 17
91054 Erlangen
Germany

Gerald Thiel
Department of Medical Biochemistry and
Molecular Biology
University of the Saarland Medical Center
Building 44
66421 Homburg
Germany

Fan Wang
Department of Cell Biology
Duke University Medical Center
308 Nanaline Duke Bldg, Box 3709
Durham, North Carolina 27710
USA

Xuemin Wang
Department of Pharmacology
Center for Neurodegenerative Disease
Emory University
615 Michael Street
Atlanta, Georgia 30329

Michael Wegner
Department of Biochemistry
University Erlangen-Nuremberg
Fahrstrasse 17
91054 Erlangen
Germany

Heiner Westphal
Laboratory of Mammalian Genes and
Development
National Institute of Child Health and
Human Development
6 Center Drive
Bethesda, Maryland 20892–2790
USA

Darius Widera
Institute of Neurobiochemistry
University Witten/Herdecke
Stockumer Strasse 10
58448 Witten
Germany

Yangu Zhao
Laboratory of Mammalian Genes and
Development
National Institute of Child Health and
National Institutes of Health
6 Center Drive
Bethesda, Maryland 20892–2790
USA

Color Plates

Transcription Factors. Edited by Gerald Thiel
Copyright © 2006 WILEY-VCH Verlag GmbH & Co. KGaA, Weinheim
ISBN 3-527-31285-4

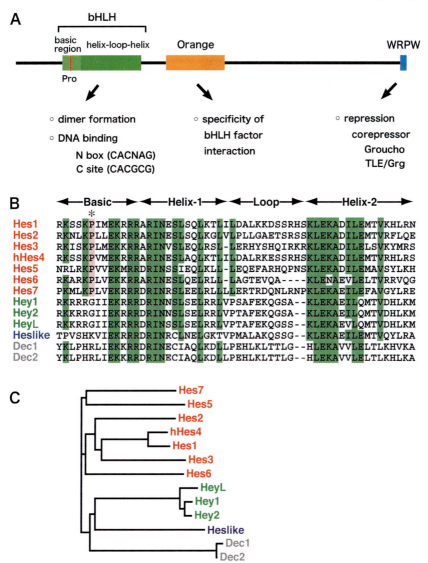

Fig. 1.2 Features of Hes bHLH factors. (A) Three conserved domains of Hes factors, the bHLH, Orange and WRPW domains. (B) Sequence alignment of the bHLH domain of Hes and related factors. Proline is conserved in the middle of the basic region of Hes factors (asterisk). (C) Phylogenetic tree of Hes and related factors. This figure also appears on page 6.

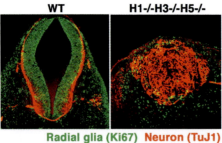

Fig. 1.7 Premature neuronal differentiation in *Hes1:Hes3:Hes5* triple knockout mice. The horizontal sections of the neural tube of mouse embryos at day 10. In the wild type, cell bodies of radial glia (Ki67$^+$) are located in the ventricular zone while neurons (TuJ1$^+$) reside in the outer layers. In the absence of *Hes1*, *Hes3* and *Hes5*, neuronal differentiation is severely accelerated. As a result, virtually all cells become neurons and neural stem cells are depleted. Adopted from Hatakeyama et al. (2004). This figure also appears on page 12.

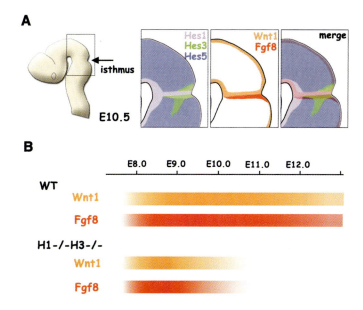

Fig. 1.11 Roles of *Hes1* and *Hes3* in the isthmic organizer. (A) Expression patterns of *Hes*, *Wnt1* and *Fgf8* genes. *Hes1* and *Hes3* are expressed in the isthmic organizer, which secrets Wnt1 and Fgf8 and specifies the midbrain and hindbrain. (B) *Wnt1* and *Fgf8* expression in the isthmic organizer. In the absence of *Hes1* and *Hes3*, the isthmic cells prematurely lose Wnt1 and Fgf8 expression and differentiate into neurons. This figure also appears on page 17.

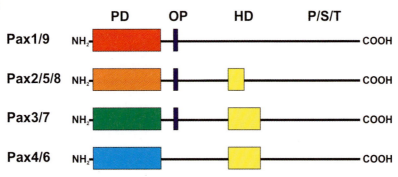

Fig. 2.1 Schematic representation of the four different vertebrate Pax gene classes. Pax1/Pax9 consist of a paired domain (PD) and an octapeptide (OP), a PST-rich transactivating domain (TAD), and are lacking the homeobox domain (HD). Pax2/5/8 contain a PD, OP and a partial HD followed by a PST-rich TAD. Pax3/7 are characterized by the presence of a PD, OP, HD and a PST-rich TAD. Pax4/6 consist of a PD followed by a complete HD and a PST-rich TAD. This figure also appears on page 24.

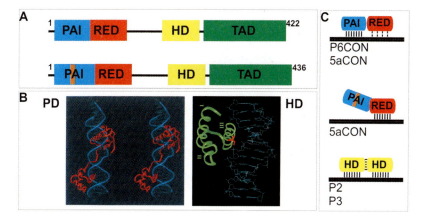

Fig. 2.2 (A) The canonical Pax6 form depicted on top (422 AA) consists of a PD, which is subdivided in an N-terminal ‚PAI' and a C-terminal ‚RED' subdomain (PAI-RED) linked to a HD followed by a TAD, whereas the Pax6(5a) isoform (436 AA) is characterized by a 14-AA insert into the PAI domain. (B) Overview of the Pax6 PD-DNA complex. Left: PD-DNA binding [DNA (blue), protein (red)]. Right: HD-DNA binding [DNA (blue), protein (green), numbers indicate the helices, red: critical AA residue at position 50 of HD] (Reproduced from [7, 42].) (C) DNA binding of the PD of the canonical Pax6 form occurs predominantly via the N-terminal PAI (blue). In the Pax6(5a) isoform, DNA-binding of the PAI is abolished and occurs exclusively via the RED domain to 5aCON sites (middle panel). The HD (yellow) binds preferentially as dimer to palindromic P2 or P3 sites containing a TAAT core sequence. (Modified after [29–31,176,177].) This figure also appears on page 26.

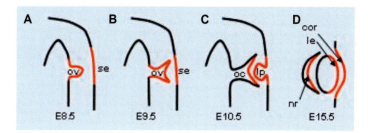

Fig. 2.3 *Pax6* expression in the developing eye (red) of the mouse [45, 78]. At earlier stages (E8.5–E9.5), *Pax6* expression is seen in the optic vesicle (ov) and the surface ectoderm (se) in the head region. As development of the eye progresses (E10.5), *Pax6* expression becomes confined to the inner layer of the optic cup (oc), the lens placode cells which form the lens pit (lp), and the immediately adjacent regions of the surface ectoderm. At E15.5, *Pax6* expression remains in the neural retina (nr), the lens epithelium (le) and the cornea (cor). This figure also appears on page 30.

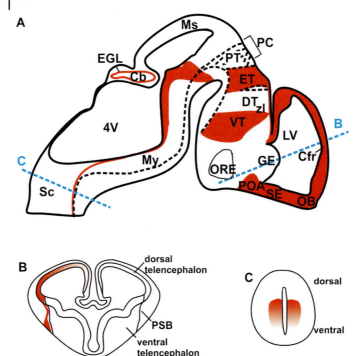

Fig. 2.5 (A) Sagittal section through the developing brain at E13.5 showing *Pax6* expression in red. The blue dashed lines indicate the planes of sections B and C, respectively. (B) Frontal section through the telencephalon showing *Pax6* expression in the cerebral cortex and in the pallial-subpallial boundary (PSB). (C) *Pax6* expression in the spinal cord occurs in ventral low to dorsal high gradient. Frontal section of the developing spinal cord with *Pax6* expression (red) in the ventral portion in a ventral^{low}-dorsal^{high} gradient. (Modified after [162].) Abbreviations: Cb = cerebellum; Cfr = frontal cortex; DT = dorsal thalamus; EGL = external granule layer; ET = epithalamus; GE = ganglionic eminence; LV = lateral ventricle; Ms = mesencephalon; My = myelencephalon; OB = olfactory bulb; ORE = optic recess; PC = posterior commissure; Pn = pons; PT = pretectum; Sc = spinal cord; SE = septum; 4V = fourth ventricle; zl = zona limitans intrathalamica. This figure also appears on page 34.

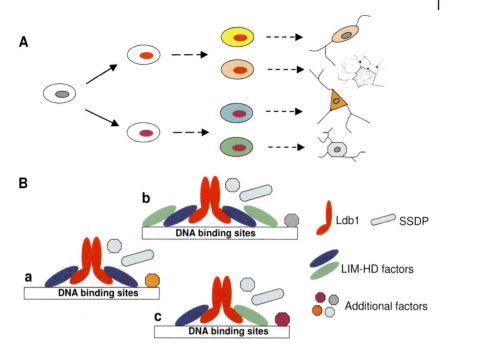

Fig. 4.2 Multiple complexes composed of LIM-HD factors, Ldb1, SSDP, and other nuclear factors are involved in the specification of diverse neuronal cell types. (A) A schematic illustration of pathways that lead from a single neural progenitor cell to various types of neurons. (B) Mediated by Ldb1, the LIM-HD factors form multiple types of complexes (a, b, c) that also include SSDP and possibly additional unidentified factors. These complexes are involved in controlling transcription by binding to specific sites in the regulatory regions of their downstream target genes. The products of these target genes are thought to specify diverse neuronal cell types in the developing nervous system. This figure also appears on page 86.

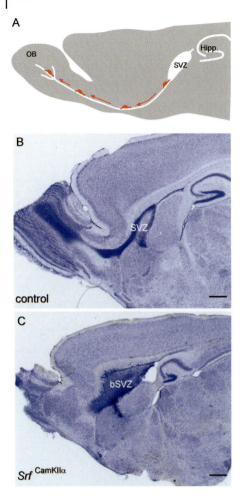

Fig. 5.2 The role of SRF in neuronal migration is best understood in the so-called rostral migratory stream (RMS), which replenishes the olfactory bulb (OB) with neurons derived from a stem cell pool localized in the subventricular zone (SVZ). (A) Migration of these neurons (shown by arrows) persists throughout the entire lifespan of organisms. (B,C) Sagittal brain sections (Nissl staining) derived from control (B) and forebrain-specific SRF-deficient mutants (C). In (B) the entire length of the RMS, from the SVZ to the olfactory bulb, is visible. By contrast, in (C) SRF ablation led to the retention of neurons in a migratory status in the SVZ, giving this brain structure an inflated or broadened (bSVZ) appearance. Consequently the number of migrating neurons entering the OB was dramatically reduced in SRF mutants. Hipp. = hippocampus. This figure also appears on page 101.

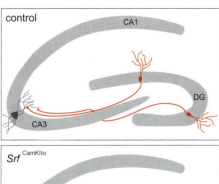
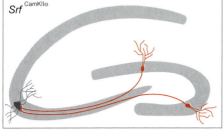

Fig. 5.3 SRF controls guidance and synaptic targeting of hippocampal mossy fibers. In control mice, mossy fibers emanating dentate gyrus (DG) granule cells bifurcate in a supra- and infrapyramidal tract. Both branches of the mossy fibers navigate precisely on either side outside the band formed by CA3 pyramidal neurons. The infrapyramidal branch crosses the CA3 stratum pyramidale at some point and joins the main suprapyramidal branch. Control mossy fiber terminals synapse with dendrites outside the layer of CA3 cell bodies. Conditional ablation of SRF function results in severe axon guidance defects. Here, mossy fibers, instead of bifurcating, grow preferentially inside the CA3 layer between individual CA3 somata. Synaptic targeting of *Srf*-deficient mossy fibers occurs aberrantly at CA3 somata and somatic protrusions (B. Knöll et al., unpublished results). This figure also appears on page 102.

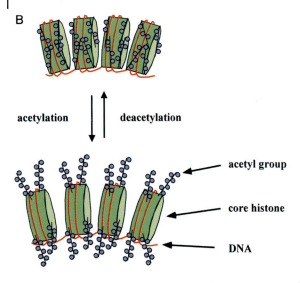

Fig. 6.3 Acetylation and deacetylation of histones determine the chromatin structure. (A) Chemical composition of the side chains of lysine and acetyl-lysine. Histone acetyltransferases (HAT) catalyze the transfer of an acetyl group from acetyl coenzyme A to the ε-amino group of internal lysine residues of histone N-terminal domains, removing the positive charge of the ε-amino group at physiological pH. Histone deacetylases (HDAC) catalyze the removal of the acetyl groups. (B) Acetylation of histones loosens the contact between DNA and the histone octamer, thus generating an open configuration of the chromatin. Deacetylation, in contrast, stabilizes the DNA/histone binding, leading to chromatin compaction. [Reproduced with modifications from Thiel and Lietz (2004) with copyright permission of Wiley-VCH.] This figure also appears on page 118.

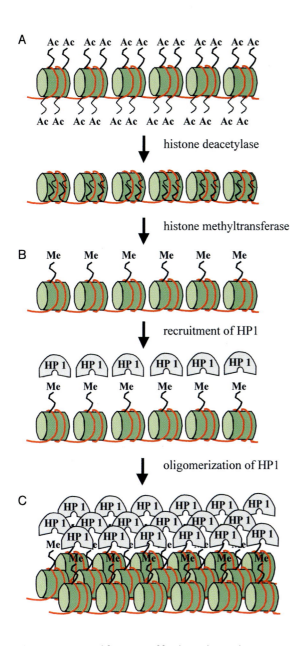

Fig. 6.5 Sequential formation of facultative heterochromatin. (A) Recruitment of histone deacetylases to the transcription unit by transcriptional repressors such as REST induces deacetylation of histone tails, thus making them suitable substrates for histone methyltransferases such as SUV39H1 and G9a. These enzymes transfer methyl groups to the ε-nitrogen of lysine residue 9 of histone H3 using S-adenosyl-L-methionine as methyl donor. (B) The methylated lysine 9 of histone H3 provides a high-affinity binding site for HP1. (C) Dimerization and oligomerization of HP1 proteins spreads the compaction of nucleosomes, forming facultative heterochromatin. [Reproduced from Thiel et al. (2004) with copyright permission of Blackwell Publishing, Oxford.] This figure also appears on page 121.

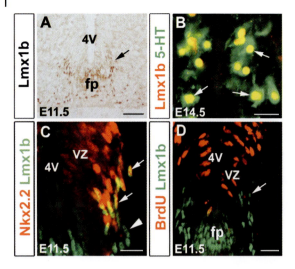

Fig. 8.2 Double staining of *Lmx1b* and 5–HT in embryonic mouse hindbrain. (A) *Lmx1b* staining in r5 detected with immunocytochemical staining. (B) *Lmx1b* (red) and 5–HT (green) double staining. Arrows indicate double-stained cells. (C) *Nkx2.2* (red) and *Lmx1b* (green) double staining. *Nkx2.2* is mainly detected in the VZ, whereas *Lmx1b* is found in postmitotic cells. Arrows indicate double-stained cells, whereas arrowhead indicates *Lmx1b*-expressing cells only. (D) BrdU (red) and *Lmx1b* (green) double staining, indicating postmitotic expression of *Lmx1b* (arrow). Scale bars: 100 µm (A); 25 µm (B); and 50 µm (C, D). Abbreviations: fp = floor plate; 4V = fourth ventricle; VZ = ventricular zone. This figure also appears on page 151.

A

Homeodomains

```
mNkx2.1   RRKRRVLFSQ AQVYELERRF KQQKYLSAPE REHLASMIHL TPTQVKIWFQ NHRYKMKRQA
mNkx2.2   KRKRRVLFSK AQTYELERRF RQQRYLSAPE REHLASLIRL TPTQVKIWFQ NHRYKMKRAR
mNkx2.9   RRKRRVLFSK AQTLELERRF RQQRYLSAPE REQLARLLRL TPTQVKIWFQ NHRYKLKRGR

mNkx6.1   RKHTRPTFSG QQIFALEKTF EQTKYLAGPE RARLAYSLGM TESQVKVWFQ NRRTKWRKKH
mNkx6.2   KKHSRPTFSG QQIFALEKTF EQTKYLAGPE RARLAYSLGM TESQVKVWFQ NRRTKWRKRH
mNkx6.3   KKHTRPTFTG HQIFALEKTF EQTKYLAGPE RARLAYSLGM TESQVKVWFQ NRRTKWRKKS
```

NK-2 domains   **NK-6 domains**

```
mNkx2.1   SPRRVAVPVL VKDGKPC (255-271)    mNkx6.1   DDDYNKPLDP NSDDEKI (324-340)
mNkx2.2   SPRRVAVPVL VRDGKPC (199-215)    mNkx6.2   DDEYNRPLDP NSDDEKI (235-251)
mNkx2.9   LLRRVMVPVL VHDRPPS (162-178)    mNkx6.3   DDEYNKPLDP DSDNEKI (225-241)
```

TN domains

```
mNkx2.1   TPFSVSDILS   (9-18)
mNkx2.2   TGFSVKDILD   (8-17)
mNkx2.9   LGFTVRSLLN   (7-16)
mNkx6.1   TPHGINDILS   (94-103)
mNkx6.2   TPHGISDILG   (56-65)
mNkx6.3   TPHGITDILS   (51-60)
```

B

Nkx-2 transcription factor structure

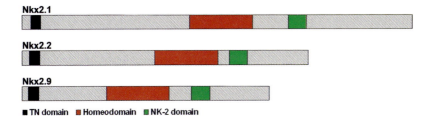

■ TN domain ■ Homeodomain ■ NK-2 domain

Nkx-6 transcription factor structure

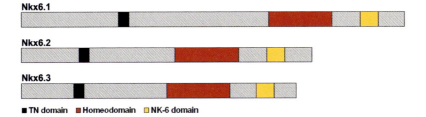

■ TN domain ■ Homeodomain ☐ NK-6 domain

Fig. 9.1 (A) Sequence similarities of homeodomains and other conserved motifs of the vertebrate *Nkx2* and *Nkx6* homeobox genes. (B) Positions and relative lengths of the homeodomains and other conserved motifs in *Nkx* genes. This figure also appears on page 165.

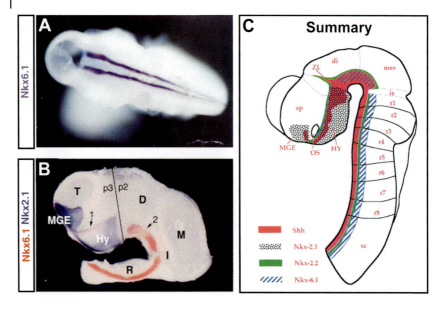

Fig. 9.2 (A,B) Exemplary expression of *Nkx* genes in the ventral neural tube. Neural tube tissue from E4 chicken embryos were subjected to whole-mount *in-situ* RNA hybridization with *Nkx6.1* (A) or simultaneously with *Nkx6.1* and *Nkx2.1* (B). (C) Schematic representation of *Nkx* expression in the developing central nervous system. Abbreviations: D (di) = diencephalon; Hy (HY) = hypothalamus; is = isthmus; M (mes) = mesencephalon; MGE = medial ganglionic eminence; OS = optic stalk; R (r) = rhombomere; sc = spinal cord; sp = secondary prosemere; T = telencephalon; ZL = zona limitans intrathalamica; p2, p3 = progenitor domains. This figure also appears on page 167.

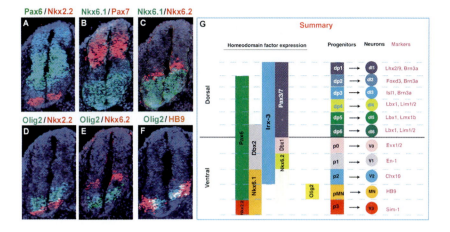

Fig. 9.3 (A-F) Expression of *Nkx* genes in relation to other transcription factors in E10.5 mouse spinal cord. Tissues were subjected to double immunofluorescent staining with antibodies against Nkx proteins and other transcriptions factors. Nuclei were counterstained with DAPI (in purple). (G) Schematic illustration of the homeodomain code that specifies the identity of spinal neural progenitor cells. Nested expression of homeodomain transcription factors subdivides the ventral neuroepithelium into five distinct progenitor domains (p0–p3, pMN) and the dorsal neuroepithelium into six domains (dp1–6). While the ventral progenitor domains give rise to five classes of neurons (V0–V3 ventral interneurons and motor neurons), the dorsal progenitors generate six classes of dorsal interneurons (dI1–6). Different classes of postmitotic neurons can be readily identified by their expression of unique combination of other transcription factors, mostly homeodomain proteins. This figure also appears on page 169.

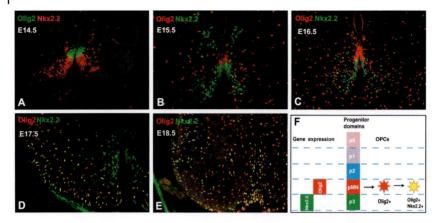

Fig. 9.4 Up-regulation of *Nkx2.2* in *Olig2*+ oligodendrocyte progenitor cells (OPCs) in the spinal cord. (A-E) Spinal cord sections from various stages of rat embryos were subjected to double immunofluorescence staining with anti-Olig2 and anti-Nkx2.2. Prior to E17.5, Olig2 is expressed in OPCs, whereas Nkx2.2 labels p3 progenitor cells and possibly V3 interneurons in the ventral gray matter. Starting at E17.5, Olig2+ OPCs in the white matter start to co-express Nkx2.2 (the double-positive cells are labeled as yellow). (F) Schematic representation of the origin and gene expression profile of OPCs in the ventral spinal cord. This figure also appears on page 175.

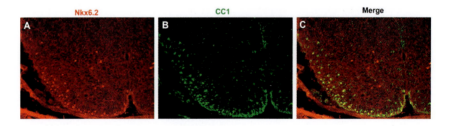

Fig. 9.5 *Nkx6.2* is expressed in differentiated oligodendrocytes in postnatal spinal cord. P4 mouse spinal cord was subjected to double immunostaining with (A) anti-Nkx6.2 and (B) anti-APC antibody (CC1). In the white matter region, all Nkx6.2+ cells co-express APC which specifically labels differentiated oligodendrocytes. This figure also appears on page 176.

Molecular pathways in MN and OL development

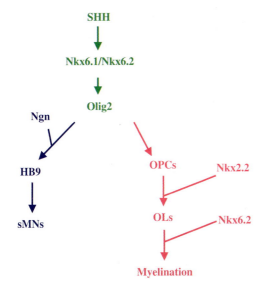

Fig. 9.6 A summary of the major molecular pathways in the specification and differentiation of motor neurons (MNs) and oligodendrocytes (OLs) generated in the ventral spinal cord. OPCs = oligodendrocyte progenitor cells. This figure also appears on page 177.

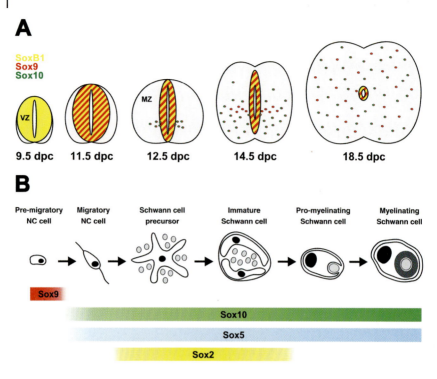

Fig. 10.2 Sox protein expression in the developing mouse nervous system. (A) Spinal cord. Areas expressing Sox2 are marked in yellow; regions or cells expressing Sox9 are highlighted in red. Green indicates Sox10–expressing cells. Areas labeled in yellow (or yellow and red) correspond to the ventricular zone (VZ) with its neuroepithelial progenitors; labeled cells in the mantle zone (MZ) correspond to astrocytes (red) or oligodendrocytes (red and green). Time points correspond to days of mouse embryogenesis post-coitum (dpc). (B) Schwann cell lineage. Various phases of Schwann cell development from pre-migratory neural crest (NC) stem cell to terminally differentiated myelinating Schwann cell are indicated. Bars indicate expression periods for several important Sox proteins. This figure also appears on page 184.

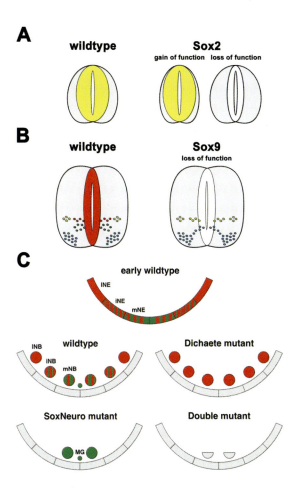

Fig. 10.3 Sox protein function in nervous system development. (A) Sox2 overexpression in the chicken neural tube leads to expansion of the ventricular zone, whereas loss of Sox2 function causes premature differentiation and a reduction of pluripotent neuroepithelial progenitors. (B) Loss of Sox9 in the mouse spinal cord reduces generation of oligodendrocytes (green dots) and astrocytes (red dots). Instead, motor neurons (blue dots) and V2 interneurons (yellow dots) are increased in numbers. (C) The *Drosophila* SoxB proteins SoxNeuro (red) and Dichaete/fish-hook (green) show an overlapping expression in the early neuroectoderm (NE) from which neuroblasts (NB) arise. In the SoxNeuro mutant, many lateral (lNB) and intermediate (iNB) neuroblasts are missing, whereas only midline glia (MG) are lost in the Dichaete/fish-hook mutant. Deletion of both SoxNeuro and Dichaete/fish-hook in the double mutant additionally leads to an increased deletion of medial neuroblasts (mNB) pointing to functional redundancy of both SoxB proteins. Abbreviations: lNE = lateral neuroectoderm; iNE = intermediate neuroectoderm; mNE = medial neuroectoderm. This figure also appears on page 186.

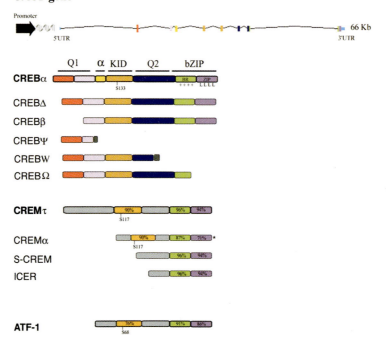

Fig. 11.1 *Creb1* gene structure and domain organization of the CREB family of transcription factors. The members of the CREB family of transcription factors have a highly conserved leucine zipper (ZIP) and adjacent basic region responsible for DNA-binding (BR), a regulatory kinase inducible domain (KID), and two glutamine-rich regions, Q1 and Q2, which contribute to constitutive transcription activation and are less conserved among different family members. The percentages of similar amino acids in CREM and ATF1 with the corresponding bZIP and KID domains of CREB are indicated. The most relevant isoforms of these proteins and represented and the locations of some important sites are indicated and discussed in the text. The *Creb* and *Crem* genes encode both activator and repressor variants. The upper part of the figure shows the exonic organization of the *Creb* gene, only the exons encoding domains present in the most relevant forms of CREB are highlighted in color. This figure also appears on page 209.

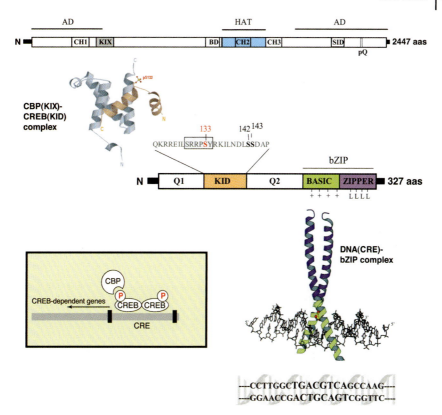

Fig. 11.2 Critical molecular interactions in the CREB activation pathway. CREB has a highly conserved leucine zipper and adjacent basic region responsible for binding to CRE sites and a regulatory kinase inducible domain (KID) that, once phosphorylated, interacts with the KIX domain of CBP. The interactions between the KID and KIX domain and the bZIP domain and the CRE sequence are known with atomic details and are represented here using ribbon structural models. The location of some important domains and sites in CREB and CBP structure are labeled and discussed in the text. This figure also appears on page 210.

A. A bridge between transcription factors that bind to DNA and the RNApol II complex.

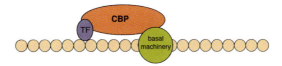

B. A scaffold bringing together different proteins to the promoter.

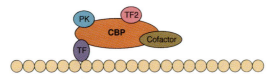

C. Histone acetyltransferase that adds Acetyl groups to histones in nucleosomes.

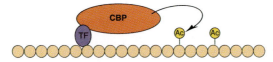

D. Acetyl transferase that adds Acetyl groups to interacting transcription factors.

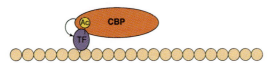

Fig. 11.3 The multiple functions of CBP. The capability of CBP and p300 to co-activate transcription depends on four different activities. TF = transcription factor; PK = protein kinase; Ac = acetyl group. This figure also appears on page 213.

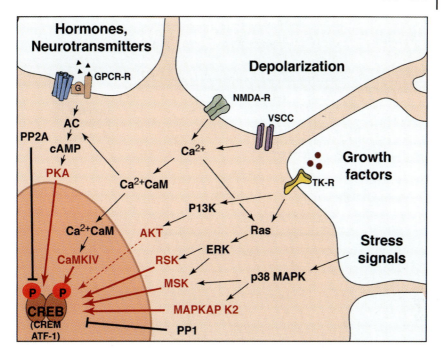

Fig. 11.4 Activation of the CREB signaling pathway. Diverse external stimuli, such as activation of receptors coupled to adenylyl cyclase (AC), such as the G protein-coupled receptors (GPC-R), or opening of Ca^{2+} channels (NMDA-R and VSCC, voltage-sensitive calcium channels), activate protein kinases pathways that converge on CREB phosphorylation at Ser133. Phosphorylation of this residue promotes the recruitment of the co-activator CBP and initiates the transcription of targets genes. However, this is an extremely simplified vision of these regulatory processes. Multiple layers of complexity in the CREB pathway allow the integration of diverse cytoplasmic signals and the divergence of nuclear responses (cartoon adapted from Lonze and Ginty (2002) [9]). This figure also appears on page 215.

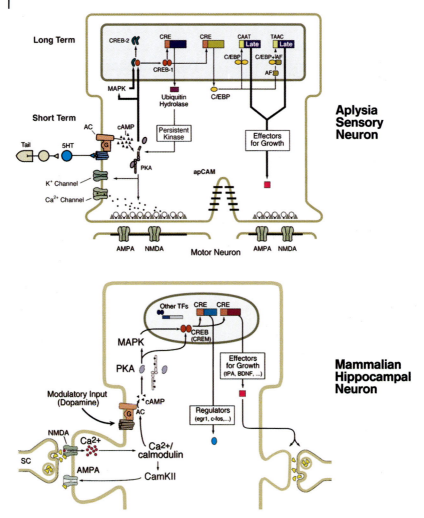

Fig. 11.6 Molecular signaling for short-term and long-term synaptic facilitation in *Aplysia* and mammalian hippocampal neurons. Molecular details are discussed in the text. This figure also appears on page 225.

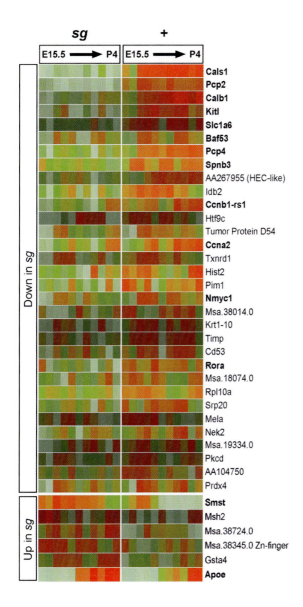

Fig. 15.4 RORα-dependent gene expression in developing cerebellum. Microarray data from Gold et al. [24] are shown here in an alternate view. Whole-cerebellum RNA from paired *staggerer* (sg) and nonmutant littermates of several developmental stages was analyzed on GeneChip arrays (Affymetrix). Data are normalized across all hybridizations for each gene. Red indicates a relative increase and green a relative decrease in expression compared to the mean of all measurements. (See Gold et al. [24] for details of data handling and rank ordering of RORα-responsive genes.) This figure also appears on page 319.

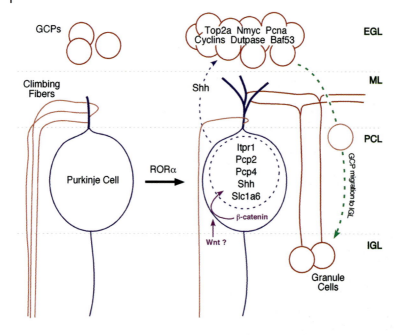

Fig. 15.5 RORα coordinates reciprocal signaling between Purkinje cells and afferent neurons. Shh signaling from Purkinje cells stimulates proliferation of granule cell precursors (GPCs) in the external granule cell layers. Granule cells migrate through the molecular layer (ML) and Purkinje cell layer (PCL) to the internal granule cell layer (IGL). At the same time, RORα activates genes necessary for receiving innervation from granule cells and reduction of supernumerary synapses from climbing fibers. This figure also appears on page 320.

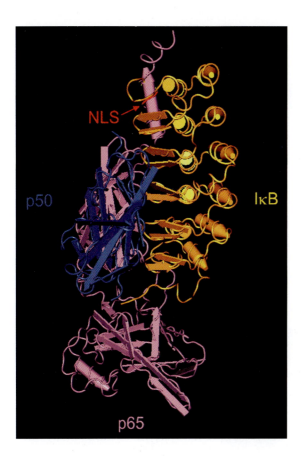

Fig. 16.2 Three-dimensional model of a co-crystal containing IκB, p50, and p65. The alpha-helical conformation of the p65 NLS (marked by an arrow) is due to an interaction with IκB-α. After degradation of IκB-α, the NLS loses its alpha-helical conformation and can be recognized by the nuclear import machinery via interaction with importin-α. Drawn after pbd: molecule 1IKN (Huxford et al., 1998). This figure also appears on page 330.

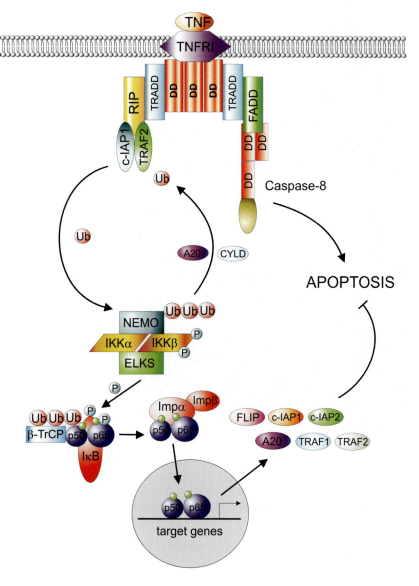

Fig. 16.3 Canonical pathway of NF-κB activation by tumor necrosis factor (TNF). The anti-apoptosis (NF-κB) and caspase-mediated apoptosis pathway is shown. Activation of the TNF receptor by ligand binding is transmitted to the IKK complex, which phosphorylates IκB family inhibitory molecules (see text for details). This targets IκB for degradation within the 26S proteasome, freeing nuclear localization signals on the DNA-binding p65/p50 subunits. After nuclear import, target gene transcription is initiated. Not all signaling components depicted in the canonical pathway have been investigated in the nervous system, but appear to be present in neurons and glia. Proteins are depicted as icons which illustrate a functional category (receptor, enzyme etc.) as suggested by the Alliance for Signalling convention (www.signaling-gateway.org). Ub = ubiquitination; P = phosphorylation. This figure also appears on page 332.

Color Plates | LV

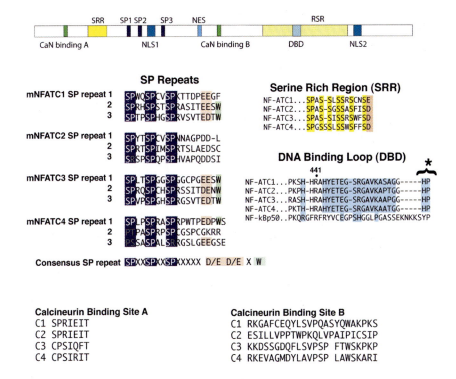

Fig. 17.2 The critical sequences within the NFATc family of proteins that mediate its response to calcineurin and its rapid export from the nucleus. The N-terminal domain of the protein is necessary and sufficient to allow Ca^{2+}/calcineurin-dependent nuclear import and GSK3–dependent export and functions as a potent dominant negative of NFAT function [8]. The SP-repeat and serine-rich regions [24] contain most of the phosphorylation sites for GSK3 and PKA, which are in turn dephosphorylated by calcineurin. The two calcineurin binding sites that probably account for the dominant negative effects of the N-termini are shown as sites A and B. This figure also appears on page 357.

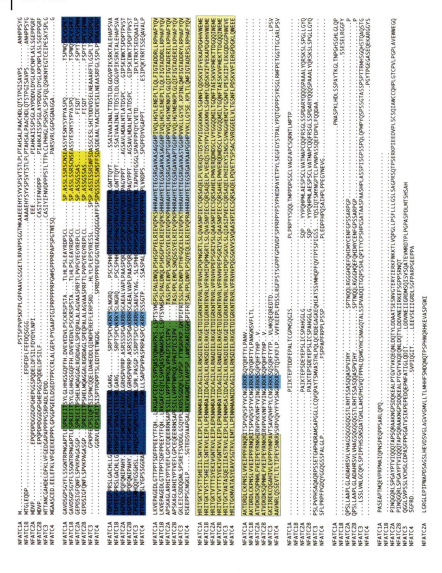

Fig. 17.3 Alignment of the sequences of the four NFATc family members and their splice products. The coloring of the different domains in the proteins are taken from Fig. 17.2. This figure also appears on page 358.

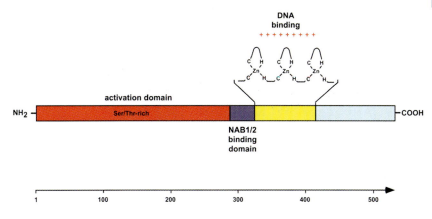

Fig. 18.1 Modular structure of the zinc finger transcription factor Egr–1. The Egr–1 protein contains an extended transcriptional activation domain on the N-terminus and a DNA-binding domain, consisting of three zinc finger motifs. Additionally, an inhibitory domain has been mapped between the activation and DNA-binding domains that functions as a binding site for the transcriptional co-repressor proteins NAB1 and NAB2. This figure also appears on page 381.

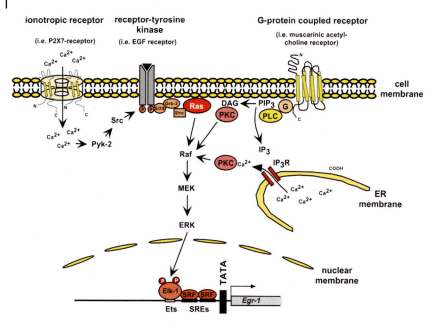

Fig. 18.2 Intracellular signaling pathways leading to Egr–1 biosynthesis. Ligand binding to receptor tyrosine kinases leads to receptor dimerization and intracellular *trans*-autophosphorylation of key tyrosine residues. The phosphotyrosyl residues function as docking sites for SH2–containing adapter proteins such as Grb2 (growth factor-receptor-bound 2). The nucleotide exchange factor Sos (son-of-sevenless) is recruited and activates the G-protein Ras. GTP-bound Ras in turn activates the protein kinase Raf via recruitment to the plasma membrane, leading to the sequential phosphorylation and activation of the protein kinases MEK and ERK. Ligands that bind to G-protein-coupled receptors (GPCR) stimulate ERK activation via activation of protein kinase C or transactivation of the EGF receptor. Protein kinase C can directly or indirectly stimulate the activity of Raf via phosphorylation. Transactivation of the EGF receptor may be accomplished by cytosolic tyrosine kinases of the src-family or via the activation of membrane-bound metalloproteinases. Likewise, an increase in the intracellular Ca^{2+}-concentration as a result of $P2X_7$-receptor stimulation triggers Egr–1 biosynthesis via transactivation of the EGF receptor and activation of ERK. This figure also appears on page 382.

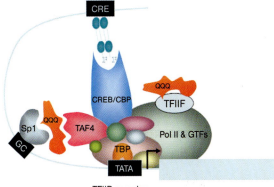

Fig. 20.1 Model of potential mechanisms used by mutant huntingtin (QQQ) to disrupt Sp1– and CRE-mediated transcription. In normal cells, transcription factors Sp1 and CREB bind to GC-box and CRE sequences, respectively. Sp1 and CREB/CBP target TAF4 and recruit TFIID and other components of the general transcription machinery to form a productive preinitiation complex. In HD cells, mutant huntingtin may target multiple components of the general transcription machinery for repression. First, mutant huntingtin can sequester Sp1 and prevent it from binding to GC-box sequences in the promoter. Second, mutant huntingtin can target TAF4 in the TFIID complex, and therefore impair the recruitment of TFIID by Sp1 and CREB. Third, mutant huntingtin disrupts the TFIIF complex formation and thus interferes with transcription initiation, promoter escape, and elongation. This figure also appears on page 425.

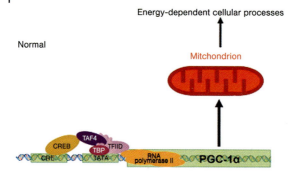
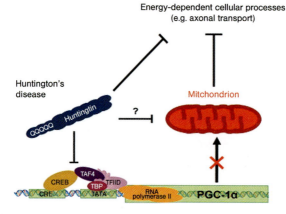

Fig. 20.2 Model for regulation of PGC–1α in Huntington's disease (HD). Upper panel: In a normal state, PGC–1α regulates metabolic programs and maintains energy homeostasis in the CNS. Lower panel: In HD, mutant huntingtin interferes with CREB and TAF4 regulation of PGC–1α transcription that leads to inhibited expression of PGC–1α. Inhibition of PGC–1α expression limits the ability of the vulnerable neurons to adequately respond to energy demands in HD. Direct interactions of mutant huntingtin with mitochondria may also contribute to defects in energy metabolism in HD. This figure also appears on page 434.

Part I
Transcription Factors in Neural Development

1
Roles of Hes bHLH Factors in Neural Development

Ryoichiro Kageyama, Jun Hatakeyama, and Toshiyuki Ohtsuka

Abstract

Hes genes, mammalian homologues of *Drosophila hairy* and *Enhancer of split* genes, encode basic helix-loop-helix (bHLH) transcriptional repressors. There are seven members in the *Hes* family, among which *Hes1*, *Hes3*, and *Hes5* are expressed by embryonic neural stem cells. Mutations in these *Hes* genes lead to up-regulation of proneural bHLH gene expression and concomitantly premature neuronal differentiation. As a result, neural stem cells are prematurely depleted without proliferating sufficiently, and without generating later born cell types such as astrocytes and ependymal cells. In addition, premature depletion of neural stem cells leads to the disruption of brain structures, because these cells constitute a framework by forming the inner and outer barriers. Thus, *Hes* genes regulate the generation of cells not only in the correct number but also in their full diversity and in an organized manner by maintaining neural stem cells. At later stages, *Hes* genes promote gliogenesis. In contrast, proneural bHLH genes induce *Hes6*, which antagonizes *Hes1* activity and promotes neuronal specification. This antagonistic regulation between *Hes1/3/5* and proneural bHLH genes is important for the normal timing of neural stem cell differentiation. *Hes* genes are also required for maintenance of the isthmic organizer, which specifies the midbrain and hindbrain by secreting morphogens. Thus, *Hes* genes regulate formation of complex brain structures with appropriate size, shape, cytoarchitecture and specification by controlling neural stem cells and the organizing center.

1.1
Introduction

The neural plate consists of neuroepithelial cells, and these cells divide symmetrically to produce more neuroepithelial cells (Fig. 1.1). This cell type is the earliest form of embryonic neural stem cells (Alvarez-Buylla et al., 2001; Fujita, 2003). After neural tube formation, neuroepithelial cells become radial glial cells, which have a cell body in the ventricular zone and long radial fibers extending from the internal surface to the pial (outer) surface of the neural tube (Fig. 1.1). This cell type was long thought of as a specialized glial cell that guides neuronal migration along the radial fibers, but it has been recently shown that radial glia are embryonic neural stem cells. Radial glial cells divide asymmetrically, forming one radial glial cell and one neuron (or a neuronal precursor) from each division. Neurons migrate along the radial fibers to the outer layers. Radial glial cells are later differentiated into ependymal cells, which form the internal lining of the neural tube (Fig. 1.1) (Spassky et al., 2005). After production of neurons, radial glial cells give rise to oligodendrocytes and finally to astrocytes (Fig. 1.1). At around the time of birth, the radial glial cells disappear, but recent studies have shown that some astrocytes or astrocyte-like cells are neural stem cells, which remain in the adult brain. Thus, neural stem cells change their characteristics in both morphology and competency during development (Alvarez-Buylla et al., 2001; Fujita, 2003). Because it takes a certain period of time for neural stem cells to change their characteristics, maintenance of these cells until later stages is required for generation of cells not only in the correct number but also in their full diversity.

Another important aspect of neural development is that the nervous system is partitioned into several compartments such as the midbrain and hindbrain. These compartments are divided by specialized boundary cells, which secrete morphogens and specify the adjacent compartments, thus acting as the organizing center. For example, the isthmus, the boundary demarcating the midbrain and hindbrain, secretes Wnt1 and Fgf8 and thereby regulates midbrain and hindbrain development (the isthmic organizer, see Fig. 1.11) (Lumsden and Krumlauf, 1996; Joyner et al., 2000; Mason et al., 2000; Wurst and Bally-Cuif, 2001). Premature loss of the isthmus leads to mis-specification of the midbrain and hindbrain neurons. Thus, maintenance of the boundary cells is very important for development of region-specific neurons.

Recent studies have shown that *Hes* genes, which encode basic helix-loop-helix (bHLH) transcriptional repressors, play a critical role in maintenance of both neural stem cells and boundary cells in the developing nervous system. In the absence of *Hes* genes, neural stem cells and boundary cells are prematurely lost, leading to severe impairment of neural development. In this chapter, we describe the structures, expression, regulation and functions of Hes factors in neural development.

1.2
Structure and Transcriptional Activities of Hes Factors

Hes genes are mammalian homologues of *Drosophila hairy* and *Enhancer of split* [*E(spl)*] genes, which negatively regulate neural development (Akazawa et al., 1992; Sasai et al., 1992; Feder et al., 1993). There are seven members in the *Hes* family (Fig. 1.2B), among which *Hes1* and *Hes4* are more similar to *hairy* in structure while the other members are more similar to *E(spl)*. There are also several *Hes*-related bHLH genes such as *Hesr/Hey/HRT/Herp/CHF/Gridlock* (Iso et al., 2001) and *Heslike* (Miyoshi et al., 2004), which form distinct subfamilies (Fig. 1.2B,C). Among the *Hes*

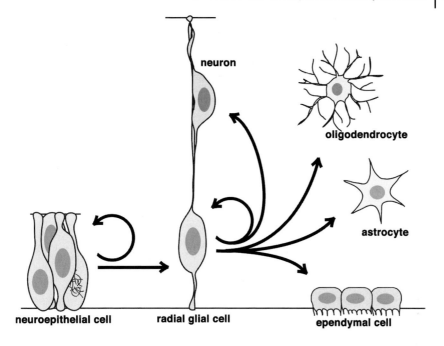

Fig. 1.1 Change of morphology and competency of neural stem cells during development. Neuroepithelial cells divide symmetrically to increase the cell number. After neural tube formation, neuroepithelial cells become radial glial cells, which have a cell body in the ventricular zone and long processes extending from the internal to the outer surface. Radial glial cells divide asymmetrically to produce one radial glial cell and one neuron (or neuronal precursor) from each division. Neurons migrate along the radial fibers to the outer layers. After neurogenesis, radial glial cells give rise to oligodendrocytes, ependymal cells and finally astrocytes.

family members, *Hes1*, *Hes3* and *Hes5* are expressed by neural stem cells in mouse developing nervous system. Each Hes factor has the following three conserved domains: the bHLH domain; the Orange domain (the helix 3–helix 4 domain); and the WRPW (Trp-Arg-Pro-Trp) domain, which are essential for transcriptional activities (Fig. 1.2A).

The bHLH domain involves dimer formation and DNA binding. bHLH factors form homodimers and heterodimers through the HLH domain while binding to DNA targets via their basic regions. Strikingly, a proline residue is conserved in the middle of the basic region of all Hes factors as well as of *Drosophila* Hairy and E(spl) proteins (Fig. 1.2A,B, asterisk). It has been suggested that this proline residue may be involved in the specificity of the target DNA sequences, although the exact significance of this conservation remains to be determined. Hes1 binds to the N box (CACNAG) and the class C site (CACGCG) with a higher affinity than to the E box (CANNTG) (Sasai et al., 1992; Chen et al., 1997), unlike most other bHLH factors,

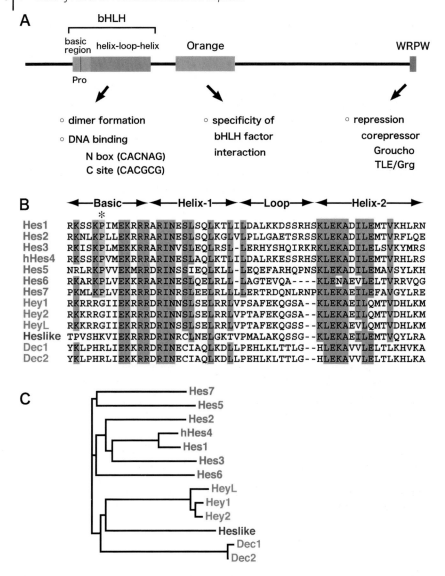

Fig. 1.2 Features of Hes bHLH factors. (A) Three conserved domains of Hes factors, the bHLH, Orange and WRPW domains. (B) Sequence alignment of the bHLH domain of Hes and related factors. Proline is conserved in the middle of the basic region of Hes factors (asterisk). (C) Phylogenetic tree of Hes and related factors. (This figure also appears with the color plates.)

which bind to the E box with a higher affinity. Hes1 can bind to these sites not only as a homodimer (Fig. 1.3A) but also as a heterodimer with Hes-related bHLH factors such as Hesr (Iso et al., 2001). In contrast, when Hes1 forms a heterodimer with other bHLH factors such as Mash1 and E47, these heterodimers do not bind to DNA (non-functional heterodimers) (Fig. 1.3B) (Sasai et al., 1992). The HLH factor Id, which lacks the basic region (Benezra et al., 1990), also forms a heterodimer with Hes1, but this heterodimer does not bind to DNA either (Jögi et al., 2002).

The Orange domain, located just downstream of the bHLH domain, is suggested to consist of two amphipathic helices (see Fig. 1.2A). This domain is shown to confer specificity for bHLH factor interactions (Dawson et al., 1995; Taelman et al., 2004). For example, the Hes-related bHLH factor Hairy interacts with the bHLH factor Scute efficiently, while another Hes-related bHLH factor, E(spl)m8, does not, and this difference in the interaction specificity maps to the Orange domain (Dawson et al., 1995). This domain is also known to mediate transcriptional repression (Castella et al., 2000), although a corepressor interacting with this domain is not known.

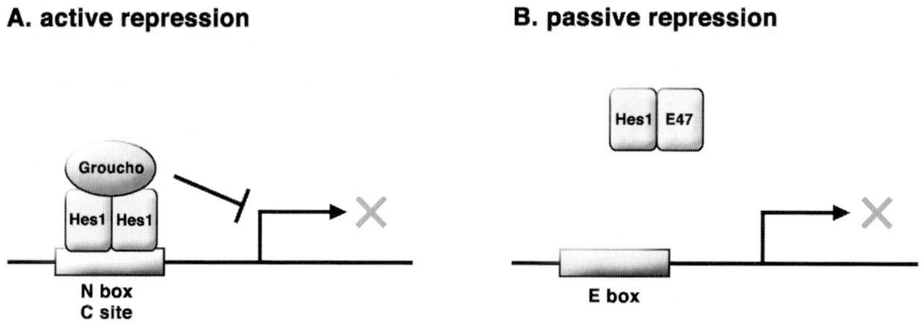

Fig. 1.3 Two different mechanisms of repression by Hes factors. (A) Active repression. Hes binds to the N box and the class C site. The corepressor Groucho/TLE/Grg interacts with the WRPW domain of Hes factors and actively inhibits the chromatin by recruiting the histone deacetylase. (B) Passive repression. Many bHLH activators form a heterodimer with E47 and activate gene expression. Hes inhibits bHLH activators by sequestering E47.

The WRPW domain is located at or near the carboxyl terminus (see Fig. 1.2A). This domain acts as a repression domain and interacts with the corepressor TLE/Grg, a homologue of *Drosophila* Groucho (Paroush et al., 1994; Fisher et al., 1996; Grbavec and Stifani, 1996). It is suggested that Groucho mediates long-range transcriptional repression that can function over distances of several kilobases in *Drosophila* embryos (Zhang and Levine, 1999). Groucho modifies the chromatin structure by recruiting the histone deacetylase Rpd3 and thereby actively represses transcription (called "active repression"; Fig. 1.3A) (Chen et al., 1999). In addition to Groucho-mediated "active repression", Hes1 represses transcription by forming nonfunctional heterodimers with bHLH activators such as Mash1 and E47, as described above (called "passive repression"; Fig. 1.3B). Thus, Hes1 represses transcription by two different mechanisms.

Hes1 activity is controlled post-translationally. The serine residues in the basic region can be phosphorylated by protein kinase C (PKC) in a phosphatidyl serine- and ATP-dependent manner, and the phosphorylated Hes1 cannot bind to DNA (Ström et al., 1997). A similar effect is also induced by protein kinase A. This inhibition of the DNA-binding activity is reversible by treatment with phosphatase. Another important feature for post-translational modification is polyubiquitination (Hirata et al., 2002). Polyubiquitinated Hes1 is soon degraded by the proteasome system, and thus Hes1 protein is very unstable (the half-life is about 22 min) (Hirata et al., 2002).

Hes1 and Hes5 have the same conserved domains described above. However, Hes3 has two different forms generated by alternative splicing (Hirata et al., 2000). One (Hes3b) has a complete basic region like Hes1 and Hes5, while the other (Hes3a) lacks the amino-terminal half of the basic region. Hes3a thus cannot bind to DNA, but can inhibit other bHLH factors by forming nonfunctional heterodimers, whereas Hes3b can do both. Like Hes1 and Hes5, Hes3b is expressed by neural stem cells while Hes3a is expressed by differentiating or mature Purkinje neurons in the cerebellum, indicating that Hes3a has a different role in neuronal differentiation from Hes1, Hes3b and Hes5 (Sasai et al., 1992; Hirata et al., 2001). In this chapter, Hes3b is designated as Hes3.

1.3
Regulation of *Hes* Gene Expression

One of the best characterized signaling pathways that regulate *Hes* gene expression is Notch signaling (Honjo, 1996; Kageyama and Nakanishi, 1997; Artavanis-Tsakonas et al., 1999; Gaiano and Fishell, 2002). Notch, a transmembrane protein, is activated by the ligands Delta and Jagged, which are also transmembrane proteins expressed by neighboring cells (Fig. 1.4). Upon activation, Notch is processed to release the intracellular domain (ICD), which is transferred into the nucleus and forms a complex with the DNA-binding protein RBP-J. In the *Hes1* promoter, there are two tandem repeats of the RBP-J sites (the core sequence: TGGGAA) at nucleotide positions −70 and −84 (relative to the transcription start site). The *Hes5* promoter also has two RBP-J sites at nucleotide positions −77 and −293. RBP-J itself is a transcriptional repressor and represses *Hes1* and *Hes5* expression by binding to their promoters (Fig. 1.4). However, when RBP-J forms a complex with Notch ICD, this complex becomes a transcriptional activator and induces *Hes1* and *Hes5* expression (Fig. 1.4). Thus, Notch activation leads to up-regulation of *Hes1* and *Hes5* expression (Jarriault et al., 1995; Nishimura et al., 1998). Notch is known to inhibit neuronal differentiation and to maintain neural stem cells (Gaiano et al., 2000). In the absence of *Hes1* and *Hes5*, however, Notch fails to inhibit neuronal differentiation, indicating that *Hes1* and *Hes5* are essential effectors of Notch signaling (Ohtsuka et al., 1999).

1.3 Regulation of Hes Gene Expression

In contrast to *Hes1* and *Hes5*, there is no evidence that *Hes3* expression is controlled by Notch signaling (Nishimura et al., 1998). Moreover, initial *Hes1* expression occurs before *Notch* and *Delta* are expressed, indicating that Notch signaling is not the sole regulator of *Hes* expression. Interestingly, the core sequence of the RBP-J site (TGGGAA) is identical to that of the NF-κB half-site (Brou et al., 1994), and *Hes1* expression is activated by NF-κB signaling while being repressed by IκB, an endogenous inhibitor of NF-κB (Aguilera et al., 2004). In addition, *Hes1* expression is regulated by cAMP. There are several half-sites (TGAC) of the cAMP-responsive element (CRE) in the *Hes1* promoter, and CRE-binding protein phosphorylated by cAMP-dependent protein kinase induces *Hes1* expression (Herzig et al., 2003). BMP, Shh and Wnt signaling pathways have also been shown to induce *Hes* expression (Issack and Ziff, 1998; Nakashima et al., 2001; Solecki et al., 2001). Thus, *Hes* expression is controlled by multiple signaling pathways, in addition to Notch signaling.

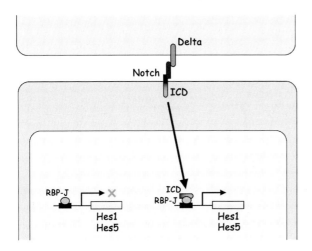

Fig. 1.4 Notch signaling. Delta expressed by neighboring cells activates Notch. Upon activation, the intracellular domain (ICD) of Notch is cleaved off the membrane portion and transferred into the nucleus. In the nucleus, the ICD forms a complex with RBP-J. RBP-J alone represses *Hes1* and *Hes5* expression by binding to their promoters, but when it forms a complex with the ICD, this complex is a transcriptional activator and induces *Hes1* and *Hes5* expression.

Another striking feature of the regulation of *Hes* expression is that Hes1 forms a negative feedback loop. There are multiple N box sequences in the *Hes1* promoter, and Hes1 can repress its own transcription by directly binding to these N boxes (Fig. 1.5) (Takebayashi et al., 1994). *Hes1* promoter activation leads to the synthesis of Hes1 protein, which in turn represses its own transcription by binding to the N boxes. When *Hes1* transcription is repressed, Hes1 protein disappears rapidly because it is very unstable (see above), allowing the next round of *Hes1* promoter activation. As a result, Hes1 autonomously exhibits oscillatory expression with a periodicity of 2 h (Fig. 1.5), indicating that Hes1 functions as a biological clock with a 2–h cycle (Hirata et al., 2002). This oscillatory expression of Hes1 is widely observed in many cell types, including neural progenitors. In the presomitic mesoderm, Hes7 expression oscillates with a 2–h periodicity and regulates somite segmentation, which occurs every 2 h in mouse embryos. In the absence of *Hes7*, somites are severely fused, indicating that Hes7 is an essential component of the somite segmentation clock (Bessho et al., 2001, 2003; Hirata et al., 2004). The significance of the Hes1 clock in neural development, however, remains to be determined.

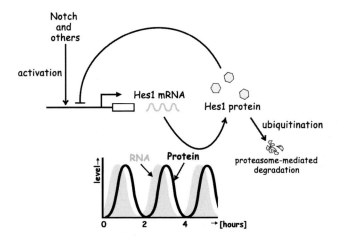

Fig. 1.5 Oscillatory expression of Hes1 by a negative feedback loop. Hes1 seems to act as a 2–h cycle biological clock in many systems. Activation of *Hes1* promoter leads to synthesis of *Hes1* mRNA and Hes1 protein, which in turn represses its own transcription by binding to the *Hes1* promoter (negative feedback loop). Because Hes1 protein is polyubiquitinated and degraded by the proteasome system, Hes1 protein disappears rapidly when the transcription is repressed. Disappearance of Hes1 protein allows the next round of the *Hes1* promoter activation. As a result, Hes1 autonomously exhibits oscillatory expression with a periodicity of 2 h.

1.4
Expression of *Hes* Genes in the Developing Nervous System

At the initial neuroepithelial cell stage, *Hes1* and *Hes3* are widely expressed by neural stem cells (Allen and Lobe, 1999; Hatakeyama et al., 2004). However, *Hes3* expression is gradually down-regulated in the ventral part of the neural tube (Fig. 1.6) and later disappears from most regions at the radial glial stage except for the isthmus, the boundary between the midbrain and hindbrain (Allen and Lobe, 1999; Hirata et al., 2001). As *Hes3* expression is down-regulated, *Hes5* expression is up-regulated (Fig.

1.6). Up-regulation of *Hes5* expression coincides with that of *Notch* and *Delta* expression, suggesting that *Hes5* expression is controlled by Notch signaling while initial *Hes1* and *Hes3* expression is not. *Hes1* expression is maintained even after *Hes3* expression is repressed, and it is likely that *Hes1* expression at later stages may depend on Notch signaling.

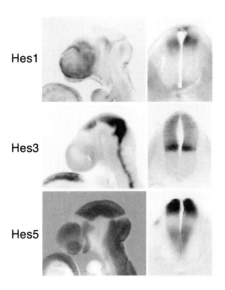

Fig. 1.6 Expression patterns of *Hes1*, *Hes3* and *Hes5* in the developing nervous system. *Hes1* and *Hes3* expression domains are overlapped, while *Hes1* and *Hes5* expression domains are mostly complementary to each other. Adopted from Hatakeyama et al. (2004).

During the early stages of radial glial cells, *Hes1* and *Hes5* expression is mostly complementary to each other (Fig. 1.6) (Hatakeyama et al., 2004). For example, *Hes5* is strongly expressed in the midbrain and hindbrain but not in the isthmus, while *Hes1* is expressed in the isthmus (Fig. 1.6; see also Fig. 1.11). In the optic vesicle at early stages, *Hes1* but not *Hes5* is expressed. Interestingly, in the absence of either *Hes1* or *Hes5*, expression of the remaining *Hes* genes is up-regulated in many regions. For example, *Hes5* expression occurs ectopically in both the isthmus and the optic vesicles of *Hes1*–null embryos. Similarly, *Hes1* and *Hes5* expression domains are expanded in the spinal cord of *Hes5*–null and *Hes1*–null embryos, respectively. These results suggest that *Hes1* and *Hes5* may functionally compensate for each other. At later stages, the apparent complementary expression patterns of *Hes1* and *Hes5* are lost, and these genes seem to be coexpressed by many neural stem cells.

1.5
Maintenance of Neural Stem Cells by *Hes* Genes

Roles for *Hes* genes in neural stem cells have been investigated by gain-of-function and loss-of-function experiments. Mis-expression of *Hes1*, *Hes3* or *Hes5* in the embryonic brain inhibits neuronal differentiation and maintains radial glial cells (Ishibashi et al., 1994; Hirata et al., 2000; Ohtsuka et al., 2001). Hes1 is known to repress expression of the proneural gene *Mash1* by binding to the class C site of the *Mash1* promoter (Chen et al., 1997). Conversely, in the absence of *Hes1* and *Hes5*, many radial glial cells are not maintained and prematurely differentiate into neurons (Ishibashi et al., 1995; Tomita et al., 1996; Ohtsuka et al., 1999; Cau et al., 2000; Nakamura et al., 2000; Hatakeyama et al., 2004). In *Hes1:Hes5* double-mutant embryos, expression of the proneural bHLH genes *Mash1* and *Math3* is highly up-regulated, which may lead to premature neuronal differentiation (Hatakeyama et al., 2004). Furthermore, $Hes1^{(-/-)}:Hes5^{(-/-)}$ neurospheres do not expand properly even in the presence of bFGF and EGF, in contrast to the wild-type neurospheres, which proliferate extensively (Ohtsuka et al., 2001). Thus, *Hes1* and *Hes5* regulate maintenance of neural stem cells by preventing premature onset of the proneural bHLH gene expression in the embryonic brain.

In *Hes1:Hes5* double-mutant embryos, some radial glial cells are still maintained, suggesting that *Hes3* may compensate for *Hes1* and *Hes5* deficiency. Agreeing with this notion, in the absence of *Hes1*, *Hes3* and *Hes5*, even neuroepithelial cells are prematurely differentiated into neurons, in contrast to the wild-type neuroepithelial cells, which are never differentiated into neurons (Hatakeyama et al., 2004). Furthermore, virtually all radial glial cells are prematurely differentiated into neurons and become depleted without generating the later-born cell types (later born neurons, oligodendrocytes, astrocytes and ependymal cells) (Fig. 1.7) (Hatakeyama et al., 2004). Thus, *Hes1*, *Hes3* and *Hes5* are essential to generate cells in correct numbers and in their full diversity by maintaining neural stem cells until later stages.

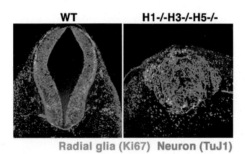

Fig. 1.7 Premature neuronal differentiation in *Hes1:Hes3:Hes5* triple knock-out mice. The horizontal sections of the neural tube of mouse embryos at day 10. In the wild type, cell bodies of radial glia (Ki67$^+$) are located in the ventricular zone while neurons (TuJ1$^+$) reside in the outer layers. In the absence of *Hes1*, *Hes3* and *Hes5*, neuronal differentiation is severely accelerated. As a result, virtually all cells become neurons and neural stem cells are depleted. Adopted from Hatakeyama et al. (2004). (This figure also appears with the color plates.)

Strikingly, even in *Hes1:Hes3:Hes5* triple-mutant mice, the neuroepithelial cells are initially formed, indicating that formation of neural stem cells is independent of *Hes* gene activities (Fig. 1.8). However, in the absence of *Hes* genes, neuroepithelial cells and radial glial cells are prematurely differentiated, indicating that their maintenance depends on *Hes* gene activities (Hatakeyama et al., 2004). Neural stem cells thus change their characteristics over time as follows: *Hes*-independent neuroepithelial cells, *Hes*-dependent neuroepithelial cells, and *Hes*-dependent radial glial cells (Fig. 1.8). Based on their expression patterns, *Hes1* and *Hes3* are important for the maintenance of neuroepithelial cells, while *Hes1* and *Hes5* are required for most radial glial cells.

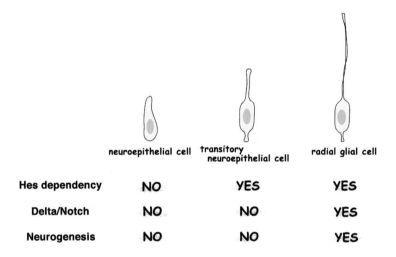

Fig. 1.8 Change of characteristics of neural stem cells. Neuroepithelial cells are formed independently of *Hes* genes, but their maintenance critically depends on *Hes* genes and not on Delta/Notch. Radial glial cells depend on Delta, Notch and Hes activities. Thus, neural stem cells change their characteristics as follows: *Hes*-independent neuroepithelial cells, *Hes*-dependent neuroepithelial cells, and *Hes*-dependent radial glial cells.

It has been shown that there are at least two types of neural stem cells depending on the developmental stage, namely primitive and definitive (Hitoshi et al., 2004). Definitive neural stem cells are derived from later stages and depend on Notch signaling, while primitive neural stem cells are from earlier stages and do not depend on Notch signaling: rather, they depend on LIF signaling (Hitoshi et al., 2004). Because initial *Hes1* and *Hes3* expression in neuroepithelial cells is not controlled by Notch signaling, this expression could be controlled by LIF signaling or by a related signaling pathway. Primitive neural stem cells thus could be *Hes*-dependent or *Hes*-independent neuroepithelial cells, while definitive neural stem cells could be *Hes*-dependent radial glial cells.

Hes-related bHLH genes, *Hesr1* and *Hesr2*, are also expressed by neural stem cells in the embryonic brain, and mis-expression of *Hesr1* or *Hesr2* promotes maintenance of neural stem cells (Sakamoto et al., 2003). *Hesr* expression is also controlled by Notch signaling, and Hesr and Hes proteins form heterodimers and act as repressors (Iso et al., 2001). Thus, it is possible that Hesr and Hes cooperatively regulate maintenance of neural stem cells.

Cytokine signaling is known to regulate neural stem cells. In response to the activation of cytokine receptors, JAK2 phosphorylates tyrosine residues of STAT3, and this phosphorylated STAT3 can promote maintenance of neural stem cells. Interestingly, JAK2–STAT3 signaling depends on Notch signaling (Kamakura et al., 2004). The Notch effectors Hes1 and Hes5 physically interact with both JAK2 and STAT3, and this complex facilitates the phosphorylation and activation of STAT3 by JAK2 (Kamakura et al., 2004), thus highlighting the significance of the cross-talk between the Notch-Hes and JAK-STAT pathways in neural stem cells.

Neural stem cells are essential for neural development because they give rise to all cell types, but the analysis of *Hes*-mutant mice has revealed another important function of this cell type. Both neuroepithelial cells and radial glial cells have epithelial features: they carry the tight junction and adherens junction at the apical side and form the basal lamina at the basal side (Fig. 1.9). These apical and basal structures constitute the inner and outer barriers of the neural tube, respectively. In the absence of *Hes* genes, both the apical and basal structures are disrupted due to the premature loss of neural stem cells, leading to spilling of neurons into the lumen as well as into the surrounding tissues (Fig. 1.9) (Hatakeyama et al., 2004). Therefore, neural stem cells are essential for the structural integrity of the nervous system. In wild-type embryos, by the time neural stem cells disappear, the ependymal cells are differentiated at the apical side and form the apical junctional complex, while astrocytes are differentiated and contribute to the basal lamina formation at the basal side. *Hes* genes are required to maintain neural stem cells until formation of ependymal cells and astrocytes and are thus essential for their structural integrity.

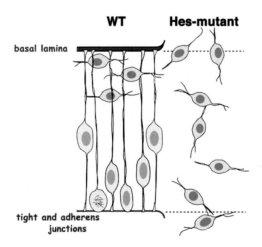

Fig. 1.9 Epithelial features of radial glia. Radial glial cells carry the tight junction and adherens junction at the apical side and form the basal lamina at the basal side. These apical and basal structures constitute the inner and outer barriers of the neural tube, respectively. In the absence of *Hes* genes, both the apical and basal structures are disrupted due to premature loss of neural stem cells, leading to spilling of neurons into the lumen as well as into the surrounding tissues.

Proneural bHLH genes such as *Mash1* override the inhibitory activities of *Hes* genes and promote neuronal differentiation. This process involves another member of the *Hes* family, *Hes6*. Hes6 can form a heterodimer with Hes1, but this complex does not bind to DNA (Bae et al., 2000). Furthermore, Hes6 was shown to inhibit the interaction between Hes1 and Groucho/TLE/Grg and induce degradation of Hes1 protein (Gratton et al., 2003). As a result, Mash1 is relieved from Hes-induced inhibition. Thus, Hes6 inhibits Hes1 but supports Mash1 and promotes neuronal differentiation in the developing brain and retina (Fig. 1.10) (Bae et al., 2000; Koyano-Nakagawa et al., 2000; Gratton et al., 2003). *Hes6* expression is induced by proneural bHLH genes such as *Neurogenin* (Fig. 1.10) (Koyano-Nakagawa et al., 2000). Thus, the proneural bHLH genes inhibit *Hes1/3/5* genes by inducing *Hes6*, while *Hes1/3/5* genes inhibit the proneural bHLH genes, indicating that these bHLH genes regulate each other in a mutually antagonistic manner.

1.6
Promotion of Gliogenesis by *Hes* Genes

At later stages, when gliogenesis occurs, *Hes1* and *Hes5* are transiently expressed by astrocytes in the developing brain (Nakashima et al., 2001; Wu et al., 2003) and by Müller glial cells in the developing retina (Hojo et al., 2000; Furukawa et al., 2000). Mis-expression of *Hes1* and *Hes5* at later stages promotes generation of astrocytes in the brain and Müller glial cells in the retina (Fig. 1.10) (Hojo et al., 2000; Furukawa et al., 2000; Ohtsuka et al., 2001; Takatsuka et al., 2004). Conversely, in the absence of *Hes1* or *Hes5*, production of Müller glial cells is decreased (Hojo et al., 2000; Furukawa et al., 2000; Takatsuka et al., 2004). Thus, *Hes1* and *Hes5* are involved in gliogenesis at later stages, indicating that *Hes* genes exhibit different activities depending on their developmental stage: maintenance of neural stem cells at early stages and promotion of gliogenesis at later stages. It remains to be determined whether *Hes1* and *Hes5* instruct neural stem cells to adopt a glial fate at later stages, or whether *Hes1* and *Hes5* just maintain neural stem cells until the gliogenic phase. Interestingly, it has been shown that the proneural bHLH gene *Neurogenin1* (*Ngn1*) has two activities: promotion of neurogenesis and inhibition of gliogenesis (Sun et al., 2001). Ngn1 sequesters the CBP-Smad1 transcriptional complex away from the glial-specific promoters and recruits the complex to the neuronal-specific promoters, thereby promoting neurogenesis while inhibiting alternative fates. Conversely, inactivation of the proneural genes *Mash1*, *Ngn2* and *Math3* blocks neurogenesis while enhancing gliogenesis (Tomita et al., 2000; Nieto et al., 2001). Thus, suppression of the proneural genes could be one of the major mechanisms for *Hes1*– and *Hes5*–induced gliogenesis.

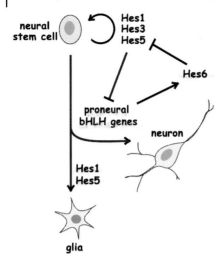

Fig. 1.10 The bHLH gene network in neural development. *Hes1*, *Hes3* and *Hes5* repress proneural bHLH gene expression and maintain neural stem cells. In contrast, proneural bHLH genes induce Hes6, which inhibits Hes1 and promotes neuronal differentiation. Hes1/Hes5–expressing cells finally become glial cells.

In the postnatal retina, *Hesr2*, but not *Hesr1* or *Hesr3*, is specifically expressed by Müller glial cells (Satow et al., 2001). Furthermore, mis-expression of *Hesr2* promotes generation of Müller glial cells (Satow et al., 2001). It is thus possible that Hesr2 regulates gliogenesis by forming a heterodimer with Hes1 or Hes5.

1.7
Maintenance of the Isthmic Organizer by *Hes* Genes

As development proceeds, the nervous system is partitioned into several compartments, which are demarcated by boundary cells. These cells secrete morphogens and specify the adjacent compartments, thus acting as organizing centers. One such example is the isthmus, the boundary between the midbrain and hindbrain (Lumsden and Krumlauf, 1996; Joyner et al., 2000; Mason et al., 2000; Wurst and Bally-Cuif, 2001). The isthmic cells secrete Wnt1 and Fgf8 and regulate development of the midbrain and hindbrain (Fig. 1.11). The isthmic cells express *Hes1* and *Hes3* (Fig. 1.11A) and do not give rise to any neurons. In the absence of *Hes1* and *Hes3*, however, the isthmic cells prematurely lose Wnt1 and Fgf8 expression (Fig. 1.11B) and are ectopically differentiated into neurons (Hirata et al., 2001). As a result, the midbrain and hindbrain neurons are not properly specified. For example, oculomotor and trochlear nuclei and dopaminergic neurons of the midbrain and locus ceruleus neurons of the hindbrain are missing in *Hes1:Hes3* double-mutant embryos (Hirata et al., 2001). Thus, *Hes1* and *Hes3* are essential for maintenance of the isthmic organizer and development of the midbrain and hindbrain. Similar functions have been reported for *Hes*-related bHLH genes *Her5* and *Her11/Him* in zebrafish isthmus (Ninkovic et al., 2005).

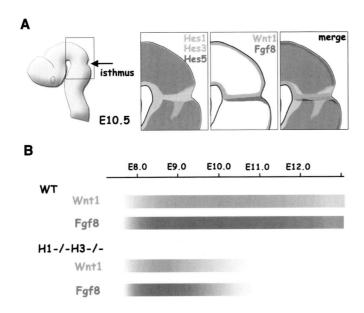

Fig. 1.11 Roles of *Hes1* and *Hes3* in the isthmic organizer. (A) Expression patterns of *Hes*, *Wnt1* and *Fgf8* genes. *Hes1* and *Hes3* are expressed in the isthmic organizer, which secrets Wnt1 and Fgf8 and specifies the midbrain and hindbrain. (B) *Wnt1* and *Fgf8* expression in the isthmic organizer. In the absence of *Hes1* and *Hes3*, the isthmic cells prematurely lose Wnt1 and Fgf8 expression and differentiate into neurons. (This figure also appears with the color plates.)

1.8
Perspective

It has by now become clear that the bHLH genes *Hes1*, *Hes3* and *Hes5* regulate the maintenance of virtually all neural stem cells. However, many questions are still unaddressed, regarding Hes functions in neural stem cells. First, neural stem cells have two important activities – self-renewal and differentiation – but it is not known how coordinately neural stem cells undertake these two different activities. The persistent expression of Hes1 induces self-renewal but inhibits differentiation, while loss of Hes1 induces differentiation but inhibits self-renewal. Thus, both persistent expression and loss of expression of *Hes1* impairs neural stem cell activities, suggesting that dynamic changes of Hes1 expression seem to be important for neural stem cells. Because Hes1 expression oscillates with a 2–h periodicity, this oscillatory expression might help perform the self-renewal and differentiation activities coordinately. For example, at the cell cycle checkpoint for self-renewal versus differentiation, the cells at high levels in Hes1 oscillation could adopt self-renewal, while those at low levels could adopt differentiation. Clearly, further studies are required to substantiate this model.

Second, while *Hes1*, *Hes3* and *Hes5* functionally compensate for each other, it remains to be determined whether these three *Hes* genes have the same activities in neural stem cells. During early stages, *Hes1* and *Hes3* are expressed, but at later stages, *Hes3* is down-regulated while *Hes5* is up-regulated. It is not known why *Hes3* and *Hes5* expression is switched during development. Expression of *Hes1* and *Hes3* at early stages is independent of Notch signaling while that of *Hes5* (and probably *Hes1* also) is dependent on Notch signaling at later stages. Furthermore, *Hes1/Hes5*–expressing radial glial cells give rise to neurons while *Hes1/Hes3*–expressing neuroepithelial cells do not, raising the possibility that *Hes1/Hes5* at later stages allow neurogenesis while *Hes1/Hes3* at early stages do not. Thus, *Hes* genes could have different activities.

During embryogenesis, *Hes*-expressing cells remain as neural stem cells but finally become astrocytes in the brain and Müller glia in the retina, indicating that *Hes* genes exhibit different activities depending on developmental stages. These different activities could be due to other factors that are coexpressed with Hes, but those factors specific for neural stem cells and glial cells remain to be analyzed.

Another important issue is that although *Hes* genes are essential for the maintenance of neural stem cells, they are not required for the initial formation of neural stem cells. The initial formation of neural stem cells might be regulated by other *Hes*-related bHLH genes or by totally different factors. Further analysis of *Hes* and related bHLH genes is definitely required to understand the molecular dynamics of the regulation of neural stem cells.

Acknowledgments

These studies were supported by research grants from the Ministry of Education, Culture, Sports, Science and Technology of Japan and Japan Society for the Promotion of Science. J.H. was supported by Research Fellowships of the Japan Society for the Promotion of Science for Young Scientists.

Abbreviations

bHLH	basic helix-loop-helix
CRE	cAMP-responsive element
ICD	intracellular domain
PKC	protein kinase C

References

Aguilera C, Hoya-Arias R, Haegeman G, Espinosa L, Bigas A. 2004. Recruitment of IκBα to the *hes1* promoter is associated with transcriptional repression. *Proc Natl Acad Sci USA* 101:16537–16542.

Akazawa C, Sasai Y, Nakanishi S, Kageyama R. 1992. Molecular characterization of a rat negative regulator with a basic helix-loop-helix structure predominantly expressed in the developing nervous system. *J Biol Chem* 267:21879–21885.

Allen T, Lobe CG. 1999. A comparison of *Notch*, *Hes* and *Grg* expression during murine embryonic and post-natal development. *Cell Mol Biol* 45:687–708.

Alvarez-Buylla A, Garcia-Verdugo JM, Tramontin AD. 2001. A unified hypothesis on the lineage of neural stem cells. *Nat Rev Neurosci* 2:287–293.

Artavanis-Tsakonas S, Rand MD, lake RJ. 1999. Notch signaling: cell fate control and signal integration in development. *Science* 284:770–776.

Bae S, Bessho Y, Hojo M, Kageyama R. 2000. The bHLH gene *Hes6*, an inhibitor of *Hes1*, promotes neuronal differentiation. *Development* 127:2933–2943.

Benezra R, Davis RL, Lockshon D, Turner DL, Weintraub H. 1990. The protein Id: a negative regulator of helix-loop-helix DNA binding proteins. *Cell* 61:49–59.

Bessho Y, Sakata R, Komatsu S, Shiota K, Yamada S, Kageyama R. 2001. Dynamic expression and essential functions of *Hes7* in somite segmentation. *Genes Dev* 15:2642–2647.

Bessho Y, Hirata H, Masamizu Y, Kageyama R. 2003. Periodic repression by the bHLH factor Hes7 is an essential mechanism for the somite segmentation clock. *Genes Dev* 17:1451–1456.

Brou C, Logeat F, Lecourtois M, Vandekerckhove J, Kourilsky P, Schweisguth F, Israël A. 1994. Inhibition of the DNA-binding activity of *Drosophila* Suppressor of Hairless and its human homolog, KBF2/RBP-Jκ, by direct protein-protein interaction with *Drosophila* Hairless. *Genes Dev* 8:2491–2503.

Castella P, Sawai S, Nakao K, Wagner JA, Caudy M. 2000. HES–1 repression of differentiation and proliferation in PC12 cells: role for the helix 3–helix 4 domain in transcription repression. *Mol Cell Biol* 20:6170–6183.

Cau E, Gradwohl G, Casarosa S, Kageyama R, Guillemot F. 2000. *Hes* genes regulate sequential stages of neurogenesis in the olfactory epithelium. *Development* 127:2323–2332.

Chen H, Thiagalingam A, Chopra H, Borges MW, Feder JN, Nelkin BD, Baylin SB, Ball DW. 1997. Conservation of the *Drosophila* lateral inhibition pathway in human lung cancer: A hairy-related protein (HES–1) directly represses achaete-scute homolog–1 expression. *Proc Natl Acad Sci USA* 94:5355–5360.

Chen G, Fernandez J, Mische S, Courey AJ. 1999. A functional interaction between the histone deacetylase Rpd3 and the corepressor Groucho in *Drosophila* development. *Genes Dev* 13:2218–2230.

Dawson SR, Turner DL, Weintraub H, Parkhurst SM. 1995. Specificity for the hairy/enhancer of split basic helix-loop-helix (bHLH) proteins maps outside the bHLH domain and suggests two separable modes of transcriptional repression. *Mol Cell Biol* 15:6923–6931.

Feder JN, Jan LY, Jan Y-N. 1993. A rat gene with sequence homology to the *Drosophila* gene hairy is rapidly induced by growth factors known to influence neuronal differentiation. *Mol Cell Biol* 13:105–113.

Fisher AL, Ohsako S, Caudy M. 1996. The WRPW motif of the Hairy-related basic helix-loop-helix repressor proteins acts as a 4–amino-acid transcription repression and protein-protein interaction domain. *Mol Cell Biol* 16:2670–2677.

Fujita S. 2003. The discovery of the matrix cell, the identification of the multipotent neural stem cell and the development of the central nervous system. *Cell Struct Funct* 28:205–228.

Furukawa T, Mukherjee S, Bao ZZ, Morrow EM, Cepko CL. 2000. rax, Hes1, and notch1 promote the formation of Müller glia by postnatal retinal progenitor cells. *Neuron* 26:383–394.

Gaiano, N, Fishell G. **2002**. The role of Notch in promoting glial and neural stem cell fates. *Annu Rev Neurosci* 25:471–490.

Gaiano N, Nye JS, Fishell G. **2000**. Radial glial identity is promoted by Notch1 signaling in the murine forebrain. *Neuron* 26:395–404.

Gratton M-O, Torban E, Jasmin SB, Theriault FM, German MS, Stifani S. **2003**. Hes6 promotes cortical neurogenesis and inhibits Hes1 transcription repression activity by multiple mechanisms. *Mol Cell Biol* 23:6922–6935.

Grbavec D, Stifani S. **1996**. Molecular interaction between TLE1 and the carboxyl-terminal domain of HES–1 containing the WRPW motif. *Biochem Biophys Res Commun* 223:701–705.

Hatakeyama J, Bessho Y, Katoh K, Ookawara S, Fujioka M, Guillemot F, Kageyama R. **2004**. Hes genes regulate size, shape and histogenesis of the nervous system by control of the timing of neural stem cell differentiation. *Development* 131:5539–5550.

Herzig S, Hedrick S, Morantte I, Koo S-H, Galimi F, Montminy M. **2003**. CREB controls hepatic lipid metabolism through nuclear hormone receptor PPAR-γ. *Nature* 426:190–193.

Hirata H, Ohtsuka T, Bessho Y, Kageyama R. **2000**. Generation of structurally and functionally distinct factors from the basic helix-loop-helix gene *Hes3* by alternative first exons. *J Biol Chem* 275:19083–19089.

Hirata H, Tomita K, Bessho Y, Kageyama R. **2001**. *Hes1* and *Hes3* regulate maintenance of the isthmic organizer and development of the mid/hindbrain. *EMBO J* 20:4454–4466.

Hirata H, Yoshiura S, Ohtsuka T, Bessho Y, Harada T, Yoshikawa K, Kageyama R. **2002**. Oscillatory expression of the bHLH factor Hes1 regulated by a negative feedback loop. *Science* 298:840–843.

Hirata H, Bessho Y, Masamizu Y, Yamada S, Lewis J, Kageyama R. **2004**. Instability of Hes7 protein is critical for the somite segmentation clock. *Nat Genet* 36:750–754.

Hitoshi S. Seaberg RM, Koscik C, Alexson T, Kusunoki S, Kanazawa I, Tsuji S, van der Kooy D. **2004**. Primitive neural stem cells from the mammalian epiblast differentiate to definitive neural stem cells under the control of Notch signaling. *Genes Dev* 18:1806–1811.

Hojo M, Ohtsuka T, Hashimoto N, Gradwohl G, Guillemot F, Kageyama R. **2000**. Glial cell fate specification modulated by the bHLH gene *Hes5* in mouse retina. *Development* 127:2515–2522.

Honjo T. **1996**. The shortest path from the surface to the nucleus: RBP-J $\varkappa$/Su(H) transcription factor. *Genes Cells* 1:1–9.

Ishibashi M, Moriyoshi K, Sasai Y, Shiota K, Nakanishi S, Kageyama R. **1994**. Persistent expression of helix-loop-helix factor HES–1 prevents mammalian neural differentiation in the central nervous system. *EMBO J* 13:1799–1805.

Ishibashi M, Ang S-L, Shiota K, Nakanishi S, Kageyama R, Guillemot F. **1995**. Targeted disruption of mammalian *hairy* and *Enhancer of split* homolog–1 (*HES–1*) leads to up-regulation of neural helix-loop-helix factors, premature neurogenesis, and severe neural tube defects. *Genes Dev* 9:3136–3148.

Iso T, Sartorelli V, Poizat C, Iezzi S, Wu H, Chung G, Kedes L, Hamamori Y. **2001**. HERP, a novel heterodimer partner of HES/E(spl) in Notch signaling. *Mol Cell Biol* 21:6080–6089.

Issack PS, Ziff EB. **1998**. Genetic elements regulating HES–1 induction in Wnt-1–transformed PC12 cells. *Cell Growth Differ* 9:827–836.

Jarriault S, Brou C, Logeat F, Schroeter EH, Kopan R, Israel A. **1995**. Signalling downstream of activated mammalian Notch. *Nature* 377:355–358.

Jögi A, Persson P, Grynfeld A, Påhlman S, Axelson H. **2002**. Modulation of basic helix-loop-helix transcription complex formation by Id proteins during neuronal differentiation. *J Biol Chem* 277:9118–9126.

Joyner AL, Liu A, Millet S. **2000**. *Otx2, Gbx2* and *Fgf8* interact to position and maintain a mid-hindbrain organizer. *Curr Opin Cell Biol* 12:736–741.

Kageyama R, Nakanishi S. **1997**. Helix-loop-helix factors in growth and differentiation of the nervous system. *Curr Opin Genet Dev* 7:659–665.

Kamakura S, Oishi K, Yoshimatsu T, Nakafuku M, Masuyama N, Gotoh Y. **2004**. Hes binding to STAT3 mediates crosstalk between Notch and JAK-STAT signalling. *Nat Cell Biol* 6:547–554.

Koyano-Nakagawa N, Kim J, Anderson D, Kintner C. **2000**. Hes6 acts in a positive feedback loop with the neurogenins to promote neuronal differentiation. *Development* 127:4203–4216.

Lumsden A, Krumlauf R. **1996**. Patterning the vertebrate neuraxis. *Science* 274:1109–1115.

Mason I, Chambers D, Shamim H, Waishe J, Irving C. **2000**. Regulation and function of FGF8 in patterning of midbrain and anterior hindbrain. *Biochem Cell Biol* 78:577–584.

Miyoshi G, Bessho Y, Yamada S, Kageyama R. **2004**. Identification of a novel basic helix-loop-helix gene, *Heslike*, and its role in GABAergic neurogenesis. *J Neurosci* 24:3672–3682.

Nakamura Y, Sakakibara S, Miyata T, Ogawa M, Shimazaki T, Weiss S, Kageyama R, Okano H. **2000**. The bHLH gene *Hes1* as a repressor of the neuronal commitment of CNS stem cells. *J Neurosci* 20:283–293.

Nakashima K, Takizawa T, Ochiai W, Yanagisawa M, Hisatsune T, Nakafuku M, Miyazono K, Kishimoto T, Kageyama R, Taga T. **2001**. BMP2-mediated alteration in the developmental pathway of fetal mouse brain cells from neurogenesis to astrocytogenesis. *Proc Natl Acad Sci USA* 98:5868–5873.

Nieto M, Schuurmans S, Britz O, Guillemot F. **2000**. Neural bHLH genes control the neuronal versus glial fate decision in cortical progenitors. *Neuron* 29:401–413.

Ninkovic J, Tallafuss A, Leucht C, Topczewski J, Tannhauser B, Solnica-Krezel L, Bally-Cuif L. **2005**. Inhibition of neurogenesis at the zebrafish midbrain-hindbrain boundary by the combined and dose-dependent activity of a new *hairy/E(spl)* gene pair. *Development*. 132:75–88.

Nishimura M, Isaka F, Ishibashi M, Tomita K, Tsuda H, Nakanishi S, Kageyama R. **1998**. Structure, chromosomal locus, and promoter of mouse *Hes2* gene, a homologue of *Drosophila hairy* and *Enhancer of split*. *Genomics* 49:69–75.

Ohtsuka T, Ishibashi M, Gradwohl G, Nakanishi S, Guillemot F, Kageyama R. **1999**. Hes1 and Hes5 as Notch effectors in mammalian neuronal differentiation. *EMBO J* 18:2196–2207.

Ohtsuka T, Sakamoto M, Guillemot F, Kageyama R. **2001**. Roles of the basic helix-loop-helix genes *Hes1* and *Hes5* in expansion of neural stem cells of the developing brain. *J Biol Chem* 276:30467–30474.

Paroush Z, Finley Jr RL, Kidd T, Wainwright SM, Ingham PW, Brent R, Ish-Horowicz D. **1994**. Groucho is required for Drosophila neurogenesis, segmentation, and sex determination and interacts directly with hairy-related bHLH proteins. *Cell* 79:805–815.

Sakamoto M. Hirata H, Ohtsuka T, Bessho Y, Kageyama R. **2003**. The basic helix-loop-helix genes Hesr1/Hey1 and Hesr2/Hey2 regulate maintenance of neural precursor cells in the brain. *J Biol Chem* 278:44808–44815.

Sasai Y, Kageyama R, Tagawa Y, Shigemoto R, Nakanishi S. **1992**. Two mammalian helix-loop-helix factors structurally related to *Drosophila hairy* and *Enhancer of split*. *Genes Dev* 6:2620–2634.

Satow T, Bae S-K, Inoue T, Inoue C, Bessho Y, Hashimoto N, Kageyama R. **2001**. The bHLH gene *hesr2* promotes gliogenesis in mouse retina. *J Neurosci* 21:1265–1273.

Spassky N, Merkle FT, Flames N, Tramontin AD, Garcia-Verdugo JM, Alvarez-Buylla A. **2005**. Adult ependymal cells are postmitotic and are derived from radial glial cells during embryogenesis. *J Neurosci* 25:10–18.

Solecki DJ, Liu XL, Tomoda T, Fang Y, Hatten ME. **2001**. Activated Notch2 signaling inhibits differentiation of cerebellar granule neuron precursors by maintaining proliferation. *Neuron* 31:557–568.

Ström A, Castella P, Rockwood J, Wagner J, Caudy M. **1997**. Mediation of NGF signaling by post-translational inhibition of HES-1, a basic helix-loop-helix repressor of neuronal differentiation. *Genes Dev* 11:3168–3181.

Sun Y, Nadal-Vicens M, Misono S, Lin MZ, Zubiaga A, Hua X, Fan G, Greenberg ME. **2001**. Neurogenin promotes neurogenesis and inhibits glial differentiation by independent mechanisms. *Cell* 104:365–376.

Taelman V, Van Wayenbergh R, Solter M, Pichon B, Pieler T, Christophe D, Bellefroid EJ. **2004**. Sequences downstream of the bHLH domain of the *Xenopus* hairy-related transcription factor–1 act as an extended dimerization domain that contributes to the selection of the partners. *Dev Biol* 276:47–63.

Takatsuka K, Hatakeyama J, Bessho Y, Kageyama R **2004**. Roles of the bHLH gene *Hes1* in retinal morphogenesis. *Brain Res* 1004:148–155.

Takebayashi K, Sasai Y, Sakai Y, Watanabe T, Nakanishi S, Kageyama R. **1994**. Structure, chromosomal locus, and promoter analysis of the gene encoding the mouse helix-loop-helix factor HES–1: negative autoregulation through the multiple N box elements. *J Biol Chem* 269:5150–5156.

Tomita K, Ishibashi M, Nakahara K, Ang S-L, Nakanishi S, Guillemot F, Kageyama R. **1996**. Mammalian *hairy* and *Enhancer of split* homolog 1 regulates differentiation of retinal neurons and is essential for eye morphogenesis. *Neuron* 16:723–734.

Tomita K. Moriyoshi K, Nakanishi S, Guillemot F, Kageyama R. **2000**. Mammalian *achaete-scute* and *atonal* homologs regulate neuronal versus glial fate determination in the central nervous system. *EMBO J* 19:5460–5472.

Wu Y, Liu Y, Levine EM, Rao MS. **2003**. Hes1 but not Hes5 regulates an astrocyte versus oligodendrocyte fate choice in glial restricted precursors. *Dev Dyn* 226:675–689.

Wurst W, Bally-Cuif L. **2001**. Neural plate patterning: upstream and downstream of the isthmic organizer. *Nat Rev Neurosci* 2:99–108.

Zhang H, Levine M. **1999**. Groucho and dCtBP mediate separate pathways of transcriptional repression in the *Drosophila* embryo. *Proc Natl Acad Sci USA* 96:535–540.

2
The Role of Pax6 in the Nervous System during Development and in Adulthood: Master Control Regulator or Modular Function?

Nicole Haubst, Jack Favor, and Magdalena Götz

Abstract

Different paired-box transcription factors control various aspects of cell fate during organ development. Here we focus on the paired-box transcription factor Pax6 that is mostly expressed in the developing forebrain including the eye, as well as the cerebellum and spinal cord and in adult neurogenic zones of the brain. In murine CNS development Pax6 acts as a key regulator of neurogenesis, proliferation and regionalization, with the former roles largely restricted to the developing forebrain. The differential use of the two DNA-binding domains of Pax6, the paired and the homeo domain, in the developing eye and telencephalon as well as differential splicing of the paired domain that allows regulating cell proliferation with or without an effect on neurogenesis, are further discussed. Thus, this overview highlights the molecular mechanisms accounting for the multitude of functions of Pax6 in the developing nervous system.

2.1
Introduction

The development of an organism requires the appropriate specification of precursors in the different germ layers, their expansion during proliferation, and finally the correct differentiation of all the distinct cell types of specific organs. A complex network of transcription factors coordinates these developmental processes.

While some classes of transcription factors mostly act on one developmental program, either proliferation or differentiation (e.g., [1]), others also regulate several developmental aspects simultaneously such as members of the *Pax* gene family that influence proliferation, cell differentiation and even subtype specification. This is one reason why they are often considered to act as master regulators (e.g., [2]). Their broad, comprehensive function may partly be due to their different DNA-binding domains that specify different functions.

Transcription factors of the Pax family were named after one of their characteristic DNA-binding domains, the well-conserved "paired-box", that was first discovered in the *Drosophila* pair-rule gene paired [3–5]. In addition, these transcription factors also contain highly conserved DNA-binding sequences related to the homeobox, hence referred to as paired-type homeobox (HD) (Fig. 2.1). In the mouse, which is the focus of this chapter, nine members of this family are known, all of which possess the characteristic paired domain (PD) located close to the N-terminus of the protein, with the exception of some splice variants (see below). The crystal structure of the PD of *Drosophila* paired and the human PAX6 revealed two helix-turn-helix motifs, a structure also found in homeobox domains (Fig. 2.2B) [6]. Thus, the PD can be considered as two covalently associated HDs, connected by a flexible linker [7]. As described below, alternative splicing – mostly of Pax3, 4, 6, 7, and 8 – allows switching between the two DNA-binding parts of the PD.

The paired-type HD consists of 60 amino acids (AA) and contains a DNA-binding motif that recognizes unique target sequences not bound by other HD transcription factors. Thus, most Pax genes have three DNA-binding motifs [6]. However, as indicated in Fig. 2.1, Pax1 and 9 lack the paired-type HD, similar to the Pox genes in *Drosophila* [8], and Pax2, 5, and 8 only possess a partial HD lacking the DNA-binding motif, that interacts with other transcriptional regulators to regulate target genes [9]. All Pax genes except 4 and 6 possess a highly conserved eight-AA domain located between the PD and HD, the octapeptide (OP in Fig. 2.1). This octapeptide has been shown to recruit transcriptional repressors, such as members of the *groucho* family, and its deletion in the Pax2, 5, and 8 genes results in a decrease in their transcriptional repressor activity [10]. However, transcriptional co-regulators of the *groucho* family can also interact with the transactivation domain, and hence the repressor or activator function of some members of the Pax family may be context-dependent (see below). Generally, Pax2, 5, and 8 act mostly as transcriptional repressors [11],

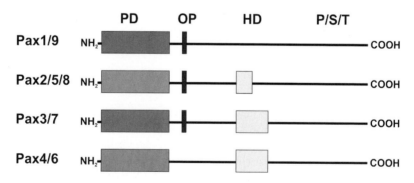

Fig. 2.1 Schematic representation of the four different vertebrate Pax gene classes. Pax1/Pax9 consist of a paired domain (PD) and an octapeptide (OP), a PST-rich transactivating domain (TAD), and lack the homeobox domain (HD). Pax2/5/8 contain a PD, OP and a partial HD followed by a PST-rich TAD. Pax3/7 are characterized by the presence of a PD, OP, HD and a PST-rich TAD. Pax4/6 consist of a PD followed by a complete HD and a PST-rich TAD. (This figure also appears with the color plates.)

while Pax4 and 6 act largely as transcriptional activators [12,13,178], consistent with the absence of the octapeptide motif in their structure. However, protein-protein interactions are important to modify their DNA-binding and transcriptional target gene regulation (e.g., [9,14,15]). For example, the partial HD of Pax2, 5, and 8 interacts with the retinoblastoma protein and TATA-binding proteins, thereby regulating homeobox domain target genes [9].

Common to all members of this family is their key role in organ development, often co-regulating cell proliferation and specification of a variety of cell fate decisions. Pax3 acts as a key determinant in premigratory neural crest cell fate as first discovered from the Splotch locus, a collection of mutations in Pax3 [6, 16]. Furthermore, Pax3 is the crucial fate determinant for cardiac neural crest specification as well as for myogenesis. An interesting switch occurs between the predominant role of Pax3 during muscle development, while Pax7 seemingly takes over this role in adult muscle regeneration [17, 18]. Similarly, essential roles for the development of entire lineages have been demonstrated for Pax2 in kidney development [19], Pax8 in thyroid development [20], and Pax5 in development of the hematopoietic B-cell lineage [11, 21]. Pax1 and Pax9 are key regulators of skeletal development [22, 23]. Finally, Pax4 and Pax6 have been identified as key regulators for endocrine cell fate in the pancreas [24], and Pax6 plays a key role in eye and central nervous system development, as will be detailed below. Interestingly, activation of the function of these Pax genes, as for example by chromosomal translocations, often results in tumor formation [25, 26]. This observation highlights their dual role in regulating cell proliferation, normally in a tightly controlled transcriptional context, and in cell fate specification.

2.2 Molecular Features of Pax6

2.2.1 The Paired Domain

Pax6 contains a PD that can recognize similar DNA sequences as the PDs of Pax2, 3 and 5 [27–31], is linked via a glycine-rich domain to the paired-like HD and followed by a proline-, serine- and threonine-rich (PST-rich) transactivating domain (TAD) at the C-terminus (e.g., [32]) (Fig. 2.2A). The murine Pax6 PD consists of 128 AA and is subdivided into the N-terminal ‚PAI' and the C-terminal ‚RED' subdomains (‚PAI-RED') (Fig. 2.2A) [27,30,33,34].

The PAI subdomain (AA 1–60) consists of an antiparallel β hairpin, a type II β turn (AA 13–16) followed by three α-helices (AA 20–60; helices 1, 2, 3) [7] (Fig. 2.2B). The β hairpin contacts the sugar-phosphate backbone of both DNA strands and spans the minor groove of the DNA [7], whereas the type II β turn contacts bases in the minor groove. The three α-helices form a helix-turn-helix (HTH) motif and contact the major groove of the DNA [7]. N-terminal PAI and C-terminal RED subdomains are connected via a linker (AA 61–76), which contacts the sugar-phosphate backbone of the minor groove over a region of 8 bp. The C-terminal RED subdomain consists of three α-helices (helices 4, 5, 6) and docks via a HTH motif (formed by helices 5 and 6) against the major groove in the distal portion of the Pax6 binding site. The recognition helix of the RED subdomain in helix 6 fits into the major groove of the DNA and contacts bases via its N-terminal part. This interaction is stabilized by the phosphate contacts of the N-terminal part of helix 5 and the C-terminal part of the linker. Deletion experiments showed that the N-terminal PAI subdomain is critical for DNA binding to Pax6 consensus sites (P6CON: ANNTTCAGCa/tTc/gANTt/ga/cAt/c [29]) (Fig. 2.2C), whereas the C-terminal part contributes to DNA binding by contacting adjacent nucleotides [30]. Transcriptional activation *in vitro* is independent of the HD, since the HD-less *Pax2* form is able to activate transcription via binding to P6CON or 5aCON [29].

2.2.2
The Paired-Type Homeodomain

The paired-type HD closely resembles other HDs, with a globular domain consisting of three α-helices (60 AA) (Fig. 2.2B, right panel). A flexible N-terminal arm (AA 1–9) is followed by helix 1 (AA 10–22), a loop structure (AA 23–27), helix 2 (AA 28–37), a

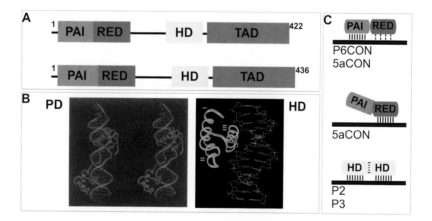

Fig. 2.2 (A) The canonical Pax6 form depicted on top (422 AA) consists of a PD, which is subdivided in an N-terminal ‚PAI' and a C-terminal ‚RED' subdomain (PAI-RED) linked to a HD followed by a TAD, whereas the Pax6(5a) isoform (436 AA) is characterized by a 14–AA insert into the PAI subdomain. (B) Overview of the Pax6 PD-DNA complex. Left: PD-DNA binding [DNA (blue), protein (red)]. Right: HD-DNA binding [DNA (blue), protein (green), Roman numbers indicate the helices, red: critical AA residue at position 50 of HD] (Reproduced from [7, 42].) (C) DNA binding of the PD of the canonical Pax6 form occurs predominantly via the N-terminal PAI (blue). In the Pax6(5a) isoform, DNA-binding of the PAI is abolished and occurs exclusively via the RED domain to 5aCON sites (middle panel). The HD (yellow) binds preferentially as dimer to palindromic P2 or P3 sites containing a TAAT core sequence. (Modified after [29–31,176,177].) (This figure also appears with the color plates.)

short turn structure (AA 38–41) and the recognition helix 3 (42–60). Helix 2 and helix 3 form a HTH motif [35]. The recognition helix 3 fits into the major groove of the DNA. The critical AA residue for DNA binding of the paired-type HD is serine located at position 50 in the recognition helix (position 9 in helix 3). Variations in the AA at position 50 can alter DNA-binding [36–39]. Interestingly, the recognition helix 3 mediates not only DNA-binding but also protein-protein interactions with the RED-subdomain of the PD, a further potential influence of transcriptional activity [40]. The Pax6 HD binds preferentially as dimer to the palindromic TAAT core motif P3 (TAAT (N)$_3$ ATTA; 3 bp spacing) [31, 41]. Loss of DNA-binding properties of the HD leads to defects in eye formation [42], but notably hardly affects forebrain development [43]. One of the most ancestral functions of the HD appears to be the regulation of rhodopsins [44]. Thus, the modular use of the PD or the HD allows Pax6 to exert specific functions during development.

2.2.3
Different Pax6 Isoforms

The PD of Pax6 is subject to alternative splicing with an insertion of 14 AA (exon 5a; Pax6(5a)) between helix 2 and helix 3 [45] in the N-terminal PAI subdomain, and thereby abolishes DNA binding of the PAI subdomain (Fig. 2.2A,C) [30] and binds specifically to the Pax6(5a) consensus site (5aCON: ATGCTCAGTGA¦ATGTT-CATTGA [30]) that consists of two 11–bp imperfect repeats and shows no significant homology to the P6CON site. While the canonical PD of Pax6 is also able to bind to the 5aCON site and activate transcription in vitro, Pax6(5a) is able to activate transcription also from 5aCON half sites (5aCON½) [30]. Therefore, the ratio of Pax6 to Pax6(5a) is crucial, and both isoforms act synergistically in the activation of transcription *in vitro* [46]. Interestingly, Pax6(5a) is expressed at later stages in development around the region of the fovea in the eye, whereas canonical Pax6 expression occurred in the entire inner neuroretina [47]. Accordingly, the Pax6(5a) form is required for aspects of postnatal differentiation in the eye [48], and an important role of the same modification of PD DNA-binding was recently discovered in the developing compound eye of *Drosophila* [49]. Gene duplication during evolution in *Drosophila* led to two homologues of the canonical Pax6 form: *eyeless* (*ey*) [50] and its paralogue twin of *eyeless* (*toy*) [51]. The *Drosophila* genes *eyegone* (*eyg*) and *twin of eyegone* (*toe*) [52, 53] are characterized by a truncated PD and have been shown to bind to target gene sequences of the 5aCON site [53], corresponding to the specific DNA-binding site of the Pax6(5a) isoform. Thus, switching between P6CON and 5aCON targets is achieved by alternative splicing in vertebrates [45,54–57] and by independent gene transcripts in other animal classes.

Other Pax6 forms lacking the PD (PD-less) have been found in mouse brain [58], the quail neuroretina [59] and in *Caenorhabditis elegans* [60]. The murine PD-less Pax6 form is generated by alternative splicing and is expressed in brain, eye, and pancreas [58], or results from a CpG island in intron 7 acting as a novel initiation site of transcription [61]. The PD-less Pax6 still binds to the HD DNA-binding consensus site but fails to activate reporter constructs and to bind the P6CON [58]. These results imply that HD can bind to the DNA, but may require the PD for transactivation. The *in-vitro* data would then predict a dominant negative function of this isoform during development, but this has not yet been tested directly [58]. However, *in-vitro* data show also that the PD-less isoform interacts with the full-length Pax6 and thereby enhances the transcriptional activation of the PD binding sites [62]. Notably, nothing is known about the specific function of the paired less isoform *in vivo*.

Additional splice variants of Pax6 (48, 46, 43, 33, and 32 kDa) have been identified in the quail neuroretina [59]. The 48–kDa form (containing an alternative paired exon 4a) and the 46–kDa form are exclusively present in the cell nucleus, whereas the 43–kDa form (lacking exon 5, due to alternative splicing), 33 kDa and 34 kDa Pax6 proteins are also found in the cytoplasm [59]. Further, three mammalian Pax6 splice variants affecting the PD have been discovered in the adult bovine eye [63]. One form (type 3) contains exon 5a, but lacks exon 6. Another splice variant lacks the HD and the TAD [63]. Clearly, the modular structure of Pax6 is further modified by splicing, but little is known so far about their specific function.

2.2.4
Protein-Protein Interactions

The binding of Pax6 to other proteins mediates transcription by other transcription factors. For example, Pax6 interacts with the TATA-box binding protein (TBP), the DNA-binding subunit of the basal transcription machinery via the N-terminal arm, and the first two helices of the Pax6 HD. Also, the C-terminal-activating domain (TAD) alone can interact with TBP *in vitro* [14]. Further, the HD of Pax6 can also bind to the complete pocket domain (A and B) of the retinoblastoma protein (pRB) *in vitro*, as shown in chicken embryonic lens nuclear extracts [14]. This mechanism may mediate the role of *Pax* genes in the regulation of proliferation (see for example Pax3 [25], Pax6 [43,64–66]). Pax6 is supposed to interact with the hypophosphorylated form of pRB, and hence should preferentially inhibit cell proliferation [14, 43]. During lens development, Pax6 interacts with Sox2 to form a co-DNA-binding complex that controls the expression of Sox2 [67]. Indeed, the interaction between *Pax* and *Sox* genes has been observed for several Pax and Sox transcription factors [6, 68]. In addition, the transcription factor *engrailed 1* (*En1*) interacts with the PD of Pax6 and prevents DNA-binding [69].

2.2.5
Post-Translational Modifications of Pax6

Finally, DNA-binding and transactivation properties of Pax6 may be altered by post-translational modifications of the Pax6 protein. The TAD of zebrafish Pax6 contains four proline-dependent kinase phosphorylation sites, three of which are phosphorylated *in vitro* by the mitogen-activated protein kinases (MAPKs), the extracellular-signal regulated kinase (ERK) and the p38 kinase, but not by Jun N-terminal kinase (JNK) [70]. One highly conserved residue is serine 413 (Ser413), that is also phosphorylated *in vivo* by p38 or ERK and is critical for the transactivation properties of Pax6 [70].

Quail neuroretina nuclear extracts have been shown to contain Pax6 with O-linked N-acetylglucosaminylation (O-GlcNAc) directly linked to serine or threonine residues [71]. However, no changes in DNA-binding were detected between the glycosylated and unglycosylated forms, which leaves the possibility that this modification may affect protein-protein interactions. Glycosylation occurs predominantly at the PD in the 46– and 48 kDa forms of Pax6 (corresponding to canonical Pax6 and Pax6(5a)), whereas the 43–, 32/33–kDa isoforms seem not to be glycosylated [71]. Since only the glycosylated isoforms were located in the nucleus, glycosylation may be a signal to keep the protein in the nucleus [71].

2.3
Function of *Pax6* in Development

2.3.1
Function of *Pax6* in the Developing Eye

The most famous role of *Pax6* is in eye development. *Pax6* is not only essential for normal eye development but is also sufficient, by ectopic expression, to induce the formation of differentiated ectopic eyes in *Drosophila* and vertebrates [50,72,73]. In the vertebrate, eye development depends upon the interaction of the optic vesicle and the surface ectoderm (Fig. 2.3). Briefly, the optic vesicle develops from an evagination of the forebrain, expands laterally, and contacts the surface ectoderm in the head region (Fig. 2.3A,B). Upon contact, the optic vesicle invaginates to form the two-layered optic cup, while a thickened region of the surface ectoderm (the lens placode) immediately adjacent to the optic vesicle invaginates to form the lens (Fig. 2.3B,C). During the early phase of eye development, *Pax6* is expressed in the optic vesicle and the surface ectoderm of the head region (red in Fig. 2.3). As development of the eye proceeds, *Pax6* expression becomes confined to the inner layer of the optic cup (the presumptive neural retina), the lens and the overlying surface ectoderm (the presumptive cornea). Upon completion of eye development, *Pax6* expression remains in the neural retina, the lens epithelium and the cornea (Fig. 2.3D). In combination with additional transcription factors, the level and spatial distribution of *Pax6* expression in the developing eye are important for the establishment of the eye axes. For example, *Pax2–Pax6* expression is critical in defining the optic stalk/optic cup boundary [74], while *Pax6–cVax* and *Tbx5* expression mediates dorsoventral patterning of the eye [75]. In the absence of *Pax6* in homozygous null mutant mice, the optic vesicle evaginates from the forebrain and expands laterally. However, the surface ectoderm fails to differentiate to the lens placode, nor does it invaginate to form the lens pit [76, 77]. The optic vesicle does not invaginate to form the optic cup [78], and has reduced proliferation and precocious differentiation of retinal neurons [79].

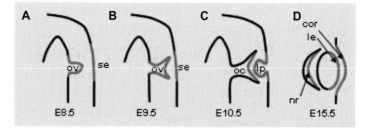

Fig. 2.3 Pax6 expression in the developing eye (red) of the mouse [45, 78]. At earlier stages (E8.5–E9.5), Pax6 expression is seen in the optic vesicle (ov) and the surface ectoderm (se) in the head region. As development of the eye progresses (E10.5), Pax6 expression becomes confined to the inner layer of the optic cup (oc), the lens placode cells which form the lens pit (lp), and the immediately adjacent regions of the surface ectoderm. At E15.5, Pax6 expression remains in the neural retina (nr), the lens epithelium (le) and the cornea (cor). (This figure also appears with the color plates.)

In *Pax6* heterozygous mouse mutants, the optic vesicle develops normally, makes proper contact with the surface ectoderm, and differentiates to an optic cup. However, the formation of the lens placode is delayed, there is a 50% reduction in the total number of lens cells, and the lens fails to detach completely from the surface ectoderm [80]. Mice lacking the Pax6(5a) isoform are viable, express iris hypoplasia, pupillated structures in the retina, and a reduced number of lens fiber cells and keratinocytes in the corneal stroma [48]. Expression of *Pax6* in the neural retina was shown to be controlled by an intronic enhancer, designated α, located between *Pax6* exons 4 and 5 [81]. In heterozygous *Pax6* knock-out mutants in which the *Pax6* exons 4, 5 and 6, including the α-enhancer, were replaced by the *lacZ* reporter gene, β-galactosidase expression was uniformly distributed throughout the neural retina. In contrast, *Pax6* expression (from the wild-type allele) in the same eyes showed the normal distal to proximal, high to low gradient in the neural retina [82]. These results indicate that the α-enhancer functions in regulating spatial-specific *Pax6* expression in the retina.

The multitude of *Pax6* functions in eye development can be elucidated by chimera analysis, or by the use of time- and space-specific conditional mutations. Analysis of *Pax6* +/+ ⟨⟩ *Pax6* −/− chimeras indicated that *Pax6* −/− cells contribute only in small numbers to the ganglion cell layer of the neural retina, and their differentiation appeared abnormal [83]. Higher numbers of *Pax6* −/− cells were found in the retinal pigmented epithelium. Their differentiation was delayed, but was observed to occur [83]. Conditional inactivation of *Pax6* via the Cre-loxP system, in which the Cre recombinase was under the control of the α-enhancer in retinal progenitor cells at E10.5 resulted in a hypocellular retina likely due to reduced retinal progenitor cell proliferation and differentiation [84]. Notably, this effect is opposite to the effect of Pax6 deletion in the developing telencephalon (see below), indicating the diverse roles of Pax6 in distinct regions of the developing central nervous system (CNS). Conditional inactivation of *Pax6* in the surface ectoderm at the stage during which lens induction occurs still allowed the development of eye structures, but arrested lens development. Multiple, fully differentiated neural retinae developed in the optic cup in the absence of the lens, indicating the importance of the lens for proper retinal morphology [85]. Determination of the retinal pigmented epithelium is dependent on *Pax2* and *Pax6* activities. Compound *Pax2;Pax6* mutants resulted in a dose-dependent reduction of *Mitf* expression [86]. The loss of *Mitf* expression in the retinal pigmented epithelium was accompanied by the formation of ectopic neural retina, a phenotype similar to that observed in $Mitf^{-/-}$ mutants [86, 87].

Mutations in *Pax6/PAX6* were first associated with the mouse *Small eye* locus [88] and in human aniridia patients [89]. Since then, a number of mouse and human *Pax6/PAX6* mutations have been molecularly characterized [42,90,91; http://pax6.hgu.mrc.ac.uk/]. Most mutant alleles in the mouse are predicted to result in a truncated gene product. Heterozygous mutants express microphthalmia and anterior segment abnormalities, including corneal opacity, corneal-lens adhesions, aniridia and anterior polar cataract. Homozygous mutants are anophthalmic (Fig. 2.4). Since heterozygotes for null mutations express a mutant phenotype, Hill et al. [88] proposed that the level of *Pax6* expression is critical for normal eye development. This was further supported by studies in which *Pax6* was over-expressed and mice also developed abnormal eye phenotypes [92]. Similar to the mutations in mice, the majority of human *PAX6* mutations [http://pax6.hgu.mrc.ac.uk/] are predicted to result in premature termination of translation. Patients heterozygous for *PAX6* mutations have been diagnosed with aniridia, Peters' anomaly, congenital cataract, keratitis, or foveal hypoplasia [30,57,93–97]. A patient expressing anophthalmia and CNS defects has been shown to be a carrier of two different *PAX6* mutant alleles inherited from the maternal and paternal germlines [93].

In the mouse, four *Pax6* missense mutations have been identified [42, 91], and two of these were shown to be in the HD. $Pax6^{4Neu}$ results in a Ser273Pro substitution at position 9 of the third α-helix, and was shown to result in greatly reduced binding activity to its P3 DNA target sequence [42]. $Pax6^{Leca1}$ is a Val270Glu substitution, also in the third α-helix of the HD, and is also predicted to affect DNA-binding activity. The remaining mouse *Pax6* missense mutations are in the PD: $Pax6^{Leca2}$ (Arg142Cys) and $Pax6^{Leca4}$ (Asn64Lys) are in the sixth and third α-helix, respectively. Both are predicted to severely disrupt DNA binding [91].

A total of 58 missense mutations are registered in the human *PAX6* mutation database [http://pax6.hgu.mrc.ac.uk/], including 35 in the paired box (at 28 codon sites), three in the linker region (at two codon sites), three in the homeobox (at three codon sites), and 11 in the PST-rich region (at 10 codon sites). In addition, there were two missense mutations at the initiation codon that likely disrupt normal translation, and four missense mutations at the termination codon. Of the 28 codon sites mutated in the paired box, the majority (15) are within AA sequences responsible for protein secondary structure (one each in the first and second β-sheets; three each in the first and second α-helices; four in the third α-helix; one in the fourth α-helix; and two in the sixth α-helix). All three missense mutations in the HD affect amino acids within the α-helix subdomains (one in the second α-helix; two in the third α-helix). Given the critical importance of the secondary structure in the PD and the HD for

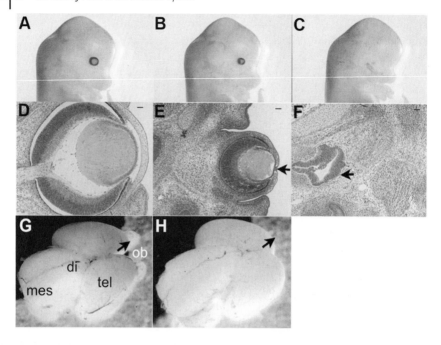

Fig. 2.4 Eye phenotype associated with mouse *Pax6* mutations. (A-C) E15 embryos: (A) *Pax6* +/+; B, *Pax6*^{3Neu} −/+ mutants express microphthalmia and triangular-shaped pupil; (C) *Pax6*^{3Neu} −/− mutants express anophthalmia. (D-F) Histological sections of eyes from E15 embryos: (D) *Pax6* +/+; (E) *Pax6*^{3Neu} −/+ mutants express anterior polar cataract with lens-corneal adhesion (arrowheads); (F) *Pax6*^{3Neu} −/− mutants express anophthalmia with remnants of the optic vesicle (arrowheads). (G) Brain morphology of an E14.5 wild-type; the arrow indicates an olfactory bulb. (H) *Pax6*$^{Sey/−}$ brain lacking the olfactory bulbs (arrow). Abbreviations: tel = telencephalon; di = diencephalon; mes = mesencephalon; ob = olfactory bulb.

PAX6 DNA-binding activity (see Pax6 PD DNA-binding, Pax6 HD DNA-binding, above), these mutated codon sites likely affect *PAX6* function. Indeed, the protein products of some of the mutant alleles were shown to reduce DNA-binding or transcriptional activation activity (see references cited for the mutant alleles in the human PAX6 mutation database). Finally, a number of deletions or cytogenetic rearrangements have been identified in human aniridia patients in whom the *PAX6* gene was not disrupted, suggesting that the level of *PAX6* expression was affected by the deletion/disorganization of cis-acting regulatory elements or by position effect [89,98–101]. These data therefore further support the critical role of Pax6 dosage for development.

2.3.2
Function of Pax6 in the Developing Brain

2.3.2.1 Telencephalon

In the forebrain, Pax6 expression starts at embryonic day (E) 8 in the mouse, just before the telencephalon and diencephalon can be distinguished [45]. Thereafter, Pax6 expression in the developing telencephalon is restricted to the dorsal telencephalon, the pallium, comprising the cerebral cortex with neo-, paleo- and archicortical regions (neocortex, piriform cortex and hippocampus, respectively (Fig. 2.5A,B). Pax6 transcription is controlled by a 5–kb fragment between P1 and P0 in the dorsal telencephalon [81], where it is expressed with a lateral high and medial (hippocampal anlage) low gradient [102], while it is weak to absent in the ventral telencephalon, the subpallium (Fig. 2.5A,B) [103–105]. Accordingly, prominent defects in the telencephalon of mice with a Pax6 truncation lacking the TAD [88, 106] affecting cell proliferation, migration, differentiation and regionalization are mainly restricted to the dorsal telencephalon Pax6 expression in the dorsal telencephalon is restricted to the ventricular zone (VZ) precursor cells, the proliferative cell population adjacent to the ventricle, whereas the secondary proliferative cell population, the subventricular zone (SVZ), located above the VZ, is devoid of Pax6 expression [65]. The reduction of neurogenesis to 50 % of normal in Pax6$^{Sey-/-}$ mice is due to a loss of the neurogenic potential in VZ cells – as shown by flourescence-activated cell sorting (FACS) [66] – while SVZ cells are less affected in their neurogenic role [107–109]. The majority of VZ cells are radial glia cells [110–113], and Pax6 proves to be important for neurogenesis from cells with radial glia or astrocyte properties. For example, neural stem cells propagated *in vitro* as neurospheres that exhibit radial glia properties [110] depend on Pax6 to generate neurons, independent of their region of origin [114]. Moreover, Pax6 is sufficient to overcome the poor neurogenic properties of these neurosphere cells and induces more than 80% to differentiate into functional neurons [114] (see also Berninger and M. Götz, unpublished observations). As described below, Pax6 also proves to be necessary and sufficient in adult neurogenesis [178] originating from astroglia-like stem cells [115]. Finally, Pax6 is even sufficient to induce neurogenesis in non-neurogenic astrocytes, such as those from the postnatal cerebral cortex [66] or in cells reacting to brain injury of the adult cortex *in vivo* (Buffo and M. Götz, unpublished observations). Thus, one prevailing function of Pax6 is its key role in neurogenesis in the telencephalon *in vivo* and in cells expanded from different regions *in vitro*.

The neurogenic role of Pax6 seems to be tightly linked to its role in proliferation. Neurons are permanently postmitotic cells, and the generation of neurons results in a reduction of the precursor pool [116, 117]. Thus, the neurogenic function of Pax6 is associated with a reduction in the progenitor pool. Indeed, the number of precursor cells in the developing cortex of Pax6$^{Sey-/-}$ mice is increased compared to wild-type [43,64–66], while overexpression of Pax6 in cortical cell culture leads to a decrease in the number of cells generated from a single infected precursor cell [43, 66]. These observations imply a role of Pax6 in reducing the precursor pool by generating postmitotic neurons. In the absence of functional Pax6 in Pax6$^{Sey-/-}$ mice, the incre-

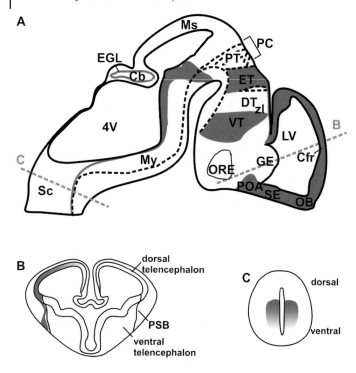

Fig. 2.5 (A) Sagittal section through the developing brain at E13.5 showing *Pax6* expression in red. The blue dashed lines indicate the planes of sections B and C, respectively. (B) Frontal section through the telencephalon showing *Pax6* expression in the cerebral cortex and in the pallial-subpallial boundary (PSB). (C) *Pax6* expression in the spinal cord occurs in ventral low to dorsal high gradient. Frontal section of the developing spinal cord with *Pax6* expression (red) in the ventral portion in a ventral^{low}-dorsal^{high} gradient. (Modified after [162].) Abbreviations: Cb = cerebellum; Cfr = frontal cortex; DT = dorsal thalamus; EGL = external granule layer; ET = epithalamus; GE = ganglionic eminence; LV = lateral ventricle; Ms = mesencephalon; My = myelencephalon; OB = olfactory bulb; ORE = optic recess; PC = posterior commissure; Pn = pons; PT = pretectum; Sc = spinal cord; SE = septum; 4V = fourth ventricle; zl = zona limitans intrathalamica. (This figure also appears with the color plates.)

ased number of precursors are located in the SVZ [43] and lack the expression of normal cortex SVZ genes such as *Svet1*, *Cux2* and *Tbr2* [118–120]. Supporting the hypothesis that SVZ precursors may generate upper layer neurons, these are mostly absent in the Pax6$^{Sey-/-}$ mice [118,121,122].

Pax6 specifies dorsal and ventral as well as lateral and medial differences in cell fate, and thereby mediates the patterning of the telencephalon [103, 123] arealization of the developing cerebral cortex [102,124–126]. Absence of functional Pax6 leads to a mis-specification of the dorsal telencephalon [127]. Severe defects in patterning are indicated by the expansion of ventrally expressed transcription factors, as for example Gsh2, Mash1 or Olig2, into the dorsal telencephalon of Pax6$^{Sey-/-}$ mice [66,103,104,123,127,128]. The expression of the proneural bHLH (basic helix-loop-helix) gene *Neurogenin2* (*Ngn2*) is strongly reduced in the cortex of Pax6$^{Sey-/-}$ mice [66, 127], as it is a direct target gene of Pax6 [129]. These defects in the region-specific transcription factor expression that is required for maintenance of the pallial-subpallial compartment boundary [130] then seem to result first in a loss of region-specific cell adhesion [128, 131]. In the absence of functional Pax6, the pallial-subpallial boundary (PSB) that delineates the dorsal and ventral telencephalon is lost or is mis-specified [128]. The PSB is formed by a fascicle of radial glia cells [110,128,132] that specifically expresses the *reticulon–1* gene [133] and high levels of soluble frizzled related protein–2 (SFRP–2) [134]. Further signaling molecules, as for example the transforming growth factor α (Tgfα), neuregulin 1 (Nrg1), Nrg3, fibroblast growth factor 7 (Fgf7), are expressed by PSB cells, this region may act as a signaling center [135]. In the Pax6$^{Sey-/-}$ mice no fasciculation of radial glia cells at this position can be observed [128], and boundary-specific expression, such as SFRP–2, Nrg1, Tgfα and reticulon–1, is lost [133–135]. An analysis of Pax6$^{Sey-/-}$ and Tlx$^{-/-}$ mice led to the conclusion that Pax6 and Tlx might cooperate genetically to establish the PSB [136]. Alterations of the PSB result not only in defects of amygdala neurons that seem to be specified in this region [105], but also lead to the failure of region-specific restriction in cell migration, such that cells from the ventral telencephalon freely enter the dorsal telencephalon and some cells migrate in the opposite direction [130]. Moreover, the number of reelin/calretinin-positive cells in the marginal zone of the Pax6$^{Sey-/-}$ telencephalon is increased, most likely due to increased tangential migration of these neurons from the enlarged cortical hem structure [137, 138], while the number of calbindin-positive cells is not altered [139]. Reelin-immunoreactive, supposedly mitral cells, also seem to be misrouted and form an aberrant olfactory bulb-like structure in the ventral telencephalon of Pax6$^{Sey-/-}$ and normal olfactory bulbs are absent [140, 141].

Patterning of the developing brain specifies distinct regions that then generate distinct types of neurons. Accordingly, defects in patterning not only result in the migration of cells that would normally be restricted in a different brain region, but also causes cell-autonomous defects in neuronal specification. Precursors in the ventral telencephalon generate mostly diverse populations of GABAergic neurons, projection neurons of the basal ganglia and interneurons for all telencephalic regions (for a review, see [142]). In contrast, precursors of the dorsal telencephalon generate mostly glutamatergic projection neurons. The transcription factors Mash1, Olig2 and Dlx that normally are restricted in expression to the ventral telencephalon are expressed in precursors in the dorsal telencephalon of Pax6$^{Sey-/-}$ mice [66,127,143], and there is a switch from the generation of glutamatergic to GABAergic neurons [130, 144].

As detailed above, the lack of functional Pax6 results in a variety of defects in the developing cortex, although how exactly Pax6 regulates and coordinates these diverse effects is not clear. Recent analysis of specific mouse mutants lacking either a functional PD (Pax6$^{Aey18-/-}$) [43] or HD (Pax6$^{4Neu-/-}$) [42] showed that the regulation of neurogenesis, cell proliferation and patterning occurs mostly via target genes of the

PD, whereas the HD plays no role in that regard [43]. The loss of a functional PD, as observed in the mouse mutant Pax6$^{Aey18-/-}$ with a deletion of exons 5 and 6, exhibits all phenotypes of the Pax6$^{Sey-/-}$ mice, such as increased cell proliferation (especially in the SVZ), a decrease in neurogenesis, and defects in dorsoventral patterning as well as a lack of the OBs, while mutation of the HD has no effect [43].

As described above, the PD consists of two independent DNA-binding domains that can be altered by alternative splicing. Without the splice insert encoded in exon 5a in mice and humans, the PD of Pax6 binds predominantly to the Pax6 consensus site (P6CON), and also to the 5aCON site [30]. In contrast, upon insertion of 14 AA encoded by exon 5a, the PD binds exclusively to the 5aCON site. Retrovirally mediated overexpression of each of these forms revealed that the Pax6(5a) isoform exclusively affects cell proliferation without affecting cell fate in cortical progenitors *in vitro*, while the Pax6 form with the canonical PD affects both cell proliferation and neurogenesis simultaneously [43]. Thus, alternative splicing regulates the range of Pax6 functions. Unfortunately, little is known so far as to when and where Pax6(+5a) is expressed in the developing brain.

Given the multitude of roles that Pax6 exerts in various developmental aspects in the eye and the telencephalon, the question is whether it exerts comparable roles in other regions of the developing CNS.

2.3.2.2 Diencephalon

Pax6 expression in the developing forebrain extends from the telencephalon into the diencephalon up to the posterior commissure (see Fig. 2.5A) [45], and is regulated by several enhancer elements [61,81,145,146]. The diencephalon is subdivided by the zona limitans intrathalamica (ZLI) into a dorsal and a ventral part (Fig. 2.5A). At E10.5, strong Pax6 expression is detected in the ventral diencephalon, the presumptive thalamus, and extends into the anterior hypothalamus in the region of the preoptic recess of the third ventricle in the mouse embryo [45, 147]. Pax6 is also expressed in the epithalamus at E13 (Fig. 2.5A), and later in the nuclei habenulae [45]. Pax6 expression is low in the dorsal thalamus (E10.5–E12.5) and not detectable at E15.5, while it is high in the pineal gland at all stages (E8–E18.5) [45] and the secretory glia cells of the subcommissural organ (SCO), a circumventricular structure at the forebrain-midbrain boundary [147]. Consistent with an important role of Pax6 in areas of high-levels of expression, the SCO and pineal gland fail to form and the posterior commissure is abnormal in the Pax6$^{Sey-/-}$ mice [148]. From E 9.5, the first postmitotic neurons are detectable in the developing diencephalon. These initial neurons in the ventral thalamus are Pax6– and βIII-tubulin-positive, and presumably give rise to TH+ dopaminergic A13 neurons of the zona incerta in the ventral thalamus. In Pax6$^{Sey-/-}$ a subset of these neurons is missing [149, 150]. Besides changes in neuronal subtype specification, proliferation in the developing diencephalon of the Pax6$^{Sey-/-}$ mice is decreased at E10.5, but not at E14.5 in regions of Pax6 expression [147], apparently the opposite defect as in the Pax6$^{Sey-/-}$ telencephalon [43, 66]. The Pax6$^{Sey-/-}$ diencephalon also shows strong morphological alterations. For example, at E9.5–E10.5 there is a lack of the boundary between diencephalic prosomere 1 and the mesencephalon, implying that there is a critical function of Pax6 also in the diencephalon in boundary formation [151]. At later stages, fusion between the medial ganglionic eminence (MGE) and the anterior hypothalamus (AH) fails to occur in the Pax6$^{Sey-/-}$ mice, and this results in an enlargement of the third ventricle and a paucity of tissue within the hypothalamus, apparently due to aberrations in the dorsoventral patterning of the diencephalon [127]. Defects in axonal pathfinding in the developing diencephalon affect the tract of the postoptic commissure (tpco) (E9.5–E10.5) [151], and thalamocortical (TCA) and corticofugal axons (CFA) are aberrant in the Pax6$^{Sey-/-}$ diencephalon (E14.5–E18.5) [152]. Taken together, in the developing diencephalon Pax6 is involved in the regulation of cell proliferation and cell fate, even though its overall role in neurogenesis is less obvious.

2.3.2.3 Cerebellum

In the developing mouse cerebellum, strong Pax6 expression is detectable at E12.5–E13.5 in the rhombic lip where external granule cells are located (see Fig. 2.5A) [45, 153]. Pax6 is further expressed in the precerebellar neuroepithelium (pcn) and in two streams of migrating cells that extend from the pcn, the anterior extramural migration stream (aes) and the posterior extramural migration stream (pes) [153]. Postnatally, Pax6 is expressed in the inner and outer granule cell layer [154]. A decrease in size of the aes and pontine nucleus and cellular disorganization (lateral reticular nuclei and the external cuneate nuclei) are observed in Pax6$^{Sey-/-}$ mice while cell proliferation appears normal [153], although medulloblastoma formation seems to correlate with the progressive loss of Pax6 expression [155]. Thus, a predominant role of Pax6 in this region is on cell migration with netrin, slit and robo as some effector molecules [153, 154] and on cell morphology and polarization during fiber formation [154].

2.3.2.4 Spinal Cord

In further posterior regions, such as the developing spinal cord, Pax6 is first ubiquitously expressed shortly after neural tube closure [156], while restriction to the ventral part occurs with a ventral^{low}-dorsal^{high} gradient at later stages (Fig. 2.5C) [157]. A ventral^{high}-dorsal^{low} gradient of sonic hedgehog (Shh) signaling specifies Nkx2.2 expression that then represses Pax6 [157]. Notably, Pax6 restricts the domain of Nkx2.2 that expands into the former Pax6 domain in the developing spinal cord of Pax6$^{Sey-/-}$ mice [157] and can be repressed by Pax6 electroporation [158].

As a common feature in the developing CNS, defects in patterning are associated with alterations of the neurons generated in the mis-specified region. Accordingly, the differentiation of motor neurons (MNs) [158–160] as well as V1 interneurons that are normally derived from Pax6–expressing progenitor cells [157, 161] in the spinal cord are affected in Pax6$^{Sey-/-}$ mice. No V1 interneurons are generated, and the number of V2 interneurons is reduced [157]. The appearance of the oligodendrocyte precursor (OLP) population in the caudal hindbrain and spinal cord is delayed due to loss of Pax6, and the site of origin of these OLPs was shifted dorsally [160]. Taken together, the main function of Pax6 in the developing spinal cord is to translate patterning into neuronal subtype specification. Thus, the common role of Pax6 throughout the developing CNS is patterning and the specification of neuronal subtypes, while a generic function in neurogenesis and proliferation is exerted mostly in the developing forebrain.

2.4
Function of Pax6 in the Adult Brain

Previous expression analysis of Pax6 in the adult brain detected Pax6 mRNA in similar, but also notably different regions as during development [162]. One of the regions in the adult mouse brain with the highest number of Pax6–positive cells is the olfactory bulb (OB), where Pax6 is seen in three locations. Highest expression levels are in a subtype of olfactory bulb interneurons located in the glomerular layer, the periglomerular neurons [178]. These interneurons comprise several subsets, and Pax6 seems to be contained almost exclusively in the dopaminergic subset of these interneurons [178]. In addition, Pax6–immunoreactivity is detectable in the granular layer, where it is localized mostly in migrating double-cortin-immunoreactive neurons destined to the glomerular layer [178]. The OB is one of two regions in the adult mammalian brain where a large number of neurons are generated throughout the life of the animal (the other region is the dentate gyrus of the hippocampal formation; for a review, see [115]). Adult neurogenesis originates in a zone lining the lateral wall of the lateral ventricle, the adult subependymal zone. It has been demonstrated that a subset of astrocytes represents the progenitor cells for adult neurogenesis that generates rapidly proliferating, transit-amplifying precursors that then give rise to double-cortin-positive neuroblasts [163]. The latter migrate in chains until they disperse radially into the different layers of the OB [164]. Interestingly, Pax6 is mostly contained in double-cortin-positive neuroblasts, to a lesser degree in transit-amplifying precursors, and is absent from the astrocyte-like stem cells [114, 178]. This expression is reminiscent of the neurogenic role of Pax6 in the developing telencephalon [43,66,114]. Indeed, as described above, Pax6 plays a role in OB development since an aberrant vesicle forms within the anterior region of the basal telencephalon instead of a proper olfactory bulb [141, 165], containing reelin-, calretinin- and calbindin-positive cells; these are possibly remnants of mitral and tufted cells in the Pax6$^{Sey-/-}$ mice [139, 165]. However, hardly any TH-immunoreactive cells are present in this aberrant vesicle, suggesting that dopaminergic periglomerular interneurons fail to develop in the Pax6–mutant mice [150]. This may be due to any of the multiple functions that Pax6 serves in the developing telencephalon described above – defects in cell migration, cell fate, or neuronal subtype specification. In particular, since dorsoventral patterning of the developing telencephalon is disturbed in the Pax6–mutant mice, mis-placement of the OB may also be due to defects in regionalization. However, overexpression of Pax6 or its deletion in the adult subependymal zone (SEZ) demonstrated its necessary and sufficient role in adult neurogenesis. While overexpression of Pax6 caused virtually all adult SEZ precursors to assume a neurogenic fate, the expression of a dominant-negative form of Pax6 (replacement of the transactivator domain with a repressor domain [154]) or acute deletion of Pax6 by using a Cre/LoxP approach led to a severe reduction in neuronal precursors, and the few remaining neuroblasts did not differentiate properly or enter the OB [178]. Thus, Pax6 functions as a transactivator in adult neurogenesis and during development [43, 154]. Besides its pan-neurogenic role in adult OB neurogenesis, Pax6 also affects neuronal subtype specification [178]. Overexpression of Pax6 strongly promotes the acquisition of a dopaminergic periglomerular neuronal fate, while interference with Pax6 function at late stages of neuroblast differentiation selectively blocks the specification of this type of interneuron, but not the formation of granule neurons, the second class of interneurons generated throughout adulthood [178]. In summary, these data support a cell-autonomous role of Pax6 in the specification of some dopaminergic neurons during development and in adults [150,151,178], with Pax6 being the first key fate determinant for the generation of dopaminergic neurons in the adult mammalian brain.

Thus, the role of Pax6 in adult neurogenesis is reminiscent of its functions during development – cell fate and neuronal subtype specification. However, Pax6 is also detected in regions of the adult brain where no developmental dysfunction has been detected in Pax6–mutant mice. For example, Pax6–positive cells are prominent in the septal region, the zona incerta and entopeduncular nucleus in the diencephalon and the zona reticularis of the substantia nigra [162]. The latter point is intriguing with regard to the dopaminergic neuron specification observed in adult neurogenesis, as well as some defects in the formation of dopaminergic neurons during development of the Pax6$^{Sey-/-}$ mice [150]. Notably, however, neurons of the substantia nigra seem to form properly in the Pax6–mutant, even though their projections to the basal ganglia are severely abnormal [150]. However, as Pax6–mutant mice die at birth, it has so far been impossible to evaluate whether Pax6 plays a role in neuronal subtype survival or maintenance at later stages. Therefore, the role of Pax6 in subsets of adult neurons remains to be determined by conditional gene deletion [85].

2.5
Mechanisms of Pax6 Function

As described above, the different DNA-binding domains of Pax6 seem to act as modules dedicated to specific roles. The canonical PD affects patterning and cell fate in the developing telencephalon via P6CON-regulated targets, while proliferation is regulated via the 5aCON site. In contrast to the developing eye, HD targets seem to play a minor role in the developing telencephalon. Thus, the emerging concept is a modulator function of Pax6. However, so far little is known about the crucial effector genes of these modular transcriptional regulations. One target of the canonical Pax6 PD (P6CON) in the developing telencephalon and spinal cord regulating patterning and cell fate is the proneuronal basic helix-loop-helix (bHLH) transcription factor Ngn2 [129]. Several phenotypes in the telencephalon of Pax6$^{Sey-/-}$ mice could be explained by the loss of its target Ngn2, such as the reduction in neurogenesis and the switch from the generation of glutamatergic neurons to GABAergic neurons. However, there are also different functions of these two transcription factors. While the loss of Ngn2 affects primarily the specification of deep layer neurons in the cortex [121], mostly upper layer neurons are mis-specified in the Pax6$^{Sey-/-}$ cortex [118, 119]. The extracellular matrix glycoprotein Tenascin-C (TN-C) is expressed in the VZ of the cerebral cortex and in the region of the PSB. In the Pax6$^{Sey-/-}$ mice, TN-C expression in the dorsal VZ and the boundary region is lost [128]. Moreover, the cell surface molecule R-cadherin seems to mediate the role of Pax6 in patterning and the restriction of cell migration. R-cadherin expression is absent in the cortex of the Pax6$^{Sey-/-}$ mouse [128], and mediates cell restriction at the border between the dorsal and the ventral telencephalon [131]. At later stages, cadherin–6 and cadherin–8 are also reduced in the cortex of Pax6$^{Sey-/-}$ mice [102]. It is not clear however, whether the regulation of cadherins by Pax6 is direct or indirect. N-CAM, a member of the immunoglobulin (Ig) -superfamily proteins of cell-surface molecules, is expressed in most neural precursors and young neurons in the developing CNS, where it is thought to be important for the mediation of cell-cell interactions. The N-CAM promotor contains a putative binding site for the canonical PD of Pax6 that is able to bind and to activate transcription *in vitro* [166]. Another cell-surface molecule suggested to be regulated by Pax6 is the cell adhesion molecule L1, a cell-surface glycoprotein of the Ig-superfamily [167]. Putative binding sites for Pax6 have also been identified in the promoter of human α_5 – and β_1-integrin respectively, and overexpression of Pax6(5a) positively regulated the transcription of β_1-integrin in the lens fiber cells of the developing eye *in vivo* [168]. During eye development of *Drosophila*, the Pax6 homologue *eyeless* directly regulates the homeobox-containing gene sine occulis (*so*) [169], and its vertebrate homologue Six3 is also expressed in the eye and may be also under the direct control of Pax6 in *Xenopus* and mouse [73, 85]. Pax6 regulates crystalline gene expression in the vertebrate eye [170–172] and Maf, a transcription factor in the developing vertebrate eye, contains at least three Pax6 binding sites in its promotor region and is activated by Pax6 *in vitro* [173]. Rhodopsin expression in *Drosophila* is regulated by *eyeless* [174, 175], and by PaxB in cnidaria [44]. Taken together, Pax6 regulates a multitude of different targets including adhesion molecules, lens structural proteins as well as other transcription factors.

2.6
Conclusions and Outlook

Pax6 has been shown to be involved in the regulation of proliferation, neurogenesis, and patterning in different regions of the developing and adult brain. One intriguing question that surfaces in this context is: How can all these different functions be fulfilled in different ways in different tissues, as for example the negative regulation of proliferation in the telencephalon as opposed to the positive regulation of proliferation in the developing eye or diencephalon? One possible answer to this might be specific use of the different DNA-binding domains of Pax6, the modularity of which is further increased by alternative splicing. Moreover, different Pax6 isoforms or transcripts can interact with diverse combinations of other transcription factors to achieve region- and time-specific regulation of transcription to produce a variety of transcriptional properties. Further variety is achieved by the spatiotemporal specific use of different promoters and enhancer elements.

Abbreviations

AA	amino acid
AH	anterior hypothalamus
bHLH	basic helix-loop-helix
CFA	corticofugal axons
CNS	central nervous system
E	embryonic day
ERK	extracellular-signal regulated kinase
Fgf	fibroblast growth factor
HD	homeobox domain
HTH	helix-turn-helix
Ig	immunoglobulin
JNK	Jun N-terminal kinase
MAPK	mitogen-activated protein kinase
MGE	medial ganglionic eminence
MN	motor neuron
Nrg	neuregulin
OB	olfactory bulb
O-GlcNAc	O-linked N-acetylglucosaminylation
OLP	oligodendrocyte precursor
PD	paired domain
PSB	pallial-subpallial boundary
PST-rich	proline-, serine- and threonine-rich
SCO	subcommissural organ
SEZ	subependymal zone
SFRP–2	soluble frizzled related protein–2
SVZ	subventricular zone
TAD	transactivating domain
TBP	TATA-box binding protein
TCA	thalamocortical axons
Tgfα	transforming growth factor α
TN-C	tenascin-C
Tpco	tract of the postoptic commissure
VZ	ventricular zone
ZLI	zona limitans intrathalamica

References

1 Bertrand N, Castro DS, Guillemot F: Proneural genes and the specification of neural cell types. *Nat Rev Neurosci* 2002; 3: 517–530.
2 Gehring WJ, Ikeo K: Pax 6: mastering eye morphogenesis and eye evolution. *Trends Genet* 1999; 15: 371–377.
3 Bopp D, Burri M, Baumgartner S, Frigerio G, Noll M: Conservation of a large protein domain in the segmentation gene paired and in functionally related genes of *Drosophila*. *Cell* 1986; 47: 1033–1040.
4 Nusslein-Volhard C, Wieschaus E: Mutations affecting segment number and polarity in Drosophila. *Nature* 1980; 287: 795–801.
5 Baumgartner S, Bopp D, Burri M, Noll M: Structure of two genes at the gooseberry locus related to the paired gene and their spatial expression during *Drosophila* embryogenesis. *Genes Dev* 1987; 1: 1247–1267.
6 Chi N, Epstein JA: Getting your Pax straight: Pax proteins in development and disease. *Trends Genet* 2002; 18: 41–47.
7 Xu HE, Rould MA, Xu W, Epstein JA, Maas RL, Pabo CO: Crystal structure of the human Pax6 paired domain-DNA complex reveals specific roles for the linker region and carboxy-terminal subdomain in DNA binding. *Genes Dev* 1999; 13: 1263–1275.
8 Bopp D, Jamet E, Baumgartner S, Burri M, Noll M: Isolation of two tissue-specific *Drosophila* paired box genes, Pox meso and Pox neuro. *EMBO J* 1989; 8: 3447–3457.
9 Eberhard D, Busslinger M: The partial homeodomain of the transcription factor Pax–5 (BSAP) is an interaction motif for the retinoblastoma and TATA-binding proteins. *Cancer Res* 1999; 59: 1716s–1724s; discussion 1724s–1725s.
10 Eberhard D, Jimenez G, Heavey B, Busslinger M: Transcriptional repression by Pax5 (BSAP) through interaction with corepressors of the Groucho family. *EMBO J* 2000; 19: 2292–2303.
11 Schebesta M, Heavey B, Busslinger M: Transcriptional control of B-cell development. *Curr Opin Immunol* 2002; 14: 216–223.
12 Chauhan BK, Reed NA, Zhang W, Duncan MK, Kilimann MW, Cvekl A: Identification of genes downstream of Pax6 in the mouse lens using cDNA microarrays. *J Biol Chem* 2002; 277: 11539–11548.
13 Chauhan BK, Reed NA, Yang Y, Cermak L, Reneker L, Duncan MK, Cvekl A: A comparative cDNA microarray analysis reveals a spectrum of genes regulated by Pax6 in mouse lens. *Genes Cells* 2002; 7: 1267–1283.
14 Cvekl A, Kashanchi F, Brady JN, Piatigorsky J: Pax–6 interactions with TATA-box-binding protein and retinoblastoma protein. *Invest Ophthalmol Vis Sci* 1999; 40: 1343–1350.
15 Garvie CW, Hagman J, Wolberger C: Structural studies of Ets–1/Pax5 complex formation on DNA. *Mol Cell* 2001; 8: 1267–1276.
16 Epstein DJ, Vogan KJ, Trasler DG, Gros P: A mutation within intron 3 of the Pax–3 gene produces aberrantly spliced mRNA transcripts in the splotch (Sp) mouse mutant. *Proc Natl Acad Sci USA* 1993; 90: 532–536.
17 Seale P, Ishibashi J, Scime A, Rudnicki MA: Pax7 is necessary and sufficient for the myogenic specification of CD45+: Sca1+ stem cells from injured muscle. *PLoS Biol* 2004; 2: E130.
18 Seale P, Sabourin LA, Girgis-Gabardo A, Mansouri A, Gruss P, Rudnicki MA: Pax7 is required for the specification of myogenic satellite cells. *Cell* 2000; 102: 777–786.
19 Torres M, Gomez-Pardo E, Dressler GR, Gruss P: Pax–2 controls multiple steps of urogenital development. *Development* 1995; 121: 4057–4065.
20 Mansouri A, Chowdhury K, Gruss P: Follicular cells of the thyroid gland require Pax8 gene function. *Nat Genet* 1998; 19: 87–90.
21 Urbanek P, Wang ZQ, Fetka I, Wagner EF, Busslinger M: Complete

block of early B cell differentiation and altered patterning of the posterior midbrain in mice lacking Pax5/BSAP. *Cell* **1994**; 79: 901–912.
22. Balling R, Helwig U, Nadeau J, Neubuser A, Schmahl W, Imai K: Pax genes and skeletal development. *Ann N Y Acad Sci* **1996**; 785: 27–33.
23. Wallin J, Mizutani Y, Imai K, Miyashita N, Moriwaki K, Taniguchi M, Koseki H, Balling R: A new Pax gene, Pax–9, maps to mouse chromosome 12. *Mamm Genome* **1993**; 4: 354–358.
24. St-Onge L, Sosa-Pineda B, Chowdhury K, Mansouri A, Gruss P: Pax6 is required for differentiation of glucagon-producing alpha-cells in mouse pancreas. *Nature* **1997**; 387: 406–409.
25. Maulbecker CC, Gruss P: The oncogenic potential of Pax genes. *EMBO J* **1993**; 12: 2361–2367.
26. Stuart ET, Gruss P: PAX: developmental control genes in cell growth and differentiation. *Cell Growth Differ* **1996**; 7: 405–412.
27. Czerny T, Schaffner G, Busslinger M: DNA sequence recognition by Pax proteins: bipartite structure of the paired domain and its binding site. *Genes Dev* **1993**; 7: 2048–2061.
28. Chalepakis G, Gruss P: Identification of DNA recognition sequences for the Pax3 paired domain. *Gene* **1995**; 162: 267–270.
29. Epstein J, Cai J, Glaser T, Jepeal L, Maas R: Identification of a Pax paired domain recognition sequence and evidence for DNA-dependent conformational changes. *J Biol Chem* **1994**; 269: 8355–8361.
30. Epstein JA, Glaser T, Cai J, Jepeal L, Walton DS, Maas RL: Two independent and interactive DNA-binding subdomains of the Pax6 paired domain are regulated by alternative splicing. *Genes Dev* **1994**; 8: 2022–2034.
31. Czerny T, Busslinger M: DNA-binding and transactivation properties of Pax–6: three amino acids in the paired domain are responsible for the different sequence recognition of Pax–6 and BSAP (Pax–5). *Mol Cell Biol* **1995**; 15: 2858–2871.
32. Simpson TI, Price DJ: Pax6; a pleiotropic player in development. *BioEssays* **2002**; 24: 1041–1051.
33. Xu W, Rould MA, Jun S, Desplan C, Pabo CO: Crystal structure of a paired domain-DNA complex at 2.5 Å resolution reveals structural basis for Pax developmental mutations. *Cell* **1995**; 80: 639–650.
34. Jun S, Desplan C: Cooperative interactions between paired domain and homeodomain. *Development* **1996**; 122: 2639–2650.
35. Wilson DS, Guenther B, Desplan C, Kuriyan J: High resolution crystal structure of a paired (Pax) class cooperative homeodomain dimer on DNA. *Cell* **1995**; 82: 709–719.
36. Treisman J, Gonczy P, Vashishtha M, Harris E, Desplan C: A single amino acid can determine the DNA binding specificity of homeodomain proteins. *Cell* **1989**; 59: 553–562.
37. Hanes SD, Brent R: DNA specificity of the bicoid activator protein is determined by homeodomain recognition helix residue 9. *Cell* **1989**; 57: 1275–1283.
38. Percival-Smith A, Muller M, Affolter M, Gehring WJ: The interaction with DNA of wild-type and mutant fushi tarazu homeodomains. *EMBO J* **1990**; 9: 3967–3974.
39. Schier AF, Gehring WJ: Direct homeodomain-DNA interaction in the autoregulation of the fushi tarazu gene. *Nature* **1992**; 356: 804–807.
40. Bruun JA, Thomassen EI, Kristiansen K, Tylden G, Holm T, Mikkola I, Bjorkoy G, Johansen T: The third helix of the homeodomain of paired class homeodomain proteins acts as a recognition helix both for DNA and protein interactions. *Nucleic Acids Res* **2005**; 33: 2661–2675.
41. Wilson D, Sheng G, Lecuit T, Dostatni N, Desplan C: Cooperative dimerization of paired class homeo domains on DNA. *Genes Dev* **1993**; 7: 2120–2134.
42. Favor J, Peters H, Hermann T, Schmahl W, Chatterjee B, Neuhauser-Klaus A, Sandulache R: Molecular

characterization of Pax6(2Neu) through Pax6(10Neu): an extension of the Pax6 allelic series and the identification of two possible hypomorph alleles in the mouse *Mus musculus*. *Genetics* **2001**; 159: 1689–1700.

43 Haubst N, Berger J, Radjendirane V, Graw J, Favor J, Saunders GF, Stoykova A, Götz M: Molecular dissection of Pax6 function: the specific roles of the paired domain and homeodomain in brain development. *Development* **2004**; 131: 6131–6140.

44 Kozmik Z, Daube M, Frei E, Norman B, Kos L, Dishaw LJ, Noll M, Piatigorsky J: Role of Pax genes in eye evolution: a cnidarian PaxB gene uniting Pax2 and Pax6 functions. *Dev Cell* **2003**; 5: 773–785.

45 Walther C, Gruss P: Pax–6, a murine paired box gene, is expressed in the developing CNS. *Development* **1991**; 113: 1435–1449.

46 Chauhan BK, Yang Y, Cveklova K, Cvekl A: Functional interactions between alternatively spliced forms of Pax6 in crystallin gene regulation and in haploinsufficiency. *Nucleic Acids Res* **2004**; 32: 1696–1709.

47 Azuma N, Tadokoro K, Asaka A, Yamada M, Yamaguchi Y, Handa H, Matsushima S, Watanabe T, Kohsaka S, Kida Y, et al.: The Pax6 isoform bearing an alternative spliced exon promotes the development of the neural retinal structure. *Hum Mol Genet* **2005**; 14: 735–745.

48 Singh S, Mishra R, Arango NA, Deng JM, Behringer RR, Saunders GF: Iris hypoplasia in mice that lack the alternatively spliced Pax6(5a) isoform. *Proc Natl Acad Sci USA* **2002**; 99: 6812–6815.

49 Dominguez M, Ferres-Marco D, Gutierrez-Avino FJ, Speicher SA, Beneyto M: Growth and specification of the eye are controlled independently by Eyegone and Eyeless in *Drosophila melanogaster*. *Nat Genet* **2004**; 36: 31–39.

50 Quiring R, Walldorf U, Kloter U, Gehring WJ: Homology of the eyeless gene of *Drosophila* to the Small eye gene in mice and Aniridia in humans. *Science* **1994**; 265: 785–789.

51 Czerny T, Halder G, Kloter U, Souabni A, Gehring WJ, Busslinger M: twin of eyeless, a second Pax–6 gene of *Drosophila*, acts upstream of eyeless in the control of eye development. *Mol Cell* **1999**; 3: 297–307.

52 Jang CC, Chao JL, Jones N, Yao LC, Bessarab DA, Kuo YM, Jun S, Desplan C, Beckendorf SK, Sun YH: Two Pax genes, eye gone and eyeless, act cooperatively in promoting *Drosophila* eye development. *Development* **2003**; 130: 2939–2951.

53 Jun S, Wallen RV, Goriely A, Kalionis B, Desplan C: Lune/eye gone, a Pax-like protein, uses a partial paired domain and a homeodomain for DNA recognition. *Proc Natl Acad Sci USA* **1998**; 95: 13720–13725.

54 Puschel AW, Gruss P, Westerfield M: Sequence and expression pattern of pax–6 are highly conserved between zebrafish and mice. *Development* **1992**; 114: 643–651.

55 Nornes S, Clarkson M, Mikkola I, Pedersen M, Bardsley A, Martinez JP, Krauss S, Johansen T: Zebrafish contains two pax6 genes involved in eye development. *Mech Dev* **1998**; 77: 185–196.

56 Dozier C, Carriere C, Grevin D, Martin P, Quatannens B, Stehelin D, Saule S: Structure and DNA-binding properties of Pax-QNR, a paired box- and homeobox-containing gene. *Cell Growth Differ* **1993**; 4: 281–289.

57 Glaser T, Walton DS, Maas RL: Genomic structure, evolutionary conservation and aniridia mutations in the human PAX6 gene. *Nat Genet* **1992**; 2: 232–239.

58 Mishra R, Gorlov IP, Chao LY, Singh S, Saunders GF: PAX6, paired domain influences sequence recognition by the homeodomain. *J Biol Chem* **2002**; 277: 49488–49494.

59 Carriere C, Plaza S, Martin P, Quatannens B, Bailly M, Stehelin D, Saule S: Characterization of quail Pax–6 (Pax-QNR) proteins expressed in the neuroretina. *Mol Cell Biol* **1993**; 13: 7257–7266.

60 Zhang Y, Emmons SW: Specification of sense-organ identity by a *Caenorhabditis elegans* Pax–6 homologue. *Nature* 1995; 377: 55–59.
61 Kleinjan DA, Seawright A, Childs AJ, van Heyningen V: Conserved elements in Pax6 intron 7 involved in (auto)regulation and alternative transcription. *Dev Biol* 2004; 265: 462–477.
62 Mikkola I, Bruun JA, Holm T, Johansen T: Superactivation of Pax6-mediated transactivation from paired domain-binding sites by DNA-independent recruitment of different homeodomain proteins. *J Biol Chem* 2001; 276: 4109–4118.
63 Jaworski C, Sperbeck S, Graham C, Wistow G: Alternative splicing of Pax6 in bovine eye and evolutionary conservation of intron sequences. *Biochem Biophys Res Commun* 1997; 240: 196–202.
64 Estivill-Torrus G, Pearson H, van Heyningen V, Price DJ, Rashbass P: Pax6 is required to regulate the cell cycle and the rate of progression from symmetrical to asymmetrical division in mammalian cortical progenitors. *Development* 2002; 129: 455–466.
65 Götz M, Stoykova A, Gruss P: Pax6 controls radial glia differentiation in the cerebral cortex. *Neuron* 1998; 21: 1031–1044.
66 Heins N, Malatesta P, Cecconi F, Nakafuku M, Tucker KL, Hack MA, Chapouton P, Barde YA, Götz M: Glial cells generate neurons: the role of the transcription factor Pax6. *Nat Neurosci* 2002; 5: 308–315.
67 Kamachi Y, Uchikawa M, Tanouchi A, Sekido R, Kondoh H: Pax6 and SOX2 form a co-DNA-binding partner complex that regulates initiation of lens development. *Genes Dev* 2001; 15: 1272–1286.
68 Schilham MW, Oosterwegel MA, Moerer P, Ya J, de Boer PA, van de Wetering M, Verbeek S, Lamers WH, Kruisbeek AM, Cumano A, et al.: Defects in cardiac outflow tract formation and pro-B-lymphocyte expansion in mice lacking Sox–4. *Nature* 1996; 380: 711–714.
69 Plaza S, Langlois MC, Turque N, Le-Cornet S, Bailly M, Begue A, Quatannens B, Dozier C, Saule S: The homeobox-containing Engrailed (En–1) product down-regulates the expression of Pax–6 through a DNA binding-independent mechanism. *Cell Growth Differ* 1997; 8: 1115–1125.
70 Mikkola I, Bruun JA, Bjorkoy G, Holm T, Johansen T: Phosphorylation of the transactivation domain of Pax6 by extracellular signal-regulated kinase and p38 mitogen-activated protein kinase. *J Biol Chem* 1999; 274: 15115–15126.
71 Lefebvre T, Planque N, Leleu D, Bailly M, Caillet-Boudin ML, Saule S, Michalski JC: O-glycosylation of the nuclear forms of Pax–6 products in quail neuroretina cells. *J Cell Biochem* 2002; 85: 208–218.
72 Halder G, Callaerts P, Gehring WJ: New perspectives on eye evolution. *Curr Opin Genet Dev* 1995; 5: 602–609.
73 Chow RL, Altmann CR, Lang RA, Hemmati-Brivanlou A: Pax6 induces ectopic eyes in a vertebrate. *Development* 1999; 126: 4213–4222.
74 Schwarz M, Cecconi F, Bernier G, Andrejewski N, Kammandel B, Wagner M, Gruss P: Spatial specification of mammalian eye territories by reciprocal transcriptional repression of Pax2 and Pax6. *Development* 2000; 127: 4325–4334.
75 Leconte L, Lecoin L, Martin P, Saule S: Pax6 interacts with cVax and Tbx5 to establish the dorsoventral boundary of the developing eye. *J Biol Chem* 2004; 279: 47272–47277.
76 Hogan BL, Hirst EM, Horsburgh G, Hetherington CM: Small eye (Sey): a mouse model for the genetic analysis of craniofacial abnormalities. *Development* 1988; 103 Suppl: 115–119.
77 Hogan BL, Horsburgh G, Cohen J, Hetherington CM, Fisher G, Lyon MF: Small eyes (Sey): a homozygous lethal mutation on chromosome 2 which affects the differentiation of both lens and nasal placodes in the mouse. *J Embryol Exp Morphol* 1986; 97: 95–110.

78 Grindley JC, Davidson DR, Hill RE: The role of Pax–6 in eye and nasal development. *Development* **1995**; 121: 1433–1442.

79 Philips GT, Stair CN, Young Lee H, Wroblewski E, Berberoglu MA, Brown NL, Mastick GS: Precocious retinal neurons: Pax6 controls timing of differentiation and determination of cell type. *Dev Biol* **2005**; 279: 308–321.

80 van Raamsdonk CD, Tilghman SM: Dosage requirement and allelic expression of PAX6 during lens placode formation. *Development* **2000**; 127: 5439–5448.

81 Kammandel B, Chowdhury K, Stoykova A, Aparicio S, Brenner S, Gruss P: Distinct cis-essential modules direct the time-space pattern of the Pax6 gene activity. *Dev Biol* **1999**; 205: 79–97.

82 Bäumer N, Marquardt T, Stoykova A, Ashery-Padan R, Chowdhury K, Gruss P: Pax6 is required for establishing naso-temporal and dorsal characteristics of the optic vesicle. *Development* **2002**; 129: 4535–4545.

83 Collinson JM, Quinn JC, Hill RE, West JD: The roles of Pax6 in the cornea, retina, and olfactory epithelium of the developing mouse embryo. *Dev Biol* **2003**; 255: 303–312.

84 Marquardt T, Ashery-Padan R, Andrejewski N, Scardigli R, Guillemot F, Gruss P: Pax6 is required for the multipotent state of retinal progenitor cells. *Cell* **2001**; 105: 43–55.

85 Ashery-Padan R, Marquardt T, Zhou X, Gruss P: Pax6 activity in the lens primordium is required for lens formation and for correct placement of a single retina in the eye. *Genes Dev* **2000**; 14: 2701–2711.

86 Bäumer N, Marquardt T, Stoykova A, Spieler D, Treichel D, Ashery-Padan R, Gruss P: Retinal pigmented epithelium determination requires the redundant activities of Pax2 and Pax6. *Development* **2003**; 130: 2903–2915.

87 Planque N, Leconte L, Coquelle FM, Martin P, Saule S: Specific Pax–6/microphthalmia transcription factor interactions involve their DNA-binding domains and inhibit transcriptional properties of both proteins. *J Biol Chem* **2001**; 276: 29330–29337.

88 Hill RE, Favor J, Hogan BL, Ton CC, Saunders GF, Hanson IM, Prosser J, Jordan T, Hastie ND, van Heyningen V: Mouse small eye results from mutations in a paired-like homeobox-containing gene. *Nature* **1991**; 354: 522–525.

89 Ton CC, Hirvonen H, Miwa H, Weil MM, Monaghan P, Jordan T, van Heyningen V, Hastie ND, Meijers-Heijboer H, Drechsler M, et al.: Positional cloning and characterization of a paired box- and homeobox-containing gene from the aniridia region. *Cell* **1991**; 67: 1059–1074.

90 Lyon MF, Bogani D, Boyd Y, Guillot P, Favor J: Further genetic analysis of two autosomal dominant mouse eye defects, Ccw and Pax6(coop). *Mol Vis* **2000**; 6: 199–203.

91 Thaung C, West K, Clark BJ, McKie L, Morgan JE, Arnold K, Nolan PM, Peters J, Hunter AJ, Brown SD, et al.: Novel ENU-induced eye mutations in the mouse: models for human eye disease. *Hum Mol Genet* **2002**; 11: 755–767.

92 Schedl A, Ross A, Lee M, Engelkamp D, Rashbass P, van Heyningen V, Hastie ND: Influence of PAX6 gene dosage on development: overexpression causes severe eye abnormalities. *Cell* **1996**; 86: 71–82.

93 Glaser T, Jepeal L, Edwards JG, Young SR, Favor J, Maas RL: PAX6 gene dosage effect in a family with congenital cataracts, aniridia, anophthalmia and central nervous system defects. *Nat Genet* **1994**; 7: 463–471.

94 Jordan T, Hanson I, Zaletayev D, Hodgson S, Prosser J, Seawright A, Hastie N, van Heyningen V: The human PAX6 gene is mutated in two patients with aniridia. *Nat Genet* **1992**; 1: 328–332.

95 Hanson I, Brown A, van Heyningen V: A new PAX6 mutation in familial aniridia. *J Med Genet* **1995**; 32: 488–489.

96 Mirzayans F, Pearce WG, MacDonald IM, Walter MA: Mutation of the PAX6 gene in patients with autosomal dominant keratitis. *Am J Hum Genet* **1995**; 57: 539–548.
97 Azuma N, Nishina S, Yanagisawa H, Okuyama T, Yamada M: PAX6 missense mutation in isolated foveal hypoplasia. *Nat Genet* **1996**; 13: 141–142.
98 Fukushima Y, Hoovers J, Mannens M, Wakui K, Ohashi H, Ohno T, Ueoka Y, Niikawa N: Detection of a cryptic paracentric inversion within band 11p13 in familial aniridia by fluorescence in situ hybridization. *Hum Genet* **1993**; 91: 205–209.
99 Fantes J, Redeker B, Breen M, Boyle S, Brown J, Fletcher J, Jones S, Bickmore W, Fukushima Y, Mannens M, et al.: Aniridia-associated cytogenetic rearrangements suggest that a position effect may cause the mutant phenotype. *Hum Mol Genet* **1995**; 4: 415–422.
100 Lauderdale JD, Wilensky JS, Oliver ER, Walton DS, Glaser T: 3' deletions cause aniridia by preventing PAX6 gene expression. *Proc Natl Acad Sci USA* **2000**; 97: 13755–13759.
101 Crolla JA, van Heyningen V: Frequent chromosome aberrations revealed by molecular cytogenetic studies in patients with aniridia. *Am J Hum Genet* **2002**; 71: 1138–1149.
102 Bishop KM, Goudreau G, O'Leary DD: Regulation of area identity in the mammalian neocortex by Emx2 and Pax6. *Science* **2000**; 288: 344–349.
103 Yun K, Potter S, Rubenstein JL: Gsh2 and Pax6 play complementary roles in dorsoventral patterning of the mammalian telencephalon. *Development* **2001**; 128: 193–205.
104 Stoykova A, Treichel D, Hallonet M, Gruss P: Pax6 modulates the dorsoventral patterning of the mammalian telencephalon. *J Neurosci* **2000**; 20: 8042–8050.
105 Tole S, Remedios R, Saha B, Stoykova A: Selective requirement of Pax6, but not Emx2, in the specification and development of several nuclei of the amygdaloid complex. *J Neurosci* **2005**; 25: 2753–2760.

106 Schmahl W, Knoedlseder M, Favor J, Davidson D: Defects of neuronal migration and the pathogenesis of cortical malformations are associated with Small eye (Sey) in the mouse, a point mutation at the Pax–6–locus. *Acta Neuropathol (Berl)* **1993**; 86: 126–135.
107 Haubensak W, Attardo A, Denk W, Huttner WB: Neurons arise in the basal neuroepithelium of the early mammalian telencephalon: a major site of neurogenesis. *Proc Natl Acad Sci USA* **2004**; 101: 3196–3201.
108 Noctor SC, Martinez-Cerdeno V, Ivic L, Kriegstein AR: Cortical neurons arise in symmetric and asymmetric division zones and migrate through specific phases. *Nat Neurosci* **2004**; 7: 136–144.
109 Miyata T, Kawaguchi A, Saito K, Kawano M, Muto T, Ogawa M: Asymmetric production of surface-dividing and non-surface-dividing cortical progenitor cells. *Development* **2004**; 131: 3133–3145.
110 Hartfuss E, Galli R, Heins N, Götz M: Characterization of CNS precursor subtypes and radial glia. *Dev Biol* **2001**; 229: 15–30.
111 Hartfuss E, Forster E, Bock HH, Hack MA, Leprince P, Luque JM, Herz J, Frotscher M, Götz M: Reelin signaling directly affects radial glia morphology and biochemical maturation. *Development* **2003**; 130: 4597–4609.
112 Götz M: Glial cells generate neurons – master control within CNS regions: developmental perspectives on neural stem cells. *Neuroscientist* **2003**; 9: 379–397.
113 Noctor SC, Flint AC, Weissman TA, Wong WS, Clinton BK, Kriegstein AR: Dividing precursor cells of the embryonic cortical ventricular zone have morphological and molecular characteristics of radial glia. *J Neurosci* **2002**; 22: 3161–3173.
114 Hack MA, Sugimori M, Lundberg C, Nakafuku M, Götz M: Regionalization and fate specification in neurospheres: the role of Olig2 and Pax6. *Mol Cell Neurosci* **2004**; 25: 664–678.
115 Alvarez-Buylla A, Garcia-Verdugo JM,

Tramontin AD: A unified hypothesis on the lineage of neural stem cells. Nat Rev Neurosci 2001; 2: 287–293.
116 Caviness VS, Jr., Takahashi T: Proliferative events in the cerebral ventricular zone. Brain Dev 1995; 17: 159–163.
117 Caviness VS, Jr., Takahashi T, Nowakowski RS: The G1 restriction point as critical regulator of neocortical neuronogenesis. Neurochem Res 1999; 24: 497–506.
118 Tarabykin V, Stoykova A, Usman N, Gruss P: Cortical upper layer neurons derive from the subventricular zone as indicated by Svet1 gene expression. Development 2001; 128: 1983–1993.
119 Nieto M, Monuki ES, Tang H, Imitola J, Haubst N, Khoury SJ, Cunningham J, Götz M, Walsh CA: Expression of Cux–1 and Cux–2 in the subventricular zone and upper layers II-IV of the cerebral cortex. J Comp Neurol 2004; 479: 168–180.
120 Englund C, Fink A, Lau C, Pham D, Daza RA, Bulfone A, Kowalczyk T, Hevner RF: Pax6, Tbr2, and Tbr1 are expressed sequentially by radial glia, intermediate progenitor cells, and postmitotic neurons in developing neocortex. J Neurosci 2005; 25: 247–251.
121 Schuurmans C, Armant O, Nieto M, Stenman JM, Britz O, Klenin N, Brown C, Langevin LM, Seibt J, Tang H, et al.: Sequential phases of cortical specification involve Neurogenin-dependent and -independent pathways. EMBO J 2004; 23: 2892–2902.
122 Caric D, Gooday D, Hill RE, McConnell SK, Price DJ: Determination of the migratory capacity of embryonic cortical cells lacking the transcription factor Pax–6. Development 1997; 124: 5087–5096.
123 Toresson H, Potter SS, Campbell K: Genetic control of dorsal-ventral identity in the telencephalon: opposing roles for Pax6 and Gsh2. Development 2000; 127: 4361–4371.
124 Muzio L, DiBenedetto B, Stoykova A, Boncinelli E, Gruss P, Mallamaci A: Emx2 and Pax6 control regionalization of the pre-neuronogenic cortical primordium. Cereb Cortex 2002; 12: 129–139.
125 Muzio L, Mallamaci A: Emx1, emx2 and pax6 in specification, regionalization and arealization of the cerebral cortex. Cereb Cortex 2003; 13: 641–647.
126 Bishop KM, Rubenstein JL, O'Leary DD: Distinct actions of Emx1, Emx2, and Pax6 in regulating the specification of areas in the developing neocortex. J Neurosci 2002; 22: 7627–7638.
127 Stoykova A, Fritsch R, Walther C, Gruss P: Forebrain patterning defects in Small eye mutant mice. Development 1996; 122: 3453–3465.
128 Stoykova A, Götz M, Gruss P, Price J: Pax6–dependent regulation of adhesive patterning, R-cadherin expression and boundary formation in developing forebrain. Development 1997; 124: 3765–3777.
129 Scardigli R, Bäumer N, Gruss P, Guillemot F, Le Roux I: Direct and concentration-dependent regulation of the proneural gene Neurogenin2 by Pax6. Development 2003; 130: 3269–3281.
130 Chapouton P, Gartner A, Götz M: The role of Pax6 in restricting cell migration between developing cortex and basal ganglia. Development 1999; 126: 5569–5579.
131 Inoue T, Tanaka T, Takeichi M, Chisaka O, Nakamura S, Osumi N: Role of cadherins in maintaining the compartment boundary between the cortex and striatum during development. Development 2001; 128: 561–569.
132 Edwards MA, Yamamoto M, Caviness VS, Jr.: Organization of radial glia and related cells in the developing murine CNS. An analysis based upon a new monoclonal antibody marker. Neuroscience 1990; 36: 121–144.
133 Hirata T, Nomura T, Takagi Y, Sato Y, Tomioka N, Fujisawa H, Osumi N: Mosaic development of the olfactory cortex with Pax6–dependent and -independent components. Brain Res Dev Brain Res 2002; 136: 17–26.
134 Kim AS, Anderson SA, Rubenstein JL, Lowenstein DH, Pleasure SJ: Pax–6 regulates expression of SFRP–2 and

Wnt–7b in the developing CNS. *J Neurosci* 2001; 21: RC132.
135 Assimacopoulos S, Grove EA, Ragsdale CW: Identification of a Pax6-dependent epidermal growth factor family signaling source at the lateral edge of the embryonic cerebral cortex. *J Neurosci* 2003; 23: 6399–6403.
136 Stenman J, Yu RT, Evans RM, Campbell K: Tlx and Pax6 co-operate genetically to establish the pallio-subpallial boundary in the embryonic mouse telencephalon. *Development* 2003; 130: 1113–1122.
137 Muzio L, DiBenedetto B, Stoykova A, Boncinelli E, Gruss P, Mallamaci A: Conversion of cerebral cortex into basal ganglia in Emx2($^{-/-}$) Pax6(Sey/Sey) double-mutant mice. *Nat Neurosci* 2002; 5: 737–745.
138 Takiguchi-Hayashi K, Sekiguchi M, Ashigaki S, Takamatsu M, Hasegawa H, Suzuki-Migishima R, Yokoyama M, Nakanishi S, Tanabe Y: Generation of reelin-positive marginal zone cells from the caudomedial wall of telencephalic vesicles. *J Neurosci* 2004; 24: 2286–2295.
139 Stoykova A, Hatano O, Gruss P, Götz M: Increase in reelin-positive cells in the marginal zone of Pax6 mutant mouse cortex. *Cereb Cortex* 2003; 13: 560–571.
140 Jimenez D, Garcia C, de Castro F, Chedotal A, Sotelo C, de Carlos JA, Valverde F, Lopez-Mascaraque L: Evidence for intrinsic development of olfactory structures in Pax–6 mutant mice. *J Comp Neurol* 2000; 428: 511–526.
141 Nomura T, Osumi N: Misrouting of mitral cell progenitors in the Pax6/small eye rat telencephalon. *Development* 2004; 131: 787–796.
142 Marin O, Rubenstein JL: Cell migration in the forebrain. *Annu Rev Neurosci* 2003; 26: 441–483.
143 Fode C, Ma Q, Casarosa S, Ang SL, Anderson DJ, Guillemot F: A role for neural determination genes in specifying the dorsoventral identity of telencephalic neurons. *Genes Dev* 2000; 14: 67–80.
144 Kroll TT, O'Leary DD: Ventralized dorsal telencephalic progenitors in Pax6 mutant mice generate GABA interneurons of a lateral ganglionic eminence fate. *Proc Natl Acad Sci USA* 2005; 102: 7374–7379.
145 Kleinjan DA, Seawright A, Schedl A, Quinlan RA, Danes S, van Heyningen V: Aniridia-associated translocations, DNase hypersensitivity, sequence comparison and transgenic analysis redefine the functional domain of PAX6. *Hum Mol Genet* 2001; 10: 2049–2059.
146 Griffin C, Kleinjan DA, Doe B, van Heyningen V: New 3' elements control Pax6 expression in the developing pretectum, neural retina and olfactory region. *Mech Dev* 2002; 112: 89–100.
147 Warren N, Price DJ: Roles of Pax–6 in murine diencephalic development. *Development* 1997; 124: 1573–1582.
148 Estivill-Torrus G, Vitalis T, Fernandez-Llebrez P, Price DJ: The transcription factor Pax6 is required for development of the diencephalic dorsal midline secretory radial glia that form the subcommissural organ. *Mech Dev* 2001; 109: 215–224.
149 Mastick GS, Andrews GL: Pax6 regulates the identity of embryonic diencephalic neurons. *Mol Cell Neurosci* 2001; 17: 190–207.
150 Vitalis T, Cases O, Engelkamp D, Verney C, Price DJ: Defect of tyrosine hydroxylase-immunoreactive neurons in the brains of mice lacking the transcription factor Pax6. *J Neurosci* 2000; 20: 6501–6516.
151 Mastick GS, Davis NM, Andrew GL, Easter SS, Jr.: Pax–6 functions in boundary formation and axon guidance in the embryonic mouse forebrain. *Development* 1997; 124: 1985–1997.
152 Jones L, Lopez-Bendito G, Gruss P, Stoykova A, Molnar Z: Pax6 is required for the normal development of the forebrain axonal connections. *Development* 2002; 129: 5041–5052.
153 Engelkamp D, Rashbass P, Seawright A, van Heyningen V: Role of Pax6 in development of the cerebellar system.

Development **1999**; 126: 3585–3596.

154 Yamasaki T, Kawaji K, Ono K, Bito H, Hirano T, Osumi N, Kengaku M: Pax6 regulates granule cell polarization during parallel fiber formation in the developing cerebellum. *Development* **2001**; 128: 3133–3144.

155 Oliver TG, Read TA, Kessler JD, Mehmeti A, Wells JF, Huynh TT, Lin SM, Wechsler-Reya RJ: Loss of patched and disruption of granule cell development in a pre-neoplastic stage of medulloblastoma. *Development* **2005**; 132: 2425–2439.

156 Pituello F, Yamada G, Gruss P: Activin A inhibits Pax–6 expression and perturbs cell differentiation in the developing spinal cord in vitro. *Proc Natl Acad Sci USA* **1995**; 92: 6952–6956.

157 Ericson J, Rashbass P, Schedl A, Brenner-Morton S, Kawakami A, van Heyningen V, Jessell TM, Briscoe J: Pax6 controls progenitor cell identity and neuronal fate in response to graded Shh signaling. *Cell* **1997**; 90: 169–180.

158 Takahashi M, Osumi N: Pax6 regulates specification of ventral neurone subtypes in the hindbrain by establishing progenitor domains. *Development* **2002**; 129: 1327–1338.

159 Osumi N, Hirota A, Ohuchi H, Nakafuku M, Iimura T, Kuratani S, Fujiwara M, Noji S, Eto K: Pax–6 is involved in the specification of hindbrain motor neuron subtype. *Development* **1997**; 124: 2961–2972.

160 Sun T, Pringle NP, Hardy AP, Richardson WD, Smith HK: Pax6 influences the time and site of origin of glial precursors in the ventral neural tube. *Mol Cell Neurosci* **1998**; 12: 228–239.

161 Burrill JD, Moran L, Goulding MD, Saueressig H: PAX2 is expressed in multiple spinal cord interneurons, including a population of EN1+ interneurons that require PAX6 for their development. *Development* **1997**; 124: 4493–4503.

162 Stoykova A, Gruss P: Roles of Pax-genes in developing and adult brain as suggested by expression patterns. *J Neurosci* **1994**; 14: 1395–1412.

163 Doetsch F, Caille I, Lim DA, Garcia-Verdugo JM, Alvarez-Buylla A: Subventricular zone astrocytes are neural stem cells in the adult mammalian brain. *Cell* **1999**; 97: 703–716.

164 Carleton A, Petreanu LT, Lansford R, Alvarez-Buylla A, Lledo PM: Becoming a new neuron in the adult olfactory bulb. *Nat Neurosci* **2003**; 6: 507–518.

165 Jimenez D, Lopez-Mascaraque L, de Carlos JA, Valverde F: Further studies on cortical tangential migration in wild type and Pax–6 mutant mice. *J Neurocytol* **2002**; 31: 719–728.

166 Holst BD, Wang Y, Jones FS, Edelman GM: A binding site for Pax proteins regulates expression of the gene for the neural cell adhesion molecule in the embryonic spinal cord. *Proc Natl Acad Sci USA* **1997**; 94: 1465–1470.

167 Meech R, Kallunki P, Edelman GM, Jones FS: A binding site for homeodomain and Pax proteins is necessary for L1 cell adhesion molecule gene expression by Pax–6 and bone morphogenetic proteins. *Proc Natl Acad Sci USA* **1999**; 96: 2420–2425.

168 Duncan MK, Kozmik Z, Cveklova K, Piatigorsky J, Cvekl A: Overexpression of PAX6(5a) in lens fiber cells results in cataract and upregulation of (alpha)5(beta)1 integrin expression. *J Cell Sci* **2000**; 113 (Pt 18): 3173–3185.

169 Niimi T, Seimiya M, Kloter U, Flister S, Gehring WJ: Direct regulatory interaction of the eyeless protein with an eye-specific enhancer in the sine oculis gene during eye induction in *Drosophila*. *Development* **1999**; 126: 2253–2260.

170 Cvekl A, Kashanchi F, Sax CM, Brady JN, Piatigorsky J: Transcriptional regulation of the mouse alpha A-crystallin gene: activation dependent on a cyclic AMP-responsive element (DE1/CRE) and a Pax–6–binding site. *Mol Cell Biol* **1995**; 15: 653–660.

171 Duncan MK, Haynes JI, 2nd, Cvekl A, Piatigorsky J: Dual roles for Pax–6: a transcriptional repressor of lens fiber cell-specific beta-crystallin genes. *Mol Cell Biol* **1998**; 18: 5579–5586.

172 Richardson J, Cvekl A, Wistow G:

Pax–6 is essential for lens-specific expression of zeta-crystallin. *Proc Natl Acad Sci USA* **1995**; 92: 4676–4680.

173 Sakai M, Serria MS, Ikeda H, Yoshida K, Imaki J, Nishi S: Regulation of c-maf gene expression by Pax6 in cultured cells. *Nucleic Acids Res* **2001**; 29: 1228–1237.

174 Papatsenko D, Nazina A, Desplan C: A conserved regulatory element present in all *Drosophila* rhodopsin genes mediates Pax6 functions and participates in the fine-tuning of cell-specific expression. *Mech Dev* **2001**; 101: 143–153.

175 Sheng G, Thouvenot E, Schmucker D, Wilson DS, Desplan C: Direct regulation of rhodopsin 1 by Pax–6/eyeless in *Drosophila*: evidence for a conserved function in photoreceptors. *Genes Dev* **1997**; 11: 1122–1131.

176 Kozmik Z, Czerny T, Busslinger M: Alternatively spliced insertions in the paired domain restrict the DNA sequence specificity of Pax6 and Pax8. *EMBO J* **1997**; 16: 6793–6803.

177 Chauhan BK, Yang Y, Cveklova K, Cvekl A: Functional properties of natural human PAX6 and PAX6(5a) mutants. *Invest Ophthalmol Vis Sci* **2004**; 45: 385–392.

178 Hack MA, Saghatelyan A, de Chevigny A, Pfeifer A, Ashery-Padan R, Lledo PM, Götz M: Neuronal fate determinants of adult olfactory bulb neurogenesis. *Nature Neuroscience* **2005** Jul; 8(7): 865–72.

3
Phox2a and Phox2b: Essential Transcription Factors for Neuron Specification and Differentiation

Uwe Ernsberger and Hermann Rohrer

Abstract

The paired-homeodomain transcription factors Phox2a and Phox2b are selectively expressed in restricted parts of the peripheral and central nervous systems. Phox2a and Phox2b are required for neuron development, affecting cell cycle exit, subtype-specific and generic neuronal differentiation. The specification of noradrenergic neurons is nicely explained by their direct action on the promoter of the dopamine beta-hydroxylase gene, whereas other Phox2 targets remain to be characterized. The multiple functions of Phox2 transcription factors in different neuronal lineages implicate specific co-determinants, acting together with Phox2 in neuron subtype specification and differentiation. The expression of Phox2a and Phox2b in autonomic visceral reflex circuits suggests that circuit formation is also included amongst the many important functions controlled by these transcription factors.

3.1
Introduction

The Phox2 transcription factors Phox2a/Arix and Phox2b/NBPhox are important regulators of cell fate and differentiation in the nervous system. They are expressed in virtually all neurons of autonomic visceral reflex circuits in both the peripheral nervous system (PNS) and central nervous system (CNS), and are essential for the development of these neurons. They were discovered during the search for proteins which regulated expression from the neural cell adhesion molecule (NCAM) (Valarché et al., 1993) and the dopamine-β-hydroxylase (DBH) (Zellmer et al., 1995) promoter. Characterization of a nuclear protein that binds to a consensus site for homeobox-containing transcription factors in both promoters and activates expression from corresponding promoter/reporter constructs led to the identification of Phox2, alternatively called Arix. With the cloning by homology screen of a related gene product (Pattyn et al., 1997), the initially identified family member was called Phox2a/Arix, and the newly described factor Phox2b/NBPhox. They are characterized by a paired class homeodomain (HD) with the amino acid Q at position 50,

---

Transcription Factors. Edited by Gerald Thiel
Copyright © 2006 WILEY-VCH Verlag GmbH & Co. KGaA, Weinheim
ISBN 3-527-31285-4

classifying the Phox2 proteins as Q50 paired-like proteins (Galliot et al., 1999). For simplicity, we use the terms Phox2a and Phox2b in this chapter, which focuses mainly on the physiological functions of Phox2 signaling. Reference is also made to a previous excellent review detailing Phox2 functions (Brunet and Pattyn, 2002)

3.2
Molecular Characteristics of Phox2 Genes and Proteins

3.2.1
Sequence and Gene Structure Conservation in the Animal Kingdom

Mouse Phox2a and Phox2b share identical homeodomain sequences (Pattyn et al., 1997). The N-terminal domains are 57% identical, whilst the C-termini are highly divergent. Zebrafish Phox2a is 66% identical at the amino acid level to mouse Phox2a (Guo et al., 1999). CEH17 in *Caenorhabditis elegans*, which is discussed as nematode Phox2 orthologue, shows 88% identity to the mouse Phox2 homeodomain (Pujol et al., 2000). Outside the homeodomain, however, no significant similarities are detected.

The genes for Phox2a and Phox2b are unlinked on mouse and human chromosomes (Adachi et al., 2000). The gene structure is similar, with each gene containing three exons and two introns. The same gene structure is reported for *C. elegans* CEH17 (Pujol et al., 2000). In *Xenopus*, a Phox2a splice variant, XPhox2a.2 has been characterized that diverges at the exon1/intron1 boundary from XPhox2a (Talikka et al., 2004).

3.2.2
Transcriptional Activation by Phox2 Proteins

RGS4 and gustucin were identified as Phox2b target genes in a screen for genes differentially expressed in hindbrain branchial motor neuron precursors of Phox2b heterozygous or homozygous mutant embryos (Grillet et al., 2003). Although Phox2b, in collaboration with Mash1, is able to induce RGS4 expression in chick spinal cord, it is unclear whether Phox2b is activating gene expression directly or indirectly. Evidence for the direct transcriptional action by Phox2 proteins has been obtained with promoter constructs derived from DBH, NCAM, Phox2a, and NET genes. As DBH, Phox2a and NET genes are expressed in a small selected group of neurons, while NCAM expression is widely detectable throughout the nervous system, this set of target genes points at the exciting question of how transcription factors coordinate expression of pan-neuronal and population-specific genes.

The most detailed analysis of Phox2 action is available for the rat and human DBH promoter. The DB1 enhancer fragment derived from the rat DBH promoter has been used to characterize Phox2a (Zellmer et al., 1995). The DB1 fragment contains two Phox2 binding sites, and mutation of both reduces gene expression from a DB1 reporter construct in PC12 cells. The corresponding region in the human DBH promoter, called domain IV, is considered a noradrenergic-specific cis-acting element, since mutation of the Phox2 binding sites does not diminish reporter gene expression in non-noradrenergic cell lines (Yang et al., 1998). Both, Phox2a and Phox2b can activate expression from a domain IV reporter construct. Another Phox2 binding region, domain II, is found between domain IV and the TATA box and mediates transactivation of reporter constructs by Phox2a and Phox2b (Kim et al., 1998). Electrophoretic mobility shift analysis indicates that Phox2a dimers can bind to the domain II sites while monomers bind the domain IV sites (Seo et al., 2002).

In order to understand the regulatory cascades leading to cell type-specific and physiologically appropriate gene activation, Phox2 interaction with other transcription factors and signaling pathways has been analyzed. In non-neuronal cell cultures, transcription from a rat DBH promoter is substantially activated when Phox2a is present together with cAMP (Swanson et al., 1997). Decrease of the phosphorylation status of Phox2a upon stimulation of cAMP-dependent protein kinase (PKA), coinciding with enhancement of Phox2a DNA binding, is necessary to fully activate DBH expression (Adachi and Lewis, 2002). In addition, homeodomain binding motives and cAMP-responsive element (CRE) are found adjacent in domain IV of the DBH promoter. Phox2a can interact via its N-terminal activation domain with cAMP-response element-binding protein-binding protein (CBP) to potentiate transcription from the DBH promoter (Swanson et al., 2000). Since co-stimulation with the cAMP/PKA system is necessary for noradrenergic differentiation in neural crest cultures (Lo et al., 1999), convergence of Phox2 and CBP action may be required to induce expression of the DBH gene during neuronal differentiation.

A second protein interacting with Phox2a at the rat DBH promoter is the basic helix-loop-helix (bHLH) transcription factor Hand2 (Rychlik et al., 2003; Xu et al., 2003). This stimulates expression from a DBH promoter reporter construct in the presence of Phox2a independently of its basic DNA-binding domain. Activation is lost in coexpression assays when the homeodomain sites in the DBH promoter are removed, demonstrating dependence upon Phox2a-DNA interaction. The issue of direct protein-protein interaction between Phox2a and Hand2 is controversial (Rychlik et al., 2003; Xu et al., 2003). Sequential expression of Phox2s and Hand2 during the development of noradrenergic neurons (Howard et al., 2000) points at the successive recruitment of transcription factors required for sympathetic neuron specification and noradrenergic differentiation.

Phox2a was originally characterized by its binding to a promoter fragment of the gene coding for mouse neural cell adhesion molecule (NCAM) (Valarché et al., 1993). In N2a cells, Phox2a can relieve the inhibitory action of the homeodomain protein Cux on a NCAM promoter-reporter construct. The importance of Phox2a for NCAM expression *in vivo* has still to be clarified. As Phox2a is expressed in a restricted number of neuronal subpopulations, while NCAM is much more widely detectable, the question arises of how such population-specific transcription factors regulate widely expressed neuronal genes (Ernsberger, 2004). It is important to note that inactivation of Phox2 transcription factors in mice affects not only the population-specific, but also the generic neuronal differentiation program (Dubreuil et al., 2000). Correspondingly, overexpression of Phox2 transcription factors induces ectopic differentiation of neurons expressing population-specific and general neuronal properties (Stanke et al., 1999; Dubreuil et al., 2000; Patzke et al., 2001). How this coordinated regulation of different sets of neuronally expressed genes is achieved in molecular detail, remains to be elucidated.

Phox2a and Phox2b show a remarkably similar expression pattern in space (Pattyn et al., 1997). The onset of expression, however, differs slightly – but significantly – from either Phox2b expression or Phox2a expression preceding the other, depending on the neuronal lineage (Pattyn et al., 1997; Ernsberger et al., 2000; Howard et al., 2000). The presence of a putative binding site for HD transcription factors in the human Phox2a promoter, the demonstration of Phox2b interaction with this site, and the transactivation of Phox2a promoter-reporter constructs by Phox2b in HeLa cells (Flora et al., 2001; Hong et al., 2001) suggests that transactivation of Phox2a expression by Phox2b may be one denominator of the Phox2a expression pattern. Despite the similarity of expression patterns, Phox2a and Phox2b promoters appear remarkably different (Hong et al., 2001, 2004). However, similar to the regulatory regions in the *Phox2a* gene, the Phox2b promoter contains a Phox2 binding motive and is weakly transactivated by Phox2b. This may stabilize Phox2b expression in an autoregulatory loop.

DBH and tyrosine hydroxylase (TH) – both of which are enzymes in the norepinephrine biosynthesis cascade – are coexpressed in time and space during sympathetic neuron development (Ernsberger et al., 2000). If genes that code for functionally interrelated gene products are also regulated by common mechanisms, they qualify as members of a synexpression group (Niehrs and Pollet, 1999; Ernsberger, 2004). This is supported by the common induction of DBH and TH by bone morphogenetic protein (BMP) growth factors and Phox2 transcription factors. The observation of direct transactivation of the TH promoter by Phox2 transcription factors, however, remains controversial (compare Zellmer et al., 1995, and Yang et al., 1998). Whether the norepinephrine transporter (NET) is also co-regulated by Phox2 transcription factors is presently unclear. The human NET promoter contains a homeodomain-binding motif which interacts with Phox2a (Kim et al., 2002), although transactivation was not observed. In addition, there is no correlation between the onset of NET expression and TH/DBH expression in sympathetic neurons (M. Stanke and H. Rohrer, unpublished results).

Taken together, there is evidence for the direct transactivation by Phox2 transcription factors of genes specifically expressed in noradrenergic neurons. Transmitter phenotype-related genes may be regulated as a synexpression group. Whether generic neuronal genes may also be directly transactivated by Phox2 transcription factors has to be examined. Mutual transactivation of Phox2 transcription factors may serve as a means of establishing transcriptional cascades that stabilize specific gene expression patterns.

3.3
Physiological Relevance of Phox2 Transcription Factors

3.3.1
Expression Pattern

Phox2a and Phox2b are selectively expressed in the developing nervous system in sets of neuronal subtypes that are functionally linked in several ways.

A striking correlation observed during the initial analysis of Phox2a expression sites was that Phox2a is present in all neurons that permanently or transiently express noradrenergic properties, in particular DBH (Valarché et al., 1993; Tiveron et al., 1996; Morin et al., 1997). This includes all noradrenergic centers of the brain – that is, the locus coeruleus (LC) and the neurons of the A1, A2, A5, and A7 regions. There is, however, no exclusive correlation with the noradrenergic phenotype in the CNS as cholinergic branchiomotor and visceromotor neurons in the brainstem and interneurons in the spinal cord express Phox2a and/or Phox2b (Tiveron et al., 1996; Pattyn et al., 1997, 2000b; Dubreuil et al., 2000). In the PNS, Phox2a and Phox2b are expressed by the entire autonomic nervous system – that is, by sympathetic, parasympathetic and enteric neurons, including noradrenergic and cholinergic cells (Tiveron et al., 1996; Pattyn et al., 1997, 1999; Morin et al., 1997; Ernsberger et al., 2000). In addition, the epibranchial placode-derived cranial sensory ganglia, the geniculate, petrosal and nodose ganglia, which also contain DBH-positive neurons, express Phox2a and Phox2b (Tiveron et al., 1996; Pattyn et al., 1997). Thus, Phox2 transcription factors are consistently expressed in DBH-expressing noradrenergic neurons. In addition, Phox2a and Phox2b may be used in defined groups of non-noradrenergic neurons such as certain sets of cholinergic motor neurons.

The detection of Phox2 transcription factors at different levels of autonomic reflex pathways marks another intriguing observation. Autonomic neurons and the placode-derived components of the cranial sensory ganglia represent efferent and afferent parts of visceral reflex pathways. In addition, Phox2a/b are also expressed in visceral motor neurons of the hindbrain (nucleus ambiguus, salivatory nuclei, dorsal nucleus of the vagus nerve), which relay visceral sensory input and autonomic targets in the parasympathetic and enteric system (for a summary, see Brunet and Pattyn, 2002). For example, the neurons of the three-relay visceral sensory pathway comprising the carotid body, petrosal ganglia and the nucleus of the solitary tract (nTS) express Phox2a/b from the onset of their differentiation (Dauger et al., 2003). The Phox2a/b-expressing motor neurons of the trigeminal and facial nuclei can be considered as parts of autonomic reflex circuits when phylogenetic aspects are considered (Fritzsch and Nortcutt, 1993). The only visceral circuit neurons devoid of Phox2 expression are the preganglionic neurons of the spinal cord providing input to Phox2–positive neurons of pre- and paravertebral sympathetic ganglia. Phox2–expressing neurons that do not participate in autonomic circuits are the cranial motor neurons of the oculomotor and trochlear nuclei, the dorsal interneurons in the spinal cord, and the noradrenergic centers LC and A7. But even for the LC there is evidence supporting the idea that it represents a response system to external stimuli and changes in the state of autonomic functions (Moore and Bloom, 1979; Viemari et al., 2004). Taken together, visceral reflex circuits with the exception of preganglionic sympathetic neurons express Phox2 transcription factors. This observation provokes the questions of whether and how Phox2a and Phox2b may regulate neuronal connectivity during the development of these circuits.

3.3.2
Effects of Phox2 Gene Mutations

The physiological role of Phox2 transcription factors has been investigated in the respective knockouts in mice (Morin et al., 1997; Pattyn et al., 1999, 2000a, 2000b, 2003; Dauger et al., 2003; Cross et al., 2004; Viemari et al., 2004) and in Phox2 zebrafish mutants (Guo et al., 1999). Mice deficient for Phox2a survive until birth. Phox2b knockout embryos die shortly after midgestation, but can be maintained by the application of noradrenergic agonists until E18.5 (Pattyn et al., 2000b). Thus, the role of Phox2 factors in the embryonic development of different neuronal lineages could be analyzed and the results are summarized in the following section.

3.3.2.1 Autonomic Neural Crest Derivatives and Visceral Sensory Ganglia

The lack of Phox2a results in the absence of cranial parasympathetic ganglia and produces a minor migratory defect in the rostral sympathetic chain (Morin et al., 1997). In visceral sensory ganglia, the transient expression of DBH is completely blocked. This result provided the first evidence that *Phox2* genes are physiologically important for noradrenergic differentiation – that is, for DBH expression *in vivo*. In addition, the expression of c-*ret*, the signaling receptor subunit of GDNF-related ligands, is strongly down-regulated, while the geniculate, petrosal and nodose cranial sensory neurons die.

In the *Phox2b* knockout an even more dramatic phenotype was observed, with a complete lack of sympathetic, parasympathetic and enteric neurons, as well as of visceral sensory neurons and one of their peripheral targets, the carotid body (Pattyn et al., 1999; Dauger et al., 2003). The development of the ciliary ganglion is already severely impaired in *Phox2b* heterozygotes (Cross et al., 2004). The common finding for autonomic neural crest derivatives in the *Phox2b* knockout is that the cells migrate to their correct positions, form ganglia, but do not differentiate and subsequently die. This has been observed for precursors of the sympathetic ganglia, enteric ganglia and the carotid body (Pattyn et al., 1999; Dauger et al., 2003). Enteric precursors populate only the foregut, and fail to migrate further.

Noradrenergic differentiation of petrosal neurons is also affected in *Phox2b* heterozygous mutant embryos, which explains the physiological defects in *Phox2b* heterozygous mice, an altered response to hypoxemia and hypercapnia at birth (Dauger et al., 2003).

3.3.2.2 Central Noradrenergic Neurons

In the absence of Phox2a, the LC never forms, whereas the other noradrenergic centers in the hindbrain are unaffected (Morin et al., 1997). In contrast, in the *Phox2b* knockout all noradrenergic neurons of the brain are missing (Pattyn et al., 2000a). Recent evidence suggests an essential role of LC neurons in the maturation of the hindbrain respiratory rhythm generator, thus providing an explanation for the death of *Phox2a* knockout mice at birth (Viemari et al., 2004).

3.3.2.3 Autonomic Centers in the Hindbrain

In homozygous *Phox2b* mutants, the central target of visceral sensory ganglia – the nTS, which integrates all visceral information – never forms, although nTS precursors expressing Lmx1b and Rnx are initially born (Dauger et al., 2003). Neurons of the area postrema (AP) that are located dorsal to the nTS, are also missing. Not only the afferent but also all efferent visceral and branchial motor neurons require Phox2b for their formation (Dubreuil et al., 2000; Pattyn et al., 2000b). Branchial motor neurons (bm) innervate muscles derived from the branchial arches, visceral motor neurons (vm) innervate parasympathetic ganglia. Both bm and vm differentiate from progenitors expressing the homeobox genes *Nkx2.2* and *Nkx2.9*, and are located ventral to the domain that gives rise to the somatic motor neurons. In mice lacking Phox2b both the generic and subtype-specific programs of bm/vm motor neuron differentiation are disrupted at an early stage.

Phox2a is essential for the generation of the rostral sites of *Phox2* gene expression – that is, oculomotor (nIII) and trochlear (nIV) nuclei – but is dispensable for the formation of bm and vm (Morin et al., 1997).

3.3.3
Human Mutations

The functions shown for Phox2a and Phox2b in mutant mice are to some extent – but apparently not completely – conserved in man, as revealed by the effects in *Phox2* mutations leading to human congenital diseases. Homozygous mutations in *PHOX2A* were shown to result in congenital fibrosis of the extraocular muscle type 2 (CFEOM2) (Nakano et al., 2001; Yazdani et al., 2003). Congenital fibrosis syndromes were once believed to result from extraocular muscle fibrosis, but are now known to result from aberrant development of the oculomotor (nIII), trochlear (nIV) and abducens (nVI) cranial nerve nuclei. The effects observed confirm the critical role of Phox2a for the development of midbrain motor nuclei. However, it is surprising that the phenotype resulting from these *Phox2a* loss-of-function mutations (Yazdani et al., 2003) is restricted to the ocular and pupillar phenotype, whereas *Phox2a* in the mouse is essential for the survival of parasympathetic and sensory neurons and for survival after birth. This suggests that there are functional differences between the highly related human and mouse genes (Yazdani et al., 2003).

The importance of Phox2b for autonomic nervous system development and function in man has been revealed by the demonstration that *PHOX2B* is causally involved in the congenital central hypoventilation syndrome (CCHS; Ondine's curse) (Amiel et al., 2003; Weese-Mayer et al., 2003; Sasaki et al., 2003). CCHS is a rare disorder characterized by persistent hypoventilation during sleep, beginning during the neonatal period. The core phenotype is associated with lower-penetrance anomalies of the autonomic nervous system, including Hirschsprung disease. The most likely disease mechanism resulting in hypoventilation is an altered integration of afferent inputs from central and peripheral chemoreceptors in the brainstem (Gozal, 1998). The cellular and molecular basis of the ventilatory control anomalies in CCHS are still unclear. The RET-GDNF signaling pathway and the upstream regulatory genes *HASH1* and *PHOX2B* were strongly implicated due to their importance in the development of visceral circuits. Mutations in several of these genes were observed in CCHS, with *PHOX2B* as the major disease-causing gene (more than 86% of cases) (Katz, 2003; Sasaki et al., 2003; Huang et al., 2005). Most *PHOX2B* mutations lead to an expansion of two polyalanine repeats, and there is a correlation between the severity of the phenotype and the length of the repeat (Weese-Mayer et al., 2003). As the phenotype is evident in heterozygous mutations, and is inherited in an autosomal-dominant manner, the mutations suggest that a dominant-negative variant of the Phox2b protein, blocking the normal protein, is encoded by the mutated gene. This conclusion is also supported by recent comparisons of respiratory and pupillary phenotype in *Phox2b* heterozygous mice and CCHS patients. Although $Phox2b^{+/-}$ mice have an altered response to hypoxia and hypercapnia, most likely due to a reduced TH expression in chemoafferent petrosal neurons, this phenotype is transient and thus only partially models the CCHS phenotype (Dauger et al., 2003). Haploinsufficiency for *Phox2b* in mice results in dilated pupils, caused by a highly atrophic ciliary ganglion (Cross et al., 2004), whereas CCHS patients display constricted, rather than dilated pupils. Thus, mutations in CCHS lead to a much stronger phenotype than observed in *Phox2b* heterozygous mice, including effects on the superior cervical ganglion that result in pupil constriction.

3.4
Molecular Mechanism of Action in Different Lineages

The Phox2 transcription factors are expressed at the stage of proliferating progenitor cells, but are mostly maintained in mature, differentiated neurons. They display distinct functions in different stages. In addition, they act in a context-dependent manner and affect the development of different characteristics in different lineages. In the following section, findings on the molecular mechanism of Phox2 action are described for the individual neuronal subtypes and their progenitors.

3.4.1
Sympathetic Neurons

The development of sympathetic neurons is controlled by a group of transcription factors, including Mash1, Phox2b, Phox2a, Gata2/3, and Hand2 that are induced by BMPs (Goridis and Rohrer, 2002). Mash1 and Phox2b are the first markers of sympathetic neuron development, and are initially expressed in parallel, independently of each other, as shown by the individual knockouts (Hirsch et al., 1998; Pattyn et al., 1999). Phox2a, Hand2, and Gata2/3 are more downstream members of this network. The transcription factor Sox10 is required for Mash1 and Phox2b expression, but represses Phox2a and sympathetic neuronal differentiation, thus inducing neurogenic potential and simultaneously delaying differentiation (Kim et al., 2003). In the absence of Phox2b, Mash1–expressing cells are detectable in the ganglion primordia, but this expression is rapidly lost and further differentiation to a sympathetic neuron phenotype (Phox2a, TH, DBH, neurofilament) is prevented (Pattyn et al., 1999). The undifferentiated precursor cells subsequently die. The lack of noradrenergic differentiation of sympathetic precursors (and of noradrenergic precursors in the hindbrain) in Phox2b-deficient mice, together with the ability of both Phox2a and Phox2b to transactivate the DBH promotor (see above), provides a direct link between cell-type specification and neuron subtype differentiation. The function of Phox2 transcription factors in the specification of noradrenergic sympathetic neurons was also confirmed by overexpression of Phox2a and Phox2b in neural crest precursor cells, both *in vitro* and *in vivo* (Stanke et al., 1999; Patzke et al., 2001). As Phox2 factors are essential also for the development of other, non-noradrenergic phenotypes, additional co-determinants are implicated that interact with Phox2 in the positive control of noradrenergic marker genes in peripheral and central noradrenergic neurons. Gata2/3 and Hand2 act together with Phox2a/b in the control of noradrenergic differentiation in the PNS (Howard et al., 2000; Lim et al., 2000; Xu et al., 2003; Rychlik et al., 2003; Tsarovina et al., 2004). It should be pointed out, however, that the action of Phox2 transcription factors in sympathetic precursors is not restricted to noradrenergic genes, as generic neuronal genes are also affected in both loss-of-function and gain-of-function experiments (Pattyn et al., 1999; Stanke et al., 1999; 2004; Patzke et al., 2001). There is both evidence for a direct effect of Phox2 on the expression of generic neuronal properties (Valarché et al., 1993) and for an indirect action involving Mash1 (Stanke et al., 2004).

3.4.2
Parasympathetic Neurons

Parasympathetic neuron development shares with sympathetic neuron development the dependence on Mash1 (Guillemot et al., 1993; Hirsch et al., 1998), and Phox2b (Pattyn et al., 1999), with an additional requirement for Phox2a for rostral parasympathetic ganglia (Morin et al., 1997). In addition, BMPs are required for the development of parasympathetic neurons, at least in avian ciliary ganglia (Müller and Rohrer, 2002). In the absence of BMPs, neural crest cells migrate normally and form ganglion aggregates, but do not express the downstream transcription factors Mash1, Phox2b and Phox2a, nor other differentiation markers. In *Phox2b* knockout mice, parasympathetic ganglia are undetectable at E13.5, and thus it is assumed that they never form. In the *Mash1* knockout, Phox2b-positive parasympathetic ganglia are present at E13.5 (with the exception of the ciliary ganglion that could never be detected), but they lack Phox2a, never express any noradrenergic markers, and subsequently disappear (Guillemot et al., 1993; Hirsch et al., 1998).

Parasympathetic neurons display mostly a cholinergic phenotype in the mature state, but during development they transiently express noradrenergic properties, as shown in rat (Leblanc and Landis, 1989), mouse (Hirsch et al., 1998) and chick (Müller and Rohrer, 2002). As TH expression in the chick ciliary ganglion is maintained by ectopic expression of the bHLH transcription factor Hand2, it has been suggested that Hand2 may be responsible for the maintenance of noradrenergic differentiation in peripheral autonomic neurons (Müller and Rohrer, 2002). This hypothesis is supported by the finding that Hand2 is expressed in the chick sphenopalatine ganglion where TH expression is maintained (Dai et al., 2004; F. Müller and H. Rohrer, unpublished results), but not in the ciliary ganglion, where TH and DBH are expressed only transiently (Müller and Rohrer, 2002). These data emphasize the involvement of Phox2 transcription factors in noradrenergic as well as non-noradrenergic neuron development, and highlight the importance of additional transcriptional regulators in the final decision on transmitter phenotype.

3.4.3
Enteric Neurons

The enteric nervous system is completely missing in homozygous *Phox2b* mutant mice at E13.5. The enteric *Phox2b* phenotype is characterized by the initial population of the foregut by enteric precursors at E10.5, the lack of further caudal migration, and differentiation. Subsequent cell death is explained by the loss of c-*ret* expression (Pattyn et al., 1999). As this phenotype closely resembles the c-*ret* knockout at that stage (Durbec et al., 1996), and since c-*ret* expression is lost in Phox2–deficient mice, the phenotype can fully be accounted for by its effect on c-*ret* expression. The lack of differentiation also includes the absence of TH and DBH, that are transiently expressed in a subpopulation of enteric neurons in the esophagus (Baetge et al., 1990; Pattyn et al., 1999). In comparison with sympathetic and parasympathetic neurons, much less is known about the initial stages of enteric neuron development. It remains to be shown whether BMPs are also essential for enteric neuron specification, which proneural genes (in addition to Mash1) cooperate with Phox2b, and which co-determinants are involved in the generation of the different enteric neuron subtypes. The *Mash1* knockout affects only a subpopulation of transiently adrenergic neurons located in the foregut (Blaugrund et al., 1996).

3.4.4
Visceral Sensory Neurons of the Geniculate, Petrosal and Nodose Ganglia

The generation of epibranchial placodes seems to be initiated by BMP-7 derived from the endoderm (Begbie et al., 1999), and controlled by the winged helix domain-containing transcription factor Foxi1 (Lee et al., 2003), acting upstream of the proneural genes *Ngn2* and *Phox2a*. The sensory neuron precursors express Phox2a in the placodes, and slightly later Phox2b, as they delaminate and form ganglion primordia (Tiveron et al., 1996; Pattyn et al., 1997). In the *Phox2a* mutant, cranial placode-derived ganglia are morphologically normal at E10.5. The precursors have initiated generic neuronal differentiation (b-tubulin, peripherin), but are completely devoid of subtype-specific differentiation (i.e., DBH expression; Morin et al., 1997). The converse phenotype is observed in the *Ngn2* knockout, with a block in pan-neuronal differentiation without affecting Phox2a expression (Fode et al., 1998). Thus, subtype-specific and generic neuronal differentiation seem to be controlled separately in this lineage, involving *Phox2* and *Ngn2*, respectively. This resembles to some extent the situation in sympathetic precursor cells, with *Mash1* preferentially controlling aspects of generic neuronal differentiation (Parras et al., 2002; Stanke et al., 2004; see also Sommer et al., 1995). Phox2a-deficient mice display a reduced expression of c-*ret* in the cranial sensory ganglia. As the survival of placode-derived cranial sensory neurons depends on the neurotrophic action of the GDNF family (Moore et al., 1996; Baloh et al., 2000), their death in *Phox2a* mutants may be explained by reduced c-*ret* expression. The *Phox2b* knockout displays a very similar phenotype. Since Phox2b is controlled by Phox2a, and since Phox2a is unable to compensate for the lack of Phox2b with respect to noradrenergic differentiation, Phox2b is the downstream effector in this lineage (Pattyn et al., 1999). As cranial sensory neurons depend on *Ngn2* as a proneural gene, while sympathetic neurons require *Mash1*, Phox2b can act within the context of either *Mash1* or *Ngn2* to induce the generation of a noradrenergic phenotype.

Besides acting in the early and transient noradrenergic differentiation of epibranchial sensory neurons, Phox2 transcription factors seem also to be required for the later differentiation of dopaminergic chemoafferent neurons in the petrosal ganglion (Brosenitsch and Katz, 2002). This differentiation process requires membrane depolarization during a critical time period, in addition to Phox2 expression. Interestingly, Phox2 is only able to induce TH, but not DBH in these cells. The peripheral target of dopaminergic chemoafferent neurons in the petrosal ganglion are the glomus cells in the carotid body, which also depend on Phox2b for their survival. The molecular and cellular mechanism involved in the action of Phox2 in this context are not well understood, besides the finding that differentiation of glomus cells, as reflected by the expression of Phox2a and TH, is blocked in the absence of Phox2b (Dauger et al., 2003).

3.4.5
Central Noradrenergic Neurons

The LC, which is the major noradrenergic center of the brain, is generated in the dorsal rhombombere 1, involving BMP-induced dorsoventral patterning (Vogel-Höpker and Rohrer, 2002). The initial steps of LC specification are controlled by *Mash1, Phox2a* and *Phox2b*, with *Mash1* upstream of *Phox2a* and *Phox2b* (Hirsch et al., 1998; Lo et al., 1998). The similarity to the situation in sympathetic neurons lead to the hypothesis that the molecular control of noradrenergic differentiation may be conserved throughout the nervous system. However, this idea cannot be maintained in view of the many differences in the way that different noradrenergic neurons acquired their fate and characteristics:

- LC neurons, in contrast to sympathetic neurons depend on both Phox2b and Phox2a.
- Noradrenergic differentiation of sympathetic precursors depends on Gata2/3 transcription factors (Gata2 in the chick, Gata3 in the mouse) and presumably also on Hand2, whereas these transcription factors are not expressed in central noradrenergic neurons (Howard et al., 2000; Lim et al., 2000; Tsarovina et al., 2004).
- The noradrenergic centers A1, A2, A5, and A7 depend only on Phox2b, not on Phox2a, like sympathetic neurons. As Hand2 and Gata2/3 are thought to represent noradrenergic co-determinants, acting together with Phox2a/b in the control of peripheral noradrenergic gene expression (; Howard et al., 2000; Müller and Rohrer, 2002; Tsarovina et al., 2004), additional unknown co-determinants are implicated in central noradrenergic neurons. Thus, the sequence of Phox2a and Phox2b expression as well as co-regulatory transcription factors may differ between noradrenergic neuron populations.

3.4.6
Autonomic Centers in the Hindbrain

3.4.6.1 Afferent Visceral Centers

The nTS is generated from dorsal precursors that start to express Phox2b as postmitotic cells in the mantle zone (Dauger et al., 2003). In the *Phox2b* knockout, nTS cells coexpressing Rnx and Lmx1b are born as in wild-type embryos and are present in a dorsal position at E11.5. Subsequently, they do not migrate ventrally, are lost, and nTS and AP never form. The development of the nTS is also dependent on Rnx (Qian et al., 2001), the action of which, however, is not restricted to visceral reflex circuits but also includes the development of somatic sensory neurons (Qian et al., 2002).

3.4.6.2 Efferent Visceral Centers

Phox2b is required for the formation of all bm and cranial vm, the nuclei of the trigeminal and facial nerves, the nucleus ambiguus (nA) and the dorsal motor nucleus of the vagus nerve (dmnX) (Pattyn et al., 2000b). These neurons develop normally in the absence of Phox2a. The bm display voluntary functions in head and jaw musculature in higher vertebrates, but fulfill visceral functions in fish and amphibia by controlling breathing through innervation of the gill muscles (Fritzsch and Northcutt, 1993). For this reason, bm and vm are discussed together in this section. In the *Phox2b* knockout, precursors of bm and vm are generated on schedule but the majority do not differentiate, reflected by the lack of Phox2a, Islet1, Math3, Ebf2, and Ebf3 expression, and by a maintenance of neuroepithelial markers (i.e., Nkx2.2 and Mash1; Pattyn et al., 2000b). Most precursors die before they migrate to the mantle layer. Interestingly, some cells are able to leave the ventricular zone and become postmitotic, albeit maintaining their neuroepithelial identity. Direct evidence for a role of Phox2b in the control of cell-cycle exit has been obtained by loss- and gain-of-function approaches. In particular, forced expression of Phox2b in embryonic chick spinal cord results in differentiation of ectopic neurons with bm/vm properties (Dubreuil et al., 2000, 2002). The Phox2b-induced generation of bm/vm involves the induction of pan-neuronal properties by stimulating the expression of proneural genes and repressing the expression of inhibitors of neurogenesis (Hes5, Id2). In parallel, Phox2b affects bm/vm fate specification by controlling the expression of patterning genes that inhibit (*Pax6*, *Olig2*) or allow (*Nkx6.1*, *Nkx6.2*) bm/vm development (Dubreuil et al., 2002). The expression of Phox2b during normal hindbrain development in turn is controlled directly by dorsoventral (*Nkx2.2*) and anteroposterior (*Hoxb1* and *Hoxb2*) patterning genes (Samad et al., 2004). The induction of ectopic Phox2b-positive visceral motor neurons requires the combined actions of Hox and Nkx2.2 homeodomain proteins (Samad et al., 2004). Taken together, Phox2b displays multiple functions in bm/vm development, on the one hand acting in proliferating precursor cells towards a neuronal fate and on the other hand in controlling general and subtype-specific neuron differentiation. This led to the proposal that Phox2b coordinates cell-cycle exit and the specification of bm/vm neuronal identity.

Phox2b is not only required for the generation of vm and bm, but it also represses the generation of serotonergic neurons generated from the same Nkx2.2–positive progenitor cells. Serotonergic neuron generation correlates with the down-regulation of Phox2b and indeed, in the *Phox2b* knockout the loss of vm/bm is accompanied by premature expression of the serotonergic marker pet–1 (Pattyn et al., 2003).

3.4.7
Oculomotor (nIII) and Trochlear (nIV) Centers

The development of oculomotor and trochlear nuclei depends on Phox2a, expressed before Phox2b in this lineage (Pattyn et al., 1997). In the *Phox2a* knockout the expression of Phox2b is absent at E10.5 and the nuclei never form, as shown by the lack of Islet–1 expression at E11.5 and the absence of ChAT and peripherin expression at later stages (Pattyn et al., 1997). Whether the Edinger-Westphal nucleus that contains neurons innervating the parasympathetic ciliary ganglion, located close to nIII, also expresses Phox2a/2b is not clear. Transcription factors interacting with Phox2a in the development of nIII and nIV are not known. Oculomotor and trochlear motor neurons are usually classified as somatic motor neurons, but there are several properties (e.g., the lack of HB9 and Lim3 expression) that they share with bm/vm but not with somatic motor neurons (for a discussion, see Pattyn et al., 2000b).

3.5
Conclusions and Outlook

3.5.1
Distinct or Identical Functions for Phox2a and Phox2b?

Overexpression and promoter studies have provided minimal evidence for distinct functions of Phox2a and Phox2b. Both are sufficient to elicit the generation of sympathetic neurons from neural crest precursor cells (Stanke et al., 1999; Patzke et al., 2001). As there is cross-regulation – that is, Phox2a inducing the expression of the upstream paralogue Phox2b – it is not possible to draw conclusions on the action of the individual transcription factors in this experimental paradigm. Promotor studies have shown that Phox2a and Phox2b share identical homeodomains that bind to the same promotor sequences (Adachi et al., 2000; Seo et al., 2002) and are not distinguishable in their ability to transactivate DBH expression in cultured cells (Adachi et al., 2000; Seo et al., 2002). The only evidence for functional differences between Phox2a and Phox2b in their ability to transactivate gene expression was observed at the human Phox2a promoter which is transactivated by Phox2b, but not Phox2a (Hong et al., 2001).

Stronger evidence for distinct Phox2a and Phox2b functions is derived from studies in mutant mice. Phox2a and Phox2b are coexpressed in virtually all lineages, but differ in the order of expression. The Phox2 protein expressed first induces the expression of the second, which can explain for the most part the phenotypes observed on *Phox2a* and *Phox2b* knockout mice. An absence of the *Phox2* gene expressed first results in the lack of the downstream paralogue and in the loss or atrophy of the ganglia and nuclei. In the absence of the second *Phox2* gene, the cells develop normally due to compensation by the upstream gene in certain neuron populations. However, some lineages require the action of both Phox2a and Phox2b – that is, the LC, cranial sensory ganglia and parasympathetic ganglia of the head (Morin et al., 1997; Pattyn et al., 1997, 1999, 2000a). Thus, Phox2a and Phox2b are unable to compensate for the loss of the paralogue, and this implicates specific functions for each factor. With respect to noradrenergic differentiation, only Phox2b may be functionally relevant, whereas Phox2a may be dispensable, as DBH expression is completely absent in Phox2b-deficient LC and cranial visceral ganglia, though Phox2a expression is initially maintained (Pattyn et al., 1999, 2000a). In conclusion, whilst there is evidence suggesting functional differences between Phox2a and Phox2b proteins, this depends heavily on the neuronal lineage and may be linked to particular cellular functions (i.e., noradrenergic differentiation). In addition, gene dosage may be important in certain lineages to support maximal Phox2 effects (Dauger et al., 2003; Cross et al., 2004).

Most recently, the different roles of Phox2a and Phox2b have been addressed by the generation of knockin mutant mice in which *Phox2a* is replaced by the *Phox2b* coding sequence, and vice versa. In both cases, no full rescue of the functions lost in the respective knockout was achieved be the replacement gene. These studies have provided evidence for significant functional differences between the *Phox2* genes (Coppola et al., 2005). Thus, the *Phox2* genes that arose by gene duplication from an ancestral gene (Pujol et al., 2000) appear to have been maintained during evolution, not only because they have adopted different spatiotemporal expression patterns but also because they acquired specific properties that are essential in particular neuronal lineages.

3.5.2
Master Control Genes for Noradrenergic Differentiation

Although Phox2a and Phox2b are expressed not exclusively in cells with a noradrenergic phenotype, all cells that transiently or permanently express noradrenergic properties also express Phox2a/Phox2b (Tiveron et al., 1996; Pattyn et al., 1997). This is explained by the direct transcriptional activation of the *DBH* gene by Phox2a and Phox2b, which bind to defined response elements in the promotor regions (Tissier-Seta et al., 1993; Zellmer et al., 1995; Kim et al., 1998; Yang et al., 1998; Adachi et al., 2000). Phox2 transcription factors are essential, but are they also sufficient? They are sufficient in the sense that Phox2 overexpression in trunk neural crest precursors is able to elicit the generation of noradrenergic neurons. This action is, however, dependent on the cellular context. Phox2 overexpression in precursor cells located in peripheral nerves induces noradrenergic neuronal properties (Stanke et al., 1999; Patzke et al., 2001). In contrast, overexpression in ciliary neuron precursors does not increase the number of noradrenergic neurons in this parasympathetic ganglion (F. Müller and H. Rohrer, unpublished results), illustrating restrictions in the noradrenergic potential even within the autonomic nervous system. Phox2 overexpression in the spinal cord results in the generation of neurons that display properties of bm/vm (Dubreuil et al., 2000), but not of neurons with noradrenergic properties (U. Ernsberger and H. Rohrer, unpublished results). Thus, the function of Phox2 factors at the *DBH* promoter is essential for noradrenergic differentiation, but is not sufficient. This points to a requirement of noradrenergic co-determinants expressed in cells that acquire a noradrenergic phenotype. Hand2 and Gata2/3 are candidates for such noradrenergic co-determinants in the autonomic nervous system (Lim et al., 2000; Müller and Rohrer, 2002; Tsarovina et al., 2004). In peripheral nerve precursors, Phox2 factors induce noradrenergic neuron differentiation by eliciting expression of the entire network of transcriptional regulators involved in peripheral noradrenergic neuron development: Mash1, Hand2, and Gata2/3

(Stanke et al., 1999, 2004). As Hand2 and Gata2/3 are not expressed in central noradrenergic neurons, additional factors that interact with Phox2 signaling appear to be implicated.

3.5.3
Master Control Genes for Autonomic Reflex Circuit Generation

The expression pattern of Phox2a revealed a striking correlation with autonomic visceral reflex circuits, and this led to the proposal that Phox2a might control the development of these synaptic pathways (Tiveron et al., 1996). *Phox2* genes are thought to control the initiation or stabilization of contacts between the different cells of autonomic circuits. But how could the same transcription factor expressed by two cell types lead to the formation of specific contacts? One possibility might be that *Phox2* genes induce the expression of homophilic cell adhesion factors (e.g., cadherins) that would allow specific cell-cell interactions to be initiated and/or stabilized. This hypothesis was initially proposed to explain the specific contacts formed between proprioceptive sensory neurons and motor neurons in the ventral spinal cord which coexpress the same ETS transcription factor (Lin et al., 1998). It might be argued, however, that Phox2 expression is not essential for autonomic circuit formation, as spinal cord visceral motor neurons are integrated into the circuits without expressing Phox2. In addition, considering the differences in the phenotypes of *Phox2a* loss-of-function mutations in mice and man (Cross et al., 2004), it is conceivable that Phox2 may control circuit formation in a species-dependent manner.

Although this is one of the most important and interesting issues in the field, to date it has been impossible to test the role of *Phox2* genes in the control of cell-cell interaction during circuit formation. As *Phox2* genes are required for the specification and early differentiation of autonomic circuit neurons, these cells are either not generated at all or die before axons grow out in the absence of Phox2 transcription factors. The analysis of conditional *Phox2* knockout mice is expected to provide the answers to this important question, however. At present, the only evidence supporting a role for Phox2 in neurite outgrowth comes from a *Phox2* mutant in C. *elegans* (Pujol et al., 2000). In this model organism, the *Phox2* orthologue cePHOX2/CEH–17 Phox is expressed in a small number of head neurons that project towards the tail. In cePHOX2/CEH–17 mutants, the axon elongation is disrupted without affecting generic neuronal differentiation (Pujol et al., 2000).

The proposed function of *Phox2* genes in circuit formation implies that a set of Phox2–dependent genes is expressed in all visceral circuit neurons. In a screen for Phox2b-controlled genes, *RGS4* and *gustucin* were discovered, these being regulators of G-protein signaling and a G-protein, respectively (Grillet et al., 2003). The expression of *RGS4* in most circuit neurons and the function of *RGS*-family members in cell migration and axon navigation are in agreement with *Phox2*–controlled circuit formation. Although a *RGS4* knockout did not support the proposed role in circuit formation (Grillet et al., 2005), further studies are required to address this issue in view of the large number of *RGS* family members and the possibility of redundant functions.

Acknowledgments

The authors thank M. Studer for permission to include unpublished data, and K. Tsarovina, F. Müller and M. Stanke for helpful comments on the manuscript. The original investigations conducted by the authors were supported by the Deutsche Forschungsgemeinschaft (SFB 269, SPP 1109) and EU grants BIO4–98–0112 and QLG3–CT–2000–00072 to H.R., and DFG grant ER 145/4 to U.E.

Abbreviations

AP	area postrema
bHLH	basic helix-loop-helix
bm	branchial motor neurons
BMP	bone morphogenetic protein
CBP	cAMP-response element-binding protein-binding protein
CCHS	congenital central hypoventilation syndrome
CFEOM2	congenital fibrosis of the extraocular muscle type 2
CRE	cAMP-responsive element
dmnX	dorsal motor nucleus of the vagus nerve
HD	homeodomain
LC	locus coeruleus
nA	nucleus ambiguus
NCAM	neural cell adhesion molecule
NET	norepinephrine transporter
nTS	nucleus of the solitary tract
PKA	protein kinase
TH	tyrosine hydroxylase
vm	visceral motor neurons

References

Adachi, M., Lewis, E.J. (2002). The paired-like homeodomain protein, arix, mediates protein kinase A-stimulated dopamine β-hydroxylase gene transcription through its phosphorylation status. *J. Biol. Chem.* 277, 22915–22924.

Adachi, M., Browne, D., Lewis, E.J. (2000). Paired-like homeodomain proteins Phox2a/Arix and Phox2b/NBPhox have similar genetic organization and independently regulate dopamine-β-hydroxylase gene transcription. *DNA Cell Biol.* 19, 539–554.

Amiel, J., Laudier, B., Attie-Bitach, T., Trang, H., de Pontal, L., Gener, B., Trochet, D., Etchevers, H., Ray, P., Simmonneau, M., Vekemans, M., Munnich, A., Gaultier, C., Lyonnet, S. (2003). Polyalanine expansion and frameshift mutations of the paired-like homeobox gene PHOX2B in congenital central hypoventilation syndrome. *Nature Genet.* 33, 440–442.

Baetge, G., Pintar, J.E., Gershon, M.D. (1990). Transiently catecholaminergic (TC) cells in the bowel of the fetal rat: Precursors of noncatecholaminergic enteric neurons. *Dev. Biol.* 141, 353–380.

Baloh, R.H., Enomoto, H., Johnson, E.M. Jr., Milbrandt, J. (2000) The GDNF family ligands and receptors – implications for neural development. *Curr. Opin. Neurobiol.* 10, 103–110.

Begbie, J., Brunet, J.-F., Rubenstein, J.L., Graham, A. (1999). Induction of the epibranchial placodes. *Development* 126, 895–870.

Blaugrund, E., Pham, T.D., Tennyson, V.M., Lo, L., Sommer, L., Anderson, D.J., Gershon, M.D. (1996). Distinct subpopulations of enteric neuronal progenitors defined by time of development, sympathoadrenal lineage markers and MASH–1–dependence. *Development* 122, 309–320.

Brosenitsch, T.A., Katz, D.M. (2002). Expression of Phox2 transcription factors and induction of the dopaminergic phenotype in primary sensory neurons. *Mol. Cell. Neurosci.* 20, 447–457.

Brunet, J.F., Pattyn, A. (2002). *Phox2* genes – from patterning to connectivity. *Curr. Opin. Genet. Dev.* 12, 435–440.

Coppola, E., Pattyn, A., Guthrie, S.C., Goridis, C., Studer, M. (2005) Reciprocal gene replacements reveal unique functions for Phox2 paralogous homeobox genes during neural differentiation. (submitted)

Cross, S.H., Morgan, J.E., Pattyn, A., West, K., McKie, L., Hart, A., Thaung, C., Brunet, J.-F., Jackson, I.J. (2004). Haploinsufficiency for Phox2b in mice causes dilated pupils and atrophy of the ciliary ganglion: mechanistic insights into human congenital central hypoventilation syndrome. *Hum. Mol. Genet.* 13, 1433–1439.

Dai, Y.-S., Hao, J., Bonin, C., Morikawa, Y., Cserjesi, P. (2004) JAB1 enhances Hand2 transcriptional activity by regulating Hand2 DNA binding. *J. Neurosci. Res.* 76, 613–622.

Dauger, S., Pattyn, A., Lofaso, F., Gaulthier, C., Goridis, C., Gallego, J., Brunet, J.-F. (2003). Phox2b controls the development of peripheral chemoreceptors and afferent visceral pathways. *Development* 130, 6635–6642.

Dubreuil, V., Hirsch, M.R., Pattyn, A., Brunet, J.F., Goridis, C. (2000). The Phox2b transcription factor coordinately regulates neuronal cell cycle exit and identity. *Development* 127, 5191–5201.

Dubreuil, V., Hirsch, M.R., Jouve, C., Brunet, J.F., Goridis, C. (2002). The role of Phox2b in synchronizing pan-neuronal and type-specific aspects of neurogenesis. *Development* 129, 5241–5253.

Durbec, P.L., Larsson-Blomberg, L.B., Schuchardt, A., Costantini, F., Pachnis, V. (1996). Common origin and developmental dependence on *c-ret* of subsets of enteric and sympathetic neuroblasts. *Development* 122, 349–358.

Ernsberger, U. (2004). Gene expression: from precursor to mature neuron. In: *Molecular Biology of the Neuron.* Oxford: Oxford University Press, pp. 29–73.

Ernsberger, U., Reissmann, E., Mason, I., Rohrer, H. (2000). The expression of dopamine β-hydroxylase, tyrosine hydroxylase, and Phox2 transcription factors in sympathetic neurons: evidence for common regulation during noradrenergic induction and diverging regulation later in development. *Mech. Dev.* 92, 169–177.

Flora, A., Lucchetti, H., Benfante, R., Goridis, C., Clementi, F., Fornasari, D. (2001). SP proteins and PHOX2B regulate the expression of the human *PHOX2a* gene. *J. Neurosci.* 21, 7037–7045.

Fode, C., Gradwohl, G., Morin, X., Dierich, A., LeMeur, M., Goridis, C., Guillemot, F. (1998). The bHLH protein NEUROGENIN 2 is a determination factor for epibranchial placode-derived sensory neurons. *Neuron* 20, 483–494.

Fritzsch, B., Northcutt, R.G. (1993) Cranial and spinal nerve organization in amphioxus and lampreys: evidence for an ancestral craniate pattern. *Acta Anat.* 148, 96–109.

Galliot, B., de Vargas, C., Miller, D. (1999). Evolution of homeobox genes: Q_{50} paired-like genes founded the paired class. *Dev. Genes Evol.* 209, 186–197.

Goridis, C., Rohrer, H. (2002). Specification of catecholaminergic and serotonergic neurons. *Nat. Rev. Neurosci.* 3, 531–541.

Gozal, D. (1998) Congenital central hypoventilation syndrome: an update. *Ped. Pulmonol.* 26, 273–282.

Grillet, N., Dubreuil, V., Dufour, H.D., Brunet, J.-F. (2003). Dynamic expression of RGS4 in the developing nervous system and regulation by the neural type-specific transcription factor Phox2b. *J. Neurosci.* 23, 10613–10621.

Grillet, N., Pattyn, A., Contet, C., Kieffer, B.L., Goridis, C., Brunet, J.-F. (2005) Generation and characterization of *RGS4* mutant mice. *Mol. Cell. Biol.* 25, 4221–4228.

Guillemot, F., Lo, L.-C., Johnson, J.E., Auerbach, A., Anderson, D.J., Joyner, A.L. (1993). Mammalian *achaete-scute* homolog 1 is required for the early development of olfactory and autonomic neurons. *Cell* 75, 463–476.

Guo, S., Brush, J., Teraoka, H., Goddard, A., Wilson, S.W., Mullins, M.C., Rosenthal, A. (1999). Development of noradrenergic neurons in the zebrafish hindbrain requires BMP, FGF8, and the homeodomain protein soulless/Phox2a. *Neuron* 24, 555–566.

Hirsch, M.R., Tiveron, M.C., Guillemot, F., Brunet, J.F., Goridis, C. (1998). Control of noradrenergic differentiation and Phox2a expression by MASH1 in the central and peripheral nervous system. *Development* 125, 599–608.

Hong, S.J., Kim, C.-H., Kim, K.-S. (2001). Structural and functional characterization of the 5' upstream promotor of the human Phox2a gene: possible direct transactivation by transcription factor Phox2b. *J. Neurochem.* 79, 1225–1236.

Hong S.J., Chae H., Kim K.S. (2004). Molecular cloning and characterization of the promoter region of the human Phox2b gene. *Brain Res. Mol. Brain Res.* 125, 29–39.

Howard, M.J., Stanke, M., Schneider, C., Wu, X., Rohrer, H. (2000). The transcription factor dHAND is a downstream effector of BMPs in sympathetic neuron specification. *Development* 127, 4073–4081.

Huang, L., Guo, H., Hellard, D.T., Katz, D.M. (2005). Glial cell line-derived neurotrophic factor (GDNF) is required for differentiation of pontine noradrenergic neurons and patterning of central respiratory output. *Neuroscience* 130, 95–105.

Katz, D.M. (2003). Neuronal growth factors and development of respiratory control. *Resp. Physiol. Neurobiol.* 135, 155–165.

Kim, C.-H., Hwang, D.-Y., Park, J.-J., Kim, K.-S. (2002). A proximal promotor domain containing a homeodomain-binding core motif interacts with multiple transcription factors, including HoxA5 and Phox2 proteins, and critically regulates cell type-specific transcription of the human norepinephrine transporter gene. *J. Neurochem.* 22, 2579–2589.

Kim, H.S., Seo, H., Yang, C.Y., Brunet, J.F., Kim, K.S. (1998). Noradrenergic-specific transcription of the dopamine β-hydroxylase gene requires synergy of multiple *cis*-acting elements including at least two Phox2a-binding sites. *J. Neurosci.* 18, 8247–8260.

Kim, J., Lo, L., Dormand, E., Anderson, D.J. (2003). Sox10 maintains multipotency and inhibits neuronal differentiation of neural crest stem cells. *Neuron* 28, 17–31.

Leblanc, G.G., Landis, S.C. (1989). Differentiation of noradrenergic traits in the principal neurons and small intensely fluorescent cells of the parasympathetic sphenopalatine ganglion of the rat. *Dev. Biol.* 131, 44–59.

Lee, S.A., Shen, E.L., Fiser, A., Guo, S. (2003). The zebrafish forkhead transcription factor Foxi1 specifies epibranchial placode-derived sensory neurons. *Development* 130, 2669–2679.

Lim, K.-C., Lakshmanan, G., Crawford, S.E., Gu, Y., Grosveld, F., Engel, J.D. (2000). Gata3 loss leads to embryonic lethality due to noradrenaline deficiency of the sympathetic nervous system. *Nat. Genet.* 25, 209–212.

Lin, J.H., Saito, T., Anderson, D.J., Lance-Jones, C., Jessell, T.M., Arber, S. (1998). Functionally related motor neuron pool and muscle sensory afferent subtypes defined by coordinate *ETS* gene expression. *Cell* 95, 393–407.

Lo, L.C., Tiveron, M.C., Anderson, D.J. (1998). MASH1 activates expression of the paired homeodomain transcription factor Phox2a, and couples pan-neuronal and subtype-specific components of autonomic neuronal identity. *Development* 125, 609–620.

Lo, L.C., Morin, X., Brunet, J.F., Anderson, D.J. (1999). Specification of neurotransmitter identity by Phox2 proteins in neural crest stem cells. *Neuron* 22, 693–705.

Moore, M.W., Klein, R.D., Fariñas, I., Sauer, H., Armanini, M., Phillips, H., Reichardt, L.F., Ryan, A.M., Carver-Moore, K., Rosenthal, A. (1996). Renal and neuronal abnormalities in mice lacking GDNF. *Nature* 382, 76–79.

Moore, R.Y., Bloom, F.E. (1979). Central catecholamine neuron systems: Anatomy and physiology of the norepinephrine and epinephrine systems. *Annu. Rev. Neurosci.* 2, 113–168.

Morin, X., Cremer, H., Hirsch, M.-R., Kapur, R.P., Goridis, C., Brunet, J.-F. (1997). Defects in sensory and autonomic ganglia and absence of locus coeruleus in mice deficient for the homeobox gene *Phox2*. *Neuron* 18, 411–423.

Müller, F., Rohrer, H. (2002). Molecular control of ciliary neuron development: BMPs and downstream transcriptional control in the parasympathetic lineage. *Development* 129, 5707–5717.

Nakano, M., Yamada, K., Fain, J., Sener, E.C., Selleck, C.J., Awad, A.H., Zwaan, J., Mullaney, P.B., Bosley, T.M., Engle, E.C. (2001). Homozygous mutations in ARIX (PHOX2A) result in congenital fibrosis of the extraocular muscles type 2. *Nat. Genet.* 29, 315–320.

Niehrs, C., Pollet, N. (1999). Synexpression groups in eukaryotes. *Nature* 402, 483–487.

Parras, C.M., Schuurmans, C., Scardigli, R., Kim, J., Anderson, D.J., Guillemot, F. (2002). Divergent functions of the proneural genes *Mash1* and *Ngn2* in the specification of neuronal subtype identity. *Genes Dev.* 16, 324–338.

Pattyn, A., Morin, X., Cremer, H., Goridis, C., Brunet, J.F. (1997). Expression and interactions of the two closely related homeobox genes *Phox2a* and *Phox2b* during neurogenesis. *Development* 124, 4065–4075.

Pattyn, A., Morin, X., Cremer, H., Goridis, C., Brunet, J.-F. (1999). The homeobox gene Phox2b is essential for the development of all autonomic derivatives of the neural crest. *Nature* 399, 366–370.

Pattyn, A., Goridis, C., Brunet, J.F. (2000a). Specification of the central noradrenergic phenotype by the homeobox gene *Phox2b*. *Mol. Cell. Neurosci.* 15, 235–243.

Pattyn, A., Hirsch, M.R., Goridis, C., Brunet, J.F. (2000b). Control of hindbrain motor neuron differentiation by the homeobox gene *Phox2b*. *Development* 127, 1349–1358.

Pattyn, A., Vallstedt, A., Dias, J.M., Samad, O.A., Krumlauf, R., Rijli, F., Brunet, J.-F., Ericson, J. (2003). Coordinated temporal and spatial control of motor neuron and serotonergic neuron generation from a common pool of CNS progenitors. *Genes Dev.* 17, 729–737.

Patzke, H., Reissmann, E., Stanke, M., Bixby, J.L., Ernsberger, U. (2001). BMP growth factors and Phox2 transcription factors can induce synaptotagmin I and neurexin I during sympathetic neuron development. *Mech. Dev.* 108, 149–159.

Pujol, N., Torregrossa, P., Ewbank, J.J., Brunet, J.-F. (2000). The homeodomain protein CePHOX2/CEH–17 controls anteroposterior axonal growth in *C. elegans*. *Development* 127, 3361–3371.

Qian, Y., Fritzsch, B., Shirasawa, S., Chen, C.L., Choi, Y.J., Ma, Q.F. (2001). Formation of brainstem (nor)adrenergic centers and first-order relay visceral sensory neurons is dependent on homeodomain protein Rnx/Tlx3. *Genes Dev.* 15, 2533–2545.

Qian, Y., Shirasawa, S., Chen, C.-L., Cheng, L., Ma, Q. (2002). Proper development of relay somatic sensory neurons and D2/D4 interneurons requires homeobox genes Rnx/Tlx–3 and Tlx–1. *Genes Dev.* 16, 1220–1233.

Rychlik, J.L., Gerbasi, V., Lewis, E.J. (2003). The interaction between dHAND and Arix at the dopamine β-hydroxylase promoter region is independent of direct dHAND binding to DNA. *J. Biol. Chem.* 278, 15884–15876.

Samad, O.A., Geisen, M.J., Caronia, G., Variet, I., Zappavigna, V., Ericson, J., Goridis, C., Rijli, F.M. (2004). Integration of anteroposterior and dorsoventral regulation of *Phox2b* transcription in cranial motoneuron progenitors by homeodomain proteins. *Development* 131, 4071–4083.

Sasaki, A., Kanai, M., Kijima, K., Akaba, K., Hashimoto, M., Hasegawa, H., Otaki, S., Koizumi, T., Kusada, H., Ogawa, Y., Tuchiya, K., Yamamoto, W., Nakamura, T., Hayasaka, K. (2003). Molecular analysis of congenital central hypoventilation. *Hum. Genet.* 114, 22–26.

Seo, H., Hong, S.J., Guo, S., Kim, H.-S., Kim, C.-H., Hwang, D.-Y., Isacson, O., Rosenthal, A., Kim, K.-S. (2002). A direct role of the homeoproteins Phox2a/2b in noradrenaline neurotransmitter identity determination. *J. Neurochem.* 80, 905–916.

Sommer, L., Shah, N., Rao, M., Anderson, D.J. (1995). The cellular function of MASH1 in autonomic neurogenesis. *Neuron* 15, 1245–1258.

Stanke, M., Junghans, D., Geissen, M., Goridis, C., Ernsberger, U., Rohrer, H. (1999). The Phox2 homeodomain proteins are sufficient to promote the development of sympathetic neurons. *Development* 126, 4087–4094.

Stanke, M., Stubbusch, J., Rohrer, H. (2004). Interaction of Mash1 and Phox2b in sympathetic neuron development. *Mol. Cell. Neurosci.* 25, 372–382.

Swanson, D.J., Zellmer, E., Lewis, E.J. (1997). The homeodomain protein arix interacts synergistically with cyclic AMP to regulate expression of neurotransmitter biosynthetic genes. *J. Biol. Chem.* 272, 27382–27392.

Swanson, D.J., Adachi, M., Lewis, E.J. (2000). The homeodomain protein Arix promotes protein kinase A-dependent activation of the dopamine β-hydroxylase promoter through multiple elements and interaction with the coactivator cAMP-response element-binding protein-binding protein. *J. Biol. Chem.* 275, 2911–2923.

Talikka, M., Stefani, G., Brivanou, A.H., Zimmerman, K. (2004). Characterization of *Xenopus* Phox2a and Phox2b defines expression domains within the embryonic nervous system and early heart field. *Gene Exp. Patt.* 4, 601–607.

Tissier-Seta, J.-P., Hirsch, M.-R., Valarché, I., Brunet, J.-F., Goridis, C. (1993). A possible link between cell adhesion receptors, homeodomain proteins and neuronal identity. *C. R. Acad. Sci. Paris* 316, 1305–1315.

Tiveron, M.-C., Hirsch, M.-R., Brunet, J.-F. (1996). The expression pattern of the transcription factor Phox2 delineates synaptic pathways of the autonomic nervous system. *J. Neurosci.* 16, 7649–7660.

Tsarovina, K., Pattyn, A., Stubbusch, J., Müller, F., Van der Wees, J., Schneider, C., Brunet, J.-F., Rohrer, H. (2004). Essential role of Gata transcription factors in sympathetic neuron development. *Development* 131, 4775–4786.

Valarché, I., Tissier-Seta, J.-P., Hirsch, M.-R., Martinez, S., Goridis, C., Brunet, J.-F. (1993). The mouse homeodomain protein Phox2 regulates *Ncam* promoter activity in concert with Cux/CDP and is a putative determinant of neurotransmitter phenotype. *Development* 119, 881–896.

Viemari, J.C., Bévengut, M., Burnet, H., Coulon, P., Pequignot, J.M., Tiveron, M.C., Hilaire, G. (2004). *Phox2a* gene, A6 neurons, and noradrenaline are essential for development of normal respiratory rhythm in mice. *J. Neurosci.* 24, 928–937.

Vogel-Höpker, A., Rohrer, H. (2002). The development of noradrenergic locus coeruleus (LC) neurons depends on bone morphogenetic proteins (BMPs). *Development* 129, 983–991.

Weese-Mayer, D.E., Berry-Kravis, E.M., Zhou, L., Maher, B.S., Silvestri, J.M., Curran, M.E., Marazita, M.L. (2003). Idiopathic congenital central hypoventilation syndrome: Analysis of genes pertinent to early autonomic nervous system embryologic development and identification of mutations in PHOX2B. *Am. J. Med. Genet.* 123A, 267–278.

Xu, H.M., Firulli, A.B., Zhang, X.T., Howard, M.J. (2003). HAND2 synergistically enhances transcription of dopamine-β-hydroxylase in the presence of Phox2a. *Dev. Biol.* 262, 183–193.

Yang, C., Kim, H.-S., Seo, H., Kim, C.-H., Brunet, J.-F., Kim, K-S. (1998). *Paired*-like homeodomain proteins, Phox2a and Phox2b, are responsible for noradrenergic cell-specific transcription of the dopamine β-hydroxylase gene. *J. Neurochem.* 71, 1813–1826.

Yazdani, A., Chung, D.C., Abbaszadegan, M.R., Al-Khayer, K., Chan, W.-M., Yazdani, M., Ghosdi, K., Engle, E.C., Traboulsi, E.I. (2003). A novel Phox2a/Arix mutation in an Iranian family with congenital fibrosis of extraocular muscles type 2 (CFEOM2). *Am. J. Ophthalmol.* 136, 861–865.

Zellmer, E., Zhang, Z., Greco, D., Rhodes, J., Cassel, S., Lewis, E.J. (1995). A homeodomain protein selectively expressed in noradrenergic tissue regulates transcription of neurotransmitter biosynthetic genes. *J. Neurosci.* 15, 8109–8120.

4
Functions of LIM-Homeodomain Proteins in the Development of the Nervous System

Yangu Zhao, Nasir Malik, and Heiner Westphal

Abstract

The nervous system is composed of diverse types of neurons that are precisely positioned and connected to form the highly organized neural circuitry. One great challenge in developmental neuroscience is to understand the molecular mechanisms that underlie the generation, positioning, and subsequent connection of these neurons during development. LIM-homeodomain (LIM-HD) proteins form a group of homeodomain-containing transcription factors that have been conserved in evolution. Functional analysis of most of these proteins has shown that they play prominent roles in the specification of diverse neuronal cell types by regulating various aspects of neuronal development such as precursor cell proliferation, cell migration, process outgrowth, axonal pathfinding, and synthesis of neurotransmitters. In addition, recent progress in the identification and functional characterization of interacting factors and downstream target genes of LIM-HD proteins has begun to provide valuable insights into mechanisms underlying the action of the LIM-HD proteins during development.

4.1
Introduction

LIM-homeodomain (LIM-HD) proteins are a group of phylogenetically conserved transcription factors encoded by a subfamily of homeobox genes. The functions of these transcription factors have been analyzed in a wide variety of model organisms, including worms, flies, fish, frogs, chick, and mice. These analyses have revealed important roles of these proteins in tissue patterning and cell differentiation during embryonic development. This chapter first presents a brief overview on the structure and classification of LIM-HD proteins, followed by a review of their functions in the development of the nervous system. There follows a discussion of recent progress in the identification and characterization of factors that interact with the LIM-HD proteins and of downstream target genes that are specifically controlled by these transcriptional regulators.

Transcription Factors. Edited by Gerald Thiel
Copyright © 2006 WILEY-VCH Verlag GmbH & Co. KGaA, Weinheim
ISBN 3-527-31285-4

4.2
Common Structural Features and Classification of LIM-HD Proteins

One common structural feature shared by all LIM-HD transcription factors is the presence of two unique N-terminal motifs, called LIM domains, and a DNA-binding homeodomain close to the C-terminus of the protein (Fig. 4.1A). LIM is an acronym derived from the first letters of the three founding members of this group of transcription factors, LIN–11 (Freyd et al., 1990), Islet1 (Karlsson et al., 1990), and MEC–3 (Way and Chalfie, 1988). The LIM domain is a cysteine-histidine-rich domain composed of approximately 55 amino acid residues that form double zinc fingers (Fig. 4.1B). Structural analyses using NMR spectroscopy and X-ray crystallography have shown that each of the two zinc fingers in a LIM domain consists of two orthogonally packed anti-parallel β-hairpins (Perez-Alvarado et al., 1994; Deane et al., 2003, 2004). The LIM domains are believed to be involved in mediating interactions between LIM-HD proteins and their cofactors that are required for the transcriptional regulation of downstream target genes.

The LIM-HD proteins are encoded by a large subfamily of homeobox genes that is composed of seven members in *Caenorhabditis elegans*, six in *Drosophila*, and 12 in mice. Based on amino acid sequence similarities between the homeodomains, the LIM-HD proteins have been further subdivided into six subgroups (Hobert and Westphal, 2000; see also Table 4.1). Most of these subgroups are represented by one gene in invertebrates and two closely related genes in mice, possibly reflecting a gene duplication event during evolution. Zebrafish and *Xenopus laevis* occasionally contain a third member in each subgroup of the LIM-HD proteins.

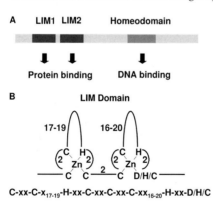

Fig. 4.1 (A) Schematic illustration showing major functional motifs of a LIM-HD protein. (B) Characteristic cysteine-histidine-rich amino acid sequence of a LIM domain that forms the double zinc finger structure.

Table 4.1 Classification of LIM-HD proteins in model organisms.

Sub-family	Mammals	Zebrafish	D. melanogaster	C. elegans
Lhx1/5	Lhx1	Lim1 (Lhx1a)	Lim1	Lin11
	Lhx5	Lim6 (Lhx1b)		
		Lim5 (Lhx5)		
Lhx2/9	Lhx2	BC093288[a]	Apterous	Ttx3
	Lhx9	33846.2[b]		
Lhx3/4	Lhx3	Lim3	Lim3	Ceh14
	Lhx4	CN05678[a]		
Lhx6/8	Lhx6	Lhx6	Arrowhead	Lim4
	Lhx8	Lhx7		
Islet	Islet1	Islet1	Islet/Tailup	Lim7
	Islet2	Islet2		
		Islet3		
Lmx	Lmx1a	BI843115[a]	CG4328[c]	Lim6
	Lmx1b	13921.1[b]	GA18112[c]	
Mec3	Mec3	None	None	None	Mec3

The list of zebrafish genes is preliminary, based upon available literature and on a UCSC Genome Bioinformatics Web Site BLAT search (http://www.genome.ucsc.edu/cgi-bin/hgBlat?command=start)
[a] GenBank Accession Number.
[b] ENSEMBL Gene Prediction Transcript Number.
[c] There may be only one active *Drosophila lmx* gene.
References for some zebrafish genes: *lim1* (Toyama et al., 1995b); *lim5* (Toyama et al., 1995a); *lim6* (Toyama and Dawid, 1997); *lim3* (Glasgow et al., 1997); *Lhx6/Lhx7* (Jackman et al., 2004).

4.3
LIM-HD Proteins and the Development of Invertebrate Nervous Systems

The function of LIM-HD proteins in the development of the nervous system has been initially analyzed in the invertebrates *C. elegans* and *Drosophila*. Studies in these organisms have shown that LIM-HD transcription factors play essential roles in controlling the generation of specific neuronal cell types and axonal pathfinding in the developing nervous system.

4.3.1
C. elegans

Whole-genome sequencing has documented the presence of seven genes encoding LIM-HD proteins in *C. elegans*. Six of these genes are known to play specific roles in the differentiation of neuronal cell types, while the seventh gene has not yet been characterized. The first LIM-HD protein ever identified was MEC–3 (Way and Chalfie, 1988). The *mec–3* gene is expressed in the six *C. elegans* neurons responsible for sensing touch, and all of these cells fail to differentiate properly in *mec–3* mutants.

Another neuronal cell type expressing *mec–3* is the PVD harsh touch sensory neuron which does not function properly in *mec–3* mutants, as the extensive dendritic branching of this neuron fails to take place in the absence of *mec–3* (Way and Chalfie, 1989; Tsalik et al., 2003).

Three other LIM-HD proteins, LIN–11, TTX–3, and CEH–14, play critical roles in the terminal differentiation and maintenance of neurons that form the thermoregulatory circuit in *C. elegans*. This circuit consists of a sensory neuron (AFD), a primary interneuron (AIY) that regulates responses to low temperature, and another primary interneuron (AIZ) that regulates responses to high temperature. The *ceh–14*, *ttx–3*, and *lin–11* genes are expressed in the AFD, AIY, and AIZ neurons, respectively, and a loss of function of each of these genes results in defects in terminal differentiation of each of these neurons and causes behavioral abnormalities indicative of dysfunctional thermoregulation (Hobert et al., 1997, 1998; Cassata et al., 2000).

The function of LIM-HD proteins has also been analyzed in development of the chemosensory neurons in *C. elegans*. There are 11 pairs of chemosensory neurons in the worm that sense the chemical environment. Among these, AWA and AWC are odor attractants, AWB is an odor repellant, and ASG is one of the neurons that sense pheromones. The AWB olfactory neuron expresses the LIM-HD gene *lim–4* (Sagasti et al., 1999). In the absence of *lim–4* function, the AWB neuron is transformed to adopt an AWC neuron phenotype, both morphologically and functionally. Conversely, an ectopic expression of *lim–4* in an AWC neuron transforms it into an AWB neuron (Sagasti et al., 1999). The *lin–11* gene is transiently expressed in the AWA neuron, and it is also expressed in AWC and ASG neurons (Sarafi-Reinach et al., 2001). In *lin–11* mutants, the AWA neuron assumes AWA/AWC hybrid morphology. Additionally, a mis-expression of *lin–11* in all ciliated neurons forces some neurons to switch to an ASG fate (Sarafi-Reinach et al., 2001). Together, these findings indicate that LIM-HD proteins play important roles in controlling the identity of distinct chemosensory neurons in *C. elegans*.

Finally in *C. elegans*, several types of sensory neurons, interneurons, and motor neurons express the Lmx sub-group LIM-HD gene *lim–6* (Hobert et al., 1999). This gene controls the specification of two classes of GABAergic motor neurons. In the absence of *lim–6*, these neurons fail to form proper axonal projection and lack expression of the enzyme required for synthesis of the neurotransmitter gamma-aminobutyric acid (GABA) (Hobert et al., 1999). The *C. elegans* nervous system shows multiple examples of neurons that show left/right asymmetry. Two sensory neurons required for taste – ASE left (ASEL) and ASE right (ASER) – are bilaterally symmetric but have distinctive functions and gene expression profiles. *Lim–6* is selectively expressed in ASEL and is required for the functional asymmetry between ASEL and ASER by turning off expression of the G-protein-coupled receptor *gcy–5* in ASEL neurons (Pierce-Shimomura et al., 2001; Chang et al., 2003). In addition, *lim–6* allows the ASEL neuron to distinguish sodium from chloride (Pierce-Shimomura et al., 2001). *Lim–6* also interacts with *ceh–14* to control specification of the PVT interneuron in the ventral nerve cord (Aurelio et al., 2003). Proteins encoded by these two genes are required for the proper expression of immunoglobulin genes that, in turn, are necessary for correct axonal patterning of the PVT neuron.

4.3.2
Drosophila

In the fruit fly Drosophila, two LIM-HD transcription factors, Lim3, and Islet, have clearly defined roles in controlling the identity and axonal projection pattern of motor neurons in the ventral nerve cord. The Drosophila ventral nerve cord contains three different subtypes of motor neurons that are characterized by their distinct projection of axons through one of the three different nerves exiting the nerve cord, the transverse nerve (TN), the intersegmental nerve b (ISNb), and the intersegmental nerve d (ISNd). The *Islet* gene is expressed in all these three types of motor neurons, whereas *Lim3* is only expressed in neurons that project axons through the TN and ISNb (Thor et al., 1999). The *Islet* gene controls axon pathfinding in all three nerves (Thor and Thomas, 1997). In mutants lacking *Lim3*, the ISNb neurons divert their axons to ISNd, and TN neurons fail to project axons to their proper target. Mis-expression of *Lim3* in ISNd motor neurons reroutes the axon of these neurons so that they project through ISNb (Thor et al., 1999). These results suggest that Lim3 and Islet function as part of a combinatorial code that controls the specificity of motor neuron axonal pathway selection.

Unlike the motor neurons, each interneuron cell type in Drosophila expresses only one LIM-HD gene. The loss of either *Apterous*, *Islet*, or *Lim3* causes axonal pathfinding defects in each type of the interneuron that expresses these genes, respectively (Lundgren et al., 1995; Thor and Thomas, 1997; Thor et al., 1999; van Meyel et al., 2000). In *Apterous* mutants, the phenotype can be rescued upon expression of a full-length *Apterous* transgene, but not by *Lim3* or an *Apterous* construct fitted with the sequence encoding the LIM domains of the Lim3 protein, indicating non-redundant function of different LIM-HD proteins in controlling axonal pathfinding in individual interneuron subtypes (O'Keefe et al., 1998).

Apterous and *Islet* are also involved in the regulation of neurotransmitter synthesis in several different neuronal cell types in the fly nervous system. *Apterous* is expressed in ventral nerve cord (VNC) neurons that use the peptide FMRF-amide as a neurotransmitter. In *Apterous* mutants, these neurons are present but fail to produce FMRF-amide (Schneider et al., 1993; Benveniste et al., 1998). *Islet* is expressed in both serotonergic and dopaminergic neurons. In mutant flies lacking *Islet* function, these neurons are impaired in the synthesis of serotonin and dopamine, respectively (Thor and Thomas, 1997).

4.4
Functions of LIM-HD Proteins in the Development of Vertebrate Nervous Systems

Compared to invertebrates, the vertebrate nervous system contains a much larger number and diversity of neurons, and a far more complex neural circuitry. In this setting, the LIM-HD proteins assume an expanding cell-autonomous role in controlling the differentiation and axonal pathfinding of developing neurons. In addition, the LIM-HD proteins can affect the patterning of broad regions of the developing vertebrate central nervous system (CNS) by controlling the formation of critical signaling centers and regulation of early neural progenitor cell specification and proliferation.

4.4.1
The Vertebrate Spinal Cord

Although the spinal cord is a less complex part of the vertebrate CNS, it too contains many different types of neurons. During development, these neurons are generated and then precisely connected, either with each other within the spinal cord or with their input and target cells outside the spinal cord, to form a circuitry that is essential for the normal sensory and motor functions. This process involves intricate transcriptional regulation by a number of LIM-HD factors.

The expression of LIM-HD factors in the developing spinal cord was initially characterized in chick embryos (Tsuchida et al., 1994). Chick Islet1, Islet2, Lhx1, and Lhx3 were found to be expressed in different combinations that correlate with both the location of the cell body and the axonal projections of the various types of motor neurons in the spinal cord. Similarly, in zebrafish, spinal cord primary motor neurons with distinct cell body positions and axonal projection patterns express unique combinations of LIM-HD genes. Furthermore, when these neurons were transplanted to a heterologous position within the spinal cord, their LIM-HD gene expression profile, morphology, and axonal projection pattern changed to resemble those of neurons residing at the implant site (Appel et al., 1995). These observations suggest that the identity and axonal projection profile of the different subclasses of spinal cord motor neurons is controlled by a combinatorial expression of different LIM-HD proteins.

The function of individual LIM-HD proteins in the development of the spinal cord has been analyzed primarily by targeted deletion or mis-expression experiments in mouse and chick embryos. The *Islet1* gene is expressed in postmitotic precursor cells that give rise to all subclasses of motor neurons soon after their exit from the cell cycle (Ericson et al., 1992). In mutant mice lacking *Islet1*, cells that are destined to become motor neurons fail to differentiate properly and instead undergo apoptosis (programmed cell death). This results in a complete absence of motor neurons in *Islet1* mutants (Pfaff et al., 1996). The closely related *Islet2* gene is also expressed in all subclasses of motor neurons (Tsuchida et al., 1994). In mutant mice lacking *Islet2*, the spinal cord motor neurons are generated. However, a subclass of visceral motor neurons fail to migrate and project their axons properly, reflecting a switch from visceral to somatic motor neuron character (Thaler et al., 2004).

Two other LIM-HD genes, *Lhx3* and *Lhx4*, have been extensively studied in this context. These two genes share a high sequence similarity (Zhadanov et al., 1995; Yamashita et al., 1997). They are coexpressed in all subclasses of motor neurons that extend their axons ventrally from the neural tube during the time window when these cells exit the cell cycle. In contrast, *Lhx3* and *Lhx4* are never expressed in motor neurons that extend their axons dorsally from the neural tube. In mutant mice with deletions of both *Lhx3* and *Lhx4*, the motor neurons are generated, but the neurons that normally extend axons ventrally switch their identity and instead extend their axons dorsally (Sharma et al., 1998). Conversely, chick hindbrain motor neurons that normally project axons dorsally change fate and project axons ventrally upon ectopic expression of *Lhx3* (Sharma et al., 1998). In motor neurons that extend axons ventrally from the neural tube, *Lhx3* and *Lhx4* are rapidly down-regulated, but their expression persists in one subclass that specifically innervates the axial muscles. This suggests that Lhx3 or Lhx4 may play additional roles in further specifying the identity and axonal projections at subsequent stages in development of this particular subclass of neurons. This has been confirmed by a gain-of-function study showing that an ectopic expression of *Lhx3* can drive all motor neurons to change their profile of molecular expression, position of the cell body, and axonal projection to adopt the identity of the subclass in which the expression of *Lhx3* is normally maintained (Sharma et al., 2000). These results strongly support the notion that the LIM-HD factors play essential and sufficient roles in sequential events that control the identity and axonal projections of specific sets of motor neurons.

Additional evidence supporting the role for LIM-HD proteins in the regulation of motor neuron axonal pathfinding has been derived from a functional analysis of the *Lhx1* and *Lmx1b* genes. *Lhx1* is specifically expressed in a subclass of motor neurons of the spinal cord that innervate the dorsal limb muscles (Tsuchida et al., 1994). In a complementary manner, *Lmx1b* is selectively expressed in cells in the dorsal limb mesenchyme (Riddle et al., 1995; Vogel et al., 1995; Chen et al., 1998). In mice lacking function of either *Lhx1* or *Lmx1b*, the motor neurons that normally express *Lhx1* are specified, but their axonal projections into the limb are randomized. Instead of projecting exclusively into the dorsal limb, the axons target the dorsal and ventral limb with equal probability (Kania et al., 2000). These results indicate that LIM-HD factors expressed in a subclass of motor neurons and their target cells coordinately regulate the formation of the precise neural circuitry of the developing vertebrate nervous system.

In the subclass of motor neurons that innervate dorsal limb muscles, the initial expression of *Islet1* is quickly down-regulated while *Lhx1* continues to be expressed. Conversely, in adjacent motor neurons that innervate ventral limb muscles, *Lhx1* is not expressed but *Islet1* remains active (Tsuchida et al., 1994). Gain-of-function experiments in chick embryos have shown that a mis-expression of either *Lhx1* or *Islet1* can cross- repress the other and direct the cells to change both their cell body position and axonal projection to adopt the identity of the cells that are appropriate for the gene that is expressed (Kania and Jessell, 2003). It thus appears that *Lhx1* and *Islet1* each specify one of these two specific subclasses of motor neurons.

LIM-HD genes are also expressed in interneurons of the spinal cord. There is evidence that they control the specification of distinct interneurons in a cell-autonomous fashion. *Lhx3* is normally expressed in V2 interneurons in the ventral spinal cord. These V2 interneurons can be induced by ectopic expression of *Lhx3* in the dorsal spinal cord (Thaler et al., 2002). Interestingly, when *Lhx3* is mis-expressed together with *Islet1* in the dorsal spinal cord, motor neurons rather than interneurons are induced (Thaler et al., 2002). These results suggest that a combinatorial expression of different LIM-HD proteins is also involved in the distinction between motor neurons and interneurons in the spinal cord. In a separate loss-of-function study, it has been shown that a targeted deletion of *Lmx1b*, a LIM-HD gene normally expressed in postmitotic interneurons in the dorsal spinal cord, results in defects that affect the differentiation and migration of interneurons in the dorsal spinal cord (Ding et al., 2004).

Interference with LIM-HD gene function in zebrafish has led to similar conclusions. Functional repression of *Islet2* results in cell-autonomous defects in the positioning, neurotransmitter expression, axonal outgrowth and pathfinding of sensory and motor neurons during development of the spinal cord in this vertebrate organism (Segawa et al., 2001).

4.4.2
The Vertebrate Brain

An incredible number of neural cell types and an elaborate circuitry are required to accommodate the complex functions of the vertebrate brain. Thus, it is not surprising that, as in the spinal cord, combinatorial LIM-HD functions appear to be involved in the control of neuronal cell differentiation and migration in discrete regions of the developing brain.

Several LIM-HD genes are essential for forebrain development. *Lhx5* is expressed in neural progenitor cells lining the medial wall of the telencephalon, the site of the developing hippocampus. In mutant mice lacking function of *Lhx5*, a large number of neural progenitor cells in the developing hippocampus fail to exit the cell cycle in time. Those that do exit show defects in neuronal differentiation and migration in the formation of distinctive layers of the developing hippocampus (Zhao et al., 1999). The two closely related genes *Lhx6* and *Lhx8* are expressed in the ventral telencephalon (Matsumoto et al., 1996; Grigoriou et al., 1998; Asbreuk et al., 2002). They appear to act downstream of the Nkx2.1 transcription factor to control the differentiation of neuronal cell types derived from the ventral telencephalon (Sussel et al., 1999; Marin et al., 2000). A loss-of-function analysis has implicated *Lhx8* in the development of many forebrain cholinergic neurons. Without the function of *Lhx8*, the forebrain cholinergic neuron progenitors still form, but their terminal differentiation is impaired (Zhao et al., 2003; Mori et al., 2004). The function of *Lhx6* in this context has been analyzed in tissue explants and dissociated neuronal cell cultures using RNA interference. After a knock-down of *Lhx6* expression, GABAergic interneurons are generated, but their tangential migration from the ventral telencephalon to the cortex is blocked (Alifragis et al., 2004).

Elsewhere in the brain, the activity of the LIM-HD gene *Lmx1b* has been associated with dopaminergic neurons in the developing and adult mesencephalon. In *Lmx1b* knockout mice, these neurons are initially generated and express the marker tyrosine hydroxylase, an enzyme required for the synthesis of dopamine. However, they fail to express *Ptx3*, a homeobox gene that is also specific for dopaminergic neurons of the mesencephalon, and are eventually lost during subsequent stages of development (Smidt et al., 2000). *Lmx1b* is thus involved in a molecular cascade that controls the development of mesencephalic dopaminergic neurons independent of the specification of the neurotransmitter phenotype of these neurons.

Lmx1b is also expressed in postmitotic serotonergic neuron precursor cells and differentiated serotonergic neurons in the developing hindbrain (Cheng et al., 2003; Ding et al., 2003). In *Lmx1b* null mutants, serotonergic neuron precursor cells are present but fail to migrate to their proper target fields and to express a series of serotonergic neuron specific markers (Cheng et al., 2003; Ding et al., 2003). Also involved in the development of hindbrain serotonergic neurons are the transcription factors Nkx2.2 and Pet1 (Briscoe et al., 1999; Hendricks et al., 2003; Pattyn et al., 2003). Coexpression of *Lmx1b* with *Nkx2.2*, and *Pet1* can induce serotonergic neurons ectopically in the spinal cord. This suggests that a combination of these transcription factors is sufficient to drive the generation of serotonergic neurons (Cheng et al., 2003). Since *Nkx2.2* is normally expressed in *Lmx1b* mutants, and conversely, the expression of *Lmx1b* is missing in *Nkx2.2* null mutants, *Lmx1b* appears to act downstream of *Nkx2.2* in a cascade that controls development of the serotonergic neurons in the hindbrain (Cheng et al., 2003; Ding et al., 2003).

Similar to their roles in the spinal cord, several LIM-HD transcription factors control development of motor neurons in the hindbrain. For example, much like their function in the spinal cord, Islet1 is required for the generation of all motor neurons in the hindbrain (Pfaff et al., 1996), and both Lhx3 and Lhx4 are required to direct proper axonal projection of those motor neurons in the hindbrain that extend their axons ventrally (Sharma et al., 1998). In addition, Lhx1/Lhx5 and Lhx3/Lhx4 are required for the correct specification of subsets of reticulospinal neurons in the chick hindbrain (Cepeda-Nieto et al., 2005).

4.4.3
The Olfactory and Visual Sensory Systems

Recent studies have uncovered important control functions of LIM-HD genes in the development of the mammalian olfactory and visual sensory systems. *Lhx2* is expressed in the olfactory sensory neurons, and the Lhx2 protein binds to the HD site present in the promoter region of an olfactory receptor gene (Hirota and Mombaerts, 2004). In *Lhx2* null mutant mice, precursor cells of the olfactory sensory neurons are properly generated. However, these cells fail to express olfactory receptors and other molecular markers characteristic of olfactory sensory neurons and are eventually eliminated by cell apoptosis (Hirota and Mombaerts, 2004; Kolterud et al., 2004). *Lhx2* thus plays a critical role in the specification of olfactory sensory neurons. The *Islet2* gene, mentioned above in the context of spinal cord development, is also expressed in a specific group of retinal ganglion cells that project their axons exclusively to the visual center on the contralateral side of the brain. Analysis of *Islet2* null mutant mice has revealed that this gene is critically involved in a negative regulation for expression of the transcription factor Zic2 and the receptor tyrosine kinase EphB1 in these cells, and prevents them from adopting a program to project their axons ipsilaterally into the brain. This finding reveals an important function of Islet2 in establishing the retinotopic map during development of the visual system (Pak et al., 2004).

4.4.4
LIM-HD Genes and Early Patterning Events in the Developing CNS

Up to this point, we have described LIM-HD gene functions involved in the specialization of neuronal precursors in discrete regions of the nervous system. In addition, some of these genes appear to affect broader patterning of the developing nervous system by controlling the formation of critical signaling centers and/or by regulating the proliferation of large populations of neural progenitor cells.

The LIM-HD gene *Lmx1a*, affected in three different alleles of the spontaneous neurological mutant mouse *dreher*, is required for proper formation of the roof plate in the developing CNS (Millonig et al., 2000). Consistent with the established role of the roof plate in mediating signals that control patterning of the dorsal neural tube, the loss of the roof plate in *Lmx1a* mutants causes non-cell-autonomous defects in the specification, migration, and axonal outgrowth of dorsal interneurons in the spinal cord (Millonig et al., 2000; Millen et al., 2004). Rostrally, at the border between the mesencephalon and metencephalon, lack of *Lmx1a* function also results in a partial loss of the choroid plexus and the rhombic lip, a germinal tissue adjacent to the roof plate that gives rise to the granule cells in the cerebellum (Millonig et al., 2000). Loss- and gain-of-function experiments in mouse and chick embryos have shown that *Lmx1a* acts cell-autonomously to withdraw neural progenitor cells from the cell cycle. It appears to play a dual role, directing the progenitors to form roof plate cells while preventing them from becoming dorsal interneurons in the developing spinal cord (Chizhikov and Millen, 2004a).

In chick embryos, the closely related *Lmx1b* gene is also expressed in the roof plate. Like *Lmx1a*, *Lmx1b* can also induce ectopic formation of functional roof plate cells. Moreover, as *Lmx1b* can activate the expression of *Lmx1a* but not vice versa, *Lmx1b* appears to act upstream of *Lmx1a* in chick embryos. Although *Lmx1b* is not expressed in the roof plate in mice, it can nevertheless partially rescue development of the roof plate in *Lmx1a* mutants, suggesting a functional redundancy between *Lmx1a* and *Lmx1b* in controlling formation of the roof plate in the developing CNS (Chizhikov and Millen, 2004b).

In addition to the roof plate, *Lmx1b* is also expressed in the isthmic organizer at the border between the mesencephalon and metencephalon of chick embryos. A gain-of-function analysis has shown that *Lmx1b* is critically involved in the formation and function of the isthmic organizer by participating in complex regulatory pathways composed of multiple signaling molecules and transcription factors (Adams et al., 2000; Matsunaga et al., 2002).

Striking morphological defects in the developing forebrain have been observed in mutant mice that lack *Lhx2* function. In those mutants, the hippocampus and the neocortex is severely reduced and replaced by excessive medial telencephalic tissues including choroid plexus and cortical hem (Porter et al., 1997; Bulchand et al., 2001; Monuki et al., 2001). More detailed marker analysis has revealed that some of the cells that normally express *Lhx2* are still present in the remnant of the dorsal telencephalon in *Lhx2* mutants. These cells retain certain characteristics of the dorsal telencephalic neural progenitors, but they fail to be normally specified to become cortical ventricular zone progenitors (Monuki et al., 2001), and their potential to proliferate is severely impaired (Porter et al., 1997). *Lhx2* thus plays an essential role in patterning of the dorsal telencephalon by controlling the specification and proliferation of the cortical ventricular zone progenitor cells. In addition, the development of the nucleus of the lateral olfactory tract is selectively curtailed in *Lhx2* mutants. This nucleus is part of the developing amygdala, thus assigning the *Lhx2* gene a hitherto unknown role in controlling the development of this part of the telencephalon (Remedios et al., 2004). Several other LIM-HD genes including *Lhx5*, *Lhx6*, *Lhx8*, and *Lhx9* are also expressed in the different nucleus of the amygdaloid complex (Remedios et al., 2004; Choi et al., 2005). The function of these factors in patterning of this complex remains to be determined.

4.5
Factors that Interact with LIM-HD Proteins

The LIM domains present in LIM-HD proteins mediate interactions with other proteins and allow for the formation of multi-protein complexes (Fig. 4.2B). So far, several proteins that are involved in formation of such complexes have been identified. These include the LIM-domain-binding protein Ldb1 (CLIM2/NLI) and its *Drosophila* counterpart Chip, the ubiquitin protein ligase RLIM, a single-stranded DNA-binding protein SSDP, an intraflagellar transport protein SLB, and the p300–associated transcriptional activator Mrg1 (Agulnick et al., 1996; Jurata et al., 1996; Bach et al., 1997, 1999; Glenn and Maurer, 1999; Howard and Maurer, 2000; Chen et al., 2002; van Meyel et al., 2003).

The best characterized of these is Ldb1, a cofactor that binds selectively to the LIM domains of the LIM-HD and LIM-only (LMO) classes of nuclear LIM-proteins. The LIM-binding domain of Ldb1 has been localized to a 38–amino acid portion near the C-terminus of the protein, and a 200–amino acid portion at the N-terminus of Ldb1 is critical for homodimerization (Jurata and Gill, 1997). These two domains allow for the assembly of tetramers composed of two molecules of Ldb1 and two molecules of the same or different LIM-HD proteins (Jurata et al., 1998) (Fig. 4.2B). In addition, direct interactions between different LIM-HD proteins have been observed which enables the formation of hexamers composed of two molecules each of Ldb1, Islet1, and Lhx3 (Thaler et al., 2002) (Fig. 4.2B). Ldb1 and Chip are essential cofactors mediating LIM-HD activity. A convincing example is the interaction of Chip and the LIM-HD gene product Apterous during *Drosophila* wing development. Mutant Chip proteins lacking either the LIM-interacting domain or the homodimerization domain interfere with Apterous function in a dominant-negative fashion, underscoring the importance of physical interaction between the LIM-HD factor and its cofactor. The LIM-only protein dLMO can act as a negative regulator of Apterous activity, possibly by competing for cofactor binding (Milan and Cohen, 1999; van Meyel et al., 1999).

As a word of caution, the function of Ldb1 and Chip may well extend beyond mediating the transcriptional regulation exerted by nuclear LIM domain-containing

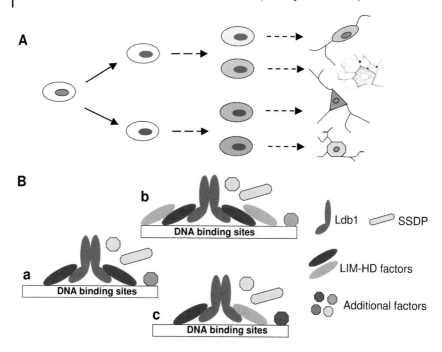

Fig. 4.2 Multiple complexes composed of LIM-HD factors, Ldb1, SSDP, and other nuclear factors are involved in the specification of diverse neuronal cell types. (A) A schematic illustration of pathways that lead from a single neural progenitor cell to various types of neurons. (B) Mediated by Ldb1, the LIM-HD factors form multiple types of complexes (a, b, c) that also include SSDP and possibly additional unidentified factors. These complexes are involved in controlling transcription by binding to specific sites in the regulatory regions of their downstream target genes. The products of these target genes are thought to specify diverse neuronal cell types in the developing nervous system. (This figure also appears with the color plates.)

proteins. The very severe and early knockout phenotype of Lbd1 null mouse mutants (Mukhopadhyay et al., 2003) is not matched by knockout phenotypes of mice lacking one or the other of these nuclear LIM proteins, suggesting possible cofactor activities in pathways unrelated to LIM-HD factors. In fact, Chip appears to interact with a number of unrelated HD proteins (Torigoi et al., 2000).

A negative control of LIM-HD activity is exerted by RLIM, a well-studied Ldb1-interacting protein. RLIM is a ubiquitin protein ligase able to target Ldb1 for degradation through the 26S proteasome pathway. Overexpression of RLIM during chick wing development results in wing defects that are similar to those caused by inhibition of the LIM-HD factor Lhx2 (Bach et al., 1999). More recent experiments have demonstrated ubiquitination-dependent association of RLIM with LIM-HD proteins in the presence of Ldb1, allowing for *in-vivo* scenarios whereby cofactor exchange takes place on DNA-bound transcription factors (Ostendorff et al., 2002).

Another protein that interacts with LIM-HD proteins, albeit indirectly via Ldb1, is the single-stranded DNA-binding protein, SSDP (Chen et al., 2002; van Meyel et al., 2003) (Fig. 4.2B). Ldb1/SSDP/LIM-HD proteins can form a ternary complex that appears to be important for the regulation of transcription (van Meyel et al., 2003). The functional significance of this interaction has been revealed through studies in both *Drosophila* and *Xenopus*. The SSDP fly mutants show defects in wing development similar to those observed in *Apterous* and *Chip* mutants (Chen et al., 2002; van Meyel et al., 2003). In *Xenopus*, co-injections of *SSDP*, *Lhx1*, and *Ldb1* mRNA in oocytes enhance axis induction compared to injections of only *Lhx1* and *Ldb1* mRNAs (Chen et al., 2002). The exact role of SSDP in the multi-protein complex that involves Ldb1 and LIM-HD proteins remains to be characterized.

Several other proteins can interact with LIM-HD transcription factors, but the functional relevance of these interactions is not well understood. The transcription factor Mrg1 can bind to Lhx2, and together they activate transcription, possibly by recruiting the p300 transcription activator and the TATA-binding protein (Glenn and Maurer, 1999). A separate report describes an interaction of the intraflagellar transport complex protein SLB with Lhx3 and Lhx4, but not with other LIM-HD proteins (Howard and Maurer, 2000). However, the phenotype of a recently described SLB mutant mouse does not support the notion that this interaction influences known functions of Lhx3 and Lhx4 in development (Huangfu et al., 2003).

4.6
Downstream Targets of LIM-HD Proteins

Extensive efforts have been undertaken to determine downstream target genes of LIM-HD factors via loss- and gain-of-function approaches. In dorsal spinal cord interneurons, the genes encoding transcription factors Drg11 and Ebf3 are positively regulated by *Lmx1b*. In mutants lacking this gene, the expression of *Drg11* and *Ebf3* is absent. When *Lmx1b* is reactivated, these genes can be turned on. In contrast, the genes encoding transcription factors Zic1 and Zic4 are up-regulated in *Lmx1b* mutants, suggesting that these genes are normally repressed by Lmx1b (Ding et al., 2004). In the retina of *Islet2* null mutant mice, the gene encoding the related transcription factor Zic2 is up-regulated, suggesting that Islet2 acts as a repressor for *Zic2* (Pak et al., 2004). In the isthmic organizer, gain-of-function studies have shown that Lmx1b can induce the expression of the signaling molecule Wnt1 (Adams et al., 2000; Matsunaga et al., 2002) and the transcription factors Otx2 and Grg4 (Matsunaga et al., 2002).

Consistent with the role of LIM-HD proteins in controlling axonal pathfinding, another set of studies has identified axon guidance molecules as targets that are regulated by these transcription factors. EphA4 is a receptor tyrosine kinase that is selectively expressed at a high level in the Lhx1–positive subclass of motor neurons that innervate the dorsal limb muscles, and at a low level in Islet1–positive neurons that innervate the ventral limb muscles. This pattern of EphA4 expression is essential for the proper formation of the topographic axonal projections of the spinal cord motor neurons to the limb muscles (Helmbacher et al., 2000; Eberhart et al., 2002; Kania and Jessell, 2003). Both loss- and gain-of-function analyses have shown that EphA4 is regulated by Lhx1 and Islet1, as Lhx1 elevates and Islet1 lowers the expression of EphA4 (Kania and Jessell, 2003). In addition, EphrinA5 – an EphA4 ligand that is expressed at a high level in the ventral and at a low level in the dorsal limb mesenchyme cells – is considered a possible regulator for axonal projections from the motor neurons to the limb muscles (Eberhart et al., 2000). The expression of EphrinA5 in limb muscle mesenchyme is regulated by Lmx1b (Kania and Jessell, 2003). Thus, the proper formation of the motor neuron axonal projections to the limb muscles is controlled by Lhx1 and Lmx1b through their regulation of Eph and Ephrin molecules (Kania and Jessell, 2003).

Several LIM-HD transcription factors bind directly and specifically to *cis*-regulatory elements of downstream target genes. There is a notable example that this interaction can result in functional target gene activation. The complex of Lhx3 and Islet1 that controls the specification of motor neurons (Thaler et al., 2002) binds to an enhancer element of the *HB9* gene (Lee and Pfaff, 2003), and thereby stimulates the enhancer activity of a transcription factor that is both required and sufficient to direct the differentiation of motor neurons in the spinal cord (Tanabe et al., 1998; Arber et al., 1999; Thaler et al., 1999). As a further example, Lhx2 binds specifically to the HD site in the promoter region of odorant receptor genes, and may be involved in regulating the expression of these genes (Hirota and Mombaerts, 2004). In *C. elegans*, the LIM-HD factor TTX–3, in conjunction with the paired- HD factor CEH–10, binds to a conserved 16–base pair *cis*-regulatory element that has been identified in a battery of genes involved in the specification of interneurons, including the *ttx–3* gene itself. Further work along similar lines of experimentation is needed to determine whether the activation of groups of target genes that together determine cell specificity might well be a general mechanism by which LIM-HD factors exert their function during development of the nervous system.

4.7
Conclusion and Future Directions

Extensive functional studies have revealed important roles for LIM-HD transcription factors in the specification of the developing nervous system. The gain or loss of individual LIM-HD gene functions can result in a complete block of differentiation and subsequent loss of an entire group of neurons, or else cause more subtle changes in various aspects of neuronal development such as cell migration, process outgrowth, axonal projection, or neurotransmitter synthesis. This variety of phenotypic responses may reflect the plethora of cell-cell interactions during development of the nervous system, allowing for interactions between LIM-HD proteins and other transcriptional cascades or signaling pathways that influence context-specific regulation of downstream target genes (Lee and Pfaff, 2003; Allan et al., 2005). The identification of direct downstream target genes that are controlled by the LIM-HD transcription factors in different fields of the developing nervous system constitutes a major challenge of future studies.

Abbreviations

LIM-HD	LIM-homeodomain
GABA	gamma-aminobutyric acid
TN	transverse nerve
ISNb	intersegmental nerve b
ISNd	intersegmental nerve d
CNS	central nervous system
SSDP	single-stranded DNA-binding protein
VNC	ventral nerve cord

References

Adams KA, Maida JM, Golden JA, Riddle RD. **2000**. The transcription factor Lmx1b maintains Wnt1 expression within the isthmic organizer. *Development* 127: 1857–1867.

Agulnick AD, Taira M, Breen JJ, Tanaka T, Dawid IB, Westphal H. **1996**. Interactions of the LIM-domain-binding factor Ldb1 with LIM homeodomain proteins. *Nature* 384: 270–272.

Alifragis P, Liapi A, Parnavelas JG. **2004**. Lhx6 regulates the migration of cortical interneurons from the ventral telencephalon but does not specify their GABA phenotype. *J. Neurosci.* 24: 5643–5648.

Allan DW, Park D, St Pierre SE, Taghert PH, Thor S. **2005**. Regulators acting in combinatorial codes also act independently in single differentiating neurons. *Neuron* 45: 689–700.

Appel B, Korzh V, Glasgow E, Thor S, Edlund T, Dawid IB, Eisen JS. **1995**. Motoneuron fate specification revealed by patterned LIM homeobox gene expression in embryonic zebrafish. *Development* 121: 4117–4125.

Arber S, Han B, Mendelsohn M, Smith M, Jessell TM, Sockanathan S. **1999**. Requirement for the homeobox gene Hb9 in the consolidation of motor neuron identity. *Neuron* 23: 659–674.

Asbreuk CH, van Schaick HS, Cox JJ, Kromkamp M, Smidt MP, Burbach JP. **2002**. The homeobox genes Lhx7 and Gbx1 are expressed in the basal forebrain cholinergic system. *Neuroscience* 109: 287–298.

Aurelio O, Boulin T, Hobert O. **2003**. Identification of spatial and temporal cues that regulate postembryonic expression of axon maintenance factors in the *C. elegans* ventral nerve cord. *Development* 130: 599–610.

Bach I, Carriere C, Ostendorff HP, Andersen B, Rosenfeld MG. **1997**. A family of LIM domain-associated cofactors confer transcriptional synergism between LIM and Otx homeodomain proteins. *Genes Dev.* 11: 1370–1380.

Bach I, Rodriguez-Esteban C, Carriere C, Bhushan A, Krones A, Rose DW, Glass CK, Andersen B, Izpisua Belmonte JC, Rosenfeld MG. **1999**. RLIM inhibits functional activity of LIM homeodomain transcription factors via recruitment of the histone deacetylase complex. *Nat. Genet.* 22: 394–399.

Benveniste RJ, Thor S, Thomas JB, Taghert PH. **1998**. Cell type-specific regulation of the *Drosophila* FMRF-NH2 neuropeptide gene by Apterous, a LIM homeodomain transcription factor. *Development* 125: 4757–4765.

Briscoe J, Sussel L, Serup P, Hartigan-O'Connor D, Jessell TM, Rubenstein JL, Ericson J. **1999**. Homeobox gene Nkx2.2 and specification of neuronal identity by graded Sonic hedgehog signalling. *Nature* 398: 622–627.

Bulchand S, Grove EA, Porter FD, Tole S. **2001**. LIM-homeodomain gene Lhx2 regulates the formation of the cortical hem. *Mech. Dev.* 100: 165–175.

Cassata G, Kagoshima H, Andachi Y, Kohara Y, Durrenberger MB, Hall DH, Burglin TR. **2000**. The LIM homeobox gene ceh–14 confers thermosensory function to the

AFD neurons in *Caenorhabditis elegans*. *Neuron* 25: 587–597.

Cepeda-Nieto AC, Pfaff SL, Varela-Echavarria A. **2005**. Homeodomain transcription factors in the development of subsets of hindbrain reticulospinal neurons. *Mol. Cell. Neurosci.* 28: 30–41.

Chang S, Johnston RJ, Jr., Hobert O. **2003**. A transcriptional regulatory cascade that controls left/right asymmetry in chemosensory neurons of *C. elegans*. *Genes Dev.* 17: 2123–2137.

Chen H, Lun Y, Ovchinnikov D, Kokubo H, Oberg KC, Pepicelli CV, Gan L, Lee B, Johnson RL. **1998**. Limb and kidney defects in Lmx1b mutant mice suggest an involvement of LMX1B in human nail patella syndrome. *Nat. Genet.* 19: 51–55.

Chen L, Segal D, Hukriede NA, Podtelejnikov AV, Bayarsaihan D, Kennison JA, Ogryzko VV, Dawid IB, Westphal H. **2002**. Ssdp proteins interact with the LIM-domain-binding protein Ldb1 to regulate development. *Proc. Natl. Acad. Sci. USA* 99: 14320–14325.

Cheng L, Chen CL, Luo P, Tan M, Qiu M, Johnson R, Ma Q. **2003**. Lmx1b, Pet–1, and Nkx2.2 coordinately specify serotonergic neurotransmitter phenotype. *J. Neurosci.* 23: 9961–9967.

Chizhikov VV, Millen KJ. **2004a**. Control of roof plate formation by Lmx1a in the developing spinal cord. *Development* 131: 2693–2705.

Chizhikov VV, Millen KJ. **2004b**. Control of roof plate development and signaling by Lmx1b in the caudal vertebrate CNS. *J. Neurosci.* 24: 5694–5703.

Choi GB, Dong H, Murphy AJ, Valenzuela DM, Yancopoulos GD, Swanson LW, Anderson DJ. **2005**. Lhx6 delineates a pathway mediating innate reproductive behaviors from the amygdala to the hypothalamus. *Neuron* 46: 647–660.

Deane JE, Mackay JP, Kwan AH, Sum EY, Visvader JE, Matthews JM. **2003**. Structural basis for the recognition of ldb1 by the N-terminal LIM domains of LMO2 and LMO4. *EMBO J.* 22: 2224–2233.

Deane JE, Ryan DP, Sunde M, Maher MJ, Guss JM, Visvader JE, Matthews JM. **2004**. Tandem LIM domains provide synergistic binding in the LMO4:Ldb1 complex. *EMBO J.* 23: 3589–3598.

Ding YQ, Marklund U, Yuan W, Yin J, Wegman L, Ericson J, Deneris E, Johnson RL, Chen ZF. **2003**. Lmx1b is essential for the development of serotonergic neurons. *Nat. Neurosci.* 6: 933–938.

Ding YQ, Yin J, Kania A, Zhao ZQ, Johnson RL, Chen ZF. **2004**. Lmx1b controls the differentiation and migration of the superficial dorsal horn neurons of the spinal cord. *Development* 131: 3693–3703.

Eberhart J, Swartz M, Koblar SA, Pasquale EB, Tanaka H, Krull CE. **2000**. Expression of EphA4, ephrin-A2 and ephrin-A5 during axon outgrowth to the hindlimb indicates potential roles in pathfinding. *Dev. Neurosci.* 22: 237–250.

Eberhart J, Swartz ME, Koblar SA, Pasquale EB, Krull CE. **2002**. EphA4 constitutes a population-specific guidance cue for motor neurons. *Dev. Biol.* 247: 89–101.

Ericson J, Thor S, Edlund T, Jessell TM, Yamada T. **1992**. Early stages of motor neuron differentiation revealed by expression of homeobox gene Islet–1. *Science* 256: 1555–1560.

Freyd G, Kim SK, Horvitz HR. **1990**. Novel cysteine-rich motif and homeodomain in the product of the *Caenorhabditis elegans* cell lineage gene lin–11. *Nature* 344: 876–879.

Glasgow E, Karavanov AA, Dawid IB. **1997**. Neuronal and neuroendocrine expression of lim3, a LIM class homeobox gene, is altered in mutant zebrafish with axial signaling defects. *Dev. Biol.* 192: 405–419.

Glenn DJ, Maurer RA. **1999**. MRG1 binds to the LIM domain of Lhx2 and may function as a coactivator to stimulate glycoprotein hormone alpha-subunit gene expression. *J. Biol. Chem.* 274: 36159–36167.

Grigoriou M, Tucker AS, Sharpe PT, Pachnis V. **1998**. Expression and regulation of Lhx6 and Lhx7, a novel subfamily of LIM homeodomain encoding genes, suggests a role in mammalian head development. *Development* 125: 2063–2074.

Helmbacher F, Schneider-Maunoury S, Topilko P, Tiret L, Charnay P. **2000**. Targeting of the EphA4 tyrosine kinase receptor affects dorsal/ventral pathfinding of limb motor axons. *Development* 127: 3313–3324.

Hendricks TJ, Fyodorov DV, Wegman LJ, Lelutiu NB, Pehek EA, Yamamoto B, Silver J, Weeber EJ, Sweatt JD, Deneris ES. **2003**. Pet-1 ETS gene plays a critical role in 5-HT neuron development and is required for normal anxiety-like and aggressive behavior. *Neuron* 37: 233–247.

Hirota J, Mombaerts P. **2004**. The LIM homeodomain protein Lhx2 is required for complete development of mouse olfactory sensory neurons. *Proc. Natl. Acad. Sci. USA* 101: 8751–8755.

Hobert O, Mori I, Yamashita Y, Honda H, Ohshima Y, Liu Y, Ruvkun G. **1997**. Regulation of interneuron function in the *C. elegans* thermoregulatory pathway by the ttx-3 LIM homeobox gene. *Neuron* 19: 345–357.

Hobert O, D'Alberti T, Liu Y, Ruvkun G. **1998**. Control of neural development and function in a thermoregulatory network by the LIM homeobox gene lin-11. *J. Neurosci.* 18: 2084–2096.

Hobert O, Tessmar K, Ruvkun G. **1999**. The *Caenorhabditis elegans* lim-6 LIM homeobox gene regulates neurite outgrowth and function of particular GABAergic neurons. *Development* 126: 1547–1562.

Hobert O, Westphal H. **2000**. Functions of LIM-homeobox genes. *Trends Genet.* 16: 75–83.

Howard PW, Maurer RA. **2000**. Identification of a conserved protein that interacts with specific LIM homeodomain transcription factors. *J. Biol. Chem.* 275: 13336–13342.

Huangfu D, Liu A, Rakeman AS, Murcia NS, Niswander L, Anderson KV. **2003**. Hedgehog signalling in the mouse requires intraflagellar transport proteins. *Nature* 426: 83–87.

Jackman WR, Draper BW, Stock DW. **2004**. Fgf signaling is required for zebrafish tooth development. *Dev. Biol.* 274: 139–157.

Jurata LW, Kenny DA, Gill GN. **1996**. Nuclear LIM interactor, a rhombotin and LIM homeodomain interacting protein, is expressed early in neuronal development. *Proc. Natl. Acad. Sci. USA* 93: 11693–11698.

Jurata LW, Gill GN. **1997**. Functional analysis of the nuclear LIM domain interactor NLI. *Mol. Cell. Biol.* 17: 5688–5698.

Jurata LW, Pfaff SL, Gill GN. **1998**. The nuclear LIM domain interactor NLI mediates homo- and heterodimerization of LIM domain transcription factors. *J. Biol. Chem.* 273: 3152–3157.

Kania A, Johnson RL, Jessell TM. **2000**. Coordinate roles for LIM homeobox genes in directing the dorsoventral trajectory of motor axons in the vertebrate limb. *Cell* 102: 161–173.

Kania A, Jessell TM. **2003**. Topographic motor projections in the limb imposed by LIM homeodomain protein regulation of ephrin-A:EphA interactions. *Neuron* 38: 581–596.

Karlsson O, Thor S, Norberg T, Ohlsson H, Edlund T. **1990**. Insulin gene enhancer binding protein Isl-1 is a member of a novel class of proteins containing both a homeo- and a Cys-His domain. *Nature* 344: 879–882.

Kolterud A, Alenius M, Carlsson L, Bohm S. **2004**. The Lim homeobox gene Lhx2 is required for olfactory sensory neuron identity. *Development* 131: 5319–5326.

Lee SK, Pfaff SL. **2003**. Synchronization of neurogenesis and motor neuron specification by direct coupling of bHLH and homeodomain transcription factors. *Neuron* 38: 731–745.

Lundgren SE, Callahan CA, Thor S, Thomas JB. **1995**. Control of neuronal pathway selection by the *Drosophila* LIM homeodomain gene apterous. *Development* 121: 1769–1773.

Marin O, Anderson SA, Rubenstein JL. **2000**. Origin and molecular specification of striatal interneurons. *J. Neurosci.* 20: 6063–6076.

Matsumoto K, Tanaka T, Furuyama T, Kashihara Y, Mori T, Ishii N, Kitanaka J, Takemura M, Tohyama M, Wanaka A. **1996**. L3, a novel murine LIM-homeodomain transcription factor expressed in the ventral telencephalon and the mesenchyme surrounding the oral cavity. *Neurosci. Lett.* 204: 113–116.

Matsunaga E, Katahira T, Nakamura H. **2002**. Role of Lmx1b and Wnt1 in mesencephalon and metencephalon development. *Development* 129: 5269–5277.

Milan M, Cohen SM. **1999**. Regulation of LIM homeodomain activity in vivo: a tetramer of dLDB and apterous confers activity and capacity for regulation by dLMO. *Mol. Cell* 4: 267–273.

Millen KJ, Millonig JH, Hatten ME. **2004**. Roof plate and dorsal spinal cord dl1 interneuron development in the *dreher* mutant mouse. *Dev. Biol.* 270: 382–392.

Millonig JH, Millen KJ, Hatten ME. **2000**. The mouse Dreher gene Lmx1a controls formation of the roof plate in the vertebrate CNS. *Nature* 403: 764–769.

Monuki ES, Porter FD, Walsh CA. **2001**. Patterning of the dorsal telencephalon and cerebral cortex by a roof plate-Lhx2 pathway. *Neuron* 32: 591–604.

Mori T, Yuxing Z, Takaki H, Takeuchi M, Iseki K, Hagino S, Kitanaka J, Takemura M, Misawa H, Ikawa M, Okabe M, Wanaka A. **2004**. The LIM homeobox gene, L3/Lhx8, is necessary for proper development of basal forebrain cholinergic neurons. *Eur. J. Neurosci.* 19: 3129–3141.

Mukhopadhyay M, Teufel A, Yamashita T, Agulnick AD, Chen L, Downs KM, Schindler A, Grinberg A, Huang SP, Dorward D, Westphal H. **2003**. Functional ablation of the mouse Ldb1 gene results in severe patterning defects during gastrulation. *Development* 130: 495–505.

O'Keefe DD, Thor S, Thomas JB. **1998**. Function and specificity of LIM domains in *Drosophila* nervous system and wing development. *Development* 125: 3915–3923.

Ostendorff HP, Peirano RI, Peters MA, Schluter A, Bossenz M, Scheffner M, Bach I. **2002**. Ubiquitination-dependent cofactor exchange on LIM homeodomain transcription factors. *Nature* 416: 99–103.

Pak W, Hindges R, Lim YS, Pfaff SL, O'Leary DD. **2004**. Magnitude of binocular vision controlled by islet–2 repression of a genetic program that specifies laterality of retinal axon pathfinding. *Cell* 119: 567–578.

Pattyn A, Vallstedt A, Dias JM, Samad OA, Krumlauf R, Rijli FM, Brunet JF, Ericson J. **2003**. Coordinated temporal and spatial control of motor neuron and serotonergic neuron generation from a common pool of CNS progenitors. *Genes Dev.* 17: 729–737.

Perez-Alvarado GC, Miles C, Michelsen JW, Louis HA, Winge DR, Beckerle MC, Summers MF. **1994**. Structure of the carboxy-terminal LIM domain from the cysteine rich protein CRP. *Nat. Struct. Biol.* 1: 388–398.

Pfaff SL, Mendelsohn M, Stewart CL, Edlund T, Jessell TM. **1996**. Requirement for LIM homeobox gene Isl1 in motor neuron generation reveals a motor neuron-dependent step in interneuron differentiation. *Cell* 84: 309–320.

Pierce-Shimomura JT, Faumont S, Gaston MR, Pearson BJ, Lockery SR. **2001**. The homeobox gene lim–6 is required for distinct chemosensory representations in C. elegans. *Nature* 410: 694–698.

Porter FD, Drago J, Xu Y, Cheema SS, Wassif C, Huang SP, Lee E, Grinberg A, Massalas JS, Bodine D, Alt F, Westphal H. **1997**. Lhx2, a LIM homeobox gene, is required for eye, forebrain, and definitive erythrocyte development. *Development* 124: 2935–2944.

Remedios R, Subramanian L, Tole S. **2004**. LIM genes parcellate the embryonic amygdala and regulate its development. *J. Neurosci.* 24: 6986–6990.

Riddle RD, Ensini M, Nelson C, Tsuchida T, Jessell TM, Tabin C. **1995**. Induction of the LIM homeobox gene Lmx1 by WNT7a establishes dorsoventral pattern in the vertebrate limb. *Cell* 83: 631–640.

Sagasti A, Hobert O, Troemel ER, Ruvkun G, Bargmann CI. **1999**. Alternative olfactory neuron fates are specified by the LIM homeobox gene lim–4. *Genes Dev.* 13: 1794–1806.

Sarafi-Reinach TR, Melkman T, Hobert O, Sengupta P. **2001**. The lin–11 LIM homeobox gene specifies olfactory and chemosensory neuron fates in C. elegans. *Development* 128: 3269–3281.

Schneider LE, Roberts MS, Taghert PH. **1993**. Cell type-specific transcriptional regulation of the *Drosophila* FMRFamide neuropeptide gene. *Neuron* 10: 279–291.

Segawa H, Miyashita T, Hirate Y, Higashijima S, Chino N, Uyemura K, Kikuchi Y, Okamoto H. **2001**. Functional repression of Islet–2 by disruption of complex with Ldb impairs peripheral axonal outgrowth in embryonic zebrafish. *Neuron* 30: 423–436.

Sharma K, Sheng HZ, Lettieri K, Li H, Karavanov A, Potter S, Westphal H, Pfaff SL.

1998. LIM homeodomain factors Lhx3 and Lhx4 assign subtype identities for motor neurons. *Cell* 95: 817–828.

Sharma K, Leonard AE, Lettieri K, Pfaff SL. **2000**. Genetic and epigenetic mechanisms contribute to motor neuron pathfinding. *Nature* 406: 515–519.

Smidt MP, Asbreuk CH, Cox JJ, Chen H, Johnson RL, Burbach JP. **2000**. A second independent pathway for development of mesencephalic dopaminergic neurons requires Lmx1b. *Nat. Neurosci.* 3: 337–341.

Sussel L, Marin O, Kimura S, Rubenstein JL. **1999**. Loss of Nkx2.1 homeobox gene function results in a ventral to dorsal molecular respecification within the basal telencephalon: evidence for a transformation of the pallidum into the striatum. *Development* 126: 3359–3370.

Tanabe Y, William C, Jessell TM. **1998**. Specification of motor neuron identity by the MNR2 homeodomain protein. *Cell* 95: 67–80.

Thaler J, Harrison K, Sharma K, Lettieri K, Kehrl J, Pfaff SL. **1999**. Active suppression of interneuron programs within developing motor neurons revealed by analysis of homeodomain factor HB9. *Neuron* 23: 675–687.

Thaler JP, Lee SK, Jurata LW, Gill GN, Pfaff SL. **2002**. LIM factor Lhx3 contributes to the specification of motor neuron and interneuron identity through cell-type-specific protein-protein interactions. *Cell* 110: 237–249.

Thaler JP, Koo SJ, Kania A, Lettieri K, Andrews S, Cox C, Jessell TM, Pfaff SL. **2004**. A postmitotic role for Isl-class LIM homeodomain proteins in the assignment of visceral spinal motor neuron identity. *Neuron* 41: 337–350.

Thor S, Thomas JB. **1997**. The *Drosophila* islet gene governs axon pathfinding and neurotransmitter identity. *Neuron* 18: 397–409.

Thor S, Andersson SG, Tomlinson A, Thomas JB. **1999**. A LIM-homeodomain combinatorial code for motor-neuron pathway selection. *Nature* 397: 76–80.

Torigoi E, Bennani-Baiti IM, Rosen C, Gonzalez K, Morcillo P, Ptashne M, Dorsett D. **2000**. Chip interacts with diverse homeodomain proteins and potentiates bicoid activity in vivo. *Proc. Natl. Acad. Sci. USA* 97: 2686–2691.

Toyama R, Curtiss PE, Otani H, Kimura M, Dawid IB, Taira M. **1995a**. The LIM class homeobox gene lim5: implied role in CNS patterning in *Xenopus* and zebrafish. *Dev. Biol.* 170: 583–593.

Toyama R, O'Connell ML, Wright CV, Kuehn MR, Dawid IB. **1995b**. Nodal induces ectopic goosecoid and lim1 expression and axis duplication in zebrafish. *Development* 121: 383–391.

Toyama R, Dawid IB. **1997**. lim6, a novel LIM homeobox gene in the zebrafish: comparison of its expression pattern with lim1. *Dev. Dyn.* 209: 406–417.

Tsalik EL, Niacaris T, Wenick AS, Pau K, Avery L, Hobert O. **2003**. LIM homeobox gene-dependent expression of biogenic amine receptors in restricted regions of the *C. elegans* nervous system. *Dev. Biol.* 263: 81–102.

Tsuchida T, Ensini M, Morton SB, Baldassare M, Edlund T, Jessell TM, Pfaff SL. **1994**. Topographic organization of embryonic motor neurons defined by expression of LIM homeobox genes. *Cell* 79: 957–970.

van Meyel DJ, O'Keefe DD, Jurata LW, Thor S, Gill GN, Thomas JB. **1999**. Chip and apterous physically interact to form a functional complex during *Drosophila* development. *Mol. Cell* 4: 259–265.

van Meyel DJ, O'Keefe DD, Thor S, Jurata LW, Gill GN, Thomas JB. **2000**. Chip is an essential cofactor for apterous in the regulation of axon guidance in *Drosophila*. *Development* 127: 1823–1831.

van Meyel DJ, Thomas JB, Agulnick AD. **2003**. Ssdp proteins bind to LIM-interacting co-factors and regulate the activity of LIM-homeodomain protein complexes in vivo. *Development* 130: 1915–1925.

Vogel A, Rodriguez C, Warnken W, Izpisua Belmonte JC. **1995**. Dorsal cell fate specified by chick Lmx1 during vertebrate limb development. *Nature* 378: 716–720.

Way JC, Chalfie M. **1988**. mec-3, a homeobox-containing gene that specifies differentiation of the touch receptor neurons in *C. elegans*. *Cell* 54: 5–16.

Way JC, Chalfie M. **1989**. The mec-3 gene of *Caenorhabditis elegans* requires its own product for maintained expression and is expressed in three neuronal cell types. *Genes Dev.* 3: 1823–1833.

Yamashita T, Moriyama K, Sheng HZ, Westphal H. **1997**. Lhx4, a LIM homeobox gene. *Genomics* 44: 144–146.

Zhadanov AB, Bertuzzi S, Taira M, Dawid IB, Westphal H. **1995**. Expression pattern of the murine LIM class homeobox gene Lhx3 in subsets of neural and neuroendocrine tissues. *Dev. Dyn.* 202: 354–364.

Zhao Y, Sheng HZ, Amini R, Grinberg A, Lee E, Huang S, Taira M, Westphal H. **1999**. Control of hippocampal morphogenesis and neuronal differentiation by the LIM homeobox gene Lhx5. *Science* 284: 1155–1158.

Zhao Y, Marin O, Hermesz E, Powell A, Flames N, Palkovits M, Rubenstein JL, Westphal H. **2003**. The LIM-homeobox gene Lhx8 is required for the development of many cholinergic neurons in the mouse forebrain. *Proc. Natl. Acad. Sci. USA* 100: 9005–9010.

5
The Roles of Serum Response Factor (SRF) in Development and Function of the Brain

Bernd Knöll and Alfred Nordheim

Abstract

The transcription factor serum response factor (SRF) regulates target genes with the help of cofactors which are responsive to either MAPK signaling (cofactors involved: TCFs) or actin signaling (cofactors involved: myocardin/MAL family proteins). In the brain, the SRF target genes identified to date belong to the class of the rapidly and transiently induced immediate-early genes (IEGs) or encode – among others – cytoskeletal proteins. Based on genetic studies generating conditional SRF deficiency in the mouse forebrain, identified functions of SRF and its cofactors in the neuronal system include: (i) activation of the IEG response in the hippocampus upon electroconvulsive shock; (ii) directing neuronal migration; (iii) regulating neurite outgrowth and axonal pathfinding; and (iv) directing features of long-term synaptic potentiation as well as long-term depression during learning and memory acquisition.

5.1
Characteristics of SRF as a Transcription Factor

Serum response factor (SRF) (Norman et al., 1988) is a widely expressed nuclear protein which acts as a transcription factor that regulates the expression of many target genes (Philippar et al., 2004; Zhang et al., 2005). SRF is essential for mammalian embryonic development (Arsenian et al., 1998) and belongs to the family of MADS-box proteins (Schwarz-Sommer et al., 1992), members of which are found in all eukaryotic systems ranging from yeast to man (Shore and Sharrocks, 1995). In size, human SRF comprises 508 amino acids, though several splice variants of SRF have been reported. SRF binds as a homodimer to DNA segments containing palindromic or near-palindromic $CC(A/T)_6GG$ sequences (called CArG boxes). The structure of a DNA-protein co-crystal, consisting of a CArG oligonucleotide and homodimeric core^{SRF}, was solved at high resolution by X-ray crystallography (Pellegrini et al., 1995). This structure revealed a novel, interdigitated protein unit which maintained specific DNA contacts to a highly bent segment of the double helix.

Transcription Factors. Edited by Gerald Thiel
Copyright © 2006 WILEY-VCH Verlag GmbH & Co. KGaA, Weinheim
ISBN 3-527-31285-4

SRF exerts its role as a transcriptional regulator largely upon recruitment of, and interaction with, partner proteins (Shaw et al., 1989; Buchwalter et al., 2004). These partner proteins – for example, the ternary complex factors (TCFs) Elk–1/Sap–1/Net1 (Hipskind et al., 1991; Dalton and Treisman, 1992; Giovane et al., 1994) and members of the myocardin/MRTF/MAL protein family (Wang et al., 2002; Miralles et al., 2003), are subject to regulation by signaling cascades (Fig. 5.1). The nuclear Ets-type TCF proteins respond with high sensitivity to incoming MAPK signals (Gille et al., 1992), thereby serving as points of convergence to at least three types of MAPK, namely Erk1/Erk2, JunK, and p38/SAPK. The myocardin/MRTF/MAL family of SRF partner proteins, on the other hand, inducibly undergo nuclear translocation for SRF activation upon induction of actin polymerization by Rho GTPase stimulation (Miralles et al., 2003). Accordingly, SRF target genes can be classified according to their selective responsiveness to either MAPK signaling (class I) or RhoA/actin signaling (class II) (Gineitis and Treisman, 2001; Murai and Treisman, 2002; Wang et al., 2004). Class I genes contain the classical immediate-early genes (IEGs) which are activated rapidly and transiently by, for example, growth factors or neurotransmitters (Greenberg and Ziff, 1984; Misra et al., 1994). Class II genes include, amongst others, cytoskeletal genes such as *actin, vinculin,* or *SM22α*.

Therefore, expression patterns and the activation states of SRF partner proteins confer temporal and spatial selectivity to SRF function with regard to cell type and target gene specificity (Buchwalter et al., 2004; Wang and Olson, 2004). Whereas TCF proteins and MAL appear to be expressed quite ubiquitously, myocardin is found selectively in muscle cells. As of yet, no brain cell-specific partner protein of SRF has been identified, although the myocardin/MAL homologue MRTF-B was shown to be expressed more restrictedly in heart and brain (Wang et al., 2002).

5.2
Neuronal Expression Patterns of SRF and Partner Proteins

The localization of RNA and protein encoded by *Srf* has been recently investigated primarily in the postnatal and adult brain of rodents (Stringer et al., 2002). These authors first described nuclear SRF localization predominantly in neurons of various brain regions including cortex, caudate-putamen, amygdala, hippocampus and cerebellar granule cells. The mesencephalon, thalamus and hypothalamus seem largely devoid of *Srf* expression. These results are in line with later reports showing localization of SRF in the postnatal and adult cornu ammonis and dentate gyrus regions of the hippocampus (Alberti et al., 2005; Ramanan et al., 2005), striatum and additionally in the olfactory bulb (Alberti et al., 2005, Knöll et al., unpublished results). Our own data (unpublished results) further revealed widespread distribution of *Srf* RNA in the embryonal brain.

Amongst the cofactors interacting with SRF, distribution of the TCF Elk–1 has been analyzed in more detail in the brain. One group (Sgambato et al., 1998) reported highest Elk–1 expression levels in the olfactory bulb, cortex (especially the pyriform cortex), dentate gyrus and cerebellum, thus largely overlapping with the ex-

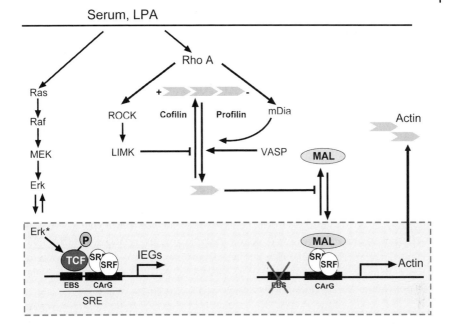

Fig. 5.1 Signaling pathways leading to stimulation of serum response factor (SRF) via activation of SRF partner proteins. Stimulation of cells with serum or lysophosphatidic acid (LPA) results in SRF activation via at least two signaling pathways. Activation of the MAP kinase cascade translocates the activated kinases Erk1/2 to the nuclear compartment and directs phosphorylation and thereby activation of ternary complex factors (TCFs) (e.g., Elk–1, Sap–1 and Net1). Hyperphosphorylation of the TCF type of SRF cofactors induces transcriptional activation of immediate-early genes (IEGs). SRF is intimately embedded in the regulation of actin dynamics in cells. Stimulation of small Rho GTPases (e.g., RhoA) triggers F-actin polymerization via ROCK-LIM kinase or mDia signaling intermediates (see text). The resulting depletion of G-actin levels in the cytoplasm releases a translocation block on the SRF cofactor MAL, enabling its nuclear accumulation. Here, MAL cooperates with SRF to stimulate expression of genes such as *actin* mainly involved in cytoskeletal dynamics. The rise of G-actin levels in the cytoplasm in turn is assumed to regulate *Srf* transcription negatively by sequestering MAL in the cytoplasm.

pression domains of its cognate partner SRF. Somewhat weaker Elk–1 expression levels were found in the cornu ammonis of the hippocampus and in some thalamic nuclei. Interestingly, and in contrast to SRF, localization of full-length Elk–1 is not exclusively nuclear, yet soma, dendrites and axon terminals also stain positively (Sgambato et al., 1998). As reported by others (Vanhoutte et al., 2001), this extranuclear localization of full-length Elk–1 might be the consequence of a truncated Elk–1 protein (sElk–1), which probably drives nuclear export. In addition, the same authors demonstrated that sElk–1 competes with, and thereby reduces, the transactivation properties of full-length Elk–1 on SRF. Functionally, this relates to the ability of sElk–1, but not full-length Elk–1, to stimulate neurite outgrowth of PC12 cells (Vanhoutte et al., 2001).

5.3
SRF Target Genes with Brain Functions

The IEG c-*fos* has been long known to be induced transcriptionally in the rodent hippocampus upon induction of seizure (Morgan et al., 1987), and IEG induction was soon appreciated to be important for brain function (Curran and Morgan, 1987; Sheng and Greenberg, 1990). Most of the candidate IEGs induced in the brain contain SRF binding sites in their promoters, although – in stringent functional terms – the contribution of SRF for neuronal IEG induction has been determined in only a few examples. Such stringent criteria would include: (i) functional proof by binding site mutation for a CArG box to be essential for transcriptional control of the gene in question; (ii) anti-SRF chromatin immune precipitations (ChIP) identifying the occupancy of an existing CArG box by SRF in neuronal tissues; and (iii) the demonstration that the putative SRF target gene is dysregulated in SRF-deficient cells. Additionally, functional CArG boxes can be expected to be conserved in the genomes of mice and man.

Our general knowledge regarding SRF target genes expanded significantly once genome-wide searches for CArG box-containing direct SRF target genes were utilized, combined with the aid of SRF-deficient ES cells (Philippar et al., 2004; Zhang et al., 2005). In the latter two studies, non-neuronal cells were used and – accordingly – only a limited number of putative SRF target genes with functions related to neuronal activities are presently known. The SRF target genes presently known to be important for brain function are summarized in Table 5.1. Most of these genes display rapid and transient transcriptional induction in an immediate-early fashion. The majority of these SRF target genes expressed in the brain encode transcription factors (c-*fos*, *FosB*, *Fra–1*, *Srf*, *Egr–1*, *Egr–2*, and *JunB*). In addition, structural proteins of the cytoskeleton (Actin, Gelsolin, Arc/Arg3.1) and extracellularly acting proteins (Sema3A, Cyr61) are encoded by SRF target genes (Zhang et al., 2005, Kim et al., 2003). It can be assumed safely that in the future many more genes will be identified whose expression in the brain is dependent upon SRF.

5.4
Essential Requirement for SRF in Neuronal Migration

Cell migration is intimately linked to dynamic changes of the actin cytoskeleton (Etienne-Manneville and Hall, 2002). The state of actin polymerization is sensed by SRF via the actin-linked transcriptional co-activator MRTF/MAL (Sotiropoulos et al., 1999; Miralles et al., 2003), which in turn permits SRF to activate direct target genes, such as *actin* itself (Fig. 5.1). Thereby, SRF influences the state of actin polymerization (Schratt et al., 2002). In so doing, SRF forms part of a feedback mechanism

Table 5.1 Serum response factor (SRF) target genes related to brain function.

Entry	Gene	CArG box		Functionality of CArG box			Reference(s)
		Consensus vs. consensus-like	Conserved vs. not-conserved	ChIP	Reporter assay	Affected by SRF deficiency	
1	c-fos	Consensus	Conserved	+	+	+	Curran and Morgan (1987); Arsenian et al. (1998); Ramanan et al. (2005)
2	fosB	Consensus	Conserved	n.d.	+	+	Lazo et al. (1992); Ramanan et al. (2005)
3	junB	Consensus	Conserved	n.d.	+	+	Perez-Albuerne et al. (1993); Philippar et al. (2004)
4	fra–1	Consensus-like	Conserved	+	+	n.d.	Adiseshaiah et al. (2005)
5	Srf	Consensus	Conserved	+	+	+	Unpublished data; Spencer and Misra (1996); Philippar et al. (2004)
6	egr–1 (krox–24, zif268)	Consensus	Conserved	+	+	+	Christy et al. (1988); Janssen-Timmen et al. (1989); Schratt et al. (2001); Philippar et al. (2004); Ramanan et al. (2005)
7	egr–2 (krox–20)	Consensus	Conserved	n.d.	+	+	Chavrier et al. (1989); Rangnekar et al. (1990); Ramanan et al. (2005); Watanabe et al. (2005)
8	Nur77	Consensus	n.d.	+	n.d.	+	Latinkic et al. (1996); Yakubov et al. (2004); Etkin et al. (submitted)
9	Actin	Consensus	Conserved	+	+	+	Minty and Kedes (1986); Arsenian et al. (1998); Manabe and Owens (2001); Alberti et al. (2005); Ramanan et al. (2005)
10	Gelsolin	Consensus	Not-conserved	-	n.d.	+	Alberti et al. (2005)
11	Arc	Consensus	Conserved	n.d.	n.d.	+	Ramanan et al. (2005)
12	Cyr61	Consensus	Conserved	+	+	+	Latinkic et al. (1991); Kim et al. (2003); unpublished data
13	Clathrin hc	n.d.	n.d.	n.d.	n.d.	+	Etkin et al. (unpublished results)
14	pmp–22	Consensus-like	Conserved	n.d.	n.d.	+	Philippar et al. (2004)
15	psd–95	-	-	n.d.	n.d.	+	Philippar et al. (2004)
16	Nestin	-	-	n.d.	n.d.	+	Philippar et al. (2004)
17	App	Consensus	n.d.	n.d.	n.d.	+	Philippar et al. (2004)
18	Prp	Consensus	Conserved	n.d.	n.d.	+	Philippar et al. (2004)
19	sema3A	Consensus	n.d.	+	+	+	Zhang et al. (2005)
20	mcl–1	Consensus	n.d.	+	+	n.d.	Townsend et al. (1999); Tullai et al. (2004)

This table lists identified SRF target genes, which are expressed in the brain. The indicated chromatin immune-precipitation (ChIP) data were all generated with non-neuronal tissues. Proof for these genes being truly regulated by SRF in the brain will require further ChIP experiments using neuronal tissue. Entries 1, 2, 6–11, and 13 are dysregulated in neurons of mouse forebrain-specific *Srf* deletion mutants; dysregulation of all other genes listed (except 4 and 20) was studied in non-neuronal SRF-deficient cells. Except for entries 15 and 16, only direct SRF target genes are listed. CArG boxes present in gene regulatory segments are either consensus (CC(A/T)GG) or consensus-like (having one nucleotide deviating from CC(A/T)GG). These boxes are either conserved or not-conserved in mouse and human genomes, as indicated.

n.d. = not determined.

regulating actin fiber assembly and disassembly. More generally, SRF represents a relay system to convert dynamical changes in actin polymerization into altered profiles of target gene expression.

SRF-deficient murine embryonic stem (ES) cells display impaired induction of c-*fos* and *Egr–1*, proving the essential contribution of SRF for mounting the cellular IEG response (Schratt et al., 2001). Importantly, SRF deficiency leads to reduced actin fiber density and an imbalanced equilibrium of globular (G) versus filamentous (F) forms of actin (Schratt et al., 2002). This is accompanied by severely compromised *in-vitro* ES cell migration and adhesion (Schratt et al., 2002).

The essential requirement of SRF for *in-vivo* neuronal cell migration in the murine brain is revealed by conditional *Srf* null mutagenesis (Alberti et al., 2005). Forebrain-specific, late-prenatal conditional deletion of the "floxed" *Srf-flex1* allele causes neurons to accumulate ectopically at the subventricular zone (Fig. 5.2), displaying impaired tangential chain migration along the rostral migratory stream (RMS) into the olfactory bulb. Intracellularly, this migratory defect is accompanied by reduced F-actin fiber density and lowered expression of the actin-severing protein gelsolin. With regard to cofilin, another key regulator of actin dynamics and cell migration, a dramatically elevated inhibitory phosphorylation at Ser 3 is observed. This study demonstrates that SRF-controlled gene expression directs the structure and dynamics of the actin microfilament, thereby determining cell-autonomous neuronal cell migration (Alberti et al., 2005).

The essential contribution of the SRF complex to cell migration was also revealed in a non-neuronal system of *Drosophila melanogaster*, where mutation of the SRF partner protein DMRTF/MAL-D led to a severe migration impairment of mesodermal cells (Han et al., 2004) and of border cells (Somogyi and Rorth, 2004).

5.5
SRF and Partner Proteins in Neurite Outgrowth and Axonal Guidance

Similar to cell migration, processes of outgrowth and guidance of newly elaborated axons rely fundamentally on the dynamic properties of cytoskeletal components. Here, the polymerization/de-polymerization status of actin filaments and networks in filopodial and lamellipodial structures of neuronal growth cones, respectively, covers pivotal roles in steering a navigating nerve fiber towards its respective target in the brain (Dent and Gertler, 2003).

At present, a possible role of SRF and its partner proteins in neurite outgrowth and guidance has only been sparsely addressed. A first hint towards SRF-mediated gene expression in growth cone turning was provided by Mehlen and co-workers (Forcet et al., 2002). Here, the authors employed the so-called turning assay, an *in-vitro* system to follow the response of individual growth cones towards artificially generated gradients of secreted guidance molecules. A gradient by netrin elicits attractive growth cone turning, a reaction which is dependent upon MAPK signaling. Importantly, the stimulation of neurons with netrin led to an increased transcription of a reporter gene driven by SREs. This suggests, that netrin-mediated axonal guidance stimulates an SRF-mediated gene expression program.

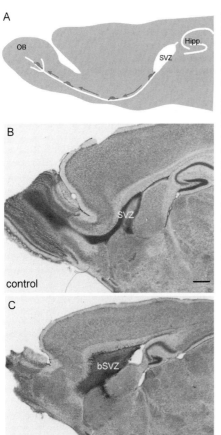

Fig. 5.2 The role of SRF in neuronal migration is best understood in the so-called rostral migratory stream (RMS), which replenishes the olfactory bulb (OB) with neurons derived from a stem cell pool localized in the subventricular zone (SVZ). (A) Migration of these neurons (shown by arrows) persists throughout the entire lifespan of organisms. (B,C) Sagittal brain sections (Nissl staining) derived from control (B) and forebrain-specific SRF-deficient mutants (C). In (B) the entire length of the RMS, from the SVZ to the olfactory bulb, is visible. By contrast, in (C) SRF ablation led to the retention of neurons in a migratory status in the SVZ, giving this brain structure an inflated or broadened (bSVZ) appearance. Consequently the number of migrating neurons entering the OB was dramatically reduced in SRF mutants. Hipp. = hippocampus. (This figure also appears with the color plates.)

Similarly, it was recently demonstrated (Tabuchi et al., 2005) that overexpression of the RhoA-GEF Tech stimulates SRF-driven reporter expression in primary cortical neurons. Additionally, the same authors demonstrated translocation of the SRF cofactor MAL from the cytoplasm to the nucleus in primary neuronal cultures.

The SRF cofactor Elk–1 has been implicated in promoting neurite outgrowth. Interestingly, in this respect, only a truncated (sElk–1) – but not full-length Elk–1 – increased the neurite-length of PC12 cells (Vanhoutte et al., 2001). Noteworthy, this transcription factor not only localized to the nuclear compartment of neurons, but was also present in the soma, dendrites, and axon terminals (Sgambato et al., 1998).

We have begun to analyze the function of SRF in neurite outgrowth, axonal guidance and synaptic targeting (B. Knöll et al., unpublished results) by forebrain-specific SRF ablation. Here, we focused on the mossy fiber pathway of the hippocampal projection, a trajectory which is closely associated with various paradigms of learning and memory, as well as circuitry aberrations described for epileptic seizures (Morimoto et al., 2004). Neurite outgrowth of these SRF-deficient neurons was severely impaired *in vivo*, but was also clearly evident by applying *in-vitro* neurite outgrowth assays. Mossy fibers are normally precisely navigated on either side outside the CA3 band formed by pyramidal neurons. In *Srf* mutants, this well-defined axon guidance process is non-functional (Fig. 5.3). Remarkably, contact-mediated axonal repulsion exerted by members of the EphA family of axon guidance molecules was strongly reduced. EphrinA triggered growth cone collapse, initiating complete breakdown of F-Actin polymers in growth cones in controls, and this resulted in *Srf* mutants in the persistence of ring-shaped structures consisting of F-Actin and also microtubules (B. Knöll et al., unpublished results).

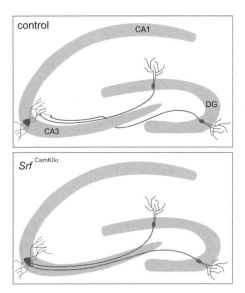

Fig. 5.3 SRF controls guidance and synaptic targeting of hippocampal mossy fibers. In control mice, mossy fibers emanating dentate gyrus (DG) granule cells bifurcate in a supra- and infrapyramidal tract. Both branches of the mossy fibers navigate precisely on either side outside the band formed by CA3 pyramidal neurons. The infrapyramidal branch crosses the CA3 stratum pyramidale at some point and joins the main suprapyramidal branch. Control mossy fiber terminals synapse with dendrites outside the layer of CA3 cell bodies. Conditional ablation of SRF function results in severe axon guidance defects. Here, mossy fibers, instead of bifurcating, grow preferentially inside the CA3 layer between individual CA3 somata. Synaptic targeting of *Srf*-deficient mossy fibers occurs aberrantly at CA3 somata and somatic protrusions (B. Knöll et al., unpublished results). (This figure also appears with the color plates.)

Interfering with the activity of the SRF cofactor MAL by electroporation of various dominant-negative mutants strongly impeded neurite outgrowth and EphA-mediated growth cone guidance. *In vivo*, mis-routed SRF-deficient mossy fibers aberrantly targeted CA3 pyramidal cell somata or somatic protrusions rather than apical dendrites for synapse formation (see Fig. 5.3) (B. Knöll et al., unpublished results).

In summary, investigations into SRF-mediated transcriptional programs in neurite outgrowth, axonal guidance and synapse formation are at their earliest stages. For many other transcriptions factors (e.g., the CREB family), interpretations of neurite outgrowth or axon guidance phenotypes are often interfered with by a well-known role of these molecules in preventing neuronal cell death. Notably, this absence of increased neuronal cell death in *Srf* mutants (Alberti et al., 2005; Ramanan et al., 2005; B. Knöll et al., unpublished results) renders SRF as an ideal candidate to explore gene expression in processes of axonal growth and pathfinding.

5.6
SRF-Mediated Gene Expression in Learning and Memory

One challenging problem in molecular neurobiology relates to the question of how the constant stimulation of organisms by diverse (sensory) inputs is manifested in learning and memory acquisition and, in the same regard, how these two processes shape or rearrange existing neuronal circuits.

Neuronal activity provided by various sensory inputs has been closely linked to signaling pathways ultimately funneling in the activation of transcriptional programs (West et al., 2002). Seminal studies conducted in the laboratory of M. Greenberg identified Ca^{2+} as a key signal transducer conveying receptor stimulation at the synapse to transcriptional activation in the nucleus, as studied in particular for genes containing SREs (Rivera et al., 1993; Misra et al., 1994; Xia et al., 1996; Johnson et al., 1997). A signaling cascade leading to the initiation of an activity-dependent transcriptional program has been described in great detail for the SRF cofactor Elk–1 (Choe and Wang, 2002) (Fig. 5.4).

The activation of both ionotropic and metabotropic subtypes of glutamate receptors engage signaling intermediates, eventually resulting in Erk1/2 translocation to the nucleus, and thereby inducing phosphorylation and activation of Elk–1 (Fig. 5.4). In turn, Elk–1–mediated IEG induction leads to a rapid increase in RNA levels from various genes (see Table 5.1). Many of the IEGs activated by SRF/Elk–1 are transcription factors themselves, yet recently, genes regulated by SRF and its partners have been identified in neurons, which code for structural components and regulators of the actin cytoskeleton (Alberti et al., 2005). In this regard, transcription of actin and the actin-severing protein gelsolin are regulated by SRF. These findings are particularly interesting in the light of morphological alterations taking place during synapse formation. In this respect, dendritic spine formation profoundly relies on dynamic features of the actin cytoskeleton (Zito et al., 2004).

The laboratory of D. Ginty (Ramanan et al., 2005) recently addressed the role of SRF in activity-dependent IEG expression and memory acquisition by conditional

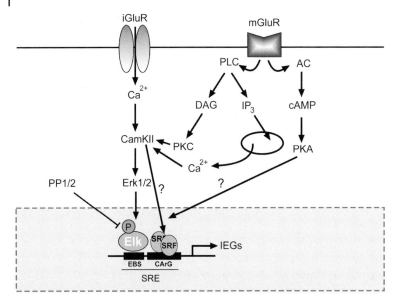

Fig. 5.4 Neuronal activity-dependent signaling leading to SRF/TCF activation. Ca^{2+} influx through ionotropic glutamate receptors (iGluR) eventually activates and translocates Erk1/2 to the nuclear compartment. Phosphorylation of the TCF Elk–1 by Erk1/2 strengthens Elk–1 interaction with SRF and stimulates the transcriptional machinery, resulting in IEG induction. Similarly, additional signaling cascades involving Protein kinases A and C (PKA, PKC, respectively) initiated by metabotropic glutamate receptors (mGluR) are also directed towards increasing phospho-Elk–1 levels. Ca^{2+}-activated kinases (CamKs) may further increase the phosphorylation levels of SRF.

SRF ablation in the hippocampus of adult mice. In these SRF mutants, IEG induction of *c-fos, fosB, egr1, egr2, arc,* and *actB* after application of electroconvulsive shocks was completely abolished. In addition, the authors further reported an intriguing lack of IEG induction in *Srf* mutants following the exploration of a novel environment. Interestingly, in this study signaling from the synapse to the nucleus (as judged by Erk1/2 activation) was not affected in *Srf* mutants. Also, neuronal cell death in *Srf* mutants was not elevated, in contrast to mice deficient for transcription factors of the CREB family. This is in line with our own observations, where conditional forebrain-specific *Srf* deletion beginning around birth, likewise did not reveal overt changes regarding neuronal apoptosis *in vivo* (Alberti et al., 2005; B. Knöll et al., unpublished results). Nonetheless, *in-vitro* experiments performed by others (Chang et al., 2004) ascribed SRF a function in neuronal survival against trophic deprivation and DNA damage.

In the studies conducted by Ginty and colleagues, SRF was shown to be dispensable for basal excitatory transmission in CA1 pyramidal neurons. However, long-term synaptic potentiation (LTP), viewed as a cellular basis for learning and memory, was severely altered (Ramanan et al., 2005). In a further investigation of SRF function in learning and memory, Dash et al. (2005) inhibited SRF target gene expression by titrating SRF with SRE-containing oligonucleotides. Using this technique, these authors described a role for SRF in long-term spatial memory, but not in the initial memory acquisition. Interestingly, by applying the same strategy to interfere with Elk–1 activity, these authors obtained no differences in learning and memory paradigms, arguing that SRF in these instances is bypassing Elk–1 as a partner protein and instead might recruit other members of the TCF family or myocardin-related proteins (e.g., MAL, MRTFs) as cofactors.

Generally, the functional significance of Elk–1 in learning and memory has not been elucidated in sufficient detail. Frequently, implications of Elk–1 in these processes rely on correlation between synapse stimulation resulting in Erk1/2 activation and eventually Elk–1 hyper-phosphorylation (often also paralleled by CREB activation). Nevertheless, using the phosphorylation status of Elk–1 as a read-out and indicator for activity, a plethora of contributions of this SRF cofactor in learning and memory has been reported (Fig. 5.4). Thus, Elk–1 has been implicated in complex behavioral paradigms such as fear conditioning (Cammarota et al., 2000; Sananbenesi et al., 2002; Ahi et al., 2004), long-term memory of taste (Berman, 2003), and photic induction involved in regulating circadian rhythms (Kaminska et al., 1999; Coogan and Piggins, 2003). When investigating LTP-dependent gene expression in the dentate gyrus of the hippocampus *in vivo*, Davis et al. (2000) reported that the induction of LTP is paralleled by the time-course of Elk–1 hyperphosphorylation.

Apart from LTP of synaptic efficacy, long-term depression (LTD) – a weakening of synaptic transmission – is the second mechanism believed to contribute towards learning and memory. During LTD, Elk–1 activation was reported in the hippocampal area CA1 (Thiels et al., 2002). A recent study in which Kandel and colleagues explored the role of LTD, the results showed unambiguously a major contribution of SRF transcriptional activation to this process (Etkin et al., unpublished results). When allowed to explore a novel environment, adult mice lacking SRF in the hippocampus have a complete loss of contextual habituation, which is a non-associative form of learning. This deficit was accompanied by an early impairment in the induction of various forms of LTD, though LTP was found to be largely unaffected – a finding somewhat at odds (though it might reflect genetic background differences) with results obtained by others (Ramanan et al., 2005).

In summary, recent reports which in particular have employed the genetic targeting of SRF have revealed a considerable role for SRF-mediated transcriptional activation in learning and in the storage of memories. In particular, the intimate link of SRF towards cytoskeletal and structural alterations in cells should allow fascinating molecular and cellular insights as to the morphological (re-)organization of synapse assembly.

5.7
SRF in Neurological Disorders

To date, as with SRF in normal brain function, few reports are published on the roles of SRF in neuronal pathology.

Temporal lobe epilepsy emerges as one neuropathological phenomenon in which SRF might be crucially involved. Morris et al. (1999) described an up-regulation of SRF protein 24 hours after pilocarpine-induced status epilepticus in rats. In agreement with these findings, kainate-induced epileptic seizures were followed by increased total and phospho-SRF protein levels (Herdegen et al., 1997). Similarly, in mice deficient for the SRF co-factor Elk–1, we described a mild impairment of IEG induction after kainate-induced seizures (Cesari et al., 2004). Along with this, Elk–1 phosphorylation in CA3 neurons was elevated in an experimental paradigm for traumatic brain injury, which is thought to be involved in rearrangements of mossy fiber projections typically found in temporal lobe epilepsy (Hu et al., 2004). In summary, reports available so far indicate a role for SRF during epileptic seizures, which might be further substantiated by addressing this issue in recently established postnatal and adult conditional *Srf* mouse models (Alberti et al., 2005; Ramanan et al., 2005).

In addition to a role in epilepsy, a possible link between Elk–1 and Alzheimer's disease (AD) was indicated. The stimulation of cortical neurons with sublethal concentrations of beta-amyloid protein interfered strongly with a neurotrophin signaling cascade and resulted in Elk–1 activation (Tong et al., 2004). Additionally, Elk–1 might function as a transcriptional repressor of the *presenilin 1* gene which, if mutated, is a major cause in early-onset AD (Pastorcic and Das, 2003).

5.8
Perspectives

Although SRF was discovered more than 15 years ago, unambiguous functions of this protein in the brain have been identified only recently, and this is due mainly to the availability of conditional, brain-specific *Srf* gene deletions. In light of the ability of SRF to integrate and convert multiple incoming signals into both rapid and long-lasting changes of gene expression profiles, this protein serves as an ideal candidate to regulate a wide range of neuronal functions. These include neuronal migration, the outgrowth and guidance of axons, and synaptic targeting. Moreover, accompanying long-term structural changes in the neuronal circuitry can be linked to SRF, affecting brain function at the anatomical and physiological levels. These initial data on the roles of SRF in learning and memory acquisition promise new general insight into both normal and pathological functions of the brain.

Acknowledgments

These studies were supported by grants from the DFG (NO 120/12–1 and SFB 446).

Abbreviations

AD	Alzheimer's disease
ChIP	chromatin immune precipitation
ES	embryonic stem
IEG	immediate-early gene
LTD	long-term depression
LTP	long-term potentiation
mDia	mDiaphanous
RMS	rostral migratory stream
SRE	serum response element
SRF	serum response factor
TCF	ternary complex factor

References

Adiseshaiah, P., Peddakama, S., Zhang, Q., Kalvakolanu, D.V., and Reddy, S. P. (2005). Mitogen regulated induction of FRA–1 proto-oncogene is controlled by the transcription factors binding to both serum and TPA response elements. *Oncogene* 24, 4193–4205.

Ahi, J., Radulovic, J., and Spiess, J. (2004). The role of hippocampal signaling cascades in consolidation of fear memory. *Behav. Brain Res.* 149, 17–31.

Alberti, S., Krause, S.M., Kretz, O., Philippar, U., Lemberger, T., Casanova, E., Wiebel, F.F., Schwarz, H., Frotscher, M., Schutz, G., and Nordheim, A. (2005). Neuronal migration in the murine rostral migratory stream requires serum response factor. *Proc. Natl. Acad. Sci. USA* 102, 6148–6153.

Arsenian, S., Weinhold, B., Oelgeschlager, M., Ruther, U., and Nordheim, A. (1998). Serum response factor is essential for mesoderm formation during mouse embryogenesis. *EMBO J.* 17, 6289–6299.

Berman, D.E. (2003). Modulation of taste-induced Elk–1 activation by identified neurotransmitter systems in the insular cortex of the behaving rat. *Neurobiol. Learn. Mem.* 79, 122–126.

Buchwalter, G., Gross, C., and Wasylyk, B. (2004). Ets ternary complex transcription factors. *Gene* 324, 1–14.

Cammarota, M., Bevilaqua, L.R., Ardenghi, P., Paratcha, G., Levi de Stein, M., Izquierdo, I., and Medina, J.H. (2000). Learning-associated activation of nuclear MAPK, CREB and Elk–1, along with Fos production, in the rat hippocampus after a one-trial avoidance learning: abolition by NMDA receptor blockade. *Brain Res. Mol. Brain Res.* 76, 36–46.

Cesari, F., Brecht, S., Vintersten, K., Vuong, L.G., Hofmann, M., Klingel, K., Schnorr, J.J., Arsenian, S., Schild, H., Herdegen, T., et al. (2004). Mice deficient for the ets transcription factor elk–1 show normal immune responses and mildly impaired neuronal gene activation. *Mol. Cell. Biol.* 24, 294–305.

Chang, S.H., Poser, S., and Xia, Z. (2004). A novel role for serum response factor in neuronal survival. *J. Neurosci.* 24, 2277–2285.

Chavrier, P., Janssen-Timmen, U., Mattei, M.G., Zerial, M., Bravo, R., and Charnay, P. (1989). Structure, chromosome location, and expression of the mouse zinc finger gene Krox–20: multiple gene products and coregulation with the proto-oncogene c-fos. *Mol. Cell. Biol.* 9, 787–797.

Choe, E.S., and Wang, J. Q. (2002). Regulation of transcription factor phosphorylation by metabotropic glutamate receptor-associated signaling pathways in rat striatal neurons. *Neuroscience* 114, 557–565.

Christy, B.A., Lau, L.F., and Nathans, D. (1988). A gene activated in mouse 3T3 cells by serum growth factors encodes a protein with "zinc finger" sequences. *Proc. Natl. Acad. Sci. USA* 85, 7857–7861.

Coogan, A.N., and Piggins, H.D. (2003). Circadian and photic regulation of phosphorylation of ERK1/2 and Elk–1 in the suprachiasmatic nuclei of the Syrian hamster. *J. Neurosci.* 23, 3085–3093.

Curran, T., and Morgan, J.I. (1987). Memories of fos. *BioEssays* 7, 255–258.

Dalton, S., and Treisman, R. (1992). Characterization of SAP–1, a protein recruited by serum response factor to the c-fos serum response element. *Cell* 68, 597–612.

Dash, P.K., Orsi, S.A., and Moore, A.N. (2005). Sequestration of serum response factor in the hippocampus impairs long-term spatial memory. *J. Neurochem.* 93, 269–278.

Davis, S., Vanhoutte, P., Pages, C., Caboche, J., and Laroche, S. (2000). The MAPK/ERK cascade targets both Elk–1 and cAMP response element-binding protein to control long-term potentiation-dependent gene expression in the dentate gyrus in vivo. *J. Neurosci.* 20, 4563–4572.

Dent, E.W., and Gertler, F.B. (2003). Cytoskeletal dynamics and transport in growth cone motility and axon guidance. *Neuron* 40, 209–227.

Etienne-Manneville, S., and Hall, A. (2002). Rho GTPases in cell biology. *Nature* 420, 629–635.

Forcet, C., Stein, E., Pays, L., Corset, V., Llambi, F., Tessier-Lavigne, M., and Mehlen, P. (2002). Netrin–1–mediated axon outgrowth requires deleted in colorectal cancer-dependent MAPK activation. *Nature* 417, 443–447.

Gille, H., Sharrocks, A.D., and Shaw, P.E. (1992). Phosphorylation of transcription factor p62TCF by MAP kinase stimulates ternary complex formation at c-fos promoter. *Nature* 358, 414–417.

Gineitis, D., and Treisman, R. (2001). Differential usage of signal transduction pathways defines two types of serum response factor target gene. *J. Biol. Chem.* 276, 24531–24539.

Giovane, A., Pintzas, A., Maira, S.M., Sobieszczuk, P., and Wasylyk, B. (1994). Net, a new ets transcription factor that is activated by Ras. *Genes Dev.* 8, 1502–1513.

Greenberg, M.E., and Ziff, E.B. (1984). Stimulation of 3T3 cells induces transcription of the c-fos proto-oncogene. *Nature* 311, 433–438.

Han, Z., Li, X., Wu, J., and Olson, E.N. (2004). A myocardin-related transcription factor regulates activity of serum response factor in Drosophila. *Proc. Natl. Acad. Sci. USA* 101, 12567–12572.

Herdegen, T., Blume, A., Buschmann, T., Georgakopoulos, E., Winter, C., Schmid, W., Hsieh, T.F., Zimmermann, M., and Gass, P. (1997). Expression of activating transcription factor–2, serum response factor and cAMP/Ca response element binding protein in the adult rat brain following generalized seizures, nerve fibre lesion and ultraviolet irradiation. *Neuroscience* 81, 199–212.

Hipskind, R.A., Rao, V.N., Mueller, C.G., Reddy, E.S., and Nordheim, A. (1991). Ets-related protein Elk–1 is homologous to the c-fos regulatory factor p62TCF. *Nature* 354, 531–534.

Hu, B., Liu, C., Bramlett, H., Sick, T.J., Alonso, O.F., Chen, S., and Dietrich, W.D. (2004). Changes in trkB-ERK1/2–CREB/Elk–1 pathways in hippocampal mossy fiber organization after traumatic brain injury. *J. Cereb. Blood Flow Metab.* 24, 934–943.

Janssen-Timmen, U., Lemaire, P., Mattei, M.G., Revelant, O., and Charnay, P. (1989). Structure, chromosome mapping and regulation of the mouse zinc-finger gene Krox–24; evidence for a common regulatory pathway for immediate-early serum-response genes. *Gene* 80, 325–336.

Johnson, C.M., Hill, C.S., Chawla, S., Treisman, R., and Bading, H. (1997). Calcium controls gene expression via three distinct pathways that can function independently of the Ras/mitogen-activated protein kinases (ERKs) signaling cascade. *J. Neurosci.* 17, 6189–6202.

Kaminska, B., Kaczmarek, L., Zangenehpour, S., and Chaudhuri, A. (1999). Rapid phosphorylation of Elk–1 transcription factor and activation of MAP kinase signal transduction pathways in response to visual stimulation. *Mol. Cell. Neurosci.* 13, 405–414.

Kim, K.H., Min, Y.K., Baik, J.H., Lau, L.F., Chaqour, B., and Chung, K.C. (2003). Expression of angiogenic factor Cyr61 during neuronal cell death via the activation of c-Jun N-terminal kinase and serum response factor. *J. Biol. Chem.* 278, 13847–13854.

Latinkic, B.V., O'Brien, T.P., and Lau, L.F. (1991). Promoter function and structure of the growth factor-inducible immediate early gene cyr61. *Nucleic Acids Res.* 19, 3261–3267.

Latinkic, B.V., Zeremski, M., and Lau, L.F. (1996). Elk–1 can recruit SRF to form a ternary complex upon the serum response element. *Nucleic Acids Res.* 24, 1345–1351.

Lazo, P.S., Dorfman, K., Noguchi, T., Mattei, M.G., and Bravo, R. (1992). Structure and mapping of the fosB gene. FosB downregulates the activity of the fosB promoter. *Nucleic Acids Res.* 20, 343–350.

Manabe, I., and Owens, G.K. (2001). Recruitment of serum response factor and hyperacetylation of histones at smooth muscle-specific regulatory regions during differentiation of a novel P19–derived in vitro smooth muscle differentiation system. *Circ. Res.* 88, 1127–1134.

Minty, A., and Kedes, L. (1986). Upstream regions of the human cardiac actin gene that modulate its transcription in muscle cells: presence of an evolutionarily conserved repeated motif. *Mol. Cell. Biol.* 6, 2125–2136.

Miralles, F., Posern, G., Zaromytidou, A.I., and Treisman, R. (2003). Actin dynamics control SRF activity by regulation of its coactivator MAL. *Cell* 113, 329–342.

Misra, R.P., Bonni, A., Miranti, C.K., Rivera, V.M., Sheng, M., and Greenberg, M.E. (1994). L-type voltage-sensitive calcium channel activation stimulates gene expression by a serum response factor-dependent pathway. *J. Biol. Chem.* 269, 25483–25493.

Morgan, J.I., Cohen, D.R., Hempstead, J.L., and Curran, T. (1987). Mapping patterns of c-fos expression in the central nervous system after seizure. *Science* 237, 192–197.

Morimoto, K., Fahnestock, M., and Racine, R.J. (2004). Kindling and status epilepticus models of epilepsy: rewiring the brain. *Prog. Neurobiol.* 73, 1–60.

Morris, T.A., Jafari, N., Rice, A.C., Vasconcelos, O., and DeLorenzo, R.J. (1999). Persistent increased DNA-binding and expression of serum response factor occur with epilepsy-associated long-term plasticity changes. *J. Neurosci.* 19, 8234–8243.

Murai, K., and Treisman, R. (2002). Interaction of serum response factor (SRF) with the Elk–1 B box inhibits RhoA-actin signaling to SRF and potentiates transcriptional activation by Elk–1. *Mol. Cell. Biol.* 22, 7083–7092.

Norman, C., Runswick, M., Pollock, R., and Treisman, R. (1988). Isolation and properties of cDNA clones encoding SRF, a transcription factor that binds to the c-fos serum response element. *Cell* 55, 989–1003.

Pastorcic, M., and Das, H.K. (2003). Ets transcription factors ER81 and Elk1 regulate the transcription of the human presenilin 1 gene promoter. *Brain Res. Mol. Brain Res.* 113, 57–66.

Pellegrini, L., Tan, S., and Richmond, T.J. (1995). Structure of serum response factor core bound to DNA. *Nature* 376, 490–498.

Perez-Albuerne, E.D., Schatteman, G., Sanders, L.K., and Nathans, D. (1993). Transcriptional regulatory elements downstream of the JunB gene. *Proc. Natl. Acad. Sci. USA* 90, 11960–11964.

Philippar, U., Schratt, G., Dieterich, C., Muller, J.M., Galgoczy, P., Engel, F.B., Keating, M.T., Gertler, F., Schule, R., Vingron, M., and Nordheim, A. (2004). The SRF target gene Fhl2 antagonizes RhoA/MAL-dependent activation of SRF. *Mol. Cell* 16, 867–880.

Ramanan, N., Shen, Y., Sarsfield, S., Lemberger, T., Schutz, G., Linden, D.J., and Ginty, D.D. (2005). SRF mediates activity-induced gene expression and synaptic plasticity but not neuronal viability. *Nat. Neurosci.* 8, 759–767.

Rangnekar, V.M., Aplin, A.C., and Sukhatme, V.P. (1990). The serum and TPA responsive promoter and intron-exon structure of EGR2, a human early growth response gene encoding a zinc finger protein. *Nucleic Acids Res.* 18, 2749–2757.

Rivera, V.M., Miranti, C.K., Misra, R.P., Ginty, D.D., Chen, R.H., Blenis, J., and Greenberg, M.E. (1993). A growth factor-induced kinase phosphorylates the serum response factor at a site that regulates its DNA-binding activity. *Mol. Cell. Biol.* 13, 6260–6273.

Sananbenesi, F., Fischer, A., Schrick, C., Spiess, J., and Radulovic, J. (2002). Phosphorylation of hippocampal Erk–1/2, Elk–1, and p90–Rsk–1 during contextual fear conditioning: interactions between Erk–1/2 and Elk–1. *Mol. Cell. Neurosci.* 21, 463–476.

Schratt, G., Philippar, U., Berger, J., Schwarz, H., Heidenreich, O., and Nordheim, A. (2002). Serum response factor is crucial for actin cytoskeletal organization and focal adhesion assembly in embryonic stem cells. *J. Cell Biol.* 156, 737–750.

Schratt, G., Weinhold, B., Lundberg, A.S., Schuck, S., Berger, J., Schwarz, H., Weinberg, R.A., Ruther, U., and Nordheim, A. (2001). Serum response factor is required for immediate-early gene activation yet is dispensable for proliferation of embryonic stem cells. *Mol. Cell. Biol.* 21, 2933–2943.

Schwarz-Sommer, Z., Hue, I., Huijser, P., Flor, P.J., Hansen, R., Tetens, F., Lonnig, W.E., Saedler, H., and Sommer, H. (1992). Characterization of the Antirrhinum floral homeotic MADS-box gene deficiencies: evidence for DNA binding and autoregulation of its persistent expression throughout flower development. *EMBO J.* 11, 251–263.

Sgambato, V., Vanhoutte, P., Pages, C., Rogard, M., Hipskind, R., Besson, M.J., and Caboche, J. (1998). In vivo expression and regulation of Elk–1, a target of the extracellular-regulated kinase signaling pathway, in the adult rat brain. *J. Neurosci.* 18, 214–226.

Shaw, P.E., Frasch, S., and Nordheim, A. (1989). Repression of c-fos transcription is mediated through p67SRF bound to the SRE. *EMBO J.* 8, 2567–2574.

Sheng, M., and Greenberg, M.E. (1990). The regulation and function of c-fos and other immediate early genes in the nervous system. *Neuron* 4, 477–485.

Shore, P., and Sharrocks, A.D. (1995). The MADS-box family of transcription factors. *Eur. J. Biochem.* 229, 1–13.

Somogyi, K., and Rorth, P. (2004). Evidence for tension-based regulation of *Drosophila* MAL and SRF during invasive cell migration. *Dev. Cell* 7, 85–93.

Sotiropoulos, A., Gineitis, D., Copeland, J., and Treisman, R. (1999). Signal-regulated activation of serum response factor is mediated by changes in actin dynamics. *Cell* 98, 159–169.

Spencer, J.A., and Misra, R.P. (1996). Expression of the serum response factor gene is regulated by serum response factor binding sites. *J. Biol. Chem.* 271, 16535–16543.

Stringer, J.L., Belaguli, N.S., Iyer, D., Schwartz, R.J., and Balasubramanyam, A. (2002). Developmental expression of serum response factor in the rat central nervous system. *Brain Res. Dev. Brain Res.* 138, 81–86.

Tabuchi, A., Estevez, M., Henderson, J.A., Marx, R., Shiota, J., Nakano, H., and Baraban, J.M. (2005). Nuclear translocation of the SRF co-activator MAL in cortical neurons: role of RhoA signalling. *J. Neurochem.* 94, 169–180.

Thiels, E., Kanterewicz, B.I., Norman, E.D., Trzaskos, J.M., and Klann, E. (2002). Long-term depression in the adult hippocampus in vivo involves activation of extracellular signal-regulated kinase and phosphorylation of Elk–1. *J. Neurosci.* 22, 2054–2062.

Tong, L., Balazs, R., Thornton, P.L., and Cotman, C.W. (2004). Beta-amyloid peptide at sublethal concentrations downregulates brain-derived neurotrophic factor functions in cultured cortical neurons. *J. Neurosci.* 24, 6799–6809.

Townsend, K.J., Zhou, P., Qian, L., Bieszczad, C.K., Lowrey, C.H., Yen, A., and Craig, R.W. (1999). Regulation of MCL1 through a serum response factor/Elk–1–mediated mechanism links expression of a viability-promoting member of the BCL2 family to the induction of hematopoietic cell differentiation. *J. Biol. Chem.* 274, 1801–1813.

Tullai, J.W., Schaffer, M.E., Mullenbrock, S.,

Kasif, S., and Cooper, G.M. (2004). Identification of transcription factor binding sites upstream of human genes regulated by the phosphatidylinositol 3–kinase and MEK/ERK signaling pathways. *J. Biol. Chem.* 279, 20167–20177.

Vanhoutte, P., Nissen, J.L., Brugg, B., Gaspera, B.D., Besson, M.J., Hipskind, R.A., and Caboche, J. (2001). Opposing roles of Elk–1 and its brain-specific isoform, short Elk–1, in nerve growth factor-induced PC12 differentiation. *J. Biol. Chem.* 276, 5189–5196.

Wang, D.Z., Li, S., Hockemeyer, D., Sutherland, L., Wang, Z., Schratt, G., Richardson, J.A., Nordheim, A., and Olson, E.N. (2002). Potentiation of serum response factor activity by a family of myocardin-related transcription factors. *Proc. Natl. Acad. Sci. USA* 99, 14855–14860.

Wang, D.Z., and Olson, E.N. (2004). Control of smooth muscle development by the myocardin family of transcriptional coactivators. *Curr. Opin. Genet. Dev.* 14, 558–566.

Wang, Z., Wang, D.Z., Hockemeyer, D., McAnally, J., Nordheim, A., and Olson, E.N. (2004). Myocardin and ternary complex factors compete for SRF to control smooth muscle gene expression. *Nature* 428, 185–189.

Watanabe, T., Hongo, I., Kidokoro, Y., and Okamoto, H. (2005). Functional role of a novel ternary complex comprising SRF and CREB in expression of Krox–20 in early embryos of *Xenopus laevis*. *Dev. Biol.* 277, 508–521.

West, A.E., Griffith, E.C., and Greenberg, M.E. (2002). Regulation of transcription factors by neuronal activity. *Nat. Rev. Neurosci.* 3, 921–931.

Xia, Z., Dudek, H., Miranti, C.K., and Greenberg, M.E. (1996). Calcium influx via the NMDA receptor induces immediate early gene transcription by a MAP kinase/ERK-dependent mechanism. *J. Neurosci.* 16, 5425–5436.

Yakubov, E., Gottlieb, M., Gil, S., Dinerman, P., Fuchs, P., and Yavin, E. (2004). Overexpression of genes in the CA1 hippocampus region of adult rat following episodes of global ischemia. *Brain Res. Mol. Brain Res.* 127, 10–26.

Zhang, S.X., Garcia-Gras, E., Wycuff, D.R., Marriot, S.J., Kadeer, N., Yu, W., Olson, E.N., Garry, D.J., Parmacek, M.S., and Schwartz, R.J. (2005). Identification of direct serum-response factor gene targets during Me_2SO-induced P19 cardiac cell differentiation. *J. Biol. Chem.* 280, 19115–19126.

Zito, K., Knott, G., Shepherd, G.M., Shenolikar, S., and Svoboda, K. (2004). Induction of spine growth and synapse formation by regulation of the spine actin cytoskeleton. *Neuron* 44, 321–334.

6
RE–1 Silencing Transcription Factor (REST): Regulation of Neuronal Gene Expression via Modification of the Chromatin Structure

Gerald Thiel and Mathias Hohl

Abstract

Humans, as multicellular organisms, contain a remarkable diversity of cell types where each cell population must fulfill a distinct function in the interest of the whole organism. The molecular basis for the variations in morphology, biochemistry, molecular biology and function of different cell types is the cell type-specific expression of genes. These genes encode proteins necessary for executing specialized functions of each cell type within an organism. The transcription factor RE–1 silencing transcription factor (REST) plays a key role in the establishment of the neuronal phenotype. REST functions as a transcriptional repressor of neuronal genes in non-neuronal tissues. Target genes of REST encode neuronal receptors, ion channels, neuropeptides, synaptic vesicle proteins, transcription factors and adhesion molecules, underlining the important role of REST in controlling the neuronal phenotype. The widespread expression of REST in non-neuronal tissues is in good agreement with the role of REST as a negative regulator of neuron-specific gene transcription. Thus, a negative regulatory mechanism, involving strong expression of REST in non-neuronal cells and marginal expression of REST in neurons, controls the establishment of the neuronal phenotype. On the molecular level, REST recruits histone deacetylases and methyltransferases to its target genes, indicating that a modulation of the chromatin structure is crucial for neuronal gene transcription.

6.1
Tissue-Specific Gene Expression: The Molecular Basis for the Function of a Multicellular Organism

The human genome contains approximately 30 000 protein-coding genes. One of the main emphases in the postgenomic era is the elucidation of control mechanisms of gene expression. This topic sheds light on the fundamental mechanisms of utilization and regulation of the genetic information. During development, many different cell types are generated that carry, with the exception of developing lymphocytes, the same genetic information. Tight control of gene expression is required for acquisition of distinct cellular identities by the differentiating cells. Only a fraction of the genes in the genome is expressed in every cell type. Those genes encode proteins necessary for the survival of the cell – that is, metabolic enzymes or structural proteins. In contrast, a portion of the genes is transcribed only in particular cell types. The tissue-specific expression of these genes is the molecular basis for the striking differences of the many cell types found in a multicellular organism. The differential expression of tissue-specific genes is thus responsible for the development of a variety of cell types such as neurons, lymphocytes, endothelial cells, hepatocytes, myocytes and astrocytes, that all together build up the organism.

Alterations in gene expression are, therefore, the basis for the understanding of multicellular organisms. Neurons differ from other cells of an organism by containing a specific set of proteins that are crucial for the execution of the specialized functions of the nervous system. These proteins are encoded by genes that must be expressed in a neuron-specific manner. The regulatory mechanisms that control gene expression in neurons are therefore fundamental for the development and function of the brain. Many neuronal genes are expressed throughout the nervous system, indicating that transcription factors are required that ensure continuous active control of neuron-specific gene transcription in every neuron throughout adulthood. One regulator protein has been discovered, in the analysis of neuronal gene expression, that may fulfill this role. The zinc finger protein, RE–1 silencing transcription factor (REST), also termed neuron-restrictive silencer factor (NRSF), functions as a negative regulator of neuron-specific gene transcription (Chong et al., 1995; Schoenherr and Anderson, 1995), in contrast to most of the transcription factors discussed in this book, which are positive regulators. Together with the expression pattern of REST in non-neuronal tissues, it was suggested that REST represses neuronal genes in non-neuronal cells.

6.2
Modular Structure of REST

A transcriptional repressor is typically composed of a DNA-tethering domain that anchors the protein to DNA, and a repressor domain. The REST protein displays a modular structure (Fig. 6.1). The DNA-binding domain has been localized within the cluster of eight zinc fingers at the N-terminus. Recently, it was shown that zinc finger 7 of REST is crucial for DNA binding (Shimojo et al., 2001). In addition, signals for nuclear targeting, nuclear entry, and for the release from the translocation machinery are embedded within zinc fingers 2 to 5 (Shimojo et al., 2001). Two repressor domains were identified, encompassing the N-terminal 83 and the C-terminal 56 amino acids, respectively (Tapia-Ramírez et al., 1997; Thiel et al., 1998; Naruse et al., 1999). The C-terminal repressor domain includes a single zinc finger motif. Both repressor domains are transferable to a heterologous DNA-binding domain and function from proximal and distal positions (Thiel et al., 1998, 2001; Leichter and Thiel, 1999). Several splice variants of REST are known (Palm et al., 1998) that encode proteins with five or four zinc finger motifs. Two variants of REST, termed REST4 and REST5, were only detected in neuronal tissues. These transcripts are generated by alternative splicing of a neuron-specific exon (exon N) located between exons V and VI. REST4 retains the N-terminal repression domain and five of the eight zinc fingers that function as a DNA-tethering domain. The neuron-specific exon adds the amino acid sequence CDLVG, followed by an in-frame stop codon, to the REST4 protein C-terminal of the fifth zinc finger (Palm et al., 1998) (Fig. 6.1). The expression levels of REST4 in the brain are low, and do not exceed 1% in comparison to the full-length REST protein (Palm et al., 1998). The neuron-specific splicing of REST is conserved in human, mouse and rat (Palm et. al., 1999). The biological function of REST4 is controversial. While transcriptional repression activity was attributed to the REST isoforms by Palm et al. (1998), Shimojo et al. (1999) described REST4 as a de-repressor. These authors suggested that REST4 binds to the full-length REST protein and silences the silencing activity of REST. Our results, obtained following expression of a "humanized" REST4 in neuroblastoma cells, revealed that hREST4 is neither a transcriptional repressor nor a de-repressor (Magin et al., 2002). Likewise, an analysis of REST function in small-cell lung cancer cells showed that none of the splice forms of REST had a significant effect on REST-mediated transcriptional repression (Gurrola-Diaz et al., 2003). The core sequence NNCAGCACCNNGCACAGNNNC has been proposed to be required for REST binding to DNA (Schoenherr et al., 1996). The binding motifs of neuronal genes known to be regulated by REST are illustrated in Fig. 6.2.

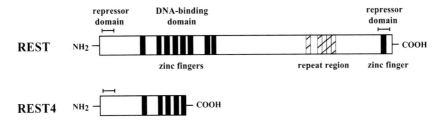

Fig. 6.1 The modular structure of REST. A cluster of eight zinc fingers in the N-terminal region functions as DNA-binding domain. Repressor domains have been mapped on the N- and C-termini of the molecule (Tapia-Ramírez et al., 1997; Thiel et al., 1998). In addition, a repeat-region was identified with so far unknown function (Chong et al., 1995; Schoenherr and Anderson, 1995). The modular structure of the REST4 splice variant is also depicted.

6.3
Biological Activity of REST

The biological activity of REST was analyzed by many investigators using transient transfections of reporter genes. The fact that transcriptional repression via REST is measurable using this approach argues for some kind of nucleosomal structure of the transfected plasmids. In line with this, it has been reported that in transiently transfected cells, nucleosomes are deposited onto non-replicated DNA, but the over-

TTCAGCACCGCGGACAGTGCC	SynapsinI (human)
TCCAGCACCGTGGACAGAGCC	Synaptophysin (human)
TTCAGCACCTTGGACAGAGCC	Brain-derived neurotrophic factor (rat)
TTCAGAACCACGGACAGCACC	Type II sodium channel (rat)
TTCAGCACCACGGAGAGTGCC	SCG10 (rat)
TTCAGCACCAGGGACAGCAAC	L1 (human)
TTTAGCACCGCGGACAGCGCT	GluR2 (mouse)
TTCAGCACCATGGGCAGCACA	Pax4 (mouse)
ACCCGAACCCCGGACGGCGCC	Connexin36 (human)

Fig. 6.2 REST binding sites in neuronal genes. The binding motif is termed repressor element–1 (RE–1) or neural-restrictive silencer element (NRSE). The sequences are taken from Schoenherr et al. (1996), Kemp et al. (2003), and Martin et al. (2003).

all structure may be incompletely organized in comparison to cellular chromatin (Smith and Hager, 1997). If the nucleosomal structure matters for transcriptional repression via REST, then naked DNA would not provide any result. However, quantitative differences between transiently transfected plasmids versus integrated transcription units are likely when the chromatin assembly is not complete. Despite the problems of transient transfections, this approach has revealed valuable information aiding in the characterization of the biological activity of REST. It has been shown, for instance, that the REST binding site within the proximal 5'-flanking region of the human synapsin I promoter is sufficient for the neuron-specific expression of a synapsin I promoter/reporter gene (Schoch et al., 1996; Thiel et al., 1998). Deletion of this site abolished neuron-specific expression mediated by the synapsin I promoter entirely, allowing constitutively acting elements of the promoter to direct expression in a non-tissue specific manner (Schoch et al., 1996). These experiments revealed that the synapsin I promoter worked perfectly in non-neuronal cells, as far as the REST binding site had been deleted. This observation indicates that REST normally represses synapsin I promoter activity. This is in agreement with the proposed function of REST as a transcriptional repressor of neuronal genes in non-neuronal tissues.

REST functions as a transcriptional repressor that is able to block the transcription of strong heterologous promoters of viral and eukaryotic origin when REST binding sites are provided in the transcription unit (Lönnerberg et al., 1996; Bessis et al., 1997; Tapia-Ramirez et al., 1997; Thiel et al., 1998). Moreover, REST functions very effectively as a transcriptional repressor at a distance; that is, REST is able to repress transcription despite the location or orientation of the binding site within a gene (Thiel et al., 1998). Thus, REST fulfils the criteria of a transcriptional silencer binding protein that functions in a similar, but opposite manner as enhancer binding proteins; that is, REST blocks transcription whether located upstream or downstream of a gene, in either orientation and in both a distance- and gene-independent manner. The biological activity of REST is therefore an "active" repression of transcription. The ability of REST to block transcription of many genes from any position within a transcription unit, fits very well with the locations of naturally occurring REST binding sites, indicating that the actual binding position of REST within a neuronal gene clearly plays no role in its biological activity.

6.4 Mechanism of Transcriptional Repression by REST: Modulation of the Chromatin Structure

A critical determinant in the regulation of eukaryotic genes is the structural organization of DNA in chromatin. The fundamental unit of chromatin is the nucleosome, with two molecules each of the histones H2A, H2B, H3 and H4 building the core histone, and approximately 146 base pairs of DNA wrapped 1.65 turns around the histone octamer. The single nucleosomes are linked by short stretches of DNA and the linker histone H1. The N-terminal regions of the core histones are often modified by acetylation, methylation or phosphorylation. The acetylation of histones involves the transfer of an acetyl group from acetyl coenzyme A to the ε-amino group of a lysine residue (Fig. 6.3A). Histone acetylation is of major importance for the regulation of gene transcription because this modification reduces the net positive charges of the core histones, leading to a decrease in their binding affinity for DNA. The termini are subsequently displaced from the nucleosome, after which the nucleosome unfolds and provides access for transcription factors. Thus, transcriptional activation occurs only after the repressive histone-DNA interaction has been destabilized by histone acetylases (Wade et al., 1997). Deacetylation of histones by histone deacetylases, in contrast, removes the acetyl group from the ε-amino group of lysine residues of histones, thereby allowing ionic interactions between the negatively charged DNA phosphate backbone and the positively charged amino termini of the core histones. This results in a more compact chromatin structure that is not easily accessible for the transcriptional machinery (Fig. 6.3B). While histone acetylation and hyperacetylation has been correlated with transcriptionally active chromatin, histone deacetylation is thought to be involved in the repression of transcription. Histone acetylation and deacetylation are major regulatory mechanisms of transcription that function by modulating the accessibility of transcription factors to their binding site on DNA. Consequently, a common feature of mammalian transcriptional repressors is the promoter-specific recruitment of histone deacetylases (Ng and Bird, 2000; Free et al., 2001).

REST is unique in having two repressor domains that are distinct in their primary structure (Tapia-Ramírez et al., 1997; Thiel et al., 1998). It has been shown that the repressor domains of REST interact with distinct nuclear factors (Leichter and Thiel, 1999), suggesting that two different modes of action are used by REST. This observation was confirmed by the finding that the N-terminal, but not the C-terminal repressor domain of REST recruits a Sin3A/histone deacetylase complex into the vicinity of the promoter (Huang et al., 1999; Naruse et al., 1999). Transcriptional

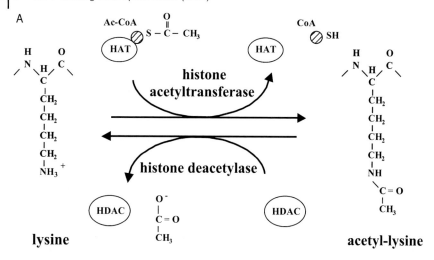
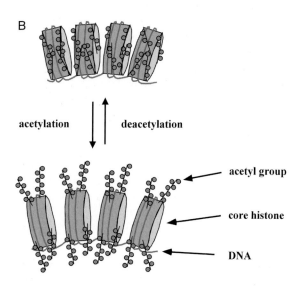

Fig. 6.3 Acetylation and deacetylation of histones determine the chromatin structure. (A) Chemical composition of the side chains of lysine and acetyl-lysine. Histone acetyltransferases (HAT) catalyze the transfer of an acetyl group from acetyl coenzyme A to the ε-amino group of internal lysine residues of histone N-terminal domains, removing the positive charge of the ε-amino group at physiological pH. Histone deacetylases (HDAC) catalyze the removal of the acetyl groups. (B) Acetylation of histones loosens the contact between DNA and the histone octamer, thus generating an open configuration of the chromatin. Deacetylation, in contrast, stabilizes the DNA/histone binding, leading to chromatin compaction. [Reproduced with modifications from Thiel and Lietz (2004) with copyright permission of Wiley-VCH.] (This figure also appears with the color plates.)

repression effected via the C-terminal repression domain of REST, however, was not impaired by the histone deacetylase inhibitor trichostatin A (Huang et al., 1999; Naruse et al., 1999). The C-terminal repression domain of REST was shown to bind to the corepressor protein CoREST (Andrés et al., 1999), a protein that has been found in a complex with the trichostatin A sensitive histone deacetylase 1 (Humphrey et al., 2001). Accordingly, it was later shown that the C-terminal repression domain of REST also functions via recruitment of histone deacetylases (Ballas et al., 2001). The recruitment of histone deacetylases to neuronal genes by REST results in the removal of acetyl groups from the core histones. Consequently, the neuronal genes are embedded into more tightly packed chromatin that is inaccessible for transcriptional activators. In neurons, REST is expressed at extremely low concentrations. As a result, the chromatin has an open configuration, allowing transcriptional activators to bind and to initiate transcription of neuronal genes. Therefore, gene transcription of neuronal genes is the result of a relief of repression. Accordingly, neuronal genes normally repressed by REST are expressed following inhibition of histone deacetylases (Huang et al., 1999; Naruse et al., 1999; Lietz et al., 2003) (Fig. 6.4). However, the de-repression of neuronal genes by histone deacetylase inhibitors depends on the cell type and the chromatin structure of the gene under investigation. The fact that REST functions via recruitment of histone deacetylases suggests that transcriptional repression via REST is independent of the nature of the activator proteins bound to the promoters of REST's target genes, thus confirming the results obtained with an activator-specific transcriptional repression assay (Lietz et al., 2001).

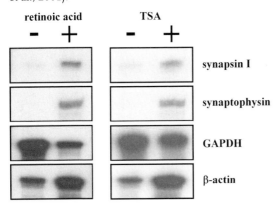

Fig. 6.4 Activation of synapsin I and synaptophysin gene transcription in P19 teratocarcinoma cells. (left panel) P19 teratocarcinoma cells were differentiated via aggregation and treatment with retinoic acid for 4 days, than plated and cultured for a further 5 days in the presence of cytosine β-D-arabinofuranoside. (right panel) P19 cells were treated for 24 h with the histone deacetylase inhibitor trichostatin A (TSA) or with the vehicle, dimethylsulfoxide (DMSO). Cytoplasmic RNA from undifferentiated (A, denoted "-"), neuronally differentiated (A, denoted "+"), DMSO-treated (B, denoted "-") and TSA-treated (B, denoted "+") P19 cells were isolated and analyzed by RNase protection mapping using cRNAs specific for synapsin I, synaptophysin, β-actin and glyceraldehyde-phosphate dehydrogenase (GAPDH), respectively. [Reproduced from Lietz et al. (2003) with copyright permission of Blackwell Publishing, Oxford.]

REST has been characterized as a dual-specific repressor (Thiel et al., 2004), that induces transcriptional repression not only via recruitment of histone deacetylases but also via gene silencing using heterochromatin protein–1 (HP1) -induced heterochromatin formation. The corepressor protein, CoREST, that binds to the C-terminal repression domain of REST and forms complexes with histone deacetylases 1 and 2, is also able to attract HP1 (Lunyak et al., 2002). DNA methylation was shown to be required to silence the voltage-gated sodium type II channel (Nav.1.2) -encoding gene in Rat–1 cells (Lunyak et al., 2002) and the GluR2 glutamate receptor and the muscarinic acetylcholine receptor type IV encoding genes in hippocampal neural stem cells (Kuwabara et al., 2004). The involvement of the methyl-CpG binding protein MeCP2 in REST-mediated gene silencing indicates that both the presence of a REST binding site and a specific DNA methylation pattern are of importance in silencing neuronal genes via the corepressor protein CoREST. In contrast, inhibition of DNA methyltransferase activity failed to de-repress muscarinic acetylcholine receptor type IV transcription in lung fibroblasts (Wood et al., 2003), suggesting that cell type-specific factors control the responsiveness of this gene to DNA methylation. Finally, histone methylation has been found to regulate the silencing activity of REST towards neuronal genes. REST recruits the histone dimethylase G9a to neuronal genes that triggers the methylation of the lysine residue 9 of histone H3 (Roopra et al., 2004). Interestingly, the recruitment of G9a by REST does not involve the CoREST protein (Roopra et al., 2004). Methylation of H3Lys9 provides a high-affinity binding site for the HP1 family of proteins. HP1 plays an essential role in the establishment and maintenance of the transcriptionally silent state of heterochromatin. HP1 binds to the H3mLys9 epitope through the "chromo domain". Subsequent oligomerization of HP1 causes the formation of a higher-ordered chromatin compaction (Fig. 6.5).

6.5
Lessons from the REST Knockout Mouse

The analysis of neuronal gene expression in transgenic mice containing an inactivated REST gene revealed that only the tubulin βIII-encoding gene was aberrantly expressed (Chen et al., 1998). In the myotome, no conversion of presumptive muscle cells into neurons was observed. The homozygous REST($^{-/-}$) knockout mice exhibited embryonic lethality that was preceded by widespread apoptotic cell death, beginning at embryonic days 9.5 and 10 (Chen et al., 1998). Unfortunately, lethality occurred at that time when neuronal genes are first expressed, making it impossible to interpret the neuronal gene expression pattern in the absence of REST. Nevertheless, the loss of REST function did not induce precocious neurogenesis in neural precursor cells or a transformation of non-neuronal cells into neurons. Thus, "... should the master regulator *rest* in peace ?" (Hemmati Brivanlou, 1998). These data obtained from the analysis of REST-knockout mice imply that REST does not control the induction of neurogenesis. Rather, REST may be required after neural versus non-neural fate determination. A conditional knockout of the REST gene in

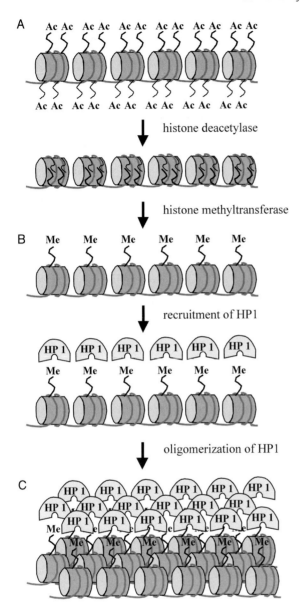

Fig. 6.5 Sequential formation of facultative heterochromatin. (A) Recruitment of histone deacetylases to the transcription unit by transcriptional repressors such as REST induces deacetylation of histone tails, thus making them suitable substrates for histone methyltransferases such as SUV39H1 and G9a. These enzymes transfer methyl groups to the ε-nitrogen of lysine residue 9 of histone H3 using S-adenosyl-L-methionine as methyl donor. (B) The methylated lysine 9 of histone H3 provides a high-affinity binding site for HP1. (C) Dimerization and oligomerization of HP1 proteins spreads the compaction of nucleosomes, forming facultative heterochromatin. [Reproduced from Thiel et al. (2004) with copyright permission of Blackwell Publishing, Oxford.] (This figure also appears with the color plates.)

mice will be helpful in analysis of the tissue-specific expression of pan-neuronal genes such as the synapsin I and synaptophysin genes in the absence of REST.

6.6
Cell Type-Specific Regulation of REST Target Genes

The analysis of three neuroblastoma cell lines revealed that the expression levels of REST and synapsin I were in a direct inverse relationship; that is, increased levels of REST mRNA were accompanied by a reduced synapsin I mRNA concentration and vice versa (Lietz et al., 1998). It was concluded that the concentration of REST in neurons and neuronal cells is of major importance for REST function and for the expression of neuronal genes. Likewise, a recent analysis of the expression levels of REST in neuronal and (neuro)endocrine cells and in fibroblasts revealed elevated levels of REST in non-neuronal cells such as fibroblasts, and low levels of REST in tissues that express REST-regulated genes. The REST-regulated target genes encoding synaptophysin and secretogranin II were found to be expressed in neuronal and (neuro)endocrine cells, but not in fibroblasts, indicating that the expression of REST was inversely proportional to the levels of synaptophysin and secretogranin II mRNA (Hohl and Thiel, 2005). A detailed analysis of REST expression in the rat nervous system, using *in-situ* hybridization and RNase protection mapping techniques, revealed that REST is expressed in neurons of the adult nervous system, although at much lower amounts than in undifferentiated neuronal progenitor cells (Palm et al., 1998). This observation indicates that low levels of REST are tolerable for allowing neuron-specific gene transcription. Higher levels of REST, however, would lead to an impairment of neuron-specific gene transcription. A constitutive expression of REST in developing neurons of the spinal cord of chicken embryos repressed the expression of two endogenous target genes of REST, N-tubulin and NgCAM, indicating that neuronal gene expression is disrupted by increasing concentration of REST. Moreover, commissural neurons expressing REST showed a significantly increased frequency of axon guidance errors, although neurogenesis was not entirely prevented by REST (Paquette et al., 2000). These data indicate that down-regulation of REST during neurogenesis is necessary for proper neuronal development. In contrast, the continuous presence of the transcriptional repressor REST in non-neuronal tissues seems to be required to prevent neuronal gene transcription in non-neuronal cells, thus ensuring continuous active control of the differentiated state (Blau, 1992).

We have tested the accessibility of REST target genes in different cell types using a dominant-positive mutant of REST that binds to the REST cognate DNA-binding site and competes with wild-type REST for DNA binding. This mutant does not contain a repression domain, and is therefore unable to recruit histone deactylases to neuronal genes. Instead, the presence of an activation domain actively promoted transcription of NRSE-containing neuronal genes in transient transfection experiments (Lietz et al., 2001, 2003; Magin et al., 2002). An analysis of REST-target genes transcription, embedded in their natural location in the chromatin, revealed that the

REST mutant strongly stimulated *synaptophysin* and *secretogranin* II gene transcription in pituitary, neuronal, and pancreatic cells (Hohl and Thiel, 2005). The results of these experiments confirm that REST is a key transcription factor controlling *synaptophysin* and *secretogranin* II gene expression. Transcription of the *connexin36* gene was strikingly enhanced in pancreatic cells, but activation of the REST mutant was not sufficient to induce connexin36 expression in neuronal and pituitary cells. These data indicate that the *connexin36* gene is a genuine target for REST in endocrine pancreatic cells, but not in neuronal cells or pituitary corticotrophs. Thus, cell type-specific factors are essential in determining whether REST is able to regulate a particular gene with REST-binding sites in the regulatory region. These data shed light on the fact that the concentration of REST is not the only determinant that controls REST target gene expression. Rather, cell type-specific modifications of the chromatin structure are most likely critical for REST to gain access to the REST-binding sites in neuronal genes. Likewise, REST is neither bound to the inactive muscarinic acetylcholine receptor type IV gene in lung fibroblasts, nor to the inactive *ANP* gene in PC12 cells (Wood et al., 2003), which suggests that repression of REST-target genes in differentiated cell types does not always require a permanent repression via REST. In this context, it is of interest that the expression of a REST-VP16 fusion protein converted C2C12 myoblasts to a physiologically active neuronal phenotype (Watanabe et al., 2004). However, expression of REST-VP16 failed to convert the fate of fully differentiated muscle fibers, suggesting that the REST target genes were not more accessible for the REST mutant. Thus, the concept that REST binds permanently to REST target genes in non-neuronal cells to allow a permanent repression of neuronal genes in these tissues requires re-evaluation. The comparable analysis of REST target gene expression in neuronal and neuroendocrine cells revealed that the actual concentration of REST is not always exactly correlated with the expression level of neuronal genes. Rather, the cell type-specific microenvironment, including the cell-specific expression of transcription regulators as well as the cell type-specific structure of the chromatin, is crucial for the ability of REST, to control gene transcription.

6.7
The Role of REST in the Differentiation of Neural Stem Cells

Establishment of the neuronal phenotype requires a down-regulation of REST expression in differentiating neuronal progenitor cells in order to switch neuronal gene transcription from a repressed state to an active state. Accordingly, the expression level of REST was shown to be largely reduced in P19 teratocarcinoma cells following neuronal differentiation (Bai et al., 2003). Likewise, forced expression of a REST mutant that activated neuronal gene transcription converted neural stem cells into mature neurons (Su et al., 2004). In addition, a small modulatory, non-coding, double-stranded RNA has been identified in neural stem cells that interacts with the REST transcriptional machinery and triggers maturation of the neuronal phenotype (Kuwabara et al., 2004). This RNA that matches to the REST binding sequence

(NRSE/RE–1) in an antisense orientation, has been proposed to convert REST from a transcriptional repressor to an activator (Kuwabara et al., 2004). The means by which a transcriptional repressor protein such as REST that contains no transcriptional activation domain, but two distinct "active" repression domains able to recruit histone deacetylases, histone methyltransferases and the heterochromatin-forming HP1 protein, could activate transcription awaits further explanation.

6.8
Involvement of REST in Brain Dysfunction and Disease

Injection of the glutamate receptor agonist kainic acid in the brain induces seizures that are accompanied by massive neuronal cell death, in particular in the hippocampus. It has been shown that this treatment is accompanied by an increase in the REST mRNA concentration in hippocampal and cortical neurons (Palm et al., 1998). This observation indicates that REST is part of the neuronal activity-implied processes in the brain that trigger a suppression of REST-controlled target genes. One of the target genes of REST encodes the GluR2 subunit of glutamate receptors that controls Ca^{2+} permeability, single channel conductance and rectification of AMPA-type glutamate receptors (Myers et al., 1998). The fact that GluR2 governs AMPA receptor Ca^{2+} permeability indicates that changes in the expression level of the receptor profoundly influence neuronal survival. Likewise, global ischemia has been shown to induce an up-regulation of REST mRNA and protein (Calderone et al., 2003), suggesting that the de-repression of REST may function as a device to counteract insult-induced neuronal death.

Huntington disease, an autosomal dominant disorder, is caused by a CAG repeat expansion that is translated into an abnormally long polyglutamine tract in the huntingtin protein. While the wild-type huntingtin protein is mainly expressed in the cytoplasm, due to a nuclear export signal on the C-terminus, the mutant accumulates in the nucleus, suggesting that alterations in gene transcription are induced as a result of huntingtin mutations. Recently, it was shown by co-immunoprecipitation assays that huntingtin interacts with REST. The huntingtin/REST complex is sequestered in the cytoplasm, thus impairing the silencing activity of REST towards neuronal genes. In contrast, mutated huntingtin binds with lower affinity to REST, leading to higher levels of REST in the nucleus and silencing of neuronal genes by REST, for instance the gene encoding brain-derived neurotrophic factor (Zuccato et al., 2003). Thus, mutations of huntingtin negatively influence neuronal gene transcription.

6.9
Conclusion and Prospects

REST target genes encode neuronal receptors and channel proteins, neuropeptides and synaptic vesicle proteins, transcription factors and adhesion molecules, underlining the important role of REST in the elaboration of the neuronal phenotype. Future analysis of REST target genes will have to combine a cell type-specific expression analysis accompanied by a cell type-specific analysis of the chromatin structure, in order to determine the impact of REST in the control of neuron-specific gene transcription. A conditional knockout of the REST gene in mice will be helpful to analyze the tissue-specific expression of pan-neuronal genes in the absence of REST. Gain-of-function experiments – that is, overexpression of REST in neurons – should provide information about the REST concentration that is tolerable by a neuron without losing its cell type-specific phenotype and function. Naturally, the elucidation of the regulatory network governed by REST will provide a handle to better understand what is going wrong in disease states. Finally, regulatory mechanisms controlling REST expression, and in particular the down-regulation of REST expression in differentiating neuronal progenitor cells, would be of special interest in order to learn how the regulator is regulated.

Note added in proof

A recent study elucidated the fundamental role of REST in orchestrating the progression of pluripotent cells to lineage-restricted neural progenitors (Ballas et al., Cell 121: 645–657).

The REST-binding protein CoREST has recently been shown to play an essential role in the demethylation of the histone 3 lysine 4 residue (Lee et al., Nature 437: 432–435).

Acknowledgments

The authors thank Libby Guethlein and Oliver Rössler for critical reading of the manuscript. The research of the laboratory concerning the modular structure and function of REST is supported by the Deutsche Forschungsgemeinschaft (grant # Th 377/6–3).

Abbreviations

HP1	heterochromatin protein–1
NRSF	neuron-restrictive silencer factor
REST	RE–1 silencing transcription factor

References

Andrés ME, Burger C, Peral-Rubio MJ, Battaglioli E, Anderson ME, Grimes J, Dallman J, Ballas N, Mandel G. **1999**. CoREST: A functional corepressor required for regulation of neural-specific gene expression. *Proc. Natl. Acad. Sci. USA* 96: 9873–9878.

Bai G, Zhuang Z, Liu A, Chai Y, Hoffman PW. **2003**. The role of the RE1 element in activation of the NR1 promoter during neuronal differentiation. *J. Neurochem.* 86: 992–1005.

Ballas N, Battaglioli E, Atouf F, Andres ME, Chenoweth J, Anderson ME, Burger C, Moniwa M, Davie JR, Bowers WJ, Federoff HJ, Rose DW, Rosenfeld MG, Brehm P, Mandel G. **2001**. Regulation of neuronal traits by a novel transcriptional complex. *Neuron* 31: 353–365.

Bessis A, Champtiaux N, Chatelin L, Changeux J-P. **1997**. The neuron-restrictive silencer element: A dual enhancer/silencer crucial for patterned expression of a nicotinic receptor gene in the brain. *Proc. Natl. Acad. Sci. USA* 94: 5906–5911.

Blau H M. **1992**. Differentiation requires continuous active control. *Annu. Rev. Biochem.* 61: 1213–1230

Calderone A, Jover T, Noh K-m, Tanaka H, Yokota H, Lin Y, Grooms SY, Regis R, Bennett MV, Zukin RS. **2003**. Ischemic insults derepress the gene silencer REST in neurons destined to die. *J. Neurosci.* 23: 2112–2121.

Chen Z-F, Paquette AJ, Anderson DJ. **1998**. NRSF/REST is required in vivo for repression of multiple neuronal target genes during embryogenesis. *Nature Genet.* 20: 136–142.

Chong JA, Tapia-Ramirez J, Kim S, Toledo-Aral JJ, Zheng Y, Boutros MC, Altshuller YM, Frohman, MA, Kraner SD, Mandel G. **1995**. REST: a mammalian silencer protein that restricts sodium channel gene expression to neurons. *Cell* 80: 949–957.

Free A, Crunotcin M, Bird A, Vogelauer M. **2001**. Histone deacetylation: mechanisms of repression. In: Elgin SCR, Workman JL (eds). *Chromatin structure and gene expression*. New York: Oxford University Press, pp. 156–181.

Gurrola-Diaz C, Lacroix J, Dihlmann S, Becker C-M, von Knebel Doeberitz M. **2003**. Reduced expression of the neuron restrictive silencer factor permits transcription of glycine receptor $\alpha 1$ in small-cell lung cancer cells. *Oncogene* 22: 5636–5645.

Hemmati Brivanlou A. **1998**. Should the master regulator *rest* in peace ? *Nature Genet.* 20: 109–110.

Hohl M. and Thiel G. **2005**. Cell type-specific regulation of RE-1 silencing transcription factor (REST) target genes. *Eur. J. Neurosci.*, in press.

Huang Y, Myers SJ, Dingledine R. **1999**. Transcriptional repression by REST: recruitment of Sin3A and histone deacetylase to neuronal genes. *Nature Neurosci.* 2: 867–872.

Humphrey GW, Wang Y, Russanova VR, Hirai T, Qin J, Nakatani Y, Howard BH. **2001**. Stable histone deacetylase complexes distinguished by the presence of SANT domain proteins CoREST/kiaa0071 and Mta-L1. *J. Biol. Chem.* 276: 6817–6824.

Kemp DM, Lin JC, Habener JF. **2003**. Regulation of pax4 paired homeodomain gene by neuron-restrictive silencer factor. *J. Biol. Chem.* 278: 35057–35062

Kuwabara T, Hsieh J, Nakashima K, Taira K, Gage FH. **2004**. A small modulatory dsRNA specifies the fate of adult neural stem cells. *Cell* 116: 779–793.

Leichter M, Thiel G. **1999**. Transcriptional repression by the zinc finger protein REST is mediated by titratable nuclear factors. *Eur. J. Neurosci.* 11: 1937–1946.

Lietz M, Bach K, Thiel G. **2001**. Biological activity of RE-1 silencing transcription factor (REST) towards distinct transcriptional activators. *Eur. J. Neurosci.* 14: 1303–1312.

Lietz M, Cicchetti P, Thiel G. **1998**. Inverse expression pattern of REST and synapsin I in human neuroblastoma cells. *Biol. Chem.* 379: 1301–1304.

Lietz M, Hohl M, Thiel G. **2003**. RE-1 silencing transcription factor (REST) regulates human synaptophysin gene transcription through an intronic sequence-specific DNA-binding site. *Eur. J. Biochem.* 270: 2–9.

Lönnerberg P, Schoenherr CJ, Anderson DJ, Ibáñez CF. **1996**. Cell type-specific regulation of choline acetyltransferase gene expression. Role of the neuron-restrictive silencer element and cholinergic-specific enhancer sequences. *J. Biol. Chem.* 271: 33358–33365.

Lunyak VV, Burgess R, Prefontaine GG, Nelson C, Sze S-H, Chenoweth J, Schwartz P, Pevzner PA, Glass C, Mandel G, Rosenfeld MG. **2002**. Corepressor-dependent silencing of chromosomal regions encoding neuronal genes. *Science* 298: 1747–1752.

Magin A, Lietz M, Cibelli G, Thiel G. **2002**. RE–1 silencing transcription factor–4 (REST4) is neither a transcriptional repressor nor a de-repressor. *Neurochem. Int.* 40: 193–200.

Martin D, Tawadros T, Meylan L, Abderrahmani A, Condorelli DF, Waeber G, Haefliger J-A. **2003**. Critical role of the transcriptional repressor neuron-restrictive silencer factor in the specific control of connexin36 in insulin-producing cell lines. *J. Biol. Chem.* 278: 53082–53089.

Myers SJ, Peters J, Huang Y, Comer MB, Barthel F, Dingledine R. **1998**. Transcriptional regulation of the GluR2 gene: neural-specific expression, multiple promoters, and regulatory elements. *J. Neurosci.* 18: 6723–6739.

Naruse Y, Aoki T, Kojima T, Mori N. **1999**. Neural restrictive silencer factor recruits mSin3 and histone deacetylase complex to repress neuron-specific target genes. *Proc. Natl. Acad. Sci. USA* 96: 13691–13696.

Ng HH, Bird A. **2000**. Histone deacetylases: silencers to hire. *Trends Biochem. Sci.* 25: 121–126.

Palm K, Belluardo N, Metsis M, Timmusk T. **1998**. Neuronal expression of zinc finger transcription factor REST/NRSF/XBR gene. *J. Neurosci.* 18: 1280–1296.

Palm K, Metsis M, Timmusk T. **1999**. Neuron-specific splicing of zinc finger transcription factor REST/NRSF/XBR is frequent in neuroblastomas and conserved in human, mouse and rat. *Mol. Brain Res.* 72: 30–39.

Paquette AJ, Perez SE, Anderson DJ. **2000**. Constitutive expression of the neuron-restrictive silencer factor (NRSF)/REST in differentiating neurons disrupts neuronal gene expression and causes axon pathfinding errors *in vivo*. *Proc. Natl. Acad. Sci. USA* 97: 12318–12323.

Roopra A, Qazi R, Schoenike B, Daley TJ, Morrison JF. **2004**. Localized domains of G9a-mediated histone methylation are required for silencing of neuronal genes. *Mol. Cell* 14: 727–738.

Schoch S, Cibelli G, Thiel G. **1996**. Neuron-specific gene expression of synapsin I: major role of a negative regulatory mechanism. *J. Biol. Chem.* 271: 3317–3323.

Schoenherr CJ, Anderson DJ. **1995**. The neuron-restrictive silencer factor (NRSF): a coordinate repressor of multiple neuron-specific genes. *Science* 267: 1360–1363.

Schoenherr CJ, Paquette AJ, Anderson DJ. **1996**. Identification of potential target genes for the neuron-restrictive silencer factor. *Proc. Natl. Acad. Sci. USA* 93: 9881–9886.

Shimojo M, Paquette AJ, Anderson DJ, Hersh LB. **1999**. Protein kinase A regulates cholinergic gene expression in PC12 cells: REST4 silences the silencing activity of neuron-restrictive silencer factor/REST. *Mol. Cell. Biol.* 19: 6788–6795.

Shimojo M, Lee JH, Hersh LB. **2001**. Role of zinc fingers of the transcription factor NRSF/REST in DNA binding and nuclear localization. *J. Biol. Chem.* 276: 13121–13126.

Smith CL, Hager GL. **1997**. Transcriptional regulation of mammalian genes *in vivo*. *J. Biol. Chem.* 272: 27493–27496.

Su X, Kameoka S, Lentz S, Majumder S. **2004**. Activation of REST/NRSF in neural stem cells is sufficient to cause neuronal differentiation. *Mol. Cell. Biol.* 24: 8018–8025.

Tapia-Ramírez J, Eggen BJ, Peral-Rubio MJ, Toledo-Aral JJ, Mandel G. **1997**. A single zinc finger motif in the silencing factor REST represses the neural-specific type II sodium channel promoter. *Proc. Natl. Acad. Sci. USA* 94: 1177–1182.

Thiel G, Lietz M. **2004**. Zinkfingerprotein REST – Regulator neuronaler Gene. *Biol. u Z.* 34: 96–101.

Thiel G, Lietz M, Bach K, Guethlein L, Cibelli G. **2001**. Biological activity of mammalian transcriptional repressors. *Biol. Chem.* 382: 891–902.

Thiel G, Lietz M, Cramer M. **1998** Biological activity and modular structure of RE–1 silencing transcription factor (REST), a repressor of neuronal genes. *J. Biol. Chem.* 273: 26891–26899.

Thiel G, Lietz M, Hohl M. **2004**. How mammalian transcriptional repressors work. *Eur. J. Biochem.* 271: 2855–2862.

Wade PA, Pruss D, Wolffe AP. **1997**. Histone acetylation: chromatin in action. *Trends Biochem. Sci.* 22: 128–132.

Watanabe Y, Kameoka S, Gopalakrishnan V, Aldape KD, Pan ZZ, Lang FF, Majumder S. **2004**. Conversion of myoblasts to physiologically active neuronal phenotype. *Genes Dev.* 18: 889–900.

Wood IC, Belyaev ND, Bruce AW, Jones C, Mistry M, Roopra A, Buckley NJ. **2003**. Interaction of the repressor element 1–silencing transcription factor (REST) with target genes. *J. Mol. Biol.* 334: 863–874.

Zuccato C, Tartari M, Crotti A, Goffredo D, Valenza M, Conti L, Cataudella T, Leavitt BR, Hayden, MR, Timmusk T, Rigamonti D, Cattaneo E. **2003**. Huntingtin interacts with REST/NRSF to modulate the transcription of NRSE-controlled neuronal genes. *Nature Genet.* 35: 76–83.

7
Roles of Tlx1 and Tlx3 and Neuronal Activity in Controlling Glutamatergic over GABAergic Cell Fates

Qiufu Ma and Le-ping Cheng

Abstract

Glutamatergic and GABAergic are two principal excitatory and inhibitory neurons, respectively, in the vertebrate nervous system. Recent studies have gained significant insights into the mechanisms by which neurons make the choice between these two fundamental neuronal cell fates. First, region-specific transcription factors such as the *Tlx* class homeobox genes are expressed in newly formed postmitotic neurons and act as binary genetic switches in selecting a glutamatergic over a GABAergic cell fate. Second, patterns of neuronal activity are able to modulate excitatory and inhibitory transmitter phenotypes in a homeostatic manner. However, the means by which neuronal activity interfaces with intrinsic switch genes to regulate transmitter phenotypes remains to be elucidated.

7.1
Introduction

Neurotransmitters are the chemicals that transmit signals from one neuron to the next across synapses. Since their discovery in the 1950s, glutamate and gamma-aminobutyric acid (GABA) have stood out as the two predominant excitatory and inhibitory neurotransmitters, respectively, in the vertebrate nervous system [1]. Most neurons release either glutamate or GABA, but rarely both, indicating that these two transmitter phenotypes define a major functional subdivision in neuronal cell type [2, 3]. Until very recently, the molecular and cellular mechanisms by which neurons choose an excitatory versus an inhibitory cell fate was poorly understood [4–6].

Neurotransmitter phenotypes are defined by expression of the enzymes responsible for transmitter synthesis and the transporters that package the transmitters to the synaptic vesicles. Glutamate is an amino acid which exists in every cell. Recent studies have led to the identification of a family of vesicular glutamate transporters, including VGLUT1, VGLUT2 and VGLUT3 [2,3,7–14]. VGLUT1 and VGLUT2 are expressed in largely non-overlapping subsets of glutamatergic neurons, thus serving

Transcription Factors. Edited by Gerald Thiel
Copyright © 2006 WILEY-VCH Verlag GmbH & Co. KGaA, Weinheim
ISBN 3-527-31285-4

as prospective markers for two separate groups of excitatory neurons [3,8–11,15,16]. VGLUT3 is expressed in cells which are generally considered to release a classical transmitter different from glutamate, including subpopulations of inhibitory neurons, cholinergic interneurons, monoamine neurons, and glia [14, 17]. GABAergic neurons are defined by the expression of two glutamate decarboxylases, GAD67/Gad1 and GAD65/Gad2, which convert glutamate into GABA and serve as the prospective markers for GABAergic neurons [18]. In addition, all GABAergic neurons express the vesicular inhibitory amino acid transporter Viaat/VGAT that packages GABA into synaptic vesicles [18].

In this chapter, recent progress in understanding how excitatory and inhibitory transmitter phenotypes are regulated is reviewed, with particular focus on the development of the dorsal spinal cord. Specifically, the roles of the *Tlx* family of homeobox genes and patterns of neuronal activity in regulating excitatory versus inhibitory cell fates will be discussed. The fundamental questions asked include:

- What are the molecular logistics underlying mutually exclusive development of the glutamatergic and GABAergic neurons?
- Are there binary genetic switches that lead to all-or-none decisions in selecting transmitter phenotypes?
- Are generic neurotransmitter phenotypes specified by universal transcriptional programs, or by distinct sets of transcription factors in different brain areas?
- How does neuronal activity interface intrinsic transcription factors to regulate transmitter phenotypes?

7.2
The Dorsal Horn of the Spinal Cord

The dorsal horn of the spinal cord is the first-order relay station that processes and transmits somatic sensory information, such as the senses of pain, temperature, touch, and itching [19, 20]. Anatomically, the dorsal horn is organized in a lamina specific fashion [20–22]. For example, nociceptive and thermoceptive afferents innervate the most superficial lamina, whereas mechanoceptive afferents project to the more ventral dorsal horn lamina [20,21,23–25]. Based on transmitter phenotypes, dorsal horn neurons can be grouped into two major categories: (a) excitatory neurons that use glutamate for fast neurotransmission; and (b) inhibitory neurons that utilize GABA (or glycine) for fast neurotransmission. All long-range ascending projection neurons are glutamatergic and excitatory, whereas local interneurons can be either excitatory or inhibitory [26–29]. The dorsal horn is not a passive relay station, but rather integrates excitatory and inhibitory synaptic inputs from peripheral sensory afferents and hindbrain descending afferents [24]. The balance of excitation and inhibition serves as a gate to determine whether the somatic sensory information – and particularly the nociceptive sensory information that leads to the perception of pain – is relayed to the brain, or not [30, 31]. Dis-inhibition under pathological conditions can lead to chronic pain disorders, the management of which remains a major medical problem [32]. An understanding of how excitatory

7.3
Neurogenesis in the Dorsal Spinal Cord

During the past decade, significant progress has been made in understanding the molecular logic that governs the generation of dorsal horn neurons. The neural precursors are patterned by signals derived from the roof plate, and along the dorsoventral axis are divided into consecutive domains distinguished by complementary expression of proneural genes, *Math1*, *Neurogenin1/Neurogenin2 (Ngn1/Ngn2)*, and *Mash1* (Fig. 7.1) [33–38]. These neuronal determination genes encode basic helix-loop-helix (bHLH) types of transcription factors that are both necessary and sufficient to promote neurogenesis [39–50]. Neurogenesis in the dorsal spinal cord then undergoes two phases. From embryonic day E10.5 to E11.5, six classes of neurons are formed along the dorsoventral axis, with the most dorsal *Math1+* precursors giving rise to DI1 cells, the dorsal *Ngn1/2+* cells to DI2 cells, the more ventral *Mash1+* precursors to DI3, DI4, and DI5 cells, and the most ventral *Ngn1/2+* domain to DI6 cells (Fig. 7.1) [34–38,51–53]. These early-born neurons tend to migrate ventrally and settle in the deep lamina of the dorsal horn, or even in the ventral spinal cord [35–38]. From E11.5 to E13.5, the remaining *Mash1+* precursors give rise to two intermingled groups of cells, DILA and DILB [51–53]. These late-born neurons migrate dorsally and occupy a major portion of the dorsal horn, including lamina 1–III [51–53]. Each of these dorsal horn neurons can be defined by a combinatorial code of transcription factors [36–38]. However, a majority of dorsal horn neurons can be distinguished by non-overlapping expression of two homeobox genes, *Pax2* (DI4, DI6 and DILA) and *Tlx3* (DI3, DI5, and DILB) [4, 52].

With recent isolation of prospective markers for glutamatergic neurons (VGLUT2 for dorsal horn excitatory neurons), the transmitter phenotypes of a variety of dorsal horn neurons at various embryonic stages (from E10.5 to E13.5) have been characterized. DI1, DI2 and Tlx3+ cells are glutamatergic, whereas Pax2+ cells are GABAergic (Fig. 7.1) [4] (also L.-P. Cheng and Q. Ma, unpublished data). The formation of GABAergic and glutamatergic neurons is, therefore, subject to complex but precise spatial and temporal control. Glutamatergic neurons (DI1–3 and DI5) and GABAergic neurons (DI4 and DI6) initially form in alternating stripes along the dorsoventral axis. During the late phase of neurogenesis, GABAergic neurons (Pax2+ DILA cells) and glutamatergic neurons (Tlx3+ DILB cells) develop sequentially from the Mash1+ domain, although there is a period during which the formation of these two classes of neurons overlaps [4] (also L.-P. Cheng and Q. Ma, unpublished data). Before discussing the function of Tlx3 in determining glutamatergic transmitter phenotype, brief overview will be provided regarding the identification, structure, and expression of the *Tlx* family of homeobox genes.

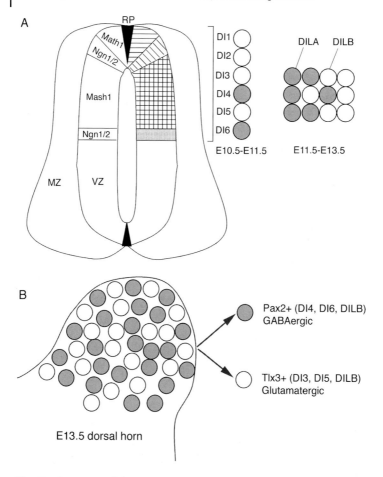

Fig. 7.1 Generation of glutamatergic and GABAergic neurons in the dorsal spinal cord.
(A) Patterns of neurogenesis in the dorsal spinal cord. Neuronal precursor cells in the dorsal ventricular zone (VZ) is patterned by signals released from the roof plate (RF), and can be distinguished by complementary expression of proneural genes: *Math1*, *Ngn1/2* and *Mash1*. Neurogenesis in the dorsal spinal cord undergoes two phases. From embryonic day E10.5 to E11.5, six classes of neurons (DI1–DI6) are formed along the dorsoventral axis. These early-born neurons first migrate to the marginal zone (MZ) and then settle in the deep lamina of the dorsal horn, or even in the ventral spinal cord. From E11.5 to E13.5, the *Mash1*⁺ precursors sequentially give rise to two intermingled groups of cells, DILA and DILB. These late-born neurons migrate dorsally and occupy a major portion of the dorsal horn, including laminae 1–III. (B) Neurotransmitter phenotypes of dorsal horn neurons. In the superficial lamina of the E13.5 dorsal horn, glutamatergic neurons can be marked by the expression of Tlx3 (DI3, DI5, and DILB), whereas GABAergic neurons can be marked by the expression of Pax2 (DI4, DI6, and DILB). Two other early born neurons (DI1 and DI2) also belong to glutamatergic neurons.

7.4
The Tlx Family of Homeobox Proteins

In the mammalian genome, the *Tlx* family of homeobox genes is composed of three members, *Tlx1* (also called *tcl3* and *Hox11*), *Tlx2* (also called *Hox11L1*, *Ncx*, *Enx*), and *Tlx3* (also called *Hox11L2* and *Rnx*) [54–67]. The founding member *Tlx1* was identified by the association of frequent *Tlx1* chromosomal translocation with the development of T-cell acute lymphoblastic leukemia (T-ALL) [54,55,58,60]. *Tlx2* and *Tlx3* were isolated by virtue of their sequence homology to *Tlx1* [56,68,69]. A close homologue called *Clawless* has also been identified in *Drosophila* [70, 71]. *Tlx* genes encode a novel family of homeobox proteins that harbors an unusual threonine at position 47 in the homeodomain instead of the more usual isoleucine or valine (Fig. 7.2) [56]. Structure-function analysis shows that Tlx1 possesses both activation and repression activities [56,72–74].

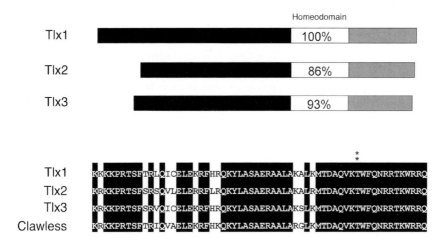

Fig. 7.2 The *Tlx* family of the homeobox genes. There are three members in mammalian genome (*Tlx1–3*), and one close homologue in the *Drosophila* genome (*Clawless*). These *Tlx* genes encode a novel family of homeobox proteins that harbors an unusual threonine at position 47 (*) in the homeodomain instead of the more usual isoleucine or valine.

In-situ hybridization showed that *Tlx* genes are dynamically expressed during mouse development. *Tlx1* is expressed in the developing embryo in the branchial arches, cranial sensory ganglia, hindbrain, spinal cord, and spleen [51,57,59,62,63,66,69]. *Tlx2* is expressed in a subset of developing neural crest-derived tissues such as the dorsal root ganglia (DRG), the cranial sensory ganglia, sympathetic ganglia, and the enteric nerve ganglia [64,66–68]. *Tlx3* is expressed in the developing cranial ganglia, DRG, sympathetic ganglia, cerebellum, hindbrain, and spinal cord [51,65,66,75,76]. *Tlx* gene knockout mice manifest different phenotypes, each reflecting the principal and unique site of expression. The Korsmeyer and Rabbitts groups have shown that *Tlx1* mutant mice do not develop spleen [57, 62]. The Korsmeyer and Hatano groups demonstrated that *Tlx2* mutant mice develop intestinal neuronal dysplasia, whereas Buchwald's group found that *Tlx2* mutant mice are embryonic lethal due to gastrulation defects [61,64,67]. The reason for the apparent discrepancy between these two independent studies remains unknown. Subsequently, the Korsmeyer and Onimaru groups showed that *Tlx3* mice die at birth due to the failure of central respiratory control [65].

7.5
Tlx Gene Expression Marks Sensory Circuits

In 1998, Lumsden's group first pointed out that *Tlx1* and *Tlx3* might be expressed in both peripheral sensory ganglia as well as various relay nuclei in the chick hindbrain [66]. Subsequent analysis in the mouse central nervous system (CNS) showed that *Tlx3* is expressed in the developing cerebellum, the nucleus of the solitary tract (NTS), the trigeminal nuclei (NTG), the noradrenergic centers, and the dorsal spinal cord, while *Tlx1* is expressed in a subset of *Tlx3*$^+$ neurons in the trigeminal nuclei and the dorsal spinal cord [51,75,76]. NTS is the major relay station for the visceral sensory afferents, and is crucial for cardiovascular and respiratory control [19, 77]. NTG and the dorsal spinal cord are the relay stations for somatic sensory information processing [19]. Collectively, *Tlx* gene expression is detected in both peripheral sensory neurons as well as the postsynaptic targets in the central relay stations. Furthermore, combinatorial expression of *Tlx3* with two other transcription factor genes, *Phox2b* [78] and *DRG11* [79], distinguishes the somatic (*Tlx3*$^+$*DRG11*$^+$) versus visceral (*Tlx3*$^+$*Phox2b*$^+$) sensory circuits [51, 75]. The Tlx proteins thus belong to those transcription factors, the expression of which marks neuronal connectivity [78–80]. In agreement with a potential role for *Tlx* genes in assembling sensory circuitry, visceral afferents fail to make a stop in the place where the presumable relay station NTS is located [75]. Similarly, nociceptive afferents fail to enter the NTG and the dorsal horn [51]. Noticeably, *Tlx3* is also required for specification of the noradrenergic neurotransmitter in the hindbrain [75]. As discussed below, within sensory relay stations, *Tlx* genes determine the excitatory over the inhibitory transmitter phenotype.

7.6
Tlx Genes Serve as Binary Switches between Glutamatergic and GABAergic Transmitter Phenotypes

There was an interesting journey before we revealed that *Tlx3* expression specifically marks the glutamatergic neurons in the dorsal spinal cord. The initial investigations were conducted to characterize the phenotypes in *Tlx1* and *Tlx3* double-mutant dorsal spinal cord. *Tlx3* single mutants showed a partial phenotype due to compensation by *Tlx1*; for clarity, *Tlx1/3* double mutants are referred to as *Tlx* mutants [51].

7.6 Tlx Genes as Binary Switches between Glutamatergic and GABAergic Transmitter Phenotypes

The journey started with the finding that the expression of *GluR2* in the dorsal spinal cord was eliminated in *Tlx* mutants. GluR2 belongs to AMPA class glutamate receptors [81], and its expression is elevated robustly in a subset of neurons in the superficial laminae of the dorsal horn [82]. GluR2 messenger RNA undergoes an editing process, such that the edited GluR2 protein replaces the glutamine residue with arginine in the fourth transmembrane domain, and renders AMPA receptors calcium-impermeable [81]. Earlier studies by Spitzer's group showed that calcium influx is able to promote the expression of GABAergic neuron markers in cultured frog spinal cord neurons [83]. It was considered that a loss of *GluR2* in *Tlx* mutants would render more neurons calcium-permeable which, according to the results of Spitzer et al., would allow more neurons to differentiate into GABAergic neurons. Indeed, a dramatic increase in GABAergic neurons was observed in the *Tlx* mutant dorsal horn [4]. *GluR2* mutant mice generated in Roder's laboratory [84] were then acquired and analyzed for GABAergic neuron development. Surprisingly, there was no increase of GABAergic neurons in *GluR2* mutants! In other words, the hypothesis that a loss of *GluR2* would lead to increased GABAergic differentiation was tested, and results according to the hypothesis – that GABAergic neuron expansion would occur – was obtained. However, the hypothesis turned out to be incorrect, that the expansion of GABAergic neurons is not involved with the loss of *GluR2*. Thus, a key function of Tlx proteins had been serendipitously discovered – the suppression of GABAergic differentiation.

The second important clue came from the finding that Tlx3$^+$ neurons are intermingled with those dorsal horn neurons marked by the expression of Pax2 [4]. Earlier studies of the cerebellum had shown that Pax2 expression is associated with GABAergic neurons [85]. It was then confirmed that Pax2$^+$ neurons in the superficial dorsal horn at embryonic stages E11.5 and E13.5 were also GABAergic [4]. At this point, the identity of Tlx3$^+$ neurons was unknown. Edwards and several other groups had identified VGLUT1–3 as being prospective markers for glutamatergic neurons, and found that glutamatergic neurons were intermingled with GABAergic neurons in most parts of the nervous system. With these lines of information, it was not difficult to postulate that *Tlx3* expression might mark glutamatergic neurons, and this hypothesis was quickly verified by the co-localization of Tlx3 with *VGLUT2* [4].

Genetic studies then demonstrated that *Tlx* genes act as a binary genetic switch that determines glutamatergic over GABAergic transmitter phenotypes [4]. Loss of *Tlx1* and *Tlx3* results in a complete elimination of *VGLUT2* expression in the superficial laminae of the dorsal horn. Meanwhile, expression of all known GABAergic neuron markers, including *Gad1/2* and *Viaat*, is dramatically expanded in *Tlx* mutant dorsal horn. Finally, overexpression of *Tlx3* in the chick neural tube is able to suppress GABAergic and promote glutamatergic differentiation. The "on and off" of *Tlx* genes are sufficient to cause a switch between excitatory versus inhibitory cell fate, implying a binary decision between these two cell fates.

The generation of excess GABAergic neurons in *Tlx* mutants also explains why *Tlx3*–deficient mice die from central respiratory control deficits [4, 65]. Electrophysiological recordings from a *Tlx3* mutant hindbrain slice preparation by Onimaru's group revealed that rhythm-generating respiratory neurons in the ventral hindbrain exhibited rapid, shallow and arrhythmic firing patterns [65]. *Tlx3* is, however, not expressed in the ventral hindbrain area, whereas respiratory neurons are located [75]. It was hypothesized that, in *Tlx3* mutants, a transformation of *Tlx3*–dependent hindbrain glutamatergic neurons into GABAergic neurons could, in principle, provide excess GABAergic inputs to the respiratory circuitry, and this in turn could disrupt the normal rhythms of respiratory neurons. In an amazing coincidence, by the time Dr. Onimaru was asked to test this possibility, his group had already independently found that the arrhythmic firing by respiratory neurons could be fully rescued after incubation with bicuculline, a potent GABA receptor antagonist [4]! These genetic and electrophysiological studies demonstrated that mutation of a region-specific switch gene which determines the glutamatergic over the GABAergic neuron cell fates can create an imbalance of excitation and inhibition and lead to profound behavioral consequences.

7.7
Binary Decision between GABAergic and Glutamatergic Cell Fates is a Common Theme

Tlx1/3 are the first transcription factors shown to be required for the specification of glutamatergic transmitter phenotype. However, an earlier study had implied that cerebral cortical precursor cells, which normally give rise to glutamatergic projection neurons, contain an intrinsic genetic program that suppresses GABAergic differentiation [86]. Cortical precursors express the neuronal determination genes *Ngn1* and *Ngn2* [33,42,86], whereas precursors in the ventral forebrain express another neuronal determination gene *Mash1* [87]. $Mash1^+$ precursors develop into GABAergic neurons, a fraction of which migrates dorsally to become cortical inhibitory interneurons [88, 89]. In *Ngn1* and *Ngn2* double mutants, *Mash1* is de-repressed in cerebral cortical precursors, which in turn leads to ectopic formation of GABAergic neurons, implying that *Ngn1* and *Ngn2* are required to inhibit a GABAergic differentiation program in cortical precursors by suppressing the expression of *Mash1* [86]. The most recent studies from the groups of Schuurmans and Guillemot demonstrated that *Ngn1* and *Ngn2* are additionally required for the specification of the glutamatergic transmitter phenotype during an early phase of cortical neurogenesis [5]. Furthermore, the homeobox gene *Pax6* and the nuclear hormone receptor gene *Tlx* (different from the *Tlx* class homeobox gene mentioned above) are required to promote glutamatergic and to suppress GABAergic differentiation during a late phase of cortical neurogenesis [5]. The binary choice between excitatory and inhibitory cell fates appears, therefore, to be a common theme in the vertebrate brain, occurring in the dorsal spinal cord, likely in the hindbrain, in the cerebral cortex and in the hippocampus. The evolvement of these binary genetic switches explains why glutamatergic and GABAergic neurons develop in a largely mutually exclusive fashion in the vertebrate brain.

7.8
Coupling of Generic Transmitter Phenotypes and Region-Specific Neuronal Identities

Glutamate and GABA are two principal excitatory and inhibitory transmitters used in most parts of the mammalian nervous system. Available data suggest that specification of these generic neurotransmitter phenotypes is not controlled by universal transcriptional programs, but rather is controlled by region-specific sets of transcription factors. First, *VGLUT1* and *VGLUT2* are expressed in complementary territories in the vertebrate brain [3,8–11,15,16]. Second, no *Tlx1* or *Tlx3* expression is detected in the forebrain areas [51,65,66,75]. Conversely, the *Dlx* class homeobox genes, which are implicated in controlling GABAergic differentiation in the forebrain, are not expressed in the hindbrain or the spinal cord [90] (also L.-P. Cheng and Q. Ma, unpublished data). Conceivably, region-specific factors might bind to distinct enhancer elements in the promoters of *Gad1/2* and *VGLUT2* genes. This control mechanism is analogous to the use of distinct neuronal determination genes (*Math1*, *Ngn1/2*, and *Mash1*) in controlling neurogenesis in different brain regions.

The involvement of region-specific transcription factors couples the following two developmental processes: specification of generic transmitter phenotypes, and the establishment of region-specific neuronal identities. For example, *Tlx1/3* are additionally required to activate many molecules associated with dorsal horn glutamatergic neurons, including the AMPA receptor GluR2, the axonal guidance molecule *Sema3C*, and multiple transcription factor genes [4, 51] (also L.-P. Cheng and Q. Ma, unpublished data). Moreover, *Tlx* genes are required to prevent dorsal horn glutamatergic neurons from expressing a set of molecules preferentially expressed in the dorsal horn GABAergic neurons, such as the kainite receptors GluR6 and GluR7, as well as the transcription factor gene *Pax2* [4]. The near-systematic cell fate transformation caused by *Tlx* mutation is remarkable, considering the fact that *Tlx1* and *Tlx3* are expressed in postmitotic neurons. The onset of *Tlx1* and *Tlx3* expression most likely represents a major readout after neuronal precursors integrate intrinsic and extrinsic signaling programs. In other words, *Tlx* genes operate at – or close to – the top of the hierarchy in controlling dorsal horn glutamatergic neuron differentiation. These findings also indicate that the distinct features of a defined group of neurons are not controlled independently, in a piecemeal fashion, but rather are at least partially controlled through coherent genetic programs.

7.9
The Plasticity of Neurotransmitter Phenotypes

The plasticity of neurotransmitter phenotypes was first observed in cultured peripheral sympathetic neurons [91, 92]. Subsequent *in-vivo* studies demonstrated that a subset of sympathetic neurons undergoes a switch of transmitter phenotypes in response to target-derived signals, from noradrenergic to cholinergic [93]. Accumulating data have also implied an activity-dependent plasticity of excitatory and inhibitory transmitter phenotypes in the vertebrate nervous system [94]. Spitzer's group showed that in the frog spinal cord at the tadpole stages, newly born neurons generate distinct patterns of spontaneous Ca^{2+} spike activity [6, 94]. Remarkably, neurons with different Ca^{2+} spike patterns express one of four classical transmitters: glutamate, glycine, GABA, and acetylcholine [6]. Early *in-vitro* culture studies showed that increased Ca^{2+} spike frequency is able to promote GABAergic neuron differentiation [83]. Most recently, Spitzer's group also showed that activity can regulate excitatory and inhibitory transmitter phenotype in a homeostatic fashion [6]. An increase in neuronal activity by the overexpression of voltage-gated rat brain Na^+ channel rNav2a leads to an increased number of neurons expressing inhibitory neurotransmitters (GABA or glycine), and a decreased number of neurons expressing excitatory neurotransmitters (glutamate or acetylcholine). Conversely, a decrease in neuronal activity by overexpression of human inward rectifier K^+ channels (hKir2.1) results in exactly inverse results: more excitatory and less inhibitory neurons. Interestingly, activity-dependent plasticity of excitatory and inhibitory transmitter phenotypes can only occur in a specific critical period after the birth of neurons [6].

Can activity regulate excitatory/inhibitory transmitter phenotype in the mammalian nervous system? Indeed, the results of several studies have suggested that this might be the case. In the rat hippocampus, the presumable "glutamatergic" granule cells in the dentate gyrus undergo a transient GABAergic phenotype during development, and in adult animals, enhanced excitability under seizures can promote GABAergic transmitter phenotype development – an effort which appears to retain the constancy of neuronal network activity [95–98]. Interestingly, neurogenesis persists to adulthood in the hippocampus [99], and it will be interesting to determine whether only immature neurons are competent for activity-dependent modulation of transmitter phenotype, in agreement with the existence of a critical period for activity-dependent modulation in the frog spinal cord. Dynamic change of inhibitory transmitter phenotype is also reported in the mammalian spinal cord. For example, there is a progressive reduction of GABAergic neurons in the rat spinal cord during prenatal and postnatal development [100, 101]. Moreover, numbers of GABAergic neurons in the dorsal horn can be dynamic under chronic pain conditions [26, 32], although it remains to be determined if the change of GABAergic neurons reflects a loss of the cells or a switch of transmitter phenotypes.

7.10
Summary and Unsolved Problems

Two important principles have emerged in regulating glutamatergic versus GABAergic transmitter phenotypes. First, there are region-specific genetic switches such as *Tlx1* and *Tlx3* in the dorsal spinal cord and *Ngn1* and *Ngn2* in the cerebral cortex that select a glutamatergic over a GABAergic cell fate. Loss or gain of these switch genes is able to cause a switch in transmitter phenotypes. Second, neuronal activity is able to modulate excitatory and inhibitory transmitter phenotypes in a homeostatic fashion, at least in the frog spinal cord and in adult mammalian hippocampus.

7.10 Summary and Unsolved Problems

One question remaining to be solved is to determine if there are switch-like genes that select a GABAergic over a glutamatergic cell fate – a function which is in opposition to that of the *Tlx* genes. The potential existence of such switch genes is indirectly implied from the finding that a decrease in neuronal activity causes a switch from a GABAergic to a glutamatergic cell fate in frog spinal cord. *Pax2* is required for the differentiation of GABAergic neurons in the mouse spinal cord, but is incapable of suppressing glutamatergic differentiation, and thus it is not involved in the cell fate choice process *per se*. The *Dlx* class homeobox genes have been implicated in controlling GABAergic neuron development in the forebrain, but their potential role in suppressing glutamatergic differentiation has not yet been examined.

The biggest unsolved question is to understand how neuronal activity interfaces with intrinsic transcription factors to control transmitter phenotypes. Our working model is that specification of transmitter phenotypes might undergo two distinct phases. The first phase is genetically "hardwired", and is controlled by the expression of intrinsic transcription factors that serve as molecular switches. In the spinal cord, expression of *Tlx1* and *Tlx3* is subject to precise spatial and temporal control. Moreover, transmitter phenotypes are among the earliest neuronal identities specified, occurring shortly after cells migrate out of the ventricular zone [4,102,103]. Even if early spontaneous neuronal activity is required for initial transmitter phenotype specification, the pattern of neuronal activity could be subject to genetic control. For example, it would be interesting to determine if *Tlx*-like switch genes could control the expression of ion channels that are responsible for the generation of spontaneous neuronal activity. During the critical period, neuronal activity might then cause the plasticity of transmitter phenotypes by modulating the expression or function of those intrinsic postmitotic switch genes, such as *Tlx1* and *Tlx3*. To test this hypothesis, it is crucial to perform genetic fate mapping to determine if there are dorsal horn neurons transiently expressing *Tlx1* and *Tlx3*, and to see if such neurons undergo a transition from a glutamatergic to a GABAergic cell fate. It is also worthwhile determining if increased neuronal activity by ectopic expression of sodium channels in the mouse spinal cord would cause a suppression of *Tlx* gene expression and a subsequent transformation from a glutamatergic to a GABAergic cell fate, as did Spitzer's group in the developing frog spinal cord.

The finding that there are region-specific molecular programs controlling the expression of excitatory and inhibitory transmitter phenotypes might be of potential clinical importance. For example, *Tlx* genes are expressed in the superficial laminae of the dorsal horn that is crucial for the transmission of nociceptive/painful sensory information. It is, therefore, important to determine if *Tlx* genes are required for the maintenance of the glutamatergic transmitter phenotype in the adult dorsal horn neurons. If they do, then a drug interfering *Tlx*-mediated transcription program could in principle be used to treat chronic pain disorders by attenuating dorsal horn neuron excitability. Such drugs would have less pronounced side effects than N-methyl-D-aspartate (NMDA) receptor antagonists or GABA receptor agonists, which affect the function of the entire nervous system [32].

Abbreviations

bHLH	basic helix-loop-helix
CNS	central nervous system
DRG	dorsal root ganglia
GABA	gamma-aminobutyric acid
NMDA	*N*-methyl-D-aspartate
NTG	trigeminal nuclei
NTS	nucleus of the solitary tract
T-ALL	T-cell acute lymphoblastic leukemia
VGLUT	vesicular glutamate transporter

References

1 M.R. Bennett, V.J. Balcar, *Neurochem. Int.* 35, 269–280 (**1999**).
2 E.E. Bellocchio, R.J. Reimer, R.T.J. Fremeau, R.H. Edwards, *Science* 289, 957–960 (**2000**).
3 R.T.J. Fremeau, et al., *Neuron* 31, 247–260 (**2001**).
4 L. Cheng, et al., *Nat. Neurosci.* 7, 510–517 (**2004**).
5 C. Schuurmans, et al., *EMBO J.* 23, 2892–1902 (**2004**).
6 L.N. Borodinsky, et al., *Nature* 429, 523–630 (**2004**).
7 F. Fujiyama, T. Furuta, T. Kaneko, *J. Comp. Neurol.* 435, 379–387 (**2001**).
8 L. Bai, H. Xu, J.F. Collins, F.K. Ghishan, *J. Biol. Chem.* 276, 36764–36769 (**2001**).
9 E. Herzog, et al., *J. Neurosci.* 21, RC181 (**2001**).
10 S. Takamori, J.S. Rhee, C. Rosenmund, R. Jahn, *J. Neurosci.* 21, RC182 (**2001**).
11 H. Varoqui, M.K. Schafer, H. Zhu, E. Weihe, J.D. Erickson, *J. Neurosci.* 22, 142–155 (**2002**).
12 C. Gras, et al., *J. Neurosci.* 22, 5442–5451 (**2002**).
13 M.K. Schafer, H. Varoqui, N. Defamie, E. Weihe, J.D. Erickson, *J. Biol. Chem.* 277, 50734–50748 (**2002**).
14 R.T.J. Fremeau, et al., *Proc. Natl. Acad. Sci. USA* 99, 14488–14493 (**2002**).
15 T. Kaneko, F. Fujiyama, H. Hioki, *J. Comp. Neurol.* 444, 39–62 (**2002**).
16 T. Kaneko, F. Fujiyama, *Neurosci. Res.* 42, 243–250 (**2002**).
17 E. Herzog, et al., *Neuroscience* 123, 983–1002 (**2004**).
18 M.G. Erlander, N.J. Tillakaratne, S. Feldblum, N. Patel, A.J. Tobin, *Neuron* 7, 91–100 (**1991**).
19 E.R. Kandel, J.H. Schwartz, T.M. Jessell, *Principles of Neural Science*. McGraw-Hill, New York, **2000**.
20 A.D. Craig, *Annu. Rev. Neurosci.* 26, 1–30 (**2003**).
21 W.D. Snider, S.B. McMahon, *Neuron* 20, 629–32 (**1998**).
22 M.J. Zylka, F.L. Rice, D.J. Anderson, *Neuron* 45, 17–25 (**2005**).
23 A.I. Basbaum, T. Jessell, in: *Principles of Neural Science*, E.R. Kandel, J.H. Schwartz, T.M. Jessell (Eds.). McGraw-Hill, **2000**, pp. 472–491.
24 S.P. Hunt, P.W. Mantyh, *Nat. Rev. Neurosci.* 2, 83–91 (**2001**).
25 A. Patapoutian, A.M. Peier, G.M. Story, V. Viswanath, *Nat. Rev. Neurosci.* 4, 529–539 (**2003**).
26 M. Malcangio, N.G. Bowery, *Trends Pharmacol. Sci.* 17, 457–462 (**1996**).
27 E. Polgar, J.H. Fowler, M.M. McGill, A.J. Todd, *Brain Res.* 833, 71–80 (**1999**).
28 J.J. Azkue, et al., *Brain Res.* 790, 74–81 (**1998**).
29 Y. Lu, E.R. Perl, *J. Neurosci.* 23, 8752–8758 (**2003**).
30 R. Melzack, P.D. Wall, *Science* 150, 971–979 (**1965**).

31 A.H. Dickenson, *Br. J. Anaesth.* 88, 755–757 (**2002**).
32 J. Scholz, C.J. Woolf, *Nat. Neurosci.* 5, Suppl:1062–1067 (**2002**).
33 Q. Ma, L. Sommer, P. Cserjesi, D.J. Anderson, *J. Neurosci.* 17, 3644–3652 (**1997**).
34 K. Gowan, et al., *Neuron* 31, 219–232 (**2001**).
35 K.J. Lee, T.M. Jessell, *Annu. Rev. Neurosci.* 22, 261–294 (**1999**).
36 T. Caspary, K.V. Anderson, *Nat. Rev. Neurosci.* 4, 289–297 (**2003**).
37 M. Goulding, G. Lanuza, T. Sapir, S. Narayan, *Curr. Opin. Neurobiol.* 12, 508–515 (**2002**).
38 A.W. Helms, J.E. Johnson, *Curr. Opin. Neurobiol.* 13, 42–49 (**2003**).
39 J.E. Johnson, S.J. Birren, D.J. Anderson, *Nature* 346, 858–861 (**1990**).
40 F. Guillemot, et al., *Cell* 75, 463–476 (**1993**).
41 Q. Ma, C. Kintner, D.J. Anderson, *Cell* 87, 43–52 (**1996**).
42 L. Sommer, Q. Ma, D.J. Anderson, *Mol. Cell. Neurosci.* 8, 221–241 (**1996**).
43 G. Gradwohl, C. Fode, F. Guillemot, *Dev. Biol.* 180, 227–241 (**1996**).
44 Q. Ma, Z.F. Chen, I.B. Barrantes, J.L. de la Pompa, D.J. Anderson, *Neuron* 20, 469–482 (**1998**).
45 C. Fode, et al., *Neuron* 20, 483–494 (**1998**).
46 N. Ben-Arie, et al., *Hum. Mol. Genet.* 5, 1207–1216 (**1996**).
47 N. Ben-Arie, et al., *Nature* 390, 169–172 (**1997**).
48 N. Ben-Arie, et al., *Development* 127, 1039–1048. (**2000**).
49 N.A. Bermingham, et al., *Neuron* 30, 411–422. (**2001**).
50 J.E. Lee, *Curr. Opin. Neurobiol.* 7, 13–20 (**1997**).
51 Y. Qian, S. Shirasawa, C.L. Chen, L. Cheng, Q. Ma, *Genes Dev.* 16, 1220–1233 (**2002**).
52 M.K. Gross, M. Dottori, M. Goulding, *Neuron* 34, 535–549 (**2002**).
53 T. Muller, et al., *Neuron* 34, 551–562 (**2002**).
54 M.A. Kennedy, et al., *Proc. Natl. Acad. Sci. USA* 88, 8900–8904 (**1991**).
55 M. Lu, Z.Y. Gong, W.F. Shen, A.D. Ho, *EMBO J.* 10, 2905–2910 (**1991**).
56 T.N. Dear, I. Sanchez-Garcia, T.H. Rabbitts, *Proc. Natl. Acad. Sci. USA* 90, 4431–4435 (**1993**).
57 T.N. Dear, et al., *Development* 121, 2909–2915 (**1995**).
58 I.D. Dube, et al., *Blood* 78, 2996–3003 (**1991**).
59 K. Raju, et al., *Mech. Dev.* 44, 51–64 (**1993**).
60 M. Hatano, C.W. Roberts, M. Minden, W.M. Crist, S.J. Korsmeyer, *Science* 253, 79–82 (**1991**).
61 M. Hatano, et al., *J. Clin. Invest.* 100, 795–801 (**1997**).
62 C.W. Roberts, J.R. Shutter, S.J. Korsmeyer, *Nature* 368, 747–749 (**1994**).
63 C.W. Roberts, A.M. Sonde, A. Lumsden, S.J. Korsmeyer, *Am. J. Pathol.* 146, 1089–1101 (**1995**).
64 S. Shirasawa, et al., *Nat. Med* 3, 646–650 (**1997**).
65 S. Shirasawa, et al., *Nat. Genet.* 24, 287–290. (**2000**).
66 C. Logan, R.J.T. Wingate, I.J. McKay, A. Lumsden, *J. Neurosci.* 18, 5389–5402 (**1998**).
67 S.J. Tang, et al., *Development* 125, 1877–1887 (**1998**).
68 M. Hatano, et al., *Anat. Embryol.* (Berl). 195, 419–425 (**1997**).
69 S.H. Cheng, T.W. Mak, *Dev. Growth Differ.* 35, 655–663 (**1993**).
70 T.N. Dear, T.H. Rabbitts, *Gene* 141, 225–229 (**1994**).
71 T. Kojima, T. Tsuji, K. Saigo, *Dev. Biol.* 279, 434–445 (**2005**).
72 N. Masson, W.K. Greene, T.H. Rabbitts, *Mol. Cell. Biol.* 18, 3502–3508 (**1998**).
73 R.L. Brake, U.R. Kees, P.M. Watt, *Oncogene* 17, 1787–1795 (**1998**).
74 B.M. Owens, et al., *Blood* 101, 4966–4974 (**2003**).
75 Y. Qian, et al., *Genes Dev.* 15, 2533–2545 (**2001**).
76 C. Logan, B.V. Millar, C. Rouleau, *J. Comp. Neurol.* 448, 138–149 (**2002**).
77 W.W. Blessing, *The Lower Brainstem and Body Homeostasis*. Oxford University Press, New York, **1997**.

78 A. Pattyn, X. Morin, H. Cremer, C. Goridis, J.-F. Brunet, *Development* 124, 4065–4075 (**1997**).
79 T. Saito, L. Lo, D.J. Anderson, K. Mikoshiba, *Dev. Biol.* 180, 143–155 (**1996**).
80 J.H. Lin, et al., *Cell* 95, 393–407 (**1998**).
81 D.S. Bredt, R.A. Nicoll, *Neuron* 40, 361–379 (**2003**).
82 R.C. Kerr, D.J. Maxwell, A.J. Todd, *Eur. J. Neurosci.* 10, 324–333 (**1998**).
83 S.D. Watt, X. Gu, R.D. Smith, N.C. Spitzer, *Mol. Cell. Neurosci.* 16, 376–387 (**2000**).
84 Z. Jia, et al., *Neuron* 17, 945–956 (**1996**).
85 S.M. Maricich, K. Herrup, *J. Neurobiol.* 41, 281–294 (**1999**).
86 C. Fode, et al., *Genes Dev.* 14, 67–80 (**2000**).
87 C. Schuurmans, F. Guillemot, *Curr. Opin. Neurobiol.* 12, 26–34 (**2002**).
88 S.A. Anderson, D.D. Eisenstat, L. Shi, J.L. Rubenstein, *Science* 278, 474–476 (**1997**).
89 N. Flames, O. Marin, *Neuron* 46, 377–381 (**2005**).
90 O. Marin, J.L. Rubenstein, *Annu. Rev. Neurosci.* 26, 441–483 (**2003**).
91 E.J. Furshpan, P.R. MacLeish, P.H. O'Lague, D.D. Potter, *Proc. Natl. Acad. Sci. USA* 73, 4225–4229 (**1976**).
92 D.D. Potter, S.G. Matsumoto, S.C. Landis, D.W. Sah, E.J. Furshpan, *Prog. Brain Res.* 68, 103–120 (**1986**).
93 N.J. Francis, S.C. Landis, *Annu. Rev. Neurosci.* 22, 541–566 (**1999**).
94 N.C. Spitzer, C.M. Root, L.N. Borodinsky, *Trends Neurosci.* 27, 415–421 (**2004**).
95 C. Schwarzer, G. Sperk, *Neuroscience* 69, 705–709 (**1995**).
96 R.S. Sloviter. et al., *J. Comp. Neurol.* 374, 593–618 (**1996**).
97 M. Ramirez, R. Gutierrez, *Brain Res.* 917, 139–146 (**2001**).
98 R. Gutierrez, et al., *J. Neurosci.* 23, 5594–5598 (**2003**).
99 D.C. Lie, H. Song, S.A. Colamarino, G.L. Ming, F.H. Gage, *Annu. Rev. Pharmacol. Toxicol.* 44, 339–421 (**2004**).
100 A.E. Schaffner, T. Behar, S. Nadi, V. Smallwood, J.L. Barker, *Brain Res. Dev. Brain Res.* 72, 265–276 (**1993**).
101 R. Somogyi, X. Wen, W. Ma, J.L. Barker, *J. Neurosci.* 15, 2575–2591 (**1995**).
102 P.E. Phelps, A. Alijani, T.S. Tran, *J. Comp. Neurol.* 409, 285–298 (**1999**).
103 C. Goridis, H. Rohrer, *Nat. Rev. Neurosci.* 3, 531–541 (**2002**).

8
Transcriptional Control of the Development of Central Serotonergic Neurons

Zhou-Feng Chen and Yu-Qiang Ding

Abstract

The central serotonergic neurons [5–hydroxytryptamine (5–HT) neurons] are several clusters of neurons located bilaterally along the midline and in the adjacent reticular formation of the brainstem. These neurons project widely to almost every part of the brain and spinal cord, and thereby modulate a variety of developmental processes and animal behaviors. A deregulation of the 5–HT level in the central nervous system might contribute to numerous psychiatric disorders, including fear, aggression, depression, and anxiety, and to pain modulation. The recent identification of 5–HT-specific transcription factors and the genetic manipulation of these factors in mice have begun to unveil molecular mechanisms underlying the specification, differentiation, survival, and maintenance of the central 5–HT neurons. This chapter summarizes some of recent advances about transcriptional control of 5–HT neuron development, and highlights the results from gene-targeting experiments in mice. The discussion concentrates on two classes of transcription factors. One class, represented by *Nkx2.2* and *Mash1*, is expressed in 5–HT progenitor cells residing within the ventricular zone. The second class, represented by *Lmx1b* and *Pet1*, is expressed in 5–HT postmitotic neurons only. The comparison of different mutant lines that lack an individual transcription factor has permitted us to speculate about their relationship during the development of 5–HT neurons. Finally, behavioral study of knockout mice with impaired 5–HT systems have shown the involvement of some transcription factors in the etiology of psychiatric abnormalities.

8.1
Introduction

The 5–hydroxytryptamine (5–HT) system in the central nervous system (CNS) comprises several groups of morphologically distinct neurons that are mainly distributed in the midline region of the brainstem along the rostrocaudal neuroaxis. They are clustered in the raphe nuclei of the brainstem, and some are dispersed in the adjacent reticular formation (Tork, 1990; Jacobs and Azmitia, 1992). One widely used nomenclature for central 5–HT neurons was originally proposed by Dahlstrom and Fuxe (1964), who classified them into nine clusters (B1–B9) on the basis of their anatomical architecture and location (Fig. 8.1). Caudal 5–HT neurons (B1–B4) are located in the medulla oblongate, whereas rostral 5–HT (B5–B9) neurons reside in the pons and the caudal-most part of the midbrain. Another commonly used nomenclature for 5–HT neurons is based on their anatomical architecture in the raphe nuclei of the brainstem (Tork, 1990; Jacobs and Azmitia, 1992) (Table 8.1). The 5–HT neurons probably possess the most complex projection network in the CNS: three major descending 5–HT efferents emanating from B1–B4 neurons project to the spinal cord, whereas B5–B9 neurons send their efferents through five major ascending routes to almost every region of the CNS (Tork, 1990; Jacobs and Azmitia, 1992). These neurons release 5–HT that interacts with at least 14 5–HT receptors, most of which are G-coupled receptors to initiate a downstream second message signal transduction pathway (Martin et al., 1998; Pauwels, 2000).

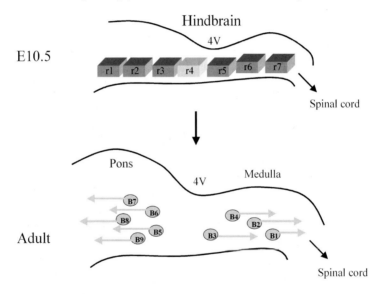

Fig. 8.1 Schematic diagram of the organization of 5–HT neurons in the brainstem during early mouse embryonic development and at adult stage. The developing hindbrain comprises a series of repeated segments called rhombomeres. Beginning at E10.5, 5–HT neurons are generated from r1–r3 and subsequently from r5–r7 (green). No 5–HT neurons are generated from r4 (pink). In the adult brainstem, nine clusters (B1–B9) of 5–HT neurons are classified in raphe nuclei according to their distinct locations and structures. Caudal 5–HT neurons (B1–B4) project to the brainstem and the spinal cord, whereas rostral 5–HT neurons (B5–B9) project to all parts of the brain. r1 may give rise to the dorsal raphe nucleus (B7/B6), whereas r2–r3 may generate the rest of rostral 5–HT neurons (B5, B8, and B9). By contrast, caudal 5–HT neurons (B1–B4) probably originate in r5–r7. The ontogenic relationship between rhombomeres, except r1, and 5–HT neurons remains unclear. Abbreviations: r = rhombomeres; 4V = fourth ventricle.

Table 8.1 Classification of 5–HT neurons and ontogenic relationship between B nuclei and rhombomeres.

Location	B nuclei	Raphe nuclei	Rhombomere
Caudal	B1	Nucleus raphes pallidus	r5–r7
Caudal	B2	Nucleus raphes obscurus	r5–r7
Caudal	B3	Nucleus raphes magnus Rostral ventrolateral medulla Lateral paragigantocellular reticular nucleus	r5–r7
Rostral	B4	Central gray of the medulla oblongata	r2–r3
Rostral	B5	Pontine median raphe nucleus	r2–r3
Rostral	B6	Pontine dorsal raphe nucleus	r1
Rostral	B7	Midbrain dorsal raphe nucleus	r1
Rostral	B8	Midbrain median raphe nucleus Caudal linear nuclei	r2–r3?
Rostral	B9	Medial lemniscus	r2–r3?

Through extensive projections, 5–HT exerts modulatory function in the neurotransmission of many types of neurons, and its dysregulation has been implicated in numerous psychiatric disorders, such as anxiety, aggression, and depression, and in pain modulation (Nelson and Chiavegatto, 2001; Millan, 2002).

During early embryonic development, the distinction among different groups of 5–HT neurons is much less clear. The 5–HT neurons are among the earliest-born neurons that are generated in the CNS. In mice, 5–HT neurons are generated between E10.5 and E12.5 as a stream of cells exits the cell cycle and migrates to settle down either near the midline of the ventral hindbrain or to the more lateral region that constitutes the reticular formation (Ding et al., 2003; Pattyn et al., 2004). The axons of 5–HT neurons do not cross the midline, and fuse until the late stages of embryonic development and early postnatal stages. Like the development of other neurons in the neural tube, the development of 5–HT neurons follows a well-defined spatiotemporal sequence. The generation of rostral 5–HT neurons precedes caudal 5–HT neurons along the brainstem axis. Moreover, before the generation of 5–HT neurons, visceral motor neurons (vMNs) are derived from the same domain of the ventral VZ; the vMNs are derived between E9.5 and E10.5 (Pattyn et al., 2003). A few 5–HT cells appear in the rostral-most part of the pons between E10.5 and E10.75. By E11.5, an increasing number of 5–HT neurons appear in the more caudal part of the pons and the rostral part of the medulla. At E12.5, almost all of distinct classes of 5–HT neurons are present in the brainstem (Ding et al., 2003). Anatomically, the embryonic hindbrain is composed of a series of segments called rhombomeres (r) (Lumsden and Krumlauf, 1996). Most 5–HT neurons derive from r2–r3 and r5–r7 (Pattyn et al., 2003). Rhombomere 1 gives rise to the dorsal raphe nuclei (presumptive B7 and B6 neurons), whereas r4, from which no 5–HT neurons are generated, serves as a spacer dividing rostral from caudal 5–HT neurons (Pattyn et al., 2003). The relationship among B clusters, 5–HT nuclei, and embryonic rhombomeres is listed in Table 8.1.

The generation of diverse ventral cell types, including somatic motor neurons and vMNs along the neural axis, requires floor plate- and notochord-derived signals (Tanabe and Jessell, 1996; Briscoe et al., 1999; Jessell, 2000). The induction and specification of 5–HT neurons may depend on Sonic hedeghog (Shh), which acts through a number of transcription factors (Hynes and Rosenthal, 1999; Goridis and Rohrer, 2002). Because Shh expression is not limited to the hindbrain, other region-specific secreted factors may also participate in the specification of 5–HT neurons. The transcription factor Nkx2.2 can specify the ventral cell types in the developing spinal cord in response to Shh signaling; thus Nkx2.2 may have an analogous role in specifying 5–HT neurons in the hindbrain (Ericson et al., 1997; Briscoe et al., 1999; Gaspar et al., 2003). However, the manner in which these signaling molecules are converted into a combinatorial transcriptional code that confers the identity of 5–HT neurons is still unknown.

8.2
Transcription Factors in the Development of 5–HT Neurons

Over the past few years, an increasing number of transcription factors have been identified in 5–HT neurons and their precursors. To date, at least six transcription factors have been identified in the development of 5–HT neurons (Hendricks et al., 1999, 2003; Cheng et al., 2003; Ding et al., 2003; Craven et al., 2004; Pattyn et al., 2003, 2004) (Table 8.2). These transcription factors can be divided into several classes according to their spatiotemporal expression profile. Nkx2.2 and Mash1 are expressed in 5–HT progenitor cells, and their expression stops as soon as 5–HT progenitors exit the cell cycle and become postmitotic. By contrast, Gata2 is expressed in both the VZ and postmitotic cells (Nardelli et al., 1999). Finally, Gata3, Pet1, and Lmx1b are mainly expressed in postmitotic 5–HT neurons. Among the six transcription factors, only Pet1 is exclusively expressed in 5–HT neurons. Researchers have shown that the other five transcription factors are important developmental regulators of the specification of distinct neuronal types in the CNS (Guillemot et al., 1993; Ericson et al., 1997; Briscoe et al., 1999; Nardelli et al., 1999; Ding et al., 2004). These results strengthen the idea that similar mechanisms used by these transcription factors in other neuronal types may be adopted in the development of 5H-T neurons, or vice versa.

8.3
Transcription Factors Expressed in 5–HT Progenitor Cells

8.3.1
Nkx2.2

Nkx2.2 is a homeodomain-containing transcription factor that plays an essential role in the specification of the ventral cell type in response to Shh signals in the spinal cord and hindbrain (Ericson et al., 1997; Briscoe et al., 1999). Nkx2.2 is the

8.3 Transcription Factors Expressed in 5–HT Progenitor Cells

Table 8.2 Summary of the phenotype of knockout mice lacking individual transcription factor in 5–HT neurons.

Genes	Expression	5–HT defects	KO lethality	5–HT-specific TFs	Reference(s)
Nkx2.2	VZ	100% except in r1	P0	Lost except in r1	Ding et al. (2003); Pattyn et al. (2003)
Mash1	VZ	Almost 100%	P0	All lost	Pattyn et al. (2004)
Gata2	VZ, postmitotic	100%	E10–12.5	All lost	Craven et al. (2004)
Lmx1b	Postmitotic	100%	P0	All lost	Cheng et al. (2003); Ding et al. (2003)
Pet1	Postmitotic	70%	Mostly viable	Unknown	Hendricks et al. (2003)
Gata3	Postmitotic	Mostly in the caudal	E11.5–13.5	Unknown	Pattyn et al. (2004); van Doorninck et al. (1999)

KO = knockout; TF = transcription factor; VZ = ventricular zone.

earliest transcription factor that has been shown to be required for the specification of 5–HT neurons (Briscoe et al., 1999; Cheng et al., 2003; Ding et al., 2003; Pattyn et al., 2003). In the developing hindbrain, *Nkx2.2* is restricted to the VZ (Pattyn et al., 2003). Between E9.5 and E10.5, it is coexpressed with two other transcription factors, *Nkx2.9* and *Phox2b*, that are important for the generation of vMNs. The extinction of *Phox2b* and *Nkx2.9* expression in the ventral-most domain of the hindbrain is accompanied by the cessation of vMN generation, and is a prerequisite for the initiation of the generation of 5–HT neurons from the same domain (Pattyn et al., 2003). In the absence of *Phox2b*, production of 5–HT neurons is premature, in addition to the failure of vMN generation. Thus, *Nkx2.2* may serve as one of the intrinsic factors upstream of *Phox2b* to instruct the switch of vMN progenitor cells into 5–HT neuronal progenitor cells.

In *Nkx2.2* mutants, *Nkx2.9* expression expands to the ventral-most part, and 5–HT neurons fail to be generated to a large degree. These findings are consistent with the idea that *Nkx2.2* is necessary for the specification of 5–HT neurons (Ding et al., 2003; Pattyn et al., 2003). *Nkx2.9* is unable to rescue the deficiency of *Nkx2.2* with regard to the generation of 5–HT neurons; this finding suggests that *Nkx2.2* may be endowed with a 5–HT-specific property. Intriguingly, not all 5–HT neurons depend on *Nkx2.2* for their specification. *Nkx2.2* is dispensable in r1 because 5–HT neurons derived from r1 are not affected by the *Nkx2.2* mutation (Ding et al., 2003; Pattyn et al., 2003). These data suggest that a discrete mechanism underlies the specification of 5–HT neurons in r1. Perhaps other unidentified r1–specific factors are involved.

In the ventral hindbrain, *Nkx2.2* possesses dual function with regard to the generation of motor neurons and 5–HT neurons. *Nkx2.2* is crucial for the generation of vMNs by initiating *Phox2b* expression that in turn represses 5–HT fate before the onset of the neurogenesis of 5–HT neurons in the ventral-most part of the hindbrain. In the absence of *Phox2b*, *Nkx2.2* adopts a default pathway to promote 5–HT fate.

8.3.2
Mash1

Mash1 is a mouse homologue of the *Drosophila* proneural genes *achaete-scute*, and a basic helix-loop-helix (bHLH) transcription factor that normally functions as either a homodimer or heterodimer (Bertrand et al., 2002). *Mash1* is one of the earliest-identified transcription factors involved in the determination of neuronal fate in vertebrates (Guillemot et al., 1993). Moreover, it has emerged as a key fate determinant for many types of neurons in the nervous system (e.g., noradrenergic neurons, 5–HT neurons, and telencephalic neurons) (Blaugrund et al., 1996; Goridis and Brunet, 1999; Fode et al., 2000; Parras et al., 2004). The function of *Mash1* in the development of central 5–HT precursor cells, however, has been explored only recently (Pattyn et al., 2004). During development, *Mash1* expression is restricted to the VZ and is coexpressed with *Nkx2.2* throughout the generation period of vMNs and 5–HT neurons (Pattyn et al., 2004). In the developing hindbrain, *Mash1* is the only known proneural bHLH transcription factor expressed in the domain of 5–HT progenitor cells.

The requirement for *Mash1* in the development of 5–HT neurons has been shown at two levels. First, in *Mash1* knockout mice, vMNs are generated normally, whereas all postmitotic transcription factors (*Lmx1b*, *Pet1*, *Gata2*, and *Gata3*) fail to be detected in the developing brainstem (Pattyn et al., 2004). Even though it is dispensable for the generation of vMNs, *Mash1* is essential for the differentiation of 5–HT progenitor cells. Second, the examination of two components of the Notch signaling pathway, *Dll* and *Hes5*, in *Mash1* mutants has revealed the loss of these two genes in the domain of 5–HT neuronal progenitor cells (Pattyn et al., 2004). Thus, *Mash1* is required for mediating the Notch signaling pathway that leads to 5–HT neurogenesis, and the absence of *Mash1* results in defective neurogenesis for 5–HT neurons.

When the *Mash1* coding sequence is replaced by another proneural gene, *Ngn2*, only about 15% of 5–HT neurons are found as compared with the wild-type control mice, despite a rescued neurogenesis as indicated by normal expression of *Dll* and *Hes5*. This finding indicates a partial block of the differentiation of 5–HT neurons (Parras et al., 2002; Pattyn et al., 2004). The observation that the function of *Mash1* in the specification of 5–HT neurons cannot be completely substituted by other bHLH factors, despite their shared similar bHLH domain, suggests that *Mash1* possesses 5–HT neuron-specific characteristics. Because the lack of *Mash1* does not lead to an alteration of *Nkx2.2* that is unable to activate *Mash1*, *Mash1* is unlikely to exert its function by regulating *Nkx2.2*. Instead, *Mash1* may act in parallel with *Nkx2.2* to specify 5–HT neurons. Other cofactors may also be required because electroporation of DNA plasmids that express either gene or both fails to induce 5–HT neurons (Pattyn et al., 2004). Therefore, although both genes are necessary, neither *Nkx2.2* nor *Mash1* is sufficient for inducing the generation of 5–HT neurons, even in the absence of *Phox2b* (Pattyn et al., 2004).

8.4 Transcription Factors Expressed in the Ventricular Zone and Postmitotic 5–HT Neurons

8.4.1 Gata2 and Gata3

Gata2 and *Gata3* are members of the GATA family that contain zinc fingers which bind to the consensus core (A/T)GATA(A/G) (Patient and McGhee, 2002). Among six GATA family members identified in vertebrates, *Gata2* and *Gata3* are important for the development of 5–HT neurons (van Doorninck et al., 1999; Craven et al., 2004; Pattyn et al., 2004). In the developing hindbrain, the expression of *Gata2* precedes that of *Gata3* (Nardelli et al., 1999). The onset of *Gata2* expression occurs at E9.0, most notably in r4 and transiently in r2. By E10.5, *Gata2* expression has expanded to all rhombomeres and is detected in progenitor cells in the VZ and in the postmitotic cells. By contrast, *Gata3* expression is weak in the VZ and mainly occupies the region outside the ventral VZ (Nardelli et al., 1999).

Gata2–null mice exhibit severe anemia and die between E10 and E11 (Tsai et al., 1994). One team has analyzed the neural development in the developing hindbrain of *Gata2* mutants and revealed several defects in neurogenesis (Nardelli et al., 1999). The team, however, did not examine the development of 5–HT neurons (Nardelli et al., 1999). The early lethality of *Gata2* mutants precluded detailed analysis of 5–HT neuronal development. For this problem to be avoided, *in-vitro* explant culture of E8 ventral hindbrain was used, and the tissue was examined at the equivalent of E13.5, when all 5–HT neurons are generated. In contrast to those in the control mice, 5–HT neurons were completely missing in *Gata2* mutants, even in the presence of *Gata3* (Craven et al., 2004). Thus, *Gata2* could be a critical factor for the specification of 5–HT neurons.

In-ovo electroporation of *Gata2* into chick embryos also suggests that the role of *Gata2* in the development of different clusters of 5–HT neurons may differ (Craven et al., 2004). Overexpression of *Gata2* in r1 induced *Pet1* and *Lmx1b*, but not in r2–r3 and r5–r7. Therefore, *Gata2* is necessary and sufficient for the development of 5–HT neurons in r1, whereas in r2–r3 and r5–r7 it is only necessary but not sufficient. Because the capacity of *Gata2* to induce other 5–HT-specific transcription factors is restricted to r1 (Craven et al., 2004), *Gata2* may be a key factor that helps to confer r1–specific 5–HT phenotype in addition to its generic function in the specification of 5–HT neurons.

$Gata3^+$ cells are almost completely co-localized with 5–HT in the caudal raphe nuclei, whereas in the rostral part of hindbrain, only 46% of $Gata3^+$ cells are overlapped with 5–HT staining (van Doorninck et al., 1999). *Gata3* mutants die between E9.5 and E12.5 and exhibit multiple defects, including brain and spinal cord abnormalities, abnormal liver hematopoiesis, and bleeding (Pandolfi et al., 1995). Two teams have rescued the early lethal phenotype to analyze the development of 5–HT neurons in the absence of Gata3. The analysis of chimeric mice composed of Gata3$^{-/-}$/Gata3$^{+/+}$ wild-type cells shows that the development of 5–HT neurons is compromised in the caudal, but not in the rostral, raphe nuclei of the hindbrain (van Doorninck et al., 1999). In line with this chimeric study, rescued Gata3 mutant embryos by noradrenergic agonists showed an 80% loss of 5–HT$^+$ neurons in the most caudal part of the hindbrain, whereas a less severe loss of 5–HT in the more rostral part of the hindbrain was found (Pattyn et al., 2004).

In Gata3 mutants rescued by a noradrenergic agonist, the expression of Pet1, Lmx1b, and Gata2 was largely normal at E13.5 (Pattyn et al., 2004). However, a partial loss of Gata3 was observed in the caudal part of the hindbrain of Lmx1b mutants (Ding et al., 2003). Although these data suggest that Lmx1b and Pet1 may lie either upstream of or parallel to Gata3, the possibility cannot be excluded that the loss of Gata3 in Lmx1b mutants could also be due to a gradual loss of 5–HT neuronal identity in general, rather than to the loss of a regulation by Lmx1b.

The reports of Gata3 expression in Gata2 mutants are conflicting. In one study, no Gata3 expression was detected in the hindbrain of Gata2 mutants (Nardelli et al., 1999), whereas in the others, Gata3 expression was not affected at a similar stage (Pata et al., 1999; Craven et al., 2004). By contrast, Gata2 expression remains unaltered in Gata3 mutants (Craven et al., 2004). Gata2 can activate Gata3 when Gata2 is overexpressed in the chick neural tube, but not vice versa. Therefore, these results suggest that Gata3 functions either downstream or independent of Gata2 and Lmx1b. Despite these studies, the epistatic relationship among these genes is still not well understood, and further work is required to resolve some of the discrepancies.

8.5
Transcription Factors Expressed in Postmitotic 5–HT Neurons

8.5.1
Lmx1b

Lmx1b is a member of the LIM (Lin–11 from *Caenorhabditis elegans*, Isl–1 from the rat, and Mec–3 from *C. elegans*) homeodomain (LIM-HD) transcription factor family that has been implicated in many aspects of developmental and biological processes (Curtiss and Heilig, 1998; Dawid et al., 1998; Bach, 2000). The LIM-HD motif comprises two zinc fingers that mediate protein-protein interaction and can bind the same domain or different class of protein; thus, diverse interactions are allowed among proteins (Bach, 2000; Matthews and Visvader, 2003). Many LIM-HD transcription factors are particularly important in the regulation of the specification and differentiation of the nervous system (Curtiss and Heilig, 1998).

Lmx1b is a mouse orthologue of the chicken Lmx1 that is required for the limb bud development (Chen et al., 1998). Its chromosome location syntenically matches a dominantly inherited human disease called nail-patella syndrome (NPS) (Chen et al., 1998; Dreyer et al., 1998). In the CNS, *Lmx1b* is widely expressed in a variety of neuronal types, including the dorsal spinal cord, dopaminergic neurons, and the eye (Kania et al., 2000; Pressman et al., 2000; Smidt et al., 2000; Asbreuk et al., 2002; Ding et al., 2004). *Lmx1b* is one of the earliest known transcription factors to be expressed in postmitotic 5–HT neurons (Cheng et al., 2003; Ding et al., 2003).

Bromodeoxyuridine (BrdU) tracing experiments show that most of *Lmx1b*-expressing cells are not stained for BrdU. Although most *Lmx1b*$^+$ cells are postmitotic, a few *Lmx1b*$^+$ cells are also stained for BrdU and co-localized with *Nkx2.2* (Ding et al., 2003). This observation raises the possibility that *Lmx1b* may serve as a "brake" signal to instruct 5–HT precursor cells to cease their proliferation. At E14.5, all 5–HT neurons are also stained for *Lmx1b* in the ventral hindbrain (Fig. 8.2). The domain of *Lmx1b* expression in the floor plate is dispensable for the development of 5–HT neurons (Ding et al., 2003). The expression of *Lmx1b* is persistent not only in all postmitotic 5–HT neurons through embryonic development but also in the adult brain (Z.-F. Chen, unpublished results). This suggests that *Lmx1b* may have multiple functions in different stages of 5–HT neuronal development.

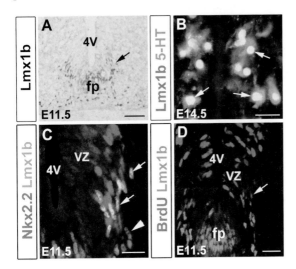

Fig. 8.2 Double staining of *Lmx1b* and 5–HT in embryonic mouse hindbrain. (A) *Lmx1b* staining in r5 detected with immunocytochemical staining. (B) *Lmx1b* (red) and 5–HT (green) double staining. Arrows indicate double-stained cells. (C) *Nkx2.2* (red) and *Lmx1b* (green) double staining. *Nkx2.2* is mainly detected in the VZ, whereas *Lmx1b* is found in postmitotic cells. Arrows indicate double-stained cells, whereas arrowhead indicates *Lmx1b*-expressing cells only. (D) BrdU (red) and *Lmx1b* (green) double staining, indicating postmitotic expression of *Lmx1b* (arrow). Scale bars: 100 μm (A); 25 μm (B); and 50 μm (C, D). Abbreviations: fp = floor plate; 4V = fourth ventricle; VZ = ventricular zone. (This figure also appears with the color plates.)

In *Lmx1b*-null mutants, most of the 5-HT neuron-specific markers are lost from the beginning (Cheng et al., 2003; Ding et al., 2003) (Fig. 8.3). Thus, *Lmx1b* could play an essential role in the specification of 5-HT neurons. The transient expression of *Pet1* in E11.5 *Lmx1b* mutants suggests that the fate of 5-HT neurons may be partially specified (Cheng et al., 2003). The overexpression of *Lmx1b* in the ventral hindbrain of embryos by *in-utero* electroporation, however, fails to induce more 5-HT neurons (Y.-Q. Ding et al., unpublished data). Thus, the early function of *Lmx1b* appears to be necessary but not sufficient for the specification of 5-HT neurons. In contrast, co-electroporation of *Lmx1b*, *Pet1*, and *Nkx2.2* together in the chick neural tube induces ectopic 5-HT neurons (Cheng et al., 2003). Although mechanisms involving 5-HT neural development between chick and mouse may vary to a certain degree (Craven et al., 2004), the previous data suggest that multiple transcription factors may work in a coordinated fashion to specify 5-HT neurons. In *Lmx1b* mutants, 5-HT neurons are eventually lost, and this loss suggests that *Lmx1b* may also be a survival factor for more differentiated 5-HT neurons.

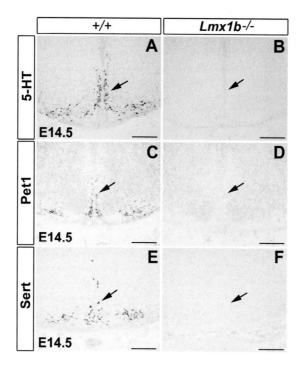

Fig. 8.3 Loss of 5-HT-specific markers in *Lmx1b* mutant embryos. (A, B) 5-HT staining in wild-type (A) and *Lmx1b* mutant embryos (B). (C, D) *Pet1* expression in wild-type (C) and mutant embryos (D). (E, F) *Sert* expression in wild-type (E) and *Lmx1b* mutant embryos (F). Scale bars: 100 μm.

Lmx1b mutants die at birth as a result of multiple defects, including kidney defects (Chen et al., 1998). In order to overcome perinatal lethality problem, we have recently deleted *Lmx1b* only in 5–HT neurons by using a conditional knockout approach (Z.-F. Chen, unpublished results). The preliminary results of these studies indicate that some 5–HT neurons are initially generated and progressively lost in *Lmx1b*-conditional knockout mice. These findings indicate a role for *Lmx1b* in maintaining the survival of 5–HT neurons. However, it has not been determined whether the loss of 5–HT neurons in the absence of *Lmx1b* could be attributed to abnormal apoptosis.

In addition to binding to proteins that interact with DNA, the LIM-HD is capable of binding to proteins that regulate the cytoskeleton, thereby mediating the morphogenesis of neurons (Bach, 2000). In the dorsal spinal cord, *Lmx1b* is critical for the migration of the dorsal horn neurons (Ding et al., 2004). *Lmx1b* most likely has a role in the morphogenesis of 5–HT neurons during development. The persistent expression of *Lmx1b* in fully differentiated 5–HT neurons suggests that *Lmx1b* may be required for maintaining the mature phenotype or synaptic activity of 5–HT neurons in the CNS. A temporal deletion of *Lmx1b* in the raphe nuclei at a postnatal stage may help to answer this question.

The downstream targets of *Lmx1b* remain unknown. In *Lmx1b* mutants, several 5–HT-specific differentiation markers such as the serotonin transporter (SERT) fail to be expressed. However, this lack of expression is more likely due to a general blockade of the differentiation program of 5–HT neurons than to a loss of direct regulation of these terminal differentiation genes by *Lmx1b*. How the loss of *Lmx1b* results in a disruption of a genetic program for 5–HT neuronal development remains to be elucidated. In the developing dorsal horn, *Lmx1b* orchestrates expression of multiple downstream transcription factors (Ding et al., 2004). Similarly, *Lmx1b* may have an analogous role in the development of 5–HT neurons.

8.5.2
Pet1

Pet1 (*p*heochromocytoma 12 *E*TS [E26 *t*ransformation-specific]) is a member of the ETS (*E*26 *t*ransformation-*s*pecific) family that consists of more than 40 members in a variety of organisms (Graves and Petersen, 1998; Wasylyk et al., 1993). The ETS transcription factors contain a DNA-binding domain of about 85 amino acids that bind to a core consensus GGAA/T (Graves and Petersen, 1998; Sharrocks, 2001; Oikawa and Yamada, 2003). Many ETS family members exhibit cell type-specific expression patterns and are required for the proliferation and differentiation of diverse cell types such as hematopoietic cells and vascular endothelial cells (Oikawa and Yamada, 2003). ETS factors are also involved in the determination and specification of neuronal connectivity during neural development (Arber et al., 2000). In the CNS, ETS proteins are components of the signal transduction pathway and are able to activate downstream effector genes (Koo and Pfaff, 2002). For example, two ETS family members, PEA3 and ER81, are expressed in proprioceptive neurons in the dorsal root ganglia and their central targets, and are important for the establishment of neuronal connectivity between primary afferents and their central targets in the spinal cord (Arber et al., 2000; Livet et al., 2002).

Pet1 was originally cloned from the adrenal chromaffin-derived phaeochromocytoma (PC12) cell line and has been found in rat, mouse, and human brains (the *fev* gene) (Fyodorov et al., 1998; Pfaar et al., 2002; Maurer et al., 2004). A *Pet1*–specific binding sequence has been identified in upstream regions of several 5–HT specific genes that are required for 5–HT synthesis, binding, or transportation: 5–HT1a receptor, serotonin transporter (Sert), tryptophan hydroxylase (TPH) gene, aromatic L-amino acid decarboxylase gene (AADC) (Hendricks et al., 1999). Co-transfection of the reporter gene that contains several *Pet1* binding sites with *Pet1*–binding domain showed a *Pet1*–dependent transcriptional activity by the reporter gene (Hendricks et al., 1999). Thus, *Pet1* appears to modulate the transcription activities of several effector genes that define the differentiated 5–HT neuronal phenotype, probably through a synergic interaction with other cofactors, which appears to be a common mechanism for many ETS factors (Oikawa and Yamada, 2003). Mechanistically, *Pet1* most likely functions as a transcriptional activator, even though its transactivation activity of downstream targets is relatively weak (Hendricks et al., 1999). On the other hand, the human homologue of *Pet1*, *Fev*, has been shown to act as a transcriptional repressor via its alanine-rich carboxy-terminal domain (Maurer et al., 2003). Whether *Pet1* contains similar repressor activity, however, is unclear.

Unlike other transcription factors, *Pet1* is found exclusively in postmitotic 5–HT neurons. Such a highly restricted expression pattern in the CNS for a transcription factor is not common, because the repeated use of the same transcription factor in distinct types of cells is a recurring theme in the nervous system. Recently, a 1.8–kb genomic fragment immediately upstream of the *Pet1* coding region has been shown to be able to direct the expression of the *LacZ* gene, which recapitulates expression of *Pet1* (Scott and Deneris, 2005). The 1.8–kb *lacZ* transgene's failure to show its activity in some of 5–HT neurons in *Pet1*–null background indicates that *Pet1* expression itself is required for maintaining the full activity of this enhancer in 5–HT neurons (Scott and Deneris, 2005). Nonetheless, the characterization of a 5–HT-specific 1.8–kb enhancer paves the way for the identification of transcription factors acting upstream of *Pet1*. Determining whether *Lmx1b*, *Gata2*, and *Gata3* interact with this element will be interesting. The unique expression pattern of *Pet1* is reminiscent of that of the homeodomain-containing transcription factor *Pitx3*, which is exclusively found in the midbrain dopaminergic neurons where *Lmx1b* expression is also found (Smidt et al., 2000; Burbach et al., 2003). Also, putative *Pitx3* binding sites are found in the promoter of tyrosine hydroxylase (TH), a enzyme which is critical for dopamine biosynthesis, and *Pitx3* can activate the TH gene through a high-affinity binding site (Lebel et al., 2001). Both *Pet1* and *Pitx3* appear to act in late steps of neuronal differentiation of two different neurotransmitter neurons. Such striking similarities between *Pitx3* and *Pet1* suggest that, in addition to shared transcription factors, tissue-specific factors might also have a key role in conferring the neurotransmitter-specific identity. *Pet1* might have been uniquely recruited by 5–HT neurons to promote 5–HT-specific characteristics during evolution.

In *Pet1* knockout mice, about 70% of 5–HT cells fail to differentiate (Hendricks et al., 2003). The differentiation capacity of the remaining 5–HT neurons is compromised because the neurons lack TPH, which may require *Pet1* for its activation

(Hendricks et al., 2003). In the absence of *Pet1*, the surviving 5–HT neurons may be attributed to some compensation effects contributed by unknown cofactors, which is in marked contrast with *Lmx1b* knockout mice in which all 5–HT neurons fail to differentiate (Ding et al., 2003). One notable observation in *Pet1* knockout mice is that *Lmx1b* expression appears normal up until the late stage of embryonic development (Z.-F. Chen, unpublished data). The developmental role of *Pet1* has not been examined in detail, and the step at which the development of 5–HT neurons is blocked remains unclear. *Pet1* is more likely to act at late steps of differentiation of 5–HT neurons.

Strikingly, *Pet1* knockout mice survive to adulthood. Recently, we have generated *Lmx1b* conditional knockout mice using the loxP-cre strategy. *Lmx1b* conditional knockout mice virtually lack 5–HT neurons after they are born. However, despite their smaller size during the first month compared with their wild-type littermates, *Lmx1b* conditional knockout mice are all viable (Z.-F. Chen, unpublished results). Therefore, 5–HT neurons are at least dispensable for prenatal development, whereas they are required for early postnatal development. These results are probably surprising given the well-documented role of 5–HT in a variety of developmental processes (Whitaker-Azmitia et al., 1996; Azmitia, 2001). Whether a compensation mechanism might have come into play in the absence of 5–HT remains unclear.

Pet1 knockout mice exhibit aggressive behaviors. Moreover, *Pet1* knockout mice show increased anxiety-like behaviors (Hendricks et al., 2003). These two abnormal phenotypes are reminiscent of those of some 5–HT receptor knockouts (Gaspar et al., 2003; Gingrich et al., 2003). Given the multiple roles of 5–HT in numerous psychiatric disorders, these animals will also likely exhibit other behavioral deficiencies. One issue that remains to be addressed is how *Pet1* contributes to the development of abnormal behaviors. This may be an indirect effect by influencing the development of the 5–HT system, or *Pet1* may consolidate 5–HT phenotypes during adulthood directly. Regardless of the underlying mechanism, the importance of *Pet1* in the development and maintenance of mature neural morphology as well as normal behaviors has been established. This opens up an exciting possibility that an abnormal transcription regulation of the 5–HT system may also contribute to the etiology of some psychiatric diseases in humans.

8.6
The Relationship between *Lmx1b* and *Pet1*

Pet1 binds to the *cis*-regulatory elements of several 5–HT effector genes, and thus probably lies at the end of the transcriptional cascade underlying the terminal differentiation and maturation of 5–HT neurons (Hendricks et al., 1999). By contrast, potential downstream targets of *Lmx1b* are unknown. *Lmx1b* may have an analogous regulatory role to that in the dorsal horn, where it controls multiple transcription factors (Ding et al., 2004), although the possibility that *Lmx1b* might regulate some 5–HT differentiation genes directly by binding to their *cis*-regulatory elements cannot be excluded. At all levels of the hindbrain, *Lmx1b* expression precedes that of

Pet1; thus, *Lmx1b* may act one or two steps earlier than *Pet1* in the genetic cascade. However, *Lmx1b* is unnecessary for the initiation of *Pet1* expression because *Pet1* is transiently expressed in *Lmx1b* mutants at E11.5 (Cheng et al., 2003). Ectopic expression of *Lmx1b* or *Pet1* consistently fails to initiate the expression of either gene. Because *Pet1* is lost after E11.5 in the absence of *Lmx1b*, *Lmx1b* may be required for maintaining expression of *Pet1*. Alternatively, the loss of *Pet1* may not be due to a lack of a regulatory relationship between *Pet1* and *Lmx1b*. Instead, it may reflect a loss of 5–HT identity in *Lmx1b*-null neurons in general or a switch of neuronal phenotype. Nevertheless, no evidence of up-regulation of any other neuronal markers is found in *Lmx1b*-null neurons (Z.-F. Chen, unpublished results), and this lack of evidence does not support any switch of neuronal type. *Lmx1b* expression is not affected in *Pet1* mutants during early development, and thus the two genes probably do not act in parallel (Z.-F. Chen, unpublished results). Taken together, the available evidence suggests that *Pet1* and *Lmx1b* function neither in a simple linear cascade nor in parallel, as was previously proposed (Cheng et al., 2003; Ding et al., 2003). *Lmx1b* may be required to maintain *Pet1* expression, but not vice versa. Given that many LIM-HDs function through protein-protein interaction, *Lmx1b* may cooperate with other LIM-interacting factor(s) to initiate *Pet1* expression.

8.7
Conclusions

Gene-targeting approaches have begun to unravel the components of a transcriptional program that dictates the specification and differentiation of 5–HT neurons during development. Figure 8.4 illustrates a tentative 5–HT-specific transcriptional cascade, although it is far from complete. In the VZ, *Gata2* and *Mash1* are important regulators of the development of 5–HT neurons in r1–r3 and r5–r7, whereas *Nkx2.2* is essential for 5–HT progenitor cells that originate only in r2–r3 and r5–r7. In the postmitotic neurons, *Gata2* appears to act upstream of *Lmx1b*, *Pet1*, and *Gata3*. *Gata3* may cross-regulate *Gata2*. *Pet1* functions probably in the last steps of the cascade that leads to the activation of terminal differentiation program of 5–HT neurons. The maintenance of *Pet1* depends on *Lmx1b*; so does *Gata3* in the caudal part of the brainstem (Fig. 8.4).

These studies have provided a strong basis for future studies, and accumulating evidence suggests that several principles exist regarding the transcriptional control of the formation of 5–HT neurons. First, transcription factors that are expressed in both 5–HT progenitor cells and postmitotic cells constitute an integral part of a genetic program that governs the specification of 5–HT progenitor cells. Second, a transcription factor may have pleiotropic functions manifested at different stages with regard to different aspects of 5–HT neuronal development. Third, although the loss of a single transcription factor could result in a loss of all 5–HT-specific markers, only combinatorial expression of a myriad of transcription factors is both necessary and sufficient to activate the generation of 5–HT neurons. Finally, the cellular diversity and complexity of 5–HT neurons may be attributed to region-specific transcription factors.

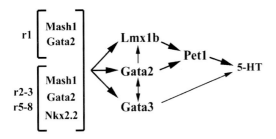

Fig. 8.4 Summary of transcriptional cascade controlling the development of 5–HT neurons. In r1, *Mash1* and *Gata2* are essential for the specification of 5–HT neurons, whereas in r2–r3 and r5–r7, *Nkx2.2* is also required. The relationship among *Mash1*, *Gata2*, and *Nkx2.2* has not yet been determined. In postmitotic progenitor cells, *Gata2* acts upstream of *Lmx1b*, *Pet1*, and *Gata3* at least in cells that originated in r1. In other rhombomeres, *Gata2* may require additional unidentified transcription factor (not shown) to activate *Lmx1b*. *Gata2* and *Gata3* are able to regulate each other. The maintenance of *Pet1* and *Gata3* expression depends on *Lmx1b*. The initiation of *Pet1* expression may require *Gata2* and other unidentified transcription factors (not shown). In addition to *Pet1*, other unidentified transcription factors may also activate some of the 5–HT terminal differentiation genes.

Among all neurotransmitter systems in the CNS, the 5–HT system is probably the best understood in terms of underlying molecular mechanisms, due mainly to the findings of genetic studies of a variety of mutant mice. Despite the rapidity of progress during the past few years, however, our understanding of the molecular machinery that functions during the development of 5–HT neurons is, at best, in its infancy. We are far from completely understanding how the combinatorial and sequential action of a myriad of transcription factors is translated into a mature 5–HT neuronal phenotype. Among the many challenges, one of particular importance and immediate significance is an elucidation of the action mechanisms of these transcription factors, in particular with respect to discrete cellular processes during early specification and differentiation of 5–HT neurons. Dissection of the genetic hierarchy of these transcription factors and assessment of their cross-regulation in the execution of a developmental program and in the acquisition of 5–HT phenotype are important. Because transcription factors often assume roles in many aspects of the neuronal phenotype, in addition to their developmental functions (Goridis and Brunet, 1999), another challenge will be to determine whether they might contribute to elaboration and maturation of 5–HT functional circuitry, such as axonal growth and synaptic activities. Region-specific factors or cofactors also need to be identified so that the complexity and heterogeneity of 5–HT neurons can be understood. Finally, temporal- and spatial-specific gene ablation strategies should be used to unravel the distinct roles of transcription factors at different stages of 5–HT neuronal development and maturation. These temporal and spatially knockout mice should eventually aid in our understanding of the transcription control of the psychiatric abnormalities involved in the 5–HT system.

Abbreviations

5–HT	5–hydroxytryptamine
AADC	aromatic L-amino acid decarboxylase gene
bHLH	basic helix-loop-helix
BrdU	bromodeoxyuridine
CNS	central nervous system
LIM-HD	LIM homeodomain
NPS	nail-patella syndrome
SERT	serotonin transporter
TH	tyrosine hydroxylase
TPH	tryptophan hydroxylase
vMN	visceral motor neuron
VZ	ventricular zone

References

Arber S, Ladle DR, Lin JH, Frank E, Jessell TM. **2000**. ETS gene Er81 controls the formation of functional connections between group Ia sensory afferents and motor neurons. *Cell* 101: 485–498.

Asbreuk CH, Vogelaar CF, Hellemons A, Smidt MP, Burbach JP. **2002**. CNS expression pattern of Lmx1b and coexpression with ptx genes suggest functional cooperativity in the development of forebrain motor control systems. *Mol. Cell. Neurosci.* 21: 410–420.

Azmitia EC. **2001**, Modern views on an ancient chemical: serotonin effects on cell proliferation, maturation, and apoptosis. *Brain Res. Bull.* 56: 413–424.

Bach I. **2000**. The LIM domain: regulation by association. *Mech. Dev.* 91: 5–17.

Bertrand N, Castro DS, Guillemot F. **2002**. Proneural genes and the specification of neural cell types. *Nat. Rev. Neurosci.* 3: 517–530.

Blaugrund E, Pham TD, Tennyson VM, Lo L, Sommer L, Anderson DJ, Gershon MD. **1996**. Distinct subpopulations of enteric neuronal progenitors defined by time of development, sympathoadrenal lineage markers and Mash 1 dependence. *Development* 122: 309–320.

Briscoe J, Sussel L, Serup P, Hartigan-O'Connor D, Jessell TM, Rubenstein JL, Ericson J. **1999**. Homeobox gene Nkx2.2 and specification of neuronal identity by graded Sonic hedgehog signalling. *Nature* **1999** 398: 622–627.

Burbach JP, Smits S, Smidt MP. **2003**. Transcription factors in the development of midbrain dopamine neurons. *Ann. N. Y. Acad. Sci.* 991: 61–68.

Chen H, Lun Y, Ovchinnikov D, Kokubo H, Oberg KC, Pepicelli CV, Gan L, Lee B, Johnson RL. **1998**. Limb and kidney defects in Lmx1b mutant mice suggest an involvement of LMX1B in human nail patella syndrome. *Nat. Genet.* 19: 51–55.

Cheng L, Chen CL, Luo P, Tan M, Qiu M, Johnson R, Ma Q. **2003**. Lmx1b, Pet-1, and Nkx2.2 coordinately specify serotonergic neurotransmitter phenotype. *J. Neurosci.* 23: 9961–9967.

Craven SE, Lim KC, Ye W, Engel JD, de Sauvage F, Rosenthal A. **2004**. Gata2 specifies serotonergic neurons downstream of sonic hedgehog. *Development* 131: 1165–1173.

Curtiss J, Heilig JS. **1998**. DeLIMiting development. *BioEssays* 20: 58–69.

Dahlstrom A, Fuxe K. **1964**. Evidence for the existence of monoamine-containing neurons in the central nervous system. I. Demonstration of monoamines in the cell bodies of brain stem neurons. *Acta Physiol. Scand.* 62: 1–55.

Dawid IB, Breen JJ, Toyama R. **1998**. LIM domains: multiple roles as adapters and functional modifiers in protein interactions. *Trends Genet.* 14: 156–162.

Ding YQ, Marklund U, Yuan W, Yin J, Wegman L, Ericson J, Deneris E, Johnson RL, Chen ZF. **2003**. Lmx1b is essential for the development of serotonergic neurons. *Nat. Neurosci.* 6: 933–938.

Ding YQ, Yin J, Kania A, Zhao ZQ, Johnson RL, Chen ZF. **2004**. Lmx1b controls the differentiation and migration of the superficial dorsal horn neurons of the spinal cord. *Development* 131: 3693–3703.

Dreyer SD, Zhou G, Baldini A, Winterpacht A, Zabel B, Cole W, Johnson RL, Lee B. **1998**. Mutations in LMX1B cause abnormal skeletal patterning and renal dysplasia in nail patella syndrome. *Nat. Genet.* 19: 47–50.

Ericson J, Rashbass P, Schedl A, Brenner-Morton S, Kawakami A, van Heyningen V, Jessell TM, Briscoe J. **1997**. Pax6 controls progenitor cell identity and neuronal fate in response to graded Shh signaling. *Cell*, 90: 169–180.

Fode C, Ma Q, Casarosa S, Ang SL, Anderson DJ, Guillemot F. **2000**. A role for neural determination genes in specifying the dorsoventral identity of telencephalic neurons. *Genes Dev.* 14: 67–80.

Fyodorov D, Nelson T, Deneris E. **1998**. Pet–1, a novel ETS domain factor that can activate neuronal nAchR gene transcription. *J. Neurobiol.* 34: 151–163.

Gaspar P, Cases O, Maroteaux L. **2003**. The developmental role of serotonin: news from mouse molecular genetics. *Nat. Rev. Neurosci.* 4: 1002–1012.

Gingrich JA, Ansorge MS, Merker R, Weisstaub N, Zhou M. **2003**. New lessons from knockout mice: the role of serotonin during development and its possible contribution to the origins of neuropsychiatric disorders. *CNS Spectr.* 8: 572–577.

Goridis C, Brunet JF. **1999**. Transcriptional control of neurotransmitter phenotype. *Curr. Opin. Neurobiol.* 9: 47–53.

Goridis C, Rohrer H. **2002**. Specification of catecholaminergic and serotonergic neurons. *Nat. Rev. Neurosci.* 3: 531–541.

Graves BJ, Petersen, JM. **1998**. Specificity within the ets family of transcription factors. *Adv. Cancer Res.* 75: 1–55.

Guillemot F, Lo LC, Johnson JE, Auerbach A, Anderson DJ, Joyner AL. **1993**. Mammalian achaete-scute homolog 1 is required for the early development of olfactory and autonomic neurons. *Cell* 75: 463–476.

Hendricks T, Francis N, Fyodorov D, Deneris ES. **1999**. The ETS domain factor Pet–1 is an early and precise marker of central serotonin neurons and interacts with a conserved element in serotonergic genes. *J. Neurosci.* 19: 10348–10356.

Hendricks TJ, Fyodorov DV, Wegman LJ, Lelutiu NB, Pehek EA, Yamamoto B, Silver J, Weeber EJ, Sweatt JD, Deneris ES. **2003**. Pet–1 ETS gene plays a critical role in 5–HT neuron development and is required for normal anxiety-like and aggressive behavior. *Neuron* 37: 233–247.

Hynes M, Rosenthal A **1999**. Specification of dopaminergic and serotonergic neurons in the vertebrate CNS. *Curr. Opin. Neurobiol.* 9: 26–36.

Jacobs BL, Azmitia EC. **1992**. Structure and function of the brain serotonin system. *Physiol. Rev.* 72: 165–229.

Jessell TM. **2000**. Neuronal specification in the spinal cord: inductive signals and transcriptional codes. *Nat. Rev. Genet.* 1: 20–29.

Kania A, Johnson RL, Jessell TM. **2000**. Coordinate roles for LIM homeobox genes in directing the dorsoventral trajectory of motor axons in the vertebrate limb. *Cell* 102: 161–173.

Koo SJ, Pfaff SL. **2002**. Fine-tuning motor neuron properties: signaling from the periphery. *Neuron* 35: 823–826.

Lebel M, Gauthier Y, Moreau A, Drouin J. **2001**. Pitx3 activates mouse tyrosine hydroxylase promoter via a high-affinity binding site. *J. Neurochem.* 77: 558–567.

Livet J, Sigrist M, Stroebel S, De Paola V, Price SR, Henderson CE, Jessell TM, Arber S. **2002**. ETS gene Pea3 controls the central position and terminal arborization of specific motor neuron pools. *Neuron* 35: 877–892.

Lumsden A, Krumlauf R. **1996**. Patterning the vertebrate neuraxis. *Science* 274: 1109–1115.

Martin GR, Eglen RM, Hamblin MW, Hoyer D, Yocca F. **1998**. The structure and signaling properties of 5–HT receptors: an endless diversity? *Trends Pharmacol. Sci.* 19: 2–4.

Matthews JM, Visvader JE. **2003**. LIM-domain-binding protein 1: a multifunctional cofactor that interacts with diverse proteins. *EMBO Rep.* 4: 1132–1137.

Maurer P, Rorive S, de Kerchove d'Exaerde A, Schiffmann SN, Salmon I, de Launoit Y. **2004**. The Ets transcription factor Fev is specifically expressed in the human central serotonergic neurons. *Neurosci. Lett.* 357: 215–218.

Maurer P, T'Sas F, Coutte L, Callens N, Brenner C, Van Lint C, de Launoit Y, Baert JL. **1999**. FEV acts as a transcriptional repressor through its DNA-binding ETS domain and alanine-rich domain. *Oncogene* 2003 22: 3319–3329.

Millan MJ. **2002**. Descending control of pain. *Prog. Neurobiol.* 66: 355–474.

Nardelli J, Thiesson D, Fujiwara Y, Tsai FY, Orkin SH. **1999**. Expression and genetic interaction of transcription factors GATA–2 and GATA–3 during development of the mouse central nervous system. *Dev. Biol.* 210: 305–321.

Nelson RJ, Chiavegatto S. **2001**. Molecular basis of aggression. *Trends Neurosci.* 24: 713–719.

Oikawa T, Yamada T. **2003**. Molecular biology of the Ets family of transcription factors. *Gene* 303: 11–34.

Pandolfi PP, Roth ME, Karis A, Leonard MW, Dzierzak E, Grosveld FG, Engel JD, Lindenbaum MH. **1995**. Targeted disruption of the GATA3 gene causes severe abnormalities in the nervous system and in fetal liver haematopoiesis. *Nat. Genet.* 11: 40–44.

Parras CM, Galli R, Britz O, Soares S, Galichet C, Battiste J, Johnson JE, Nakafuku M, Vescovi A, Guillemot F. **2004**. Mash1 specifies neurons and oligodendrocytes in the postnatal brain. *EMBO J.* 23: 4495–4505.

Parras CM, Schuurmans C, Scardigli R, Kim J, Anderson DJ, Guillemot F. **2002**. Divergent functions of the proneural genes Mash1 and Ngn2 in the specification of neuronal subtype identity. *Genes Dev.* 16: 324–338.

Pata I, Studer M, van Doorninck JH, Briscoe J, Kuuse S, Engel JD, Grosveld F, Karis A. **1999**. The transcription factor GATA3 is a downstream effector of Hoxb1 specification in rhombomere 4. *Development* 126: 5523–5531.

Patient RK, McGhee JD. **2004**. The GATA family (vertebrates and invertebrates). *Curr. Opin. Genet. Dev.* 12: 416–422.

Pattyn A, Simplicio N, van Doorninck JH, Goridis C, Guillemot F, Brunet JF. **2004**. Ascl1/Mash1 is required for the development of central serotonergic neurons. *Nat. Neurosci.* 7: 589–595.

Pattyn A, Vallstedt A, Dias JM, Samad OA, Krumlauf R, Rijli FM, Brunet JF, Ericson J. **2003**. Coordinated temporal and spatial control of motor neuron and serotonergic neuron generation from a common pool of CNS progenitors. *Genes Dev.* 17: 729–737.

Pauwels PJ. **2000**. Diverse signalling by 5–hydroxytryptamine (5–HT) receptors. *Biochem. Pharmacol.* 60: 1743–1750.

Pfaar H, von Holst A, Vogt Weisenhorn DM, Brodski C, Guimera J, Wurst W. **2002**. mPet–1, a mouse ETS-domain transcription factor, is expressed in central serotonergic neurons. *Dev. Genes Evol.* 212: 43–46.

Pressman CL, Chen H, Johnson RL. **2000**. LMX1B, a LIM homeodomain class transcription factor, is necessary for normal development of multiple tissues in the anterior segment of the murine eye. *Genesis* 26: 15–25.

Scott MM, Deneris ES. **2005**. Making and breaking serotonin neurons and autism. *Int. J. Dev. Neurosci.* 23: 277–285.

Sharrocks AD. **2001**. The ETS-domain transcription factor family. *Nat. Rev. Mol. Cell. Biol* 2: 827–837.

Smidt MP, Asbreuk CH, Cox JJ, Chen H, Johnson RL, Burbach JP. **2000**. A second independent pathway for development of mesencephalic dopaminergic neurons requires Lmx1b. *Nat. Neurosci.* 3: 337–341.

Tanabe Y, Jessell TM. **1996**. Diversity and pattern in the developing spinal cord. *Science* 274: 1115–1123.

Tork I. **1990**. Anatomy of the serotonergic system. *Ann. N. Y. Acad. Sci.* 600: 9–34; discussion 34–35.

Tsai FY, Keller G, Kuo FC, Weiss M, Chen J, Rosenblatt M, Alt FW, Orkin SH. **1994**. An early haematopoietic defect in mice lacking the transcription factor GATA–2. *Nature* 371: 221–226.

van Doorninck JH, van Der Wees J, Karis A, Goedknegt E, Engel JD, Coesmans M, Rutteman M, Grosveld F, De Zeeuw CI. **1999**. GATA–3 is involved in the development of serotonergic neurons in the caudal raphe nuclei. *J. Neurosci.* 19: RC12.

Wasylyk B, Hahn SL, Giovane A. **1993**. The Ets family of transcription factors. *Eur. J. Biochem.* 211: 7–18.

Whitaker-Azmitia PM, Druse M, Walker P, Lauder JM. **1996**. Serotonin as a developmental signal. *Behav. Brain. Res.* 73: 19–29.

9
Role of *Nkx* Homeodomain Factors in the Specification and Differentiation of Motor Neurons and Oligodendrocytes

Jun Cai and Mengsheng Qiu

Abstract

Motor neurons and oligodendrocytes are derived from the same pool of neural progenitor cells (pMN domain) in the ventral spinal cord. Homeodomain transcription factors of the *Nkx* family play important roles in the control of cell fate specification and differentiation of these two cell types. During early neural development, *Nkx6.1* and *Nkx6.2* genes promote the development of somatic motor neurons and oligodendrocytes, whereas *Nkx2.2* suppresses the development of somatic motor neurons in the spinal cord but promotes visceral motor neuron development in the hindbrain. At later stages, *Nkx6.1* and *Nkx6.2* are expressed in differentiating visceral motor neurons and control their migration and axonal projections. In addition, *Nkx2.2* and *Nkx6.2* are expressed in differentiating and mature oligodendrocytes, respectively. While *Nkx2.2* controls the terminal differentiation of oligodendrocytes, *Nkx6.2* appears to regulate the formation of paranodal structures during myelination process.

9.1
Introduction

During the past decade, major progress has been made in our understanding of the origins and molecular specification of motor neurons and oligodendrocytes in the central nervous system (CNS). Recent studies have indicated that the development of motor neurons and oligodendrocytes is under the genetic control of the homeodomain (HD) transcription factors of the *Nkx* family. During early neural development, several members of the *Nkx2* and *Nkx6* families (e.g., *Nkx2.2*, *Nkx2.9*, *Nkx6.1*, *Nkx6.2*) are specifically expressed in or immediately adjacent to the ventral neuroepithelial cells (Qiu et al., 1998) that sequentially give rise to motor neurons and oligodendrocytes. Misexpression of these *Nkx* genes can disrupt the development of both motor neurons and oligodendrocytes. In addition, at later stages, these *Nkx* genes are selectively expressed in differentiating motor neurons or oligodendrocytes and control later aspects of their differentiation. In this chapter, attention will be

Transcription Factors. Edited by Gerald Thiel
Copyright © 2006 WILEY-VCH Verlag GmbH & Co. KGaA, Weinheim
ISBN 3-527-31285-4

focused on the structural features and expression patterns of vertebrate *Nkx2* and *Nkx6* HD factors and their roles in the ventral neural patterning and the development of motor neurons and oligodendrocytes.

9.2
Structural Features of *Nkx* Homeobox Genes Involved in Ventral Neural Patterning

In the developing *Drosophila* embryo, two homeobox genes, *NK2/vnd* (ventral nervous system defective) and *NK6*, are specifically expressed in ventral neuroblasts (Nirenberg et al., 1995; Uhler et al., 2002). To date, at least eight homologues of *NK2* and three homologues of *NK6* genes are found in vertebrates, and many of these are similarly expressed in the ventral region of the developing CNS. Like other homeodomain transcription factors, all *Nkx* homeobox genes contain a 60–amino acid (aa) helix-turn-helix DNA-binding motif termed the homeodomain (HD) (Laughon and Scott, 1984). There is a high degree of sequence similarity in this motif among members of the same family. For example, HD sequences of *Nkx2* genes are highly homologous to each other and to the *Drosophila NK2* gene, with a more than 80% of sequence identity for *Nkx2.1*, *Nkx2.2*, and *Nkx2.9* (Fig. 9.1). Similarly, HDs of the *Nkx6* family have nearly identical sequences, and *Nkx6.1* and *Nkx6.2* differ from each other in this region by only three conservative amino acid changes (Fig. 9.1).

The HD structure presumably functions to recognize and bind to specific regulatory sequences of their downstream target genes. Similar to many other HD factors, *Nkx* genes bind preferentially to TATA-rich sequences. Binding site selection studies have demonstrated that the optimal binding site for *Nkx2.2* is TTAAG-TACTT (Watada et al., 2000). Similar studies showed that *Nkx6.1* binds to TTAAT-TAC (Mirmira et al., 2000) and *Nkx6.2* to TAATTA (Awatramani et al., 1997) at high affinities. These consensus binding sequences are frequently found in regulatory regions of their presumptive downstream target genes, such as myelin basic protein for *Nkx2.2* (Wei et al., 2005).

Interestingly, all six *Nkx* genes and their *Drosophila* homologues share a conserved decapeptide sequence, the TN (tinman) domain, in the N-termini of proteins. The TN domain has a high sequence similarity to the *Engrail* homology–1 (eh1) domain, which functions as a Groucho co-repressor interacting domain (Jimenez et al., 1997). Therefore, Nkx2 and Nkx6 proteins exert their effects in ventral neural patterning by functioning as transcriptional repressors via physical associations with Groucho/TLE co-repressors (Muhr et al., 2001). The TN domain is also found in many other HD transcription factors (about 36) (Muhr et al., 2001) and some basic helix-loop-helix (bHLH) factors such as *Olig2* (Mizuguchi et al., 2001; Novitch et al., 2001), suggesting that transcriptional repression involving the Groucho co-repressors is a common mechanism for regulation of cell fate specification and differentiation by many developmentally regulated transcription factors. Consistent with this notion, inhibition of the Groucho expression/activity in the spinal cord results in a deregulation of gene expression pattern in the neural tube (Muhr et al., 2001). These findings also imply that activation of downstream neural identity genes by many HD

A

Homeodomains

```
mNkx2.1   RRKRRVLFSQ AQVYELERRF KQQKYLSAPE REHLASMIHL TPTQVKIWFQ NHRYKMKRQA
mNkx2.2   RRKRRVLFSK AQTYELERRF RQQRYLSAPE REHLASLIRL TPTQVKIWFQ NHRYKMKRAR
mNkx2.9   RRKRRVLFSK AQTLELERRF RQQRYLSAPE REQLARLLRL TPTQVKIWFQ NHRYKLKRGR

mNkx6.1   RKHTRPTFSG QQIFALEKTF EQTKYLAGPE RARLAYSLGM TESQVKVWFQ NRRTKWRKKH
mNkx6.2   KKHSRPTFSG QQIFALEKTF EQTKYLAGPE RARLAYSLGM TESQVKVWFQ NRRTKWRKRH
mNkx6.3   KKHTRPTFTG HQIFALEKTF EQTKYLAGPE RARLAYSLGM TESQVKVWFQ NRRTKWRKKS
```

NK-2 domains   **NK-6 domains**

```
mNkx2.1   SPRRVAVPVL VKDGKPC (255-271)   mNkx6.1   DDDYNKPLDP NSDDEKI (324-340)
mNkx2.2   SPRRVAVPVL VRDGKPC (199-215)   mNkx6.2   DDEYNRPLDP NSDDEKI (235-251)
mNkx2.9   LLRRVMVPVL VHDRPPS (162-178)   mNkx6.3   DDEYNKPLDP DSDNEKI (225-241)
```

TN domains

```
mNkx2.1   TPFSVSDILS   (9-18)
mNkx2.2   TGFSVKDILD   (8-17)
mNkx2.9   LGFTVRSLLN   (7-16)
mNkx6.1   TPHGINDILS   (94-103)
mNkx6.2   TPHGISDILG   (56-65)
mNkx6.3   TPHGITDILS   (51-60)
```

B

Nkx-2 transcription factor structure

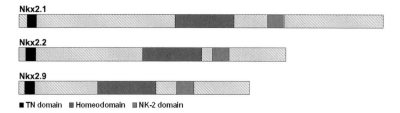

■ TN domain   ▨ Homeodomain   ▨ NK-2 domain

Nkx-6 transcription factor structure

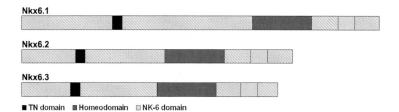

■ TN domain   ▨ Homeodomain   □ NK-6 domain

Fig. 9.1 (A) Sequence similarities of homeodomains and other conserved motifs of the vertebrate *Nkx2* and *Nkx6* homeobox genes. (B) Positions and relative lengths of the homeodomains and other conserved motifs in *Nkx* genes. (This figure also appears with the color plates.)

factors in the neural tube is achieved through the spatially controlled repression of transcriptional repressors – a derepression strategy of neuronal fate specification (Muhr et al., 2001).

In addition, *Nkx2* and *Nkx6* genes contain unique sequence domains, specifically the NK–2 domain for *Nkx2* genes and NK–6 domain for *Nkx6* genes. The NK–2 domain is a conserved 17–aa motif found in the C-termini of fly *NK2* gene and all vertebrate *Nkx2* genes (Fig. 9.1). Gene transcription studies showed that the NK–2 domain functions as an intramolecular regulator of the C-terminal activation domain in *Nkx2.2* (Watada et al., 2000). Although it has been implicated that the NK–2 domain could regulate the ability of Nkx2 proteins to activate expression of downstream genes during development, the *in vivo* role of this domain remains to be defined by further molecular and genetic studies. The NK–6 domain is present in the C-termini of vertebrate Nkx6 proteins, but not of the *Drosophila NK6* gene (Uhler et al., 2002). *In vitro* DNA binding assays have suggested that this domain functions as a mobile binding interference domain (BID), as it can dramatically reduce the DNA-binding affinity of *Nkx6.1* HD protein (Mirmira et al., 2000). This finding has raised the possibility that differences in the NK–6 domain may contribute to the differential levels of repressor activity of *Nkx6.1* and *Nkx6.2* genes in neural development. It has been shown previously that *Nkx6.2* has a weaker repressor activity than *Nkx6.1* (Vallstedt et al., 2001). The physiological importance of the NK–6 domain requires further *in vivo* functional characterization.

9.3
Selective Expression of *Nkx* Homeobox Genes in the Ventral Neural Tube

During early neural development, at least six HD factors of the *Nkx2* family (*Nkx2.1*, *Nkx2.2*, *Nkx2.9*) and *Nkx6* family (*Nkx6.1*, *Nkx6.2*, *Nkx6.3*) are selectively expressed in the ventral regions of the developing CNS (Qiu et al., 1998). Among *Nkx2* genes, *Nkx2.1* expression is first detected in the ventral forebrain (the primordial of ventral hypothalamus), but later also in the medial ganglionic eminence (MGE) of the forebrain (Fig. 9.2). *Nkx2.2* and *Nkx2.9* are expressed in the ventral neural tube along the entire neural axis, from the forebrain to the spinal cord. In the forebrain region, the *Nkx2.2*–expressing cells define the dorsal-ventral boundary (Shimamura et al., 1995). In the spinal cord and hindbrain regions, *Nkx2.2* is expressed in the ventral-most neural progenitor cells immediately dorsal to the *Shh*-expressing floor plate cells. Expression of *Nkx2.9* is more dynamic and complicated in the developing neural tube. First, it is expressed in the floor plate, but later its expression gradually shifts dorsally and begins to overlap with that of *Nkx2.2* in the spinal cord (Pabst et al., 1998; Briscoe et al., 1999).

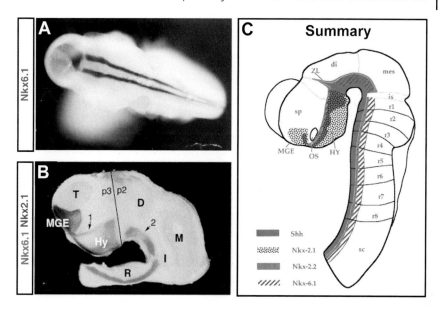

Fig. 9.2 (A,B) Exemplary expression of *Nkx* genes in the ventral neural tube. Neural tube tissue from E4 chicken embryos were subjected to whole-mount *in-situ* RNA hybridization with *Nkx6.1* (A) or simultaneously with *Nkx6.1* and *Nkx2.1* (B). (C) Schematic representation of *Nkx* expression in the developing central nervous system. Abbreviations: D (di) = diencephalon; Hy (HY) = hypothalamus; is = isthmus; M (mes) = mesencephalon; MGE = medial ganglionic eminence; OS = optic stalk; R (r) = rhombomere; sc = spinal cord; sp = secondary prosemere; T = telencephalon; ZL = zona limitans intrathalamica; p2, p3 = progenitor domains. (This figure also appears with the color plates.)

Similarly, three members of the *Nkx6* family are also expressed in the ventral neural tube with a spatially and temporally restricted pattern (Fig. 9.2). *Nkx6.1* is expressed on the ventral one-third of neuroepithelium flanking the floor plate along the entire neural axis, except for the forebrain region (Qiu et al., 1998). *Nkx6.2* HD protein, formerly named *Gtx* as glial- and testis-specific homeobox gene (Komuro et al., 1993), has pattern of expression similar to that of *Nkx6.1* in the midbrain and hindbrain (Qiu et al., 1998), but its expression in mouse spinal cord is restricted to a small number of neural progenitor cells (p1 domain) immediately dorsal to the *Nkx6.1*+ cells (Vallstedt et al., 2001). Unlike *Nkx6.1* and *Nkx6.2*, *Nkx6.3* is not expressed in neural epithelial cells in the ventricular zone. Instead, its expression is only detected in a subset of postmitotic motor neurons in the caudal hindbrain (Nelson et al., 2005).

Previous studies have demonstrated that the ventral expression of the *Nkx* genes is directly under the influence of the Shh protein secreted from the ventral midline structures, first the notochord and then the floor plate (Echelard et al., 1993). Expression of *Nkx2.1*, *Nkx2.2*, *Nkx2.9*, *Nkx6.1* and *Nkx6.2* can be induced in the undifferentiated neuroepithelial cells by the notochord tissue and the recombinant Shh protein, but inhibited by the dorsal midline signal BMP7 protein (Pabst et al., 1998; Qiu et al., 1998; Cai et al., 1999; Briscoe et al., 2000). The induction of *Nkx* gene expression by Shh is region-specific and concentration-dependent, and can also be modulated by other extracellular factors. For example, *Nkx2.1* can only be induced in the anterior neural explants, whereas *Nkx2.2* can be induced in neural tissues from the entire neural axis (Qiu et al., 1998), indicating that the posterior neural tissue is competent for the expression of *Nkx2.2*, but not of *Nkx2.1*. In spinal cord explant culture, whilst a low concentration of Shh protein can induce the expression of *Nkx6.1*, the induction of *Nkx2.2* expression requires a higher concentration of Shh activity (Briscoe et al., 2000). Moreover, the notochord – but not purified Shh protein – can induce *Nkx6.1* expression in the anterior neural plate, suggesting that the notochord produces additional factors that can regulate ventral patterning in the caudal neural tube.

9.4
Nkx Genes are Class II Components of the Homeodomain Protein Code for Ventral Neural Patterning and Cell Fate Specification

The roles of *Nkx* genes in ventral neural patterning and cell fate specification have been extensively studied in the developing spinal cord. It has been suggested that, at early stages, Shh protein forms a concentration gradient in the ventral neural tube and functions as a morphogen to either repress or induce expression of HD proteins in a concentration-dependent manner (Roelink et al., 1995; Briscoe et al., 2000). The HD factors can be subdivided into class I and class II proteins based on their differential regulation by *Shh* signaling. The class I proteins (such as HD factors of the Pax, Dbx and Irx families) are expressed by the neural progenitor cells in the absence of Shh signaling, and their expression is repressed by *Shh* activity. In contrast, expression of class II proteins (such as the *Nkx* proteins) is induced by the *Shh* signal. Since each progenitor HD protein has a different threshold response to *Shh* concentration gradient, different HD proteins are expressed at different dorsoventral positions, thus creating a nested pattern of gene expression along the dorsoventral axis (Fig. 9.3). In general, the more ventral the boundary of class I protein expression is *in vivo*, the higher the concentration of Shh protein is required for repression of gene expression *in vitro*. Conversely, the more dorsal the boundary of class II protein expression is *in vivo*, the lower the concentration of Shh protein is required for induction of expression *in vitro* (Briscoe et al., 2000).

The boundaries of HD proteins are sharply delineated and maintained by cross-regulatory interactions between complementary pairs of class I and class II proteins. For example, *Nkx2.2* and *Pax6* share a common expression boundary, and expression of *Nkx2.2* can repress that of *Pax6*, and vice versa (Briscoe et al., 2000). Cross-repressive interactions are similarly observed between other pairs of HD proteins, such as *Nkx6.1/Dbx2* and *Nkx6.2/Dbx1* (Briscoe et al., 2000; Vallstedt et al., 2001). It has been documented that mutations in dorsal misexpression of Nkx proteins can suppress *Dbx* expression, whereas mutations in *Nkx6* genes can cause the ventral

9.4 Nkx Genes are Class II Components of the Homeodomain Protein Code

expansion of the *Dbx* gene expression (Briscoe et al., 2000; Sander et al., 2000; Vallstedt et al., 2001).

Based on the nested expression profile of progenitor HD proteins, the ventral neural progenitor cells can be subdivided into five distinct progenitor domains (pMN, p0–p3), with each domain expressing a unique combination of HD factors and generating a specific neuronal subtype the identity of which can be readily identified by the expression of other transcription factors, many of which are also HD factors (Fig. 9.3G). Specifically, *Nkx2.2* is expressed in p3 progenitor domain immediately dorsal to the floor plate, from which arise the *Sim1+* V3 ventral interneurons. *Nkx6.1* is expressed in a broader region in the ventral spinal cord, including the p3, pMN, and p2 domains. While the pMN domain generates motor neurons, the p2 domain gives rise to V2 ventral interneurons (Briscoe et al., 2000). *Nkx6.2* is predominantly expressed in the p1 progenitor domain from which the *En1+* V1 ventral interneurons are produced (Vallstedt et al., 2001).

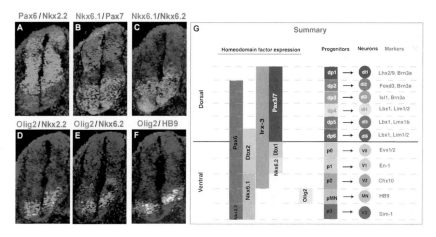

Fig. 9.3 (A-F) Expression of *Nkx* genes in relation to other transcription factors in E10.5 mouse spinal cord. Tissues were subjected to double immunofluorescent staining with antibodies against Nkx proteins and other transcriptions factors. Nuclei were counterstained with DAPI (in purple). (G) Schematic illustration of the homeodomain code that specifies the identity of spinal neural progenitor cells. Nested expression of homeodomain transcription factors subdivides the ventral neuroepithelium into five distinct progenitor domains (p0–p3, pMN) and the dorsal neuroepithelium into six domains (dp1–6). While the ventral progenitor domains give rise to five classes of neurons (V0–V3 ventral interneurons and motor neurons), the dorsal progenitors generate six classes of dorsal interneurons (dI1–6). Different classes of postmitotic neurons can be readily identified by their expression of unique combination of other transcription factors, mostly homeodomain proteins. (This figure also appears with the color plates.)

Recent studies have demonstrated the direct roles of the progenitor HD proteins in ventral neuronal fate specification. Misexpression of many HD proteins frequently alters the identities of ventral progenitor domains and their derived neuronal subtypes. For instance, dorsal expression of *Nkx2.2* induces ectopic *Sim1+* V3 interneurons (Briscoe et al., 2000), whereas loss of *Nkx2.2* expression abolishes the development of V3 neurons (Briscoe et al., 1999), indicating that *Nkx2.2* is the fate determinant of V3 interneurons. Similarly, *Nkx6.1* and *Nkx6.2* are involved in the specification of motor neurons and other ventral interneurons, as described below.

9.5
Nkx Genes Control the Fate Specification and Differentiation of Motor Neurons

9.5.1
Nkx6.1 and *Nkx6.2* have Redundant Activities in Promoting Somatic Motor Neuron Fate Specification

The role of Nkx6 proteins in motor neuron specification and differentiation has been under intensive investigation. Functional studies have shown that *Nkx6.1* specifies the pMN identity in the absence of *Irx3* but confers the p2 domain identity in combination with *Irx3*, consistent with its expression in these two ventral progenitor domains. Specifically, ectopic expression of *Nkx6.1* can induce the formation of motor neurons or V2 interneurons depending on whether *Irx3* is co-expressed (Briscoe et al., 2000). In keeping with the gain-of-function studies, loss of *Nkx6.1* function causes a marked reduction of HB9+ somatic motor neurons (sMNs) that innervate somite-derived skeletal muscles and V2 interneurons in both the spinal cord and the hindbrain regions (Sander et al., 2000; Vallstedt et al., 2001). Concurrently, expression of markers for more dorsal interneurons V1 and V0 is ventrally expanded. Thus, *Nkx6.1* functions to promote motor neuron development but to repress V0–V1 interneuron development. The *Nkx6.2* is normally expressed in the p1 progenitor domain, and its primary function is to specify the V1 interneuron fate (Vallstedt et al., 2001). *Nkx6.2*–null mutation disrupts V1 interneuron development, but has no effect on motor neuron development (Cai et al., 2001; Vallstedt et al., 2001). Interestingly, in *Nkx6.1* mutant spinal cord, *Nkx6.2* expression is up-regulated in the ventral neural progenitor cells that normally express *Nkx6.1* but not *Nkx6.2* (Vallstedt et al., 2001). The up-regulation of *Nkx6.2* in the pMN domain partially compensates for the loss of *Nkx6.1* function. Therefore, the development of sMNs is only partially inhibited in the spinal cord of *Nkx6.1*–/– single mutants (Sander et al., 2000; Vallstedt et al., 2001). In *Nkx6.1*–/–*Nkx6.2*–/–(*Nkx6* –/–) compound mutants, there is a virtual complete loss of sMNs (Vallstedt et al., 2001). Interestingly, *Drosophila Nk6* and zebrafish *Nkx6.1* proteins have similar functions in promoting motor neuron development (Cheesman et al., 2004), providing another classical example that evolutionarily conserved genes are homologous not only in sequences, expression patterns, but also in functions.

Recent studies have established that the bHLH transcription factor *Olig2* is an important downstream mediator of *Nkx6* in the control of motor neuron development. During early spinal cord development, *Olig2* is specifically expressed in the pMN domain of the ventral neuroepithelium (Lu et al., 2000; Zhou et al., 2000). Molecular and genetic studies have indicated that *Olig2* is directly responsible for sMN fate specification (Mizuguchi et al., 2001; Novitch et al., 2001). Mutation in the *Olig2* gene abolishes the development of motor neurons in mouse embryos (Lu et al., 2002; Takebayashi et al., 2002; Zhou and Anderson, 2002). Conversely, dorsal misexpression of *Olig2* in embryonic chicken spinal cord induced ectopic formation of sMNs at early stages of spinal cord development (Mizuguchi et al., 2001; Novitch et al., 2001). Therefore, the *Olig* activity is both necessary and sufficient for sMN development in the spinal cord (Rowitch et al., 2002). Although *Olig2* has been implicated as a component of the progenitor transcriptional code (Mizuguchi et al., 2001; Novitch et al., 2001), several lines of evidence suggest that *Olig2* acts downstream of *Nkx6* genes in specifying motor neuron fate. First, expression of *Olig2* in the pMN domain is slightly later than that of *Nkx6* proteins. *Nkx6.1* is among the earliest expressed transcription factors in the ventral spinal cord (Qiu et al., 1998). Second, dorsal misexpression of *Nkx6.1* in embryonic chicken spinal cord can induce *Olig2* expression, but not vice versa (Liu et al., 2003). Third, *Olig2* expression in the pMN domain is abolished in the spinal cord of *Nkx6* double mutants (Cai et al., 2005; Vallstedt et al., 2005). In contrast, mutation of *Olig2* does not affect *Nkx6.1* expression in the ventral spinal cord (Lu et al., 2002). Together, these data suggest a linear regulatory pathway from *Nkx6* to *Olig2* to sMNs in the developing spinal cord.

Intriguingly, this linear relationship does not appear to be reserved in the hindbrain region. In the caudal hindbrain of *Nkx6* compound mutant, there is an intact *Olig2* expression but a complete loss of somatic motor neurons, suggesting that *Nkx6* may function in parallel with *Olig2* to control sMN development in this region (Pattyn et al., 2003a). In addition, in the rostral hindbrain, such as at the r3 level, sMNs do not develop and *Olig2* is not expressed, despite the early expression of both Nkx6.1 and Nkx6.2 proteins in the ventral neural tube (Pattyn et al., 2003a). Therefore, the regulation of *Olig2* expression and sMN fate by Nkx6 proteins is more complicated in the rostral region of the CNS.

9.5.2
Nkx2.2 Represses Somatic Motor Neuron Fate but Promotes Visceral Motor Neuron Fate

While *Nkx6.1* and *Nkx6.2* have redundant activities in promoting motor neuron fate by activating *Olig2* expression, *Nkx2.2* acts to negatively regulate *Olig2* expression and motor neuron development in the spinal cord. Loss of *Nkx2.2* function results in a ventral expansion of *Olig2* gene expression (Qi et al., 2001), and therefore ectopic production of motor neurons at the p3 position (Briscoe et al., 1999). Conversely, dorsal misexpression of *Nkx2.2* can inhibit *Olig2* expression in the pMN domain, but promotes the expression of V3 interneuron marker *Sim1* (Novitch et al., 2001; Zhou et al., 2001). By suppressing *Olig2* expression in the ventral spinal cord, *Nkx2.2* expression acts to define the ventral boundary of *Olig2* expression and sMN generation. The dorsal boundary of *Olig2* expression appears to be defined by expression of *Irx3* homeodomain factor. *Irx3* is a potent repressor of *Olig2*, and ventral ectopic *Irx3* expression can dramatically inhibit *Olig2* expression (Novitch et al., 2001).

Unlike in the spinal cord, where the *Nkx2.2*–expressing neuroepithelial cells give rise to V3 interneurons, the *Nkx2.2+* progenitor cells in the hindbrain produce visceral motor neurons (vMNs) (Mizuguchi et al., 2001; Pattyn et al., 2003a) that innervate either autonomic ganglion (general visceral) or branchial arch-derived muscles (special visceral). While sMNs are only produced from the caudal hindbrain, vMNs are generated from the entire hindbrain ventral to sMNs. The expression of *Nkx2.2* in the hindbrain progenitor cells activates the expression of *Phox2b*, a homeodomain transcription factor required for the generation of the hindbrain vMNs (Pattyn et al., 2000). Interestingly, despite the fact that *Nkx2.2* is sufficient to induce vMN fate, the *Nkx2.2* function is not absolutely required for the development of vMNs (Pattyn et al., 2003b), possibly due to a functional redundancy from a related transcription factor, such as *Nkx2.9*. It is known that *Nkx2.9* is also expressed in vMN progenitor cells in the ventral hindbrain (Pabst et al., 1998), and mutation of *Nkx2.9* affects the development of a subset of *Phox2b+* vMNs (Pabst et al., 2003).

9.5.3
Nkx6 Proteins Control the Migration and Axonal Projection of Hindbrain vMN

Although *Nkx6.1* and *Nkx6.2* proteins are co-expressed in the vMN neural progenitor cells in the hindbrain, their redundant activities are dispensable for the vMN generation, but are instead necessary to prevent these cells from differentiating into interneurons (Pattyn et al., 2003a). In addition, expression of *Nkx6.1* and *Nkx6.2* is maintained in differentiating postmitotic vMNs, suggesting that Nkx6 proteins may direct some later aspects of their differentiation. Consistent with this idea, the migration and axonal projection properties of some vMNs (e.g., the facial branchiomotor neurons) are impaired in mice lacking *Nkx6.1* or *Nkx6.1/6.2* function (Müller et al., 2003; Pattyn et al., 2003a). The aberrant axonal growth in *Nkx6.1* mutant is associated with ectopic expression of the GDNF receptor Ret and netrin receptor Unc5h3, two molecules that have been implicated in neuronal migration or axonal guidance (Müller et al., 2003). *Nkx6.3* is also expressed in a subset of postmitotic vMNs (Nelson et al., 2005), but its role in the later aspects of neuronal differentiation remains to be examined by future studies. Interestingly, the *Drosophila* Nkx6 protein also plays a direct role in promoting axonogenesis, besides its function to specify ventrally projecting motor neuron fate (Broihier et al., 2004). *Nkx6* is necessary for the expression of the neural adhesion molecule Fasciculin III, and axon growth of *Nkx6*–expressing motor neurons is severely compromised in *Nkx6* mutant fly embryos (Broihier et al., 2004). Therefore, Nkx6 proteins have evolutionarily conserved functions in promoting motor neuron fate specification and axonogenesis.

9.6
The Role of *Nkx* Genes in Oligodendrocyte Development

9.6.1
Nkx6 Proteins Promote *Olig2* Expression and Ventral Oligodendrogenesis in the Spinal Cord

Oligodendrocytes are myelinating macroglial cells found in all regions of the CNS. Despite their wide distribution, recent studies have established that early oligodendrocyte progenitor cells (OPCs) are produced from specific loci of the ventral neural tube (Miller, 2002). In the spinal cord region, early OPCs are produced from the pMN domain after the neurogenesis stage (Sun et al., 1998; Fu et al., 2002). Therefore, oligodendrocytes and motor neurons are derived from the same pool of ventral neural progenitor cells but during different time windows (Richardson et al., 1997). It is believed that during neurogenesis stages, *Olig2* promotes sMN specification and differentiation in collaboration with other bHLH proteins, Neurogenin 1 and 2 (Ngn1 and Ngn2). However, during gliogenesis, the expression of Ngn proteins in neural progenitor cells is down-regulated in the pMN domain and in the absence of *Ngn* expression, *Olig2* expression promotes oligodendrocyte genesis instead. In support of its dual roles in motor neuron and oligodendrocyte development, *Olig2* expression is necessary (Lu et al., 2002; Takebayashi et al., 2002; Zhou et al., 2002) and sufficient (Mizuguchi et al., 2001; Novitch et al., 2001; Sun et al., 2001; Zhou et al., 2001) for induction of both cell fates.

Since *Nkx6.1* and *Nkx6.2* regulate *Olig2* expression in the ventral spinal cord and *Olig2* activity is directly responsible for oligodendrocyte development, the role of Nkx6 proteins in oligodendrogenesis has been recently investigated by our laboratory. As in motor neuron development, *Nkx6.1* and *Nkx6.2* have redundant activities in the regulation of ventral oligodendrogenesis in the spinal cord. While oligodendrocyte specification and differentiation is normal in *Nkx6.2* single mutants (Cai et al., 2001), *Olig2* expression in the pMN domain is severely reduced in *Nkx6.1* mutant embryos and as a result, the production of OPCs is markedly reduced and delayed (Liu et al., 2003). In *Nkx6.1/Nkx6.2* double mutants, OPC generation from the pMN domain is abolished (Cai et al., 2005; Vallstedt et al., 2005). Together, these data indicate that *Nkx6.1* and *Nkx6.2* have redundant activities in the regulation of *Olig2* gene expression and ventral oligodendrogenesis in the spinal cord. However, at later stages, a separate population of *Olig1/2+* OPCs arises from the dorsal neural progenitor cells in *Nkx6* double mutants. This late phase of dorsal oligodendrogenesis occurs independently of *Nkx6* regulation and *Shh* signaling (Cai et al., 2005; Vallstedt et al., 2005).

In keeping with the loss-of-function studies, ectopic expression of *Nkx6.1* in embryonic chicken spinal cord can induce a dorsal expansion of *Olig2* expression and the transient expression of OPC marker gene *Sox10* during early oligodendrogenesis stages (Liu et al., 2003). However, at later stages, forced expression of Nkx6.1 protein can repress *Olig2* gene expression and OPC production, suggesting a stage-dependent regulation of *Olig2* expression by *Nkx6* activity.

9.6.2
Nkx6 Proteins Suppress *Olig2* Expression and Ventral Oligodendrogenesis in the Rostral Hindbrain

Interestingly, in the rostral hindbrain region, ventral *Olig2* expression and OPC generation are not compromised in *Nkx6.1* single (Liu et al., 2003) and *Nkx6* double (Vallstedt et al., 2005) mutants. In this region, sMNs do not develop and vMNs are generated from *Nkx2.2*–expressing ventral neural progenitor cells (Pattyn et al., 2003a). *Olig2+* OPCs arise from the *Nkx2.2+* pMNv domain that first gives rise to vMNs and serotonergic neurons during neurogenesis stage. In the absence of *Nkx6* genes, expression of both *Nkx2.2* and *Olig2* in this region is dorsally expanded and OPC production is enhanced (Vallstedt et al., 2005). Therefore, Nkx6 proteins act to suppress *Olig* gene expression and ventral oligodendrogenesis in the anterior hindbrain, in contrast to their role in promoting *Olig2* expression and oligodendrogenesis in the spinal cord region. The molecular mechanism underlying the differential regulation of *Olig2* expression at different rostrocaudal levels is currently unknown, and it may involve the expression or activity of other region-specific cofactors.

9.6.3
Nkx2.2 Controls the Terminal Differentiation of Oligodendrocytes

Recent molecular and genetic studies have suggested that *Nkx2.2* is directly involved in oligodendrocyte differentiation and maturation. *Nkx2.2* is initially expressed in the p3 domain and its derived V3 interneurons (Briscoe et al., 1999; Liu et al., 2003). However, at later stages, its expression is up-regulated in the pMN-derived OPCs and differentiating oligodendrocytes (Fig. 9.4) (Xu et al., 2000). In the developing chicken spinal cord, the *Olig1/2+* OPCs acquire *Nkx2.2* expression before they migrate from the ventricular zone, whereas in the rodents the *Nkx2.2* up-regulation occurs after OPCs migrate from the ventricular zone (Fig. 9.4) (Fu et al., 2002). Following oligodendrocyte differentiation in the white matter, *Nkx2.2* expression is rapidly down-regulated (Xu et al., 2000). At present, the functional significance underlying the species difference in the timing of *Nkx2.2* up-regulation in OPCs is not clear. Neither is any knowledge available of the molecular mechanism underlying the up- and down-regulation of *Nkx2.2* in cells of oligodendrocyte lineage.

Nevertheless, molecular and genetic studies have shown that *Nkx2.2* activity is required for the normal differentiation of oligodendrocytes. Inhibition of *Nkx2.2* activity by antisense treatment or by targeting mutation can significantly inhibit oligodendrocyte differentiation (Qi et al., 2001; Fu et al., 2002). Conversely, overexpression of *Nkx2.2* in NIH3T3 cells can promote a weak expression of reporter gene from the proteolipid protein (PLP) promoter in co-transfection assay (Qi et al., 2001). Moreover, overexpression of *Nkx2.2* in combination with *Olig2* in embryonic chicken spinal cord is capable of inducing ectopic and precocious differentiation of oligodendrocytes and myelin gene expression (Zhou et al., 2001). Interestingly, ectopic expression of *Nkx2.2* alones is not sufficient to drive myelin basic protein

9.6 The Role of Nkx Genes in Oligodendrocyte Development

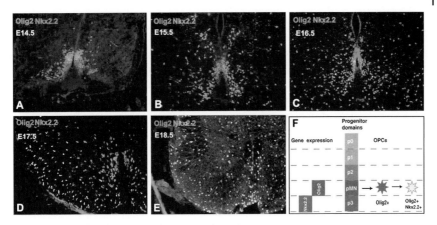

Fig. 9.4 Up-regulation of *Nkx2.2* in *Olig2+* oligodendrocyte progenitor cells (OPCs) in the spinal cord. (A-E) Spinal cord sections from various stages of rat embryos were subjected to double immunofluorescence staining with anti-Olig2 and anti-Nkx2.2. Prior to E17.5, Olig2 is expressed in OPCs, whereas Nkx2.2 labels p3 progenitor cells and possibly V3 interneurons in the ventral gray matter. Starting at E17.5, Olig2+ OPCs in the white matter start to co-express Nkx2.2 (the double-positive cells are labeled as yellow). (F) Schematic representation of the origin and gene expression profile of OPCs in the ventral spinal cord. (This figure also appears with the color plates.)

(MBP)/PLP gene expression from their natural promoters (Zhou et al., 2001), indicating that the activation of myelin gene expression by *Nkx2.2* requires a cofactor, such as *Olig2* or the oligodendrocyte-specific transcription regulator *Sox10* (Zhou et al., 2001; Stolt et al., 2002; Zhang et al., 2005).

Consistent with the concept that *Nkx2.2* regulates myelin gene expression, there are two *Nkx2.2* DNA-binding sites located in the MBP promoter region (Wei et al., 2005). However, in contrast to the earlier studies in mutant mice and in embryonic chicken spinal cord as described above, *Nkx2.2* expression in the oligodendrocyte cell line CG4 represses MBP expression (Wei et al., 2005). One plausible explanation for these conflicting observations is that *Nkx2.2* expression alone represses myelin gene expression, but activates MBP expression in collaboration with *Olig2* which is normally expressed in OPCs.

9.6.4
Nkx6.2 Homeobox Gene Regulates the Oligodendrocyte Myelination Process

Although many transcription factors that control oligodendrocyte specification and differentiation have been identified and characterized, understanding the transcriptional control of the oligodendrocyte myelination process is still in its early stages. It has been shown recently that the *Nkx6.2* molecule is involved in this later process. Earlier studies showed that the *Nkx6.2* factor was specifically expressed in differentiating OPCs and mature oligodendrocytes in postnatal CNS, and suggested that *Nkx6.2* may regulate myelin gene expression and oligodendrocyte differentiation

(Kumuro et al., 1993; Awatramani et al., 1997). However, subsequent studies indicated that *Nkx6.2* is expressed in mature oligodendrocytes (Fig. 9.5) (Southwood et al., 2004), and its activity is not required for myelin gene expression, as expression of myelin genes MBP and PLP is not affected in *Nkx6.2*–null mice (Cai et al., 2001). Instead, it is required for the normal development of myelin sheets in the white matter of the CNS. *Nkx6.2* mutant mice have abnormal paranodal structures and develop neurological deficits (Southwood et al., 2004). The paranodal defect in *Nkx6.2* mutants is associated with the abnormal expression of several cell adhesion molecules that are involved in axon-glial interactions (Southwood et al., 2004), suggesting that *Nkx6.2* regulates the myelination process by directly influencing the expression of its downstream target genes. This function is analogous to that of *Nkx6.1* in controlling cell-surface receptor expression and axonal outgrowth (Müller et al., 2003), and that of *Drosophila Nkx6* gene in regulating cell adhesion molecule expression and axonal guidance (Broihier et al., 2004). In this regard, Nkx6 proteins appear to have an evolutionarily conserved function in regulating cell-surface molecule expression and cell-matrix or cell-cell interactions.

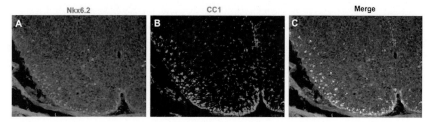

Fig. 9.5 *Nkx6.2* is expressed in differentiated oligodendrocytes in postnatal spinal cord. P4 mouse spinal cord was subjected to double immunostaining with (A) anti-Nkx6.2 and (B) anti-APC antibody (CC1). In the white matter region, all Nkx6.2+ cells co-express APC which specifically labels differentiated oligodendrocytes. (This figure also appears with the color plates.)

In summary, the related *Nkx* genes possess related and evolutionarily conserved functions in controlling the development of motor neurons in both invertebrates and vertebrates (Fig. 9.6). In addition, the *Nkx* genes play important functions in regulating the development of oligodendrocytes, ranging from their specification, differentiation to their myelin formation process (Fig. 9.6). Since the *Nkx* genes are not known to play any significant roles in glial development in invertebrates, it appears that they have acquired these novel functions in the CNS development during animal evolution.

Molecular pathways in MN and OL development

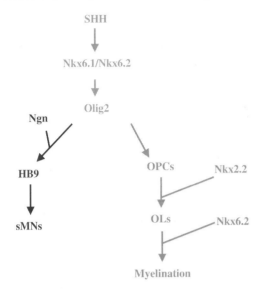

Fig. 9.6 A summary of the major molecular pathways in the specification and differentiation of motor neurons (MNs) and oligodendrocytes (OLs) generated in the ventral spinal cord. OPCs = oligodendrocyte progenitor cells. (This figure also appears with the color plates.)

Acknowledgments

These studies were supported by grants from NIH (NS37707) and NMSS (RG 3275).

Abbreviations

CNS	central nervous system
bHLH	basic helix-loop-helix
BID	binding interference domain
HD	homeodomain
MBP	myelin basic protein
PLP	proteolipid protein
sMN	somatic motor neuron
vMN	visceral motor neuron
OPC	oligodendrocyte progenitor cell

References

Awatramani, R., Scherer, S., Grinspan, J., Collarini, E., Skoff, R., O'Hagan, D., Garbern, J., Kamholz, J. (1997). Evidence that the homeodomain protein Gtx is involved in the regulation of oligodendrocyte myelination. *J. Neurosci.* 17, 6657–6668.

Briscoe, J., Sussel, L., Serup, P., Hartigan-O'Conner, D., Jessell, T.M., Rubenstein, J.L., Ericson, J. (1999). Homeobox gene Nkx–2.2 and specification of neuronal identity by graded sonic hedgehog signaling. *Nature* 398, 622–627.

Briscoe, J., Peirani, A., Jessell, T., Ericson, J. (2000). A homeodomain protein code specifies progenitor cell identity and neuronal fate in ventral neural tube. *Cell* 101, 435–445.

Broihier, H.T., Kuzin, A., Zhu, Y., Odenwald, W., Skeath, J.B. (2004). *Drosophila* homeodomain protein Nkx6 coordinates motoneuron subtype identity and axonogenesis. *Development* 131, 5233–5242.

Cai, J., St Amand, T., Yin, H., Guo, H., Li, G., Zhang, Y., Chen, Y., Qiu, M. (1999). Expression and regulation of the chicken Nkx–6.2 homeobox gene suggest its possible involvement in the ventral neural patterning and cell fate specification. *Dev. Dyn.* 216, 459–468.

Cai, J., Qi, Y., Wu, R., Modderman, G., Fu, H., Liu, R., Qiu, M. (2001). Mice lacking the Nkx6.2 (Gtx) homeodomain transcription factor develop and reproduce normally. *Mol. Cell. Biol.* 21, 4399–4403.

Cai, J., Qi, Y., Hu, X., Tan, M., Liu, Z., Zhang, J., Li, Q., Sander, M. and Qiu, M. (2005). Generation of oligodendrocyte precursor cells from mouse dorsal spinal cord independent of Nkx6–regulation and Shh signaling. *Neuron* 45, 41–53.

Cheesman, S.E., Layden, M.J., Von Ohlen, T., Doe, C.Q., Eisen, J.S. (2004). Zebrafish and fly Nkx6 proteins have similar CNS expression patterns and regulate motoneuron formation. *Development* 131, 5221–5232.

Echelard, Y., Epstein, D., St Jacques, B., Shen, L., Mohler, J., McMahon, J., McMahon, A. (1993). Sonic hedgehog, a member of a family of putative signaling molecules, is implicated in the regulation of CNS polarity. *Cell*, 75, 1417–1430.

Fu, H., Qi, Y., Tan, M., Cai, J., Takebayashi, H., Nakafuku, M., Richardson, W., Qiu, M. (2002). Dual origin of spinal oligodendrocyte progenitors and evidence for the cooperative role of Olig2 and Nkx2.2 in the control of oligodendrocyte differentiation. *Development* 129, 681–693.

Jimenez, G., Paroush, Z., Ish-Horowicz, D. (1997). Groucho acts as a corepressor for a subset of negative regulators, including Hairy and Engrailed. *Genes Dev.* 11, 3072–3082.

Komuro, I., Schalling, M., Jahn, L., Bodmer, R., Jenkins, N., Copeland, N., Izumo, S. (1993). Gtx, a novel murine homeobox-containing gene, specifically expressed in glial cells of the brain and germ cells of testis, has a transcriptional repressor activity in vitro for a serum-inducible promoter. *EMBO J.* 12, 1387–1401.

Laughon, A., Scott, MP. (1984). Sequence of a *Drosophila* segmentation gene: protein structure homology with DNA-binding proteins. *Nature* 310, 25–31.

Liu, R., Cai, J., Hu, X., Tan, M., Qi, Y., German, M., Rubenstein, J., Sander, M., Qiu, M. (2003). Region-specific and stage-dependent regulation of Olig gene expression and oligodendrogenesis by Nkx6.1 homeodomain transcription factor. *Development* 130, 6221–6231.

Lu, Q.R., Yuk, D., Alberta, J., Zhum, Z., Pawlitsky, I., Chan, J., McMahon, A., Stiles, C., Rowitch, D. (2000). Sonic Hedgehog-regulated oligodendrocyte lineage genes encoding bHLH proteins in the mammalian central nervous system. *Neuron* 25, 317–329.

Lu, Q., Sun, T., Zhu, Z., Ma, N., Garcia, M., Stiles, C., Rowitch, D. (2002). Common developmental requirement for Olig function indicates a motor neuron/oligodendrocyte connection. *Cell* 109, 75–86.

Miller, R.H. (2002). Regulation of oligodendrocyte development in the vertebrate CNS. *Prog. Neurobiol.* 67, 451–467.

Mirmira, R.G., Watada, H., German M.S. (2000). β-cell differentiation factor Nkx6.1 contains distinct DNA binding interference and transcriptional repression domains. *J. Biol. Chem.* 275, 14743–14751.

Mizuguchi, R., Sugimori, M., Takebayashi, H., Kosako, H., Nagao, M., Yoshida, S., Nabeshima, Y., Shimamura, K., Nakafuku, M. (2001). Combinatorial roles of olig2 and neurogenin 2 in the coordinated induction of pan-neuronal and subtype-specific properties of motoneurons. *Neuron* 31, 757–771.

Muhr, J., Andersson, E., Persson, M., Jessell, T.M., Ericson, J. (2001). Groucho-mediated transcriptional repression establishes progenitor cell pattern and neuronal fate in the ventral neural tube. *Cell* 104, 861–873.

Müller, M., Jabs, N., Lorke, D., Fritzsch, B., Sander, M. (2003) Nkx6.1 controls migration and axon pathfinding of cranial branchiomotor neurons. *Development* 130, 5815–5826.

Nelson, S., Janiesch, C., Sander M. (2005). Expression of Nkx6 genes in the hindbrain and gut of the developing mouse embryo. *J. Histochem. Cytochem.* 53; 787–790.

Nirenberg, M., Nakayama, K., Nakayama, N., Kim, Y., Mellerick, D., Wang, L.H., Webber, K.O., Lad, R. (1995). The NK–2 homeobox gene and the early development of the central nervous system of *Drosophila*. *Ann. N. Y. Acad. Sci.* 758, 224–242.

Novitch, B., Chen, A., Jessell, T. (2001). Coordinate regulation of motor neuron subtype identity and pan-neuronal properties by the bHLH repressor Olig2. *Neuron* 31, 773–789.

Pabst, O., Herbrand, H., Arnold, H.H. (1998). Nkx2–9 is a novel homeobox transcription factor which demarcates ventral domains in the developing mouse CNS. *Mech. Dev.* 73, 85–93.

Pabst, O., Rummelies, J., Winter, B., Arnold, H.H. (2003). Targeted disruption of the homeobox gene Nkx2.9 reveals a role in development of the spinal accessory nerve. *Development* 130, 1193–1202.

Pattyn, A., Hirsch, M., Goridis, C., Brunet, J.F. (2000). Control of hindbrain motor neuron differentiation by the homeobox gene Phox2b. *Development* 127, 1349–1358.

Pattyn, A., Vallstedt, A., Dias, J.M., Sander, M., Ericson, J. (2003a). Complementary roles for Nkx6 and Nkx2 class proteins in the establishment of motoneuron identity in the hindbrain. *Development* 130, 4149–4159.

Pattyn, A., Vallstedt, A., Dias, J.M., Samad, O.A., Krumlauf, R., Rijli, F.M., Brunet, J.F., Ericson, J. (2003b). Coordinated temporal and spatial control of motor neuron and serotonergic neuron generation from a common pool of CNS progenitors. *Genes Dev.* 17, 729–737.

Qi, Y., Cai, J., Wu, Y., Wu, R., Lee, J., Fu, H., Rao, M., Sussel, L., Rubenstein, J., Qiu, M. (2001). Control of oligodendrocyte differentiation by Nkx2.2 homeodomain transcription factor. *Development* 128, 2723–2733.

Qiu, M., Shimamura, K., Sussel, L., Chen, S., Rubenstein, J. (1998). Control of anteroposterior and dorsoventral domains of Nkx–6.1 gene expression relative to other Nkx genes during vertebrate CNS development. *Mech. Dev.* 72, 77–88.

Richardson, W., Pringle, N., Yu, W., Hall, A. (1997). Origins of spinal cord oligodendrocytes: possible developmental and evolutionary relationships with motor neurons. *Dev. Neurosci.* 19, 58–68.

Roelink, H., Porter, J., Chiang, C., Tanabe, Y., Chang, D., Beachy, P., Jessel, T. (1995). Floor plate and motor neuron induction by different concentrations of the amino-terminal cleavage product of sonic hedgehog autoproteolysis. *Cell* 81, 445–455.

Rowitch, D.H., Lu, Q.R., Kessaris, N., Richardson, W.D. (2002). An ‚oligarchy' rules neural development. *Trends Neurosci.* 25, 417–422.

Sander, M., Paydar, S., Ericson, J., Brisco, J., German, M., Jessell, T., Rubenstein, J. (2000). Ventral neural patterning by Nkx homeobox genes: Nkx6.1 controls somatic motor neuron and ventral interneuron fates. *Genes Dev.* 14, 2134–2139.

Shimamura, K., Hartigan, D.J., Martinez, S., Puelles, L., Rubenstein, J.L. (1995). Longitudinal organization of the anterior neural plate and neural tube. *Development* 121, 3923–3933.

Stolt, C., Rehberg, S., Ader, M., Lommes, P., Riethmacher, D., Schachner, M., Bartsch, U., Wegner, M. (2002). Terminal differentiation of myelin-forming oligodendrocytes depends on the transcription factor

Sox10. *Genes Dev* 16, 165–170.

Southwood, C., He, C., Garbern, J., Kamholz, J., Arroyo, E., Gow, A. (2004). CNS myelin paranodes require Nkx6-2 homeoprotein transcriptional activity for normal structure. *J. Neurosci.* 24, 11215–11225.

Sun, T., Pringle, N., Hardy, A., Richardson, W., Smith, H. (1998). Pax6 influences the time and site of origin of glial precursors in the ventral neural tube. *Mol. Cell. Neurosci.* 12, 228–239.

Sun, T., Echelard, Y., Lu, R., Yuk, D., Kaing, S., Stiles, C., Rowitch, D. (2001). Olig bHLH proteins interact with homeodomain proteins to regulate cell fate acquisition in progenitors of the ventral neural tube. *Curr. Biol.* 11, 1413–1420.

Takebayashi, H., Nabeshima, Y., Yoshida, S., Chisaka, O., Ikenaka, K., Nabeshima, Y. (2002). The basic helix-loop-helix factor olig2 is essential for the development of motoneuron and oligodendrocyte lineages. *Curr. Biol.* 12, 1157–1163.

Uhler, J., Garbern, J., Yang, L., Kamholz, J., Mellerick, D.M. (2002). Nk6, a novel Drosophila homeobox gene regulated by vnd. *Mech. Dev.* 116, 105–116.

Vallstedt, A., Muhr, J., Pattyn, A., Pierani, A., Mendelsohn, M., Sander, M., Jessell, T., Ericson, J. (2001). Different levels of repressor activity assign redundant and specific roles to Nkx6 genes in motor neuron and interneuron specification. *Neuron* 31, 743–55.

Vallstedt, A., Klos, J.M., Ericson, J. (2005). Multiple dorsoventral origins of oligodendrocyte generation in the spinal cord and hindbrain. *Neuron* 45, 55–67.

Watada, H., Mirmira, R.G., Kalamaras, J., German, M.S. (2000). Intramolecular control of transcriptional activity by the NK2–specific domain in NK-2 homeodomain proteins. *Proc. Natl. Acad. Sci. USA* 97, 9443–9448.

Wei, Q., Miskimins, W.K., Miskimins, R. (2005). Stage-specific expression of myelin basic protein in oligodendrocytes involves Nkx2.2–mediated repression that is relieved by the Sp1 transcription factor. *J. Biol. Chem.* 280, 16284–16294.

Xu, X., Cai, J., Hui, F., Qi, Y., Modderman, G., Liu, R., Qiu, M. (2000). Selective expression of Nkx-2.2 transcription factor in the migratory chicken oligodendrocyte progenitor cells and implications for the embryonic origin of oligodendrocytes. *Mol. Cell. Neurosci.* 16, 740–753.

Zhang, X., Cai, J., Klueber, K.M., Guo, Z., Lu, C., Qiu, M., Roisen, F.J. (2005). Induction of oligodendrocytes from adult human olfactory epithelial-derived progenitors by transcription factors. *Stem Cells* 23, 442–453.

Zhou, Q., Wang, S., Anderson, D.J. (2000). Identification of a novel family of oligodendrocyte lineage-specific basic helix-loop-helix transcription factors. *Neuron* 25, 331–343.

Zhou, Q., Choi, G., Anderson, D.J. (2001). The bHLH transcription factor Olig2 promotes oligodendrocyte differentiation in collaboration with Nkx2.2. *Neuron* 31, 791–807.

Zhou, Q., Anderson, D.J. (2002). The bHLH transcription factors OLIG2 and OLIG1 couple neuronal and glial subtype specification. *Cell* 109, 61–73.

10
Sox Transcription Factors in Neural Development

Michael Wegner and C. Claus Stolt

Abstract

Transcription factors of the Sox family of proteins are important regulators in both the central and the peripheral nervous systems of vertebrates. They are essential for the development of neurons and glia. During neural development, Sox proteins are required for such diverse processes as establishment and maintenance of neural stem cell characteristics, fate specifications and lineage decisions, terminal differentiation and maintenance of differentiated phenotypes. They are especially suited for participation in many tasks as they have the capacity to cooperate functionally with many different transcription factors, influence DNA conformation of many regulatory regions as architectural proteins, and recruit co-activators and chromatin-modifying complexes. Co-expression of Sox proteins during neural development sometimes allows functional compensation, whereas in other situations co-expressed Sox proteins appear to modulate each others' function. Comparison with the fruitfly *Drosophila melanogaster* furthermore reveals that some functions of Sox proteins during neural development are strongly conserved between vertebrates and invertebrates and are thus evolutionarily ancient. In other cases, vertebrate Sox proteins appear to have been newly recruited to regulate processes in neural development that are specific or specifically important to vertebrates.

10.1
The Sox Family of Transcription Factors

Sox proteins and their corresponding genes have been found throughout the animal kingdom, but are absent in plants, fungi, and protozoa (Wegner, 1999). They belong to the superfamily of HMG-group proteins, all of which are characterized by possession of a high-mobility-group domain as DNA-binding domain (Laudet et al., 1993). In many proteins, this high-mobility-group recognizes specific DNA structures, whereas in Sox proteins, it recognizes and binds specific DNA sequences. This sequence-specific variant of the high-mobility-group domain was first identified in the Sry protein, a mammalian protein required for male sex determination.

Transcription Factors. Edited by Gerald Thiel
Copyright © 2006 WILEY-VCH Verlag GmbH & Co. KGaA, Weinheim
ISBN 3-527-31285-4

Its DNA-binding domain – the so-called Sry-box – is the hallmark of all Sox proteins and, when further abbreviated to Sox, has given the family its name (Wegner, 1999). Most Sox proteins possess the typical structure of a transcription factor with separable DNA-binding and transactivation domains. Nevertheless, Sox proteins very often appear to rely for their function on the presence of additional transcription factors which function as partner proteins and jointly bind with Sox proteins to composite transcriptional regulatory elements (Kamachi et al., 2000). Upon binding to DNA, Sox proteins also introduce a strong bend in the molecule. Therefore, they have the capacity to shape the overall three-dimensional conformation of a promoter or an enhancer. Sox proteins may thus function not only as transcription factors but also as architectural proteins during transcription (Werner and Burley, 1997).

In model invertebrates such as *Caenorhabditis elegans* and *Drosophila melanogaster*, there are no more than eight *Sox* genes (Wegner, 1999; Bowles et al., 2000). This number has increased in vertebrates through gene duplications so that there are 20–24 different *Sox* genes (Schepers et al., 2002; Koopman et al., 2004). Sox proteins are particularly well studied in vertebrates. Although Sox proteins are important regulators of many developmental processes (Wegner, 1999), many Sox proteins have been detected in the developing peripheral (PNS) and central (CNS) nervous systems. These correlate with specific regulatory functions in neural development. Sox proteins are also strongly expressed in several brain tumor entities (Lee et al., 2002; Ueda et al., 2005), although their role in tumors is less well-defined than their role during development.

10.2
Sox Proteins and Neural Competence

When ectoderm is induced to become neuroectoderm at the dorsal side of the vertebrate embryo, Sox proteins are already present. Sox proteins are thus one of the earliest markers of the developing CNS. In fact, they are required for neuroectoderm induction, sometimes as instructive, but mostly as permissive factors and mediators of neural competence.

In *Xenopus*, for instance, SoxD and Sox2 are first widely expressed throughout the ectoderm and during gastrulation become restricted to the prospective neuroectoderm concomitant with an increase in expression levels (Mizuseki et al., 1998a, 1998b; Kishi et al., 2000). Sox1 and Sox3 are additionally expressed in the prospective neuroectoderm. After commitment to the neural fate, these Sox proteins remain strongly expressed in neural plate and early neural tube. Among them, only SoxD exhibits an instructive neuralizing activity upon overexpression in animal caps or embryos. Nevertheless, neural induction will not take place in *Xenopus* embryos when function of Sox1, Sox2, and Sox3 is inhibited through ectopic expression of a dominant-negative version of the Sox2 protein, thus pointing to a permissive role of these Sox proteins. Whereas Sox1, Sox2, and Sox3 are present in all vertebrates, SoxD is specific to amphibians.

Intriguingly, species-specific variations exist in the order of appearance and the exact expression domains of Sox1, Sox2, and Sox3. In the mouse, Sox2 and Sox3 are expressed pan-ectodermally until neural induction. Neural induction is then marked by the specific appearance of Sox1 in cells committed to a neural fate. Sox2 and Sox3 become concomitantly confined to the neural primordium (Collignon et al., 1996; Pevny et al., 1998; Wood and Episkopou, 1999).

In all vertebrate species so far analyzed, Sox1, Sox2, and Sox3 function as permissive factors and establish neural competence. Forced expression of each of these Sox proteins in mouse embryonic stem cells, for instance, does not interfere with self-renewal, but promotes development into neuroectoderm as soon as embryonic stem cells are released from self-renewal (Zhao et al., 2004). On the basis of these experiments, it is also assumed that Sox1, Sox2, and Sox3 perform similar functions during neural induction. These three Sox proteins are also highly related in their primary sequences and protein structures, and are jointly designated as SoxB1 proteins (Fig. 10.1).

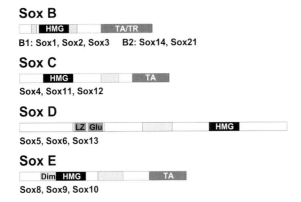

Fig. 10.1 Schematic representation of Sox proteins with importance for neural development. Sox proteins can be divided into several groups. Particularly important for neural development are SoxB (divided into SoxB1 and SoxB2), SoxC, SoxD and SoxE proteins. Within a group, Sox proteins share conserved domains (highlighted in gray) as common structural features, including transactivation domains (TA), transrepression domains (TR, only in SoxB2), dimerization domains (Dim), and coiled coil domains consisting of a leucine zipper (LZ) and an adjacent glutamine-rich region (Glu). The vertebrate members of each Sox group are listed.

The permissive role of SoxB1 proteins in establishing neural competence has not only been observed in the neuroectoderm, but also in other parts of the ectoderm which contribute as placodes to sensory organs and several ganglia of the PNS. In particular, both sensory and epibranchial placodes have been shown in fish and chicken to rely on the presence of Sox3 (Köster et al., 2000; Abu-Elmagd et al., 2001; Ishii et al., 2001). As expected for a competence factor, *Sox3* gene expression is first observed in a region broader than the later placode. Interestingly, the exact dose is critical, as too-low and too-high amounts both interfere with proper placode formation (Köster et al., 2000).

10.3
Sox Proteins and the Neuroepithelial Stem Cell

Sox1, Sox2, and Sox3 are all present throughout neural plate and early neural tube (Rex et al., 1994; Collignon et al., 1996; Wood and Episkopou, 1999). At later phases of neural tube development, when determination and differentiation processes set in, SoxB1 proteins become restricted to the cells immediately adjacent to the ventricle (Fig. 10.2A). These ventricular zone cells are highly proliferative and not specified to a certain cell fate. They give rise to all neuronal and macroglial cell types, and thus represent the stem cells of the developing CNS. Confinement of SoxB1 proteins to the ventricular zone thus suggests that SoxB1 expression correlates with the undifferentiated, pluripotent state.

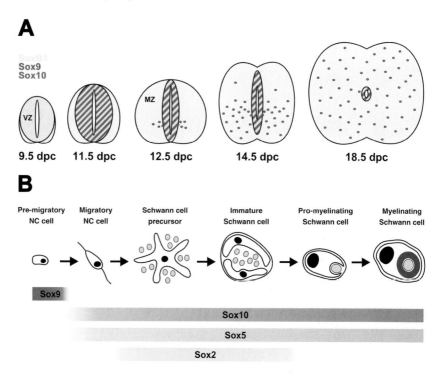

Fig. 10.2 Sox protein expression in the developing mouse nervous system. (A) Spinal cord. Areas expressing Sox2 are marked in yellow; regions or cells expressing Sox9 are highlighted in red. Green indicates Sox10–expressing cells. Areas labeled in yellow (or yellow and red) correspond to the ventricular zone (VZ) with its neuroepithelial progenitors; labeled cells in the mantle zone (MZ) correspond to astrocytes (red) or oligodendrocytes (red and green). Time points correspond to days of mouse embryogenesis post-coitum (dpc). (B) Schwann cell lineage. Various phases of Schwann cell development from pre-migratory neural crest (NC) stem cell to terminally differentiated myelinating Schwann cell are indicated. Bars indicate expression periods for several important Sox proteins. (This figure also appears with the color plates.)

An active role of SoxB1 proteins in maintaining these cells in the undifferentiated state was recently proven (Bylund et al., 2003; Graham et al., 2003). Overexpression of any of the three SoxB1 proteins in the developing chicken spinal cord led to an increase of neural progenitor cells and a concomitant decrease of neuronal differentiation (Fig. 10.3A). Inhibition of SoxB1 proteins, on the other hand, led to cell cycle arrest and depletion of the progenitor pool. Cells underwent premature neuronal differentiation without, however, obtaining a full terminally differentiated phenotype. This suggests that SoxB1 proteins during this period prevent neuronal differentiation, but that their loss alone is not sufficient for neuronal differentiation.

Expression of SoxB1 proteins and markers of the postmitotic neuronal phenotype appear mutually exclusive at early times of development. Whereas SoxB1 proteins are therefore usually not co-expressed with neuronal proteins, there is a significant overlap between SoxB1 proteins and proneural proteins which are the main factors in neuronal specification. Unlike Notch proteins, SoxB1 proteins do not regulate proneural gene expression (Bylund et al., 2003). Expression of proneural genes in the presence of SoxB1 proteins allows neural progenitor cells to be predisposed to a certain fate which may be rapidly adopted once SoxB1 expression falls below threshold levels.

Rather than interfering with proneural gene expression, SoxB1 proteins suppress the function of proneural proteins, just as proneural proteins suppress SoxB1 function. As a consequence, each group is able to repress the target genes of the other group. The underlying molecular mechanism has not been elucidated so far. However, it appears clear that SoxB1 proteins exert their effects in neural progenitors by activating rather than repressing transcription (Bylund et al., 2003). Which genes exactly are activated by SoxB1 proteins in neural progenitor cells is just beginning to be elucidated. Potential target genes include *nestin* (Tanaka et al., 2004) and the *Hes1* gene (Kan et al., 2004).

The behavior of SoxB1 proteins in overexpression studies (Bylund et al., 2003; Graham et al., 2003) and their strongly overlapping expression in the early developing neural tube indicate that these proteins exhibit significant functional redundancy and may largely compensate for each others' loss during early neural development. Accordingly, neuroepithelial progenitor cells develop normally in a hypomorphic Sox2 mutant (Ferri et al., 2004). Sox1– and Sox3–deficient mice also have only mild defects in embryonic CNS development (Malas et al., 2003; Rizzoti et al., 2004)

However, redundancy between SoxB1 proteins appears to be lost postnatally. Sox2, for instance, is not only expressed in embryonic neural progenitor cells, but is also found in all known neurogenic regions of the adult brain, including areas adjacent to the lateral ventricles, the rostral migratory stream, and the dentate gyrus of the hippocampus. In these regions, Sox2 is present in the majority of proliferating ependymal and subependymal cells as well as in migrating neuroblasts, arguing that it marks adult neural stem cells and their immediate progeny (Ellis et al., 2004; Ferri et al., 2004; Komitova and Eriksson, 2004). Whether other SoxB1 proteins are also expressed in adult neural stem cells *in situ* has not been stringently analyzed. However, strongly reduced expression of Sox2 alone is sufficient to cause profound defects in adult ventricular and hippocampal neurogenesis (Ferri et al., 2004).

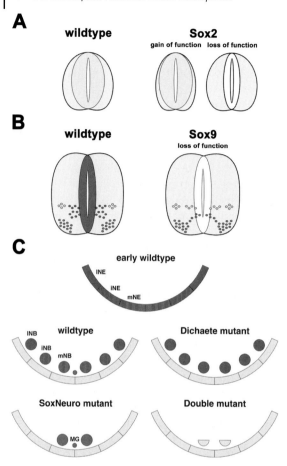

Fig. 10.3 Sox protein function in nervous system development. (A) Sox2 overexpression in the chicken neural tube leads to expansion of the ventricular zone, whereas loss of Sox2 function causes premature differentiation and a reduction of pluripotent neuroepithelial progenitors. (B) Loss of Sox9 in the mouse spinal cord reduces generation of oligodendrocytes (green dots) and astrocytes (red dots). Instead, motor neurons (blue dots) and V2 interneurons (yellow dots) are increased in numbers. (C) The *Drosophila* SoxB proteins SoxNeuro (red) and Dichaete/fish-hook (green) show an overlapping expression in the early neuroectoderm (NE) from which neuroblasts (NB) arise. In the SoxNeuro mutant, many lateral (lNB) and intermediate (iNB) neuroblasts are missing, whereas only midline glia (MG) are lost in the Dichaete/fish-hook mutant. Deletion of both SoxNeuro and Dichaete/fish-hook in the double mutant additionally leads to an increased deletion of medial neuroblasts (mNB) pointing to functional redundancy of both SoxB proteins. Abbreviations: lNE = lateral neuroectoderm; iNE = intermediate neuroectoderm; mNE = medial neuroectoderm. (This figure also appears with the color plates.)

10.4
Sox Proteins and the Neural Crest Stem Cell

In contrast to the pluripotent neuroepithelial stem cells of the CNS, neural crest stem cells which give rise to most parts of the PNS, do not express SoxB1 proteins (see Fig. 10.2B). SoxB1 proteins rather appear to be confined to later stages of PNS development (Le et al., 2005), and even prevent the acquisition of a neural crest stem cell fate as evident from in-ovo electroporation experiments of the chicken neural tube. Expansion of the Sox2 expression domain upon electroporation leads to a gain of CNS progenitors at the expense of neural crest cells (Wakamatsu et al., 2004).

This does not mean that Sox proteins do not have a role in neural crest stem cells. Here, SoxE proteins seem to be in many respects what SoxB1 proteins are for the early pluripotent CNS progenitors. SoxE proteins also consist of three highly related Sox proteins, namely Sox8, Sox9, and Sox10 (see Fig. 10.1). All SoxE proteins are expressed in the neural crest, though with different kinetics.

The first SoxE protein expressed in the neural crest is Sox9 (Fig. 10.2B). An analysis of *Xenopus* embryos indicates that Sox9 expression is induced by Wnt signals in the prospective neural crest region (Lee et al., 2004). In *Xenopus*, Sox9 is required for neural crest induction already before the neurula stage. In the absence of Sox9, the generation of neural crest cells is strongly reduced and the neuroepithelium is significantly expanded (Spokony et al., 2002). When, on the other hand, Sox9 is ectopically expressed by in-ovo electroporation in the early neural tube of chicken, neuroepithelial cells start to express neural crest markers, although they fail to delaminate as would be expected for neural crest cells (Cheung and Briscoe, 2003). Thus, Sox9 and Sox2 are not only reciprocally expressed in neural crest versus neuroepithelial cells, their presence also has opposite effects on the chosen cell fate. It is an attractive, but yet unproven assumption that both proteins influence each other's expression or activity.

Comparable to Sox2 function in the neuroepithelium, Sox9 function during neural crest induction and in the early neural crest appears to be that of a transcriptional activator rather than a transcriptional repressor, as is evident from the fact that a chimera between the DNA-binding domain of Sox9 and the VP16 transcriptional activation domain rather than a chimera with the engrailed repressor domain mimics Sox9 function.

Sox9 expression is persistent in those cranial neural crest cells that, as derivatives of pharyngeal arches 1 and 2, give rise to the craniofacial skeleton (Mori-Akiyama et al., 2003). This argues that Sox9 functions in at least some neural crest cells beyond the induction event. In most other regions of the neural crest, Sox9 expression is transient (Spokony et al., 2002), probably due to an autoregulatory feedback loop (Lee et al., 2004). In the trunk neural crest, for example, Sox9 expression is restricted to the pre-migratory stage during which it is important for survival (Fig. 10.2B). Sox9 also provides the necessary competence for the epithelial-mesenchymal transition in neural crest cells (Cheung et al., 2005). Its expression therefore is a prerequisite for delamination and migration, although Sox9 itself does not appear to be involved in these later processes. These rather seem to be under the control of additional transcription factors which coordinately function with Sox9 and include the Slug/Snail zinc finger proteins as the major regulators of epithelial-mesenchymal transition, and FoxD3 as the inductor of required changes in cell-adhesion properties. Interestingly, all these transcription factors are originally independently induced, whereas later they develop a complex cross-regulatory network where each reinforces the expression of the other (Cheung et al., 2005).

Shortly before neural crest cells delaminate, they start to express a second SoxE protein, namely Sox10 (see Fig. 10.2B). Thus, even in areas where Sox9 expression is extinguished before delamination, the presence of SoxE proteins persists. Taking into account that *in-ovo* electroporation studies in the chicken neural tube have additionally shown that all SoxE proteins are functionally similar during neural crest formation (Cheung and Briscoe, 2003), it can be assumed that Sox10 and Sox9 at least in part provide functional continuity in neural crest stem cells. The relative contribution of these two transcription factors appears to vary between species and between different regions of the body. In general, the influence of Sox10 is stronger in *Xenopus* than in mouse or in chicken (Honore et al., 2003). In the mouse, Sox9 is particularly important for early neural crest development in the trunk, and less so in cranial territories (Mori-Akiyama et al., 2003; Cheung et al., 2005). For Sox10, the situation is reversed (Herbarth et al., 1998; Southard-Smith et al., 1998).

Similar to Sox9 in the pre-migratory neural crest, Sox10 is essential for the survival of migrating neural crest stem cells (Kim et al., 2003). Sox10 expression maintains neural crest stem cells in a pluripotent state, and prevents premature neuronal differentiation (Kim et al., 2003). Interestingly, Sox10 does not simply exert its negative effect on neuronal differentiation by maintenance of the pluripotent state as these two functions possess different dosage requirements for Sox10. Whereas neural crest stem cells maintain their neurogenic potential in the presence of only one functional Sox10 allele, inhibition of neuronal differentiation shows haploinsufficiency (Kim et al., 2003). Little is known so far about the interaction of SoxE proteins with proneural genes in neural crest stem cells. However, it appears that Sox10 is necessary for the initial induction of such factors as Mash1 and Phox2b in post-migratory neural crest cells near the dorsal aorta that begin their development into autonomic neurons. These proteins in turn repress Sox10 expression and thus allow neurogenesis to proceed (Kim et al., 2003).

Overlapping with the expression of SoxE proteins in neural crest stem cells is the expression of Sox5 (see Fig. 10.2B), which belongs to the SoxD group (see Fig. 10.1, not to be confused with the previously mentioned *Xenopus* SoxD protein). Its expression slightly follows Sox9 expression, and is part of the transcriptional regulatory network that is active in neural crest stem cells (Perez-Alcala et al., 2004). As a consequence, ectopic Sox5 expression in the chicken neural tube leads to increased expression of many other neural crest transcription factors and to an expansion of the neural crest. Under native conditions, levels of Sox5 expression vary between different neural crest regions, with the largest amounts being found in the cranial neural crest (Perez-Alcala et al., 2004).

Furthermore, SoxD functions might not be restricted to neural crest cells, but may also be important for neuroectodermal development. Such a possibility is at least suggested by experiments in the P19 cell model of neuroectoderm development. This murine embryonic carcinoma cell line acquires neuroectodermal characteristics upon treatment with retinoic acid. During this treatment, expression of the SoxD protein Sox6 is transiently up-regulated, as is the expression of Sox1 (Pevny et al., 1998; Hamada-Kanazawa et al., 2004a). Intriguingly, forced expression of Sox6 induces neural determination in P19 cells in a manner similar to the known neuroectodermal competence factor Sox1, whereas down-regulation of Sox6 prevents neural determination even in the presence of retinoic acid (Hamada-Kanazawa et al., 2004b). The relationship between SoxD on the one hand, and SoxE or SoxB1 proteins on the other hand, has not yet been analyzed in the neural crest or in the neuroectoderm.

10.5
Sox Proteins in Neural Determination and Lineage Decisions

During early neural development, expression of SoxB1 and SoxE proteins counteracts specification to a neuronal fate. To allow neuronal specification, expression of SoxB1 or SoxE proteins must be extinguished, which occurs at least partly through direct inhibition of *Sox* gene expression by proneural factors.

In contrast, the presence of SoxB1 and SoxE proteins is permissive for the development of glial cells (Britsch et al., 2001; Graham et al., 2003). In the PNS, for instance, Sox10 expression is not extinguished in all neural crest-derived cells once they have reached their destination (see Fig. 10.2B). All glial cells of the PNS, including the satellite glia of peripheral ganglia, the Schwann cells and the glia of the enteric nervous system, continue to express Sox10 (Kuhlbrodt et al., 1998b; Paratore et al., 2002; Young et al., 2003; Maka et al., 2005). The importance of Sox10 for glial fate decisions in the PNS is highlighted by the loss of all peripheral glia in Sox10–deficient mice from earliest times (Britsch et al., 2001). Although neural crest cells are found at positions where they are expected to develop into glia, none of the characteristic glial markers is turned on in Sox10–deficient mouse embryos. This observation has also been reproduced in zebrafish in the Sox10–deficient *colourless* mutant (Dutton et al., 2001), thus arguing that Sox10 is likely required for glial fate specification in the PNS of all vertebrates. Sox10 exerts at least some of its effects on peripheral gliogenesis through stimulation of erbB3 expression and activation of neuregulin signaling, as evident from the absence of erbB3 expression in the developing PNS of Sox10–deficient mice and the phenotypic overlap in the PNS of erbB3– and Sox10–deficient mice (Riethmacher et al., 1997; Britsch et al., 2001).

This role of Sox10 in pan-glial fate specification is restricted to the PNS. Sox10 expression in the CNS does not occur in all glial cells (Stolt et al., 2002). Even in oligodendrocytes, where Sox10 is prominently expressed (see Fig. 10.2A), its expression follows the original specification event (Stolt et al., 2003).

Instead, Sox9 appears to possess a central role in fate specification of CNS glia (Stolt et al., 2003). Compared to Sox10, Sox9 is expressed earlier and more widely in the neural tube as it can be found in neuroepithelial progenitor cells throughout the ventricular zone from mid-embryogenesis onwards (see Fig. 10.2A). Sox9 expression thus overlaps strongly with expression of SoxB1 proteins in the neural tube. However, SoxB1 proteins significantly precede Sox9 as they are detected from earliest times, whereas Sox9 expression starts when neuronal differentiation has set in already. Sox9 expression thus slightly pre-dates the main phase of gliogenesis in the neural tube. Expression throughout the ventricular zone is also compatible with a role of Sox9 in pan-glial specification, as the different glial cell types arise from different regions along the dorsoventral axis of the spinal cord. In fact, neural tube-specific deletion of Sox9 in the mouse indicates that Sox9 might be part of a switch mechanism that diverts neuroepithelial progenitor cells from neurogenesis to gliogenesis (see Fig. 10.3B). Glial cell numbers in the CNS are strongly reduced in the absence of Sox9, whereas neuronal cell numbers are simultaneously increased (Stolt et al., 2003). This reciprocal relationship is particularly evident for oligodendrocytes and motor neurons, both of which are generated in the pMN domain as a defined ventral region of the ventricular zone. Normally, the generation of motor neurons precedes the specification of oligodendrocytes. However, in the absence of Sox9, the generation of motor neurons is prolonged and their number is increased by roughly the same amount as oligodendrocyte numbers are decreased. The loss of glial cells in the Sox9–deficient neural tube and spinal cord was, however, not complete. Several possible explanations exist for the incomplete deletion. Most likely, conditional deletion by CRE recombinase is a contributing factor, as efficiency and timing of Sox9 deletion exhibit a considerable variability and may allow Sox9 expression to be continued in some cells past the point of specification. Once specified, glial cells may then no longer rely on Sox9. Additionally, Sox9 is not the only SoxE protein expressed in the ventricular zone of the neural tube. Sox8 is also expressed (Stolt et al., 2005). Although its expression lags behind, exhibits a ventral-to-dorsal gradient, and is lower than Sox9, an analysis of compound mutant mice clearly points to an additional involvement of Sox8 in glial specification (Stolt et al., 2005). Thus, there is evidence that SoxE proteins contribute to the same function *in vivo*. Nevertheless, the situation is more complicated for SoxE proteins than for SoxB1 proteins in the developing CNS as their expression patterns and expression levels are more divergent. Sox8, in particular, appears to be only a modifying factor with little influence on its own.

10.6
Sox Proteins in Glial Differentiation

Expression of SoxE proteins continues in many glial cells past the original specification event. In the PNS, Sox10 remains expressed throughout development of Schwann cells (see Fig. 10.2B) and satellite glia, and can even be found in the terminally differentiated glial cells of the PNS (Kuhlbrodt et al., 1998b; Britsch et al., 2001). The same appears to hold true for the SoxD protein Sox5 in peripheral glia (Perez-Alcala et al., 2004). In contrast, Sox2 is transiently up-regulated in Schwann cell precursors and immature Schwann cells (see Fig. 10.2B), before it disappears again from this cell lineage upon terminal differentiation (Le et al., 2005).

Analysis of SoxE protein expression in the CNS also confirms that glial expression persists well beyond the specification event. Sox8 has been detected at later phases of mouse embryonic development in various types of glia, including oligodendrocytes and various types of radial glia, such as Bergmann glia of the cerebellum (Cheng et al., 2001; Sock et al., 2001; Stolt et al., 2004). Sox9 occurs in astrocytes and Bergmann glia of the adult CNS (Pompolo and Harley, 2001), while Sox10 continues to be expressed in the terminally differentiated oligodendrocytes (Stolt et al., 2002). Whether these later phases of SoxE protein expression correlate with specific functions in glial cells is much more difficult to analyze than the early functions, and they might be obscured by partially overlapping expression.

Studies on oligodendrocyte development of Sox10–deficient mice have been most informative (Stolt et al., 2002). In these mice, oligodendrocytes develop normally even in the absence of Sox10 until terminal differentiation. Most likely, oligodendrocytes can cope for a long time with the loss of Sox10 because they still express the related Sox9 (see Fig. 10.2A). At the time of terminal differentiation, Sox9 expression is naturally extinguished in oligodendrocyte development so that Sox10 now becomes the single decisive factor for further development (Stolt et al., 2003). Due to this constellation, a specific function of Sox10 in terminal differentiation of oligodendrocytes and particularly in myelination is revealed. At least two genes that code for major myelin proteins of the CNS are under direct transcriptional control of Sox10 (Stolt et al., 2002). These are the genes for myelin basic protein (MBP) and for proteolipid protein (PLP). In the case of the MBP gene, multiple binding sites for Sox10 are present within the promoter region that confers oligodendrocyte-specific expression of transgenes *in vivo*. Each of these binding sites contributes to promoter activation so that MBP gene expression should be exquisitely sensitive to levels of Sox10 protein.

Sox10 also appears to regulate myelin gene expression in Schwann cells, as the promoters of the genes for Myelin protein zero (MPZ) and for Connexin–32 are under direct transcriptional control of Sox10 (Peirano et al., 2000; Bondurand et al., 2001). Interestingly, a functionally relevant binding site for Sox10 within the Connexin–32 promoter is mutated in some patients with peripheral neuropathy of the Charcot-Marie-Tooth type I disease, further corroborating a role for Sox10 in terminal differentiation of Schwann cells and in maintenance of the differentiated phenotype. In agreement with such an assumption, a subset of heterozygous human *Sox10* mutations led to peripheral neuropathies and central leukodystrophy (Touraine et al., 2000; Inoue et al., 2004) in addition to the typical symptoms of a Shah-Waardenburg syndrome. The latter are observed in all patients with heterozygous *Sox10* mutations, represent neural crest defects, and likely reflect the general role of Sox10 in the early neural crest (Pingault et al., 1998; Southard-Smith et al., 1999). Whereas Sox10 supports terminal differentiation of Schwann cells and myelination, Sox2 appears to repress these processes (Le et al., 2005). From its expression in Schwann cell precursors and immature Schwann cells (see Fig. 10.2B), Sox2 is thus proposed to prevent premature Schwann cell differentiation.

10.7
Sox Proteins in Neuronal Differentiation

Apart from the established role in glial differentiation, there is evidence for an additional involvement of Sox proteins in neuronal development. In particular, all three SoxC proteins (see Fig. 10.1) are prominently expressed throughout the developing CNS and PNS (Rex et al., 1994; Hargrave et al., 1997; Jay et al., 1997; Kuhlbrodt et al., 1998a; Cheung et al., 2000). In the CNS, expression of the SoxC proteins Sox4 and Sox11 follows SoxB1 proteins. When progenitor cells leave the ventricular zone and enter neuronal lineages, they start to express SoxC proteins. This expression is transient and becomes extinguished when neurons acquire their differentiated phenotype. A similar transient peak of SoxC protein expression has also been observed in developing sensory neurons of peripheral ganglia. This has led to the hypothesis that SoxC proteins may regulate neuronal maturation throughout the nervous system. In light of this expression pattern, the detection of high levels of Sox4 and Sox11 in medulloblastomas of the cerebellum is intriguing as it might indicate that these tumors correspond to the transformed state of immature neuronal cells (Lee et al., 2002). Loss-of-function studies in mice deficient for either Sox4 or Sox11 have so far failed to corroborate an essential function for either protein in nervous system development (Cheung et al., 2000; Sock et al., 2004). Given the strongly overlapping expression of both Sox proteins (Cheung et al., 2000) and their biochemical similarity (Kuhlbrodt et al., 1998a), one SoxC protein may be able to functionally compensate the loss of the other.

Given the fact that SoxB1 proteins interfere with neuronal specification during the early phases of CNS development, it is surprising to find these proteins expressed in specific neuronal subpopulations in the postnatal mouse brain. It is currently not known whether SoxB1 expression in these neuronal subpopulations persists throughout development, or whether SoxB1 expression reappears after transient shut-down. What is clear, however, is the fact that expression of each *SoxB1* gene results from the combined action of many separate enhancers with different spatial, temporal or cell type-specific activities (Zappone et al., 2000; Brunelli et al., 2003; Uchikawa et al., 2003).

In the adult CNS, Sox2 is expressed not only in adult neural stem cells, but also in some pyramidal cells of the cerebral cortex, a subset of striatal neurons and most abundantly in thalamic neurons such as those in the periventricular nuclei (Ferri et al., 2004). Furthermore, mice with strongly reduced Sox2 levels show neurological abnormalities indicative of a dysfunctional dopaminergic system, and a neurodegenerative phenotype that includes parenchymal shrinkage, degenerating neurons and neurons with perinuclear inclusions. Taking into account that not all CNS regions are equally affected, and that affected areas correspond to those regions where Sox2 is persistently expressed in adult neurons, it is at least tempting to speculate that the observed phenotype is caused by loss of Sox2 function in postmitotic neurons.

Hints for a role of SoxB1 proteins in maintenance of mature neurons additionally come from mouse models with Sox1– or Sox3–deficiencies. Sox1, for example, is highly expressed in postmitotic GABAergic neurons of the olfactory tubercle and the nucleus accumbens of the ventral striatum (Malas et al., 2003). Sox1 deletion leads to a severe loss of the normally Sox1–expressing neurons, and consequently to a major focal neuroanatomical defect in the ventral forebrain. The resulting disruption of local neuronal circuits is the cause of enhanced synaptic excitability and spontaneous epileptiform discharges in the piriform cortex. As a result, Sox1–deficient mice suffer from epilepsy with seizure origin in the limbic forebrain.

Sox3, on the other hand, occurs in several neuronal populations of the ventral hypothalamus, including neurons of the arcuate nucleus and the median eminence. Loss of Sox3 in the mouse leads to defects in specific midline structures of the CNS including the hypothalamus and the corpus callosum (Rizzoti et al., 2004). Whereas dysgenesis of the corpus callosum is mainly a developmental defect, hypothalamic defects persist postnatally. There is not only a reduced proliferation of cells in the developing hypothalamus and infundibulum which may in turn cause defects in pituitary formation during embryogenesis. Additionally, growth hormone-releasing hormone (GHrH)-positive neurons of the adult hypothalamus are compromised in number, activity and/or connectivity. This postnatal neuronal defect strongly contributes to the hypopituitarism observed in Sox3–deficient mice and human patients with Sox3 mutations (Laumonnier et al., 2002; Rizzoti et al., 2004). The mental retardation additionally observed in affected human patients also argues in favor of defects in mature neurons.

10.8
Sox Proteins and their Molecular Mode of Action

As outlined above, Sox proteins perform many different functions in the developing nervous system. In particular, it is intriguing that one and the same Sox protein is active in different roles. Sox10, for instance, is involved in maintaining the pluripotent state of the neural crest stem cell, in glial cell fate specification, lineage progression of various glial cell types and in terminal differentiation of glia (Peirano et al., 2000; Britsch et al., 2001; Stolt et al., 2002; Kim et al., 2003). SoxB1 proteins on the other hand, are equally involved in maintaining the pluripotency of neuroepithelial progenitors and the identity of specific groups of mature neurons (Bylund et al., 2003; Graham et al., 2003; Malas et al., 2003). This does not even take into account that many Sox proteins have additional roles outside the nervous system (Wegner, 1999). Although it is formally possible that each Sox protein contributes to its many functions by always regulating the same target genes, this appears rather unlikely given the highly differing contexts. In some cases, there even exists direct evidence that different target genes are activated by the same Sox protein in a tissue- or cell type-specific manner. Sox10, for instance, activates Mitf-M expression selectively in melanocytes, but not in glial cells, whereas other targets such as myelin genes are up-regulated by Sox10 in glia, but not in melanocytes (Bondurand et al., 2000, 2001; Peirano et al., 2000; Potterf et al., 2000; Stolt et al., 2002).

Thus, it appears more likely that the function of Sox proteins is highly context-dependent and inherently flexible. An important feature of transcriptional regulation by Sox proteins is the strict requirement for partner proteins (Kamachi et al., 2000). These partner proteins vary among tissues and thus alter Sox protein activity in a tissue-specific manner.

Partner proteins have been best studied for Sox2. They include the paired domain transcription factor Pax6 (Kamachi et al., 2001) and the POU homeodomain transcription factors Oct–3/4 and Brn–2 (Yuan et al., 1995; Tanaka et al., 2004). Lens-specific activation of Sox2 target genes such as γ–*crystallin* in the chicken is achieved with Pax6 as partner, whereas Sox2–dependent gene activation in epiblast and embryonic stem cells requires cooperation with Oct–3/4 as evident from many target genes, including *fibroblast growth factor–4, upstream transcription factor, nanog* and *Sox2* itself (Yuan et al., 1995; Nishimoto et al., 1999; Tomioka et al., 2002; Catena et al., 2004; Kuroda et al., 2005). As indicated by a recent study on the *nestin* gene, Brn–2 and related POU homeodomain proteins appear to take over the role of partner proteins in neuroepithelial progenitor cells of the CNS (Tanaka et al., 2004). In the case of Sox10, identified partner proteins include Pax3 in neural crest stem cells (Bondurand et al., 2000; Potterf et al., 2000) and Mitf in neural crest-derived melanoblasts (Ludwig et al., 2004).

Very often, Sox proteins and their partners bind next to each other to composite elements in the gene regulatory region that mediates tissue-specific gene expression (Yuan et al., 1995; Nishimoto et al., 1999; Bondurand et al., 2000; Potterf et al., 2000; Kamachi et al., 2001; Tomioka et al., 2002; Tanaka et al., 2004; Kuroda et al., 2005). When bound, Sox proteins and their partners engage in additional protein-protein-contacts with glue-like surface patches that stabilize the complex and combine their transactivation domains to a super-transactivation domain (Remenyi et al., 2003).

However, there is evidence that such a requirement for adjacent binding sites is not absolute (Ludwig et al., 2004). Sox proteins have the ability to strongly bend DNA and, through their action as architectural proteins, shape the three-dimensional structure of larger regulatory regions, thereby bringing transcription factors into close contact with each other and with themselves that bind to sequence elements far apart in the primary DNA sequence of these regulatory regions.

Adding another level of complexity, Sox proteins have been observed to interact with several transcriptional co-activators. In particular CBP/p300 appears to be an important interactor as it has been identified to interact both with SoxE and SoxB1 proteins (Nowling et al., 2003; Tsuda et al., 2003). Thus, Sox proteins may not only interact directly with other transcription factors bound to the same promoter and shape the overall conformation of the enhanceosome, but might also recruit co-activator complexes, including those that modify or remodel chromatin.

Another striking feature of Sox proteins in the developing nervous system is their strongly overlapping expression. This immediately raises the question of the relationship between these co-expressed Sox proteins. In those cases where Sox proteins of the same subgroup are co-expressed, such as SoxE proteins in the oligodendrocyte lineage or SoxB1 proteins in the neuroepithelium, there is considerable evidence for at least partial functional redundancy (Bylund et al., 2003; Graham et al., 2003; Stolt et al., 2003, 2004, 2005; Maka et al., 2005). Less clear is the relationship between co-expressed Sox proteins that belong to different subgroups such as Sox2 and Sox9 in the late neuroepithelium (Collignon et al., 1996; Stolt et al., 2003) or Sox2, Sox5 and Sox10 in Schwann cell precursors and immature Schwann cells (Kuhlbrodt et al., 1998b; Perez-Alcala et al., 2004; Le et al., 2005).

Data from tissue culture experiments indicate that Sox proteins from different subgroups have biochemically distinct characteristics (Kuhlbrodt et al., 1998a, 1998b; Kamachi et al., 1999). The nuclear export sequence identified in SoxE proteins, for instance, is not conserved in most other Sox proteins (Gasca et al., 2002; Rehberg et al., 2002), raising the possibility that nucleocytoplasmic shuttling is restricted to a subset of Sox proteins. Additionally, some Sox proteins such as SoxD proteins exist as constitutive dimers (Lefebvre et al., 1998), whereas most other Sox proteins are constitutive monomers (Peirano and Wegner, 2000). SoxE proteins on the other hand seem to be able to switch between both states (Peirano and Wegner, 2000). The existence as monomers or dimers in turn impacts on the DNA-binding characteristics of Sox proteins. Additional data indicate that choice of partner proteins also differs between subgroups (Kamachi et al., 1999). Although these data do not exclude the possibility that Sox proteins from different subgroups may perform similar functions under certain circumstances, functional differences should be commonly expected for co-expressed Sox proteins from different subgroups.

These functions may in many cases be independent of each other. However, it can be imagined that Sox proteins influence each others' function when co-expressed. Evidence for such a functional interaction between Sox proteins from different subgroups has not yet been reported from the developing nervous system, but has been observed in chondrocytes where Sox9 first activates expression of SoxD proteins which then engage with Sox9 in a multiprotein complex that activates several Sox9 target genes much more strongly than Sox9 alone (Lefebvre et al., 1998; Akiyama et al., 2002).

It is also interesting that most Sox proteins have so far been primarily described as transcriptional activators, even in circumstances where they repress certain developmental pathways. The inhibitory activity of SoxB1 proteins on neuronal differentiation of neuroepithelial progenitors is, for instance, mediated by transcriptional activation (Bylund et al., 2003). The main exception to this rule so far appear to be SoxB2 proteins (see Fig. 10.1), which are also abundantly expressed in the developing nervous system along with SoxB1 proteins (Rex et al., 1997; Ohba et al., 2004). Instead of the transactivation domain present in the carboxy-terminal region of SoxB1 proteins, SoxB2 proteins carry a transrepression domain. At least part of their repressive function comes from direct interference with co-expressed SoxB1 proteins (Uchikawa et al., 1999), thus yielding further evidence for the existence of functional interactions between different Sox proteins.

10.9
Conservation of Sox Protein Function in Nervous System Development

All vertebrate Sox proteins discussed in this chapter have orthologues in the nematode *C. elegans* and the fruitfly *D. melanogaster* as model invertebrates (Schepers et al., 2002). In general, there is one single Sox protein per vertebrate Sox subgroup. For example, instead of Sox8, Sox9 and Sox10 as SoxE proteins there is only Sox100B in *D. melanogaster* and ce-SoxE in *C. elegans* (Bowles et al., 2000). Only SoxB proteins are present in higher numbers and are represented, for instance, in *D. melanogaster* by four proteins including Dichaete/fish-hook and SoxNeuro (Nambu and Nambu, 1996; Russell et al., 1996; Bowles et al., 2000; Buescher et al., 2002; Overton et al., 2002). Thus, it is interesting to ask whether Sox protein function in the nervous system is conserved between vertebrates and invertebrates. Given the fact that Sox proteins have so far been poorly studied in *C. elegans*, this comparison must be confined to *D. melanogaster*.

Both Dichaete/fish-hook and SoxNeuro are strongly expressed in the developing nervous system of the fruitfly in a highly dynamic pattern. Similar to vertebrate SoxB1 proteins, they are thus not only expressed in early neuroectodermal cells and neuroblasts, but also late in specific groups of differentiated neurons (Sanchez-Soriano and Russell, 2000). *Drosophila* SoxB proteins do not induce neuroectoderm formation, but are essential for consecutive neuroectoderm specification and neuroblast formation (Buescher et al., 2002). Correspondingly, SoxB proteins exhibit a fairly broad, and partially overlapping expression pattern during early stages of neural development. SoxNeuro is expressed throughout all three regions of the ventral neuroectoderm (i.e., the lateral, intermediate and medial neuroectoderm), whereas large amounts of Dichaete/fish-hook are restricted to the medial region (see Fig. 10.3C). Similar to their vertebrate counterparts, both *Drosophila* SoxB proteins exhibit largely redundant functions during these early phases, although there is also evidence for the existence of region-specific unique functions (Overton et al., 2002). As a consequence of functional redundancy, each SoxB protein can compensate loss of the other in those areas where they are co-expressed. Thus, neuroblast formation is severely affected in the lateral and intermediate neuroectoderm of SoxNeuro-deficient flies, but not in the medial region (Buescher et al., 2002; Overton et al., 2002). Additional strong defects in the medial neuroectoderm and therefore pan-neural hypoplasia become visible only when both SoxNeuro and Dichaete/fish-hook are deleted (see Fig. 10.3C). Loss of Dichaete/fish-hook alone, on the other hand, does not interfere strongly with neuroblast formation in any region of the CNS. Rather, Dichaete/fish-hook mutants exhibit defects in a specific cell population that selectively expresses Dichaete/fish-hook (see Fig. 10.3C) and is located in the ventral midline (Nambu and Nambu, 1996; Russell et al., 1996; Sanchez-Soriano and Russell, 1998). As a consequence of these midline glia defects, the ganglia of the ventral nerve cord fuse and axon scaffold organization is aberrant. Similar to their vertebrate counterparts, *Drosophila* SoxB proteins exert their function in the developing CNS in cooperation with other transcription factors including ventral nerve cord defective (vnd), intermediate neuroblast defective (ind) and ventral veinless (vvl)

(Sanchez-Soriano and Russell, 1998; Buescher et al., 2002; Overton et al., 2002; Zhao and Skeath, 2002). Homologues of these POU-, Nkx- and Gsh-like homeodomain proteins are known partners (or at least good candidates for partners) of mammalian SoxB1 proteins in the vertebrate nervous system. Further supporting the similarity to mammalian SoxB1 proteins, there also appears to be a functionally relevant crosstalk with proneural proteins of the Achaete-scute complex. Given this significant conservation of SoxB protein function in neural development, it is not surprising that mammalian Sox2 is able to rescue SoxB-deficiency in the fly, as is evident from ectopic overexpression studies of Sox2 in Dichaete/fish-hook mutants (Sanchez-Soriano and Russell, 1998).

In contrast to this strong conservation of SoxB function, SoxE function does not appear to be conserved. As of today, there is no evidence from gain-of-function or loss-of-function studies, nor from its expression pattern, that Sox100B plays any role in the fly's nervous system (Hui Yong Loh and Russell, 2000). Thus, neural functions may have been acquired fairly late by SoxE proteins in vertebrates during the evolution of vertebrate-specific traits such as the neural crest. Given the fact that most parts of the PNS are derived from the neural crest, its development may differ strongly from development of the PNS in flies. Another vertebrate-specific trait is the strong expansion of glial cells. Whilst there are fewer glial cells than neurons in the invertebrate nervous system, human CNS glia outnumber their neuronal counterparts by one order of magnitude. Additionally, glial functions are much more diversified in the vertebrate nervous system. Thus, it is not surprising that these hugely different demands for glial cell numbers and glial cell types may have led to the establishment of a new and efficient regulatory network that is different from that in the fly and other invertebrates, and uses as its cornerstones SoxE proteins which, in the absence of more ancient functions (Hui Yong Loh and Russell, 2000), might have been the perfect proteins to be entrusted with new and vertebrate-specific functions. Intriguingly, these new SoxE-based regulatory networks show many features which are analogous to those in the SoxB1–dependent network of early CNS development, including the requirement for similar partner proteins, the interaction with proneural genes, and the existence of functional redundancies, arguing that transcriptional regulation by Sox proteins follows general principles after all.

Acknowledgments

The research studies on Sox proteins in the Wegner laboratory are supported by grants from the Deutsche Forschungsgemeinschaft, the Thyssen-Stiftung, the Schram-Stiftung, and the Fonds der Chemischen Industrie.

Abbreviations

CNS	central nervous system
GHrH	growth hormone-releasing hormone
ind	intermediate neuroblast defective
MBP	myelin basic protein
MPZ	myelin protein zero
PLP	proteolipid protein
PNS	peripheral nervous system
vnd	ventral nerve cord defective
vvl	ventral veinless

References

Abu-Elmagd M, Ishii Y, Cheung M, Rex M, Le Rouedec D, Scotting PJ. **2001**. cSox3 expression and neurogenesis in the epibranchial placodes. *Dev. Biol.* 237: 258–269.

Akiyama H, Chaboissier M-C, Martin JF, Schedl A, de Crombrugghe B. **2002**. The transcription factor Sox9 has essential roles in successive steps of the chondrocyte differentiation pathway and is required for expression of Sox5 and Sox6. *Genes Dev.* 16: 2813–2828.

Bondurand N, Girard M, Pingault V, Lemort N, Dubourg O, Goossens M. **2001**. Human Connexin 32, a gap junction protein altered in the X-linked form of Charcot-Marie-Tooth disease, is directly regulated by the transcription factor SOX10. *Hum. Mol. Genet.* 10: 2783–2795.

Bondurand N, Pingault V, Goerich DE, Lemort N, Sock E, Le Caignec C, Wegner M, Goossens M. **2000**. Interaction between SOX10, PAX3 and MITF, three genes implicated in Waardenburg syndrome. *Hum. Mol. Genet.* 9: 1907–1917.

Bowles J, Schepers G, Koopman P. **2000**. Phylogeny of the SOX family of developmental transcription factors based on sequence and structural indicators. *Dev. Biol.* 227: 239–255.

Britsch S, Goerich DE, Riethmacher D, Peirano RI, Rossner M, Nave KA, Birchmeier C, Wegner M. **2001**. The transcription factor Sox10 is a key regulator of peripheral glial development. *Genes Dev.* 15: 66–78.

Brunelli S, Casey ES, Bell D, Harland R, Lovell-Badge R. **2003**. Expression of Sox3 throughout the developing central nervous system is dependent on the combined action of discrete, evolutionary conserved elements. *Genesis* 36: 12–24.

Buescher M, Hing FS, Chia W. **2002**. Formation of neuroblasts in the embryonic central nervous system of *Drosophila melanogaster* is controlled by SoxNeuro. *Development* 129: 4193–4203.

Bylund M, Andersson E, Novitch BG, Muhr J. **2003**. Vertebrate neurogenesis is counteracted by Sox1–3 activity. *Nat. Neurosci.* 6: 1162–1168.

Catena R, Tiveron C, Ronchi A, Porta S, Ferri A, Tatangelo L, Cavallaro M, Favaro R, Ottolenghi S, Reinbold R, Scholer H, Nicolis SK. **2004**. Conserved POU binding DNA sites in the Sox2 upstream enhancer regulate gene expression in embryonic and neural stem cells. *J. Biol. Chem.* 279: 41846–41857.

Cheng YC, Lee CJ, Badge RM, Orme AT, Scotting PJ. **2001**. Sox8 gene expression identifies immature glial cells in developing cerebellum and cerebellar tumours. *Mol. Brain Res.* 92: 193–200.

Cheung M, Abu-Elmagd M, Clevers H, Scotting PJ. **2000**. Roles of Sox4 in central nervous system development. *Mol. Brain Res.* 79: 180–191.

Cheung M, Briscoe J. **2003**. Neural crest development is regulated by the transcription factor Sox9. *Development* 130: 5681–5693.

Cheung M, Chaboissier M-C, Mynett A, Hirst E, Schedl A, Briscoe J. **2005**. The transcriptional control of trunk neural crest induction, survival, and delamination. *Dev. Cell* 8: 179–192.

Collignon J, Sockanathan S, Hacker A, Cohentannoudji M, Norris D, Rastan S, Stevanovic M, Goodfellow PN, Lovellbadge R. **1996**. A comparison of the properties of Sox–3 with Sry and two related genes, Sox–1 and Sox–2. *Development* 122: 509–520.

Dutton KA, Pauliny A, Lopes SS, Elworthy S, Carney TJ, Rauch J, Geisler R, Haffter P, Kelsh RN. **2001**. Zebrafish colourless encodes sox10 and specifies non-ectomesenchymal neural crest fates. *Development* 128: 4113–4125.

Ellis P, Fagan BM, Magness ST, Hutton S, Taranova O, Hayashi S, McMahon A, Rao M, Pevny L. **2004**. SOX2, a persistent marker for multipotential neural stem cells derived from embryonic stem cells, the embryo or the adult. *Dev Neurosci*. 26: 148–165.

Ferri ALM, Cavallaro M, Braida D, Di Cristofano A, Canta A, Vezzani A, Ottolenghi S, Pandolfi PP, Sala M, DeBiasi S, Nicolis SK. **2004**. Sox2 deficiency causes neurodegeneration and impaired neurogenesis in the adult mouse brain. *Development* 131: 3805–3819.

Gasca S, Canizares J, De Santa Barbara P, Mejean C, Poulat F, Berta P, Boizet-Bonhoure B. **2002**. A nuclear export signal within the high mobility group domain regulates the nucleocytoplasmic translocation of SOX9 during sexual determination. *Proc. Natl. Acad. Sci. USA* 99: 11199–11204.

Graham V, Khudyakov J, Ellis P, Pevny L. **2003**. SOX2 functions to maintain neural progenitor identity. *Neuron* 39: 749–765.

Hamada-Kanazawa M, Ishikawa K, Nomoto K, Uozumi T, Kawai Y, Narahara M, Miyake M. **2004a**. Sox6 overexpression causes cellular aggregation and the neuronal differentiation of P19 embryonic carcinoma cells in the absence of retinoic acid. *FEBS Lett*. 560: 192–198.

Hamada-Kanazawa M, Ishikawa K, Ogawa D, Kanai M, Kawai Y, Narahara M, Miyake M. **2004b**. Suppression of Sox6 in P19 cells leads to failure of neuronal differentiation by retinoic acid and induces retinoic acid-dependent apoptosis. *FEBS Lett*. 577: 60–66.

Hargrave M, Wright E, Kun J, Emery J, Cooper L, Koopman P. **1997**. Expression of the Sox11 gene in mouse embryos suggests roles in neuronal maturation and epithelio-mesenchymal induction. *Dev. Dyn*. 210: 79–86.

Herbarth B, Pingault V, Bondurand N, Kuhlbrodt K, Hermans-Borgmeyer I, Puliti A, Lemort N, Goossens M, Wegner M. **1998**. Mutation of the Sry-related Sox10 gene in Dominant megacolon, a mouse model for human Hirschsprung disease. *Proc. Natl. Acad. Sci. USA* 95: 5161–5165.

Honore SM, Aybar MJ, Mayor R. **2003**. Sox10 is required for the early development of the prospective neural crest in *Xenopus* embryos. *Dev. Biol*. 260: 79–96.

Hui Yong Loh S, Russell S. **2000**. A *Drosophila* group E Sox gene is dynamically expressed in the embryonic alimentary canal. *Mech. Dev*. 93: 185–188.

Inoue K, Khajavi M, Ohyama T, Hirabayashi S-i, Wilson J, Reggin JD, Mancias P, Butler IJ, Wilkinson MF, Wegner M, Lupski JR. **2004**. Molecular mechanism for distinct neurological phenotypes conveyed by allelic truncating mutations. *Nat. Genet*. 36: 361–369.

Ishii Y, Abu-Elmagd M, Scotting PJ. **2001**. Sox3 expression defines a common primordium for the epibranchial placodes in chick. *Dev. Biol*. 236: 344–353.

Jay P, Sahly I, Goze C, Taviaux S, Poulat F, Couly G, Abitbol M, Berta P. **1997**. Sox22 is a new member of the Sox gene family, mainly expressed in human nervous tissue. *Hum. Mol. Genet*. 6: 1069–1077.

Kamachi Y, Cheah KS, Kondoh H. **1999**. Mechanism of regulatory target selection by the SOX high-mobility-group domain proteins as revealed by comparison of SOX1/2/3 and SOX9. *Mol. Cell. Biol*. 19: 107–120.

Kamachi Y, Uchikawa M, Kondoh H. **2000**. Pairing SOX off: with partners in the regulation of embryonic development. *Trends Genet*. 16: 182–187.

Kamachi Y, Uchikawa M, Tanouchi A, Sekido R, Kondoh H. **2001**. Pax6 and SOX2 form a co-DNA-binding partner complex that regulates initiation of lens development. *Genes Dev.* 15: 1272–1286.

Kan L, Israsena N, Zhang Z, Hu M, Zhao LR, Jalali A, Sahni V, Kessler JA. **2004**. Sox1 acts through multiple independent pathways to promote neurogenesis. *Dev. Biol.* 269: 580–594.

Kim J, Lo L, Dormand E, Anderson DJ. **2003**. SOX10 maintains multipotency and inhibits neuronal differentiation of neural crest stem cells. *Neuron* 38: 17–31.

Kishi M, Mizuseki K, Sasai N, Yamazaki H, Shiota K, Nakanishi S, Sasai Y. **2000**. Requirement of Sox2–mediated signaling for differentiation of early *Xenopus* neuroectoderm. *Development* 127: 791–800.

Komitova M, Eriksson PS. **2004**. Sox–2 is expressed by neural progenitors and astroglia in the adult rat brain. *Neurosci. Lett.* 369: 24–27.

Koopman P, Schepers G, Brenner S, Venkatesh B. **2004**. Origin and diversity of the Sox transcription factor gene family: genome-wide analysis in *Fugu rubripes*. *Gene* 328: 177–186.

Köster RW, Kühnlein RP, Wittbrodt J. **2000**. Ectopic Sox3 activity elicits sensory placode formation. *Mech. Dev.* 95: 175–187.

Kuhlbrodt K, Herbarth B, Sock E, Enderich J, Hermans-Borgmeyer I, Wegner M. **1998**a. Cooperative function of POU proteins and Sox proteins in glial cells. *J. Biol. Chem.* 273: 16050–16057.

Kuhlbrodt K, Herbarth B, Sock E, Hermans-Borgmeyer I, Wegner M. **1998**b. Sox10, a novel transcriptional modulator in glial cells. *J. Neurosci.* 18: 237–250.

Kuroda T, Tada M, Kubota H, Kimura H, Hatano SY, Suemori H, Nakatsuji N, Tada T. **2005**. Octamer and Sox elements are required for transcriptional cis regulation of Nanog gene expression. *Mol. Cell. Biol.* 25: 2475–2485.

Laudet V, Stehelin D, Clevers H. **1993**. Ancestry and diversity of the HMG box superfamily. *Nucleic Acids Res.* 21: 2493–2501.

Laumonnier F, Ronce N, Hamel BC, Thomas P, Lespinasse J, Raynaud M, Paringaux C, Van Bokhoven H, Kalscheuer V, Fryns JP, Chelly J, Moraine C, Briault S. **2002**. Transcription factor SOX3 is involved in X-linked mental retardation with growth hormone deficiency. *Am. J. Hum. Genet.* 71: 1450–1455.

Le N, Nagarajan R, Wang JY, Araki T, Schmidt RE, Milbrandt J. **2005**. Analysis of congenital hypomyelinating Egr2Lo/Lo nerves identifies Sox2 as an inhibitor of Schwann cell differentiation and myelination. *Proc. Natl. Acad. Sci. USA* 102: 2596–2601.

Lee CJ, Appleby VJ, Orme AT, Chan WI, Scotting PJ. **2002**. Differential expression of SOX4 and SOX11 in medulloblastoma. *J. Neurooncol.* 57: 201–214.

Lee YH, Aoki Y, Hong CS, Saint-Germain N, Credidio C, Saint-Jeannet JP. **2004**. Early requirement of the transcriptional activator Sox9 for neural crest specification in *Xenopus*. *Dev. Biol.* 275: 93–103.

Lefebvre V, Li P, de Crombrugghe B. **1998**. A new long form of Sox5 (L-Sox5), Sox6 and Sox9 are coexpressed in chondrogenesis and cooperatively activate the type II collagen gene. *EMBO J.* 17: 5718–5733.

Ludwig A, Rehberg S, Wegner M. **2004**. Melanocyte-specific expression of dopachrome tautomerase is dependent on synergistic gene activation by the Sox10 and Mitf transcription factors. *FEBS Lett.* 556: 236–244.

Maka M, Stolt CC, Wegner M. **2005**. Identification of Sox8 as a modifier gene in a mouse model of Hirschsprung disease reveals underlying molecular defect. *Dev. Biol.* 277: 155–169.

Malas S, Postlethwaite M, Ekonomou A, Whalley B, Nishiguchi S, Wood H, Meldrum B, Constanti A, Episkopou V. **2003**. Sox1–deficient mice suffer from epilepsy associated with abnormal ventral forebrain development and olfactory cortex hyperexcitability. *Neuroscience* 119: 421–432.

Mizuseki K, Kishi M, Matsui M, Nakanishi S, Sasai Y. **1998**a. *Xenopus* Zic-related–1 and Sox–2, two factors induced by chordin, have distinct activities in the initiation of neural induction. *Development* 125: 579–587.

Mizuseki K, Kishi M, Shiota K, Nakanishi S, Sasai Y. **1998**b. SoxD: an essential mediator of induction of anterior neural tissues in *Xenopus* embryos. *Neuron* 21: 77–85.

Mori-Akiyama Y, Akiyama H, Rowitch DH, de Crombrugghe B. **2003**. Sox9 is required for determination of the chondrogenic cell lineage in the cranial neural crest. *Proc. Natl. Acad. Sci. USA* 100: 9360–9365.

Nambu PA, Nambu JR. **1996**. The Drosophila fish-hook gene encodes a HMG domain protein essential for segmentation and CNS development. *Development* 122: 3467–3475.

Nishimoto M, Fukushima A, Okuda A, Muramatsu M. **1999**. The gene for the embryonic stem cell coactivator UTF1 carries a regulatory element which selectively interacts with a complex composed of Oct–3/4 and Sox–2. *Mol. Cell. Biol.* 19: 5453–5465.

Nowling T, Bernadt C, Johnson L, Desler M, Rizzino A. **2003**. The co-activator p300 associates physically with and can mediate the action of the distal enhancer of the FGF–4 gene. *J. Biol. Chem.* 278: 13696–13705.

Ohba H, Chiyoda T, Endo E, Yano M, Hayakawa Y, Sakaguchi M, Darnell RB, Okano HJ, Okano H. **2004**. Sox21 is a repressor of neuronal differentiation and is antagonized by YB–1. *Neurosci. Lett.* 358: 157–160.

Overton PM, Meadows LA, Urban J, Russell S. **2002**. Evidence for differential and redundant function of the Sox genes Dichaete and SoxN during CNS development in *Drosophila*. *Development* 129: 4219–4228.

Paratore C, Eichenberger C, Suter U, Sommer L. **2002**. Sox10 haploinsufficiency affects maintenance of progenitor cells in a mouse model of Hirschsprung disease. *Hum. Mol. Genet.* 11: 3075–3085.

Peirano RI, Goerich DE, Riethmacher D, Wegner M. **2000**. Protein zero expression is regulated by the glial transcription factor Sox10. *Mol. Cell. Biol.* 20: 3198–3209.

Peirano RI, Wegner M. **2000**. The glial transcription factor Sox10 binds to DNA both as monomer and dimer with different functional consequences. *Nucleic Acids Res.* 28: 3047–3055.

Perez-Alcala S, Nieto MA, Barbas JA. **2004**. LSox5 regulates RhoB expression in the neural tube and promotes generation of the neural crest. *Development* 131: 4455–4465.

Pevny LH, Sockanathan S, Placzek M, Lovell-Badge R. **1998**. A role for SOX1 in neural determination. *Development* 125: 1967–1978.

Pingault V, Bondurand N, Kuhlbrodt K, Goerich DE, Prehu M-O, Puliti A, Herbarth B, Hermans-Borgmeyer I, Legius E, Matthijs G, Amiel J, Lyonnet S, Ceccherini I, Romeo G, Smith JC, Read AP, Wegner M, Goossens M. **1998**. Sox10 mutations in patients with Waardenburg-Hirschsprung disease. *Nat. Genet.* 18: 171–173.

Pompolo S, Harley VR. **2001**. Localisation of the SRY-related HMG box protein, SOX9, in rodent brain. *Brain Res.* 906: 143–148.

Potterf BS, Furumura M, Dunn KJ, Arnheiter H, Pavan WJ. **2000**. Transcription factor hierarchy in Waardenburg syndrome: regulation of MITF expression by SOX10 and PAX3. *Hum. Genet.* 107: 1–6.

Rehberg S, Lischka P, Glaser G, Stamminger S, Wegner M, Rosorius O. **2002**. Sox10 is an active nucleocytoplasmic shuttling protein and shuttling is required for Sox10–mediated transactivation. *Mol. Cell. Biol.* 22: 5826–5834.

Remenyi A, Lins K, Nissen LJ, Reinbold R, Schöler HR, Wilmanns M. **2003**. Crystal structure of a POU/HMG/DNA ternary complex suggests differential assembly of Oct4 and Sox2 on two enhancers. *Genes Dev.* 17: 2048–2059.

Rex M, Uwanogho D, Cartwright E, Pearl G, Sharpe PT, Scotting PJ. **1994**. Sox gene expression during neuronal development. *Biochem. Soc. Transactions* 22.

Rex M, Uwanogho DA, Orme A, Scotting PJ, Sharpe PT. **1997**. cSox21 exhibits a complex and dynamic pattern of transcription during embryonic development of the chick central nervous system. *Mech. Dev.* 66: 39–53.

Riethmacher D, Sonnenberg-Riethmacher E, Brinkmann V, Yamaai T, Lewin GR, Birchmeier C. **1997**. Severe neuropathies in mice with targeted mutations in the ErbB3 receptor. *Nature* 389: 725–730.

Rizzoti K, Brunelli S, Carmignac D, Thomas PQ, Robinson IC, Lovell-Badge R. **2004**. SOX3 is required during the formation of the hypothalamo-pituitary axis. *Nat. Genet.* 36: 247–255.

Russell SRH, Sanchez-Sorinao N, Wright CR, Ashburner M. **1996**. The Dichaete gene of *Drosophila melanogaster* encodes a SOX-domain protein required for embryonic segmentation. *Development* 122: 3669–3676.

Sanchez-Soriano N, Russell S. **1998**. The *Drosophila* Sox-domain protein Dichaete is required for the development of the central nervous system. *Development* 125: 3989–3996.

Sanchez-Soriano N, Russell S. **2000**. Regulatory mutations of the *Drosophila* Sox gene Dichaete reveal new functions in embryonic brain and hindgut development. *Dev. Biol.* 220: 307–321.

Schepers GE, Taesdale RD, Koopman P. **2002**. Twenty pairs of Sox: extent, homology, and nomenclature of the mouse and human Sox transcription factor families. *Dev. Cell* 3: 167–170.

Sock E, Rettig SD, Enderich J, Bösl MR, Tamm ER, Wegner M. **2004**. Gene targeting reveals a widespread role for the high-mobility-group transcription factor Sox11 in tissue remodeling. *Mol. Cell. Biol.* 24: 6635–6644.

Sock E, Schmidt K, Hermanns-Borgmeyer I, Bösl MR, Wegner M. **2001**. Idiopathic weight reduction in mice deficient in the high-mobility-group transcription factor Sox8. *Mol. Cell. Biol.* 21: 6951–6959.

Southard-Smith EM, Angrist M, Ellison JS, Agarwala R, Baxevanis AD, Chakravarti A, Pavan WJ. **1999**. The Sox10(Dom) mouse: modeling the genetic variation of Waardenburg-Shah (WS4) syndrome. *Genome Res.* 9: 215–225.

Southard-Smith EM, Kos L, Pavan WJ. **1998**. Sox10 mutation disrupts neural crest development in Dom Hirschsprung mouse model. *Nat. Genet.* 18: 60–64.

Spokony RF, Aoki Y, Saint-Germain N, Magner-Fink E, Saint-Jeannet JP. **2002**. The transcription factor Sox9 is required for cranial neural crest development in *Xenopus*. *Development* 129: 421–432.

Stolt CC, Lommes P, Friedrich RP, Wegner M. **2004**. Transcription factors Sox8 and Sox10 perform non-equivalent roles during oligodendrocyte development despite functional redundancy. *Development* 131: 2349–2358.

Stolt CC, Lommes P, Sock E, Chaboissier M-C, Schedl A, Wegner M. **2003**. The Sox9 transcription factor determines glial fate choice in the developing spinal cord. *Genes Dev.* 17: 1677–1689.

Stolt CC, Rehberg S, Ader M, Lommes P, Riethmacher D, Schachner M, Bartsch U, Wegner M. **2002**. Terminal differentiation of myelin-forming oligodendrocytes depends on the transcription factor Sox10. *Genes Dev.* 16: 165–170.

Stolt CC, Schmitt S, Lommes P, Sock E, Wegner M. **2005**. Impact of transcription factor Sox8 on oligodendrocyte specification in the mouse embryonic spinal cord. *Dev. Biol.* 281: 309–317.

Tanaka S, Kamachi Y, Tanouchi A, Hamada H, Jing N, Kondoh H. **2004**. Interplay of SOX and POU factors in regulation of the nestin gene in neural primordial cells. *Mol. Cell. Biol.* 24: 8834–8846.

Tomioka M, Nishimoto M, Miyagi S, Katayanagi T, Fukui N, Niwa H, Muramatsu M, Okuda A. **2002**. Identification of Sox–2 regulatory region which is under the control of Oct–3/4–Sox–2 complex. *Nucleic Acids Res.* 30: 3202–3213.

Touraine RL, Attie-Bitach T, Manceau E, Korsch E, Sarda P, Pingault V, Encha-Razavi F, Pelet A, Auge J, Nivelon-Chevallier A, Holschneider AM, Munnes M, Doerfler W, Goossens M, Munnich A, Vekemans M, Lyonnet S. **2000**. Neurological phenotype in Waardenburg syndrome type 4 correlates with novel SOX10 truncating mutations and expression in developing brain. *Am. J. Hum. Genet.* 66: 1496–1503.

Tsuda M, Takahashi S, Takahashi Y, Asahara H. **2003**. Transcriptional co-activators CREB-binding protein and p300 regulate chondrocyte-specific gene expression via association with Sox9. *J. Biol. Chem.* 278: 27224–27229.

Uchikawa M, Ishida Y, Takemoto T, Kamachi Y, Kondoh H. **2003**. Functional analysis of chicken Sox2 enhancers highlights an array of diverse regulatory elements that are conserved in mammals. *Dev. Cell* 4: 509–519.

Uchikawa M, Kamachi Y, Kondoh H. **1999**. Two distinct subgroups of Group B Sox genes for transcriptional activators and repressors: their expression during embryonic organogenesis of the chicken. *Mech. Dev.* 84: 103–120.

Ueda R, Yoshida K, Kawakami Y, Kawase T, Toda M. **2005**. Immunohistochemical analysis of SOX6 expression in human brain tumors. *Brain Tumor Pathol.* 21: 117–120.

Wakamatsu Y, Endo Y, Osumi N, Weston JA. **2004**. Multiple roles of Sox2, an HMG-box transcription factor in avian neural crest development. *Dev. Dyn.* 229: 74–86.

Wegner M. **1999**. From head to toes: the multiple facets of Sox proteins. *Nucleic Acids Res.* 27: 1409–1420.

Werner MH, Burley SK. **1997**. Architectural transcription factors: proteins that remodel DNA. *Cell* 88: 733–736.

Wood H, Episkopou V. **1999**. Comparative expression of the mouse Sox1, Sox2 and Sox3 genes from pre-gastrulation to early somite stages. *Mech. Dev.* 86: 197–201.

Young HM, Bergner AJ, Müller T. **2003**. Acquisition of neuronal and glial markers by neural crest-derived cells in the mouse intestine. *J. Comp. Neurol.* 456: 1–11.

Yuan HB, Corbi N, Basilico C, Dailey L. **1995**. Developmental-specific activity of the FGF–4 enhancer requires the synergistic action of Sox2 and Oct–3. *Genes Dev.* 9: 2635–2645.

Zappone MV, Galli R, Catena R, Meani N, De Biasi S, Mattei E, Tiveron C, Vescovi AL, Lovell-Badge R, Ottolenghi S, Nicolis SK. **2000**. Sox2 regulatory sequences direct expression of a (beta)-geo transgene to telencephalic neural stem cells and precursors of the mouse embryo, revealing regionalization of gene expression in CNS stem cells. *Development* 127: 2367–2382.

Zhao G, Skeath JB. **2002**. The Sox-domain containing gene Dichaete/fish-hook acts in concert with vnd and ind to regulate cell fate in the *Drosophila* neuroectoderm. *Development* 129: 1165–1174.

Zhao S, Nichols J, Smith AG, Li M. **2004**. SoxB transcription factors specify neuroectodermal lineage choice in ES cells. *Mol. Cell. Neurosci.* 27: 332–342.

Part II
Transcription Factors in Brain Function

11
The Role of CREB and CBP in Brain Function

Angel Barco and Eric R. Kandel

Abstract

In this chapter, the various functions in the nervous system of the CRE-binding protein (CREB), its paralogues CREM and ATF1, and its co-activator the CREB-binding protein (CBP), are reviewed. A wide array of techniques, including mouse transgenesis, gene targeting, genome-wide analyses and behavioral and electrophysiological studies, have revealed an important role for the CREB activation pathway in different brain functions. These studies have clarified our understanding of the mechanisms whereby CREB activity is regulated and have provided essential clues about how the same family of transcription factors can mediate such apparently distinct biological functions as memory storage, neuroprotection, and drug addiction.

11.1
Introduction

The intracellular second messenger cAMP mediates signal transduction in response to a variety of external stimuli and can exert important long-term effects by regulating gene expression. This regulation is achieved through the activation of DNA-binding proteins that bind to a specific DNA sequence called the cAMP-responsive-element (CRE). These DNA elements are present in one or more copies in the promoter of target genes. The prototypic member of this family of regulatory molecules is the CRE-binding protein (CREB). Although CREB was first identified in 1987 and cloned through studies investigating the expression and regulation of the neuropeptide hormone gene somatostatin [1], it was later found to contribute to the regulation of many other cellular responses, including metabolism control, cell proliferation, cell differentiation, apoptosis, and spermatogenesis. CREB also participates in many processes related to signaling in the nervous system. Indeed, the CREB pathway for regulation of transcription probably represents the transcriptional cascade whose function in the adult brain has been most extensively studied.

The CREB-binding protein (CBP) was initially described in 1993 as a co-activator of CREB required for induction of CRE-driven transcription [2]. Subsequent studies demonstrated that CBP also interacts with other transcription factors, and thereby participates in other essential biological processes, including the control of a variety of neuronal responses.

In this chapter, we describe selected aspects of our understanding of the functions of the CREB activation pathway in the nervous system, focusing in particular on the roles of CREB and CBP proteins. A number of excellent articles have reviewed this area [3–9]. Here, we will focus on the central nervous system, where CREB has been implicated, in the mechanisms of activity-dependent synaptic plasticity [10, 11], neurogenesis [12], neuronal survival [13, 14], synaptic refinement during development [15], drug addiction [16, 17], circadian rhythm [18], and depression [19].

11.2
The CREB Family of Transcription Factors

11.2.1
CREB Family Members and Close Friends

The genes *Creb*, *Crem*, and *Atf–1* encode for a group of highly homologous proteins that is frequently referred as the CREB family of transcription factors. This family is characterized by a conserved basic region/leucine zipper (bZIP) domain that bind to CRE sites found in one or several copies in the promoters of many genes (Figs. 11.1 and 11.2). CREB and ATF1 are the two most abundant CRE-binding proteins expressed in neurons, and are expressed ubiquitously throughout the nervous system and elsewhere. In contrast, CREM is expressed primarily in the neuroendocrine system [3]. However, these three proteins are not the only transcription factors known to bind to CRE sites. Other factors also bind to these sites and contribute to regulate cAMP-mediated gene expression. These other factors share many structural features with the CREB family, and the larger group is frequently referred to as the ATF/CREB family of transcription factors. This larger family includes not only the CREB family, as a subgroup, but also other bZIP proteins such as ATF2, ATF3, and ATF4. Different ATF/CREB proteins can form selective heterodimers with each other, and also with other transcription factors such as AP–1 and C/EBP that do not belong to this family but share a bZIP DNA-binding domain. Members of different families may compete for the same DNA sites and form heterodimers with members of other families. As has occurred in other fields of biology, the nomenclature and organization of these families was often dictated for the history of their discovery rather than for their sequence homology. An exhaustive discussion on the nomenclature of the ATF/CREB family members has been prepared by Hai and Hartman [20].

Creb1 gene

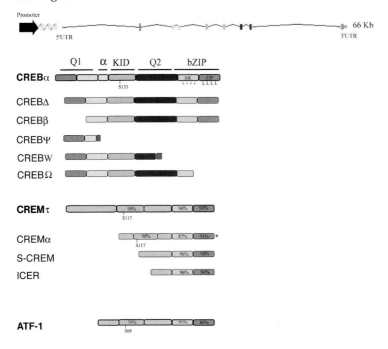

Fig. 11.1 *Creb1* gene structure and domain organization of the CREB family of transcription factors. The members of the CREB family of transcription factors have a highly conserved leucine zipper (ZIP) and adjacent basic region responsible for DNA-binding (BR), a regulatory kinase inducible domain (KID), and two glutamine-rich regions, Q1 and Q2, which contribute to constitutive transcription activation and are less conserved among different family members. The percentages of similar amino acids in CREM and ATF1 with the corresponding bZIP and KID domains of CREB are indicated. The most relevant isoforms of these proteins and represented and the locations of some important sites are indicated and discussed in the text. The *Creb* and *Crem* genes encode both activator and repressor variants. The upper part of the figure shows the exonic organization of the *Creb* gene, only the exons encoding domains present in the most relevant forms of CREB are highlighted in color. (This figure also appears with the color plates.)

11.2.2
Structural Features of the CREB Family of Transcription Factors

A number of structural features can be recognized in most members of the CREB family of transcription factors [6] (see Fig. 11.1). A well-conserved bZIP domain at the carboxy-terminus enables the dimerization between different family members

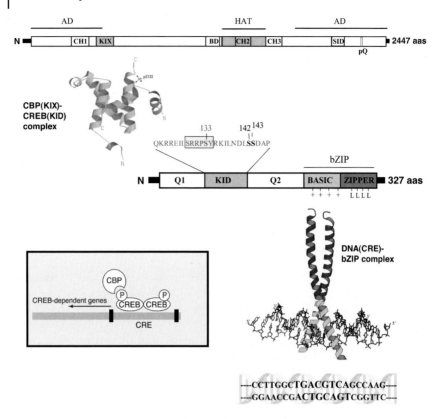

Fig. 11.2 Critical molecular interactions in the CREB activation pathway. CREB has a highly conserved leucine zipper and adjacent basic region responsible for binding to CRE sites and a regulatory kinase inducible domain (KID) that, once phosphorylated, interacts with the KIX domain of CBP. The interactions between the KID and KIX domain and the bZIP domain and the CRE sequence are known with atomic details and are represented here using ribbon structural models. The location of some important domains and sites in CREB and CBP structure are labeled and discussed in the text. (This figure also appears with the color plates.)

and the specific binding to CRE sites. Other well-conserved domains mediate the interaction with the RNApol II complex and with modulators and co-activators of the transactivation activity (see Fig. 11.2). Thus, the central kinase-inducible domain (KID) contains sites for phosphorylation by protein kinase A (PKA) and other kinases, and is flanked by two glutamine-rich domains, designated Q1 and Q2 that contribute to basal transactivation activity. The Q2 domain interacts with the $TAF_{II}130$ subunit of TFIID, recruits the transcription machinery to the promoter, and stabilizes the binding to CRE sites. The mode of action of Q1 remains unclear. The KID contains several serine residues that are critical for the regulation of CREB activity, since their phosphorylation state depends upon neuronal stimulation and regulates the binding of the transcriptional co-activator CBP. The interaction between these two proteins is believed to trigger the inducible transcriptional activity of CREB, as will be discussed later in detail. Constitutive (Q1 and Q2) and inducible (KID) domains act synergistically in response to the wide array of stimuli that trigger CREB-dependent gene expression.

The structures of the two most functionally relevant domains of CREB – the KID and bZIP domains – have been solved respectively by NMR [21, 22] and X-ray diffraction techniques [23], and are now known in atomic detail. This structural knowledge has enabled a better understanding of CREB function and regulation, and may allow the rational design of mutant variants of CREB or drugs targeted to this protein (Fig. 11.2).

11.2.3
Gene Structure and the Regulation of Expression of CREB Family Members

The *Creb*, *Crem*, and *Atf–1* genes have a complex structure, with multiple exons and introns [3, 24]. As observed in many other proteins, the exons that make up these genes are modular in nature, encoding domains with differentiated biological function, such as binding to DNA, transactivation or interaction with co-activators (Fig. 11.1). Whereas *Atf–1* encodes only one major protein product, the *Creb* and *Crem* genes support the expression of multiple splice isoforms. The alternative splicing of CREB and CREM RNAs generates transcripts encoding variants that exhibit distinct capabilities to activate or repress transcription [3, 6]. This is achieved by splicing out domains necessary for transactivation but which are not required for dimerization or DNA binding. Thus, these shorter variants act as repressors that compete for CRE sites as homodimers or that form less active heterodimers. Most alternatively spliced exons in the CREB gene introduce stop codons that result in the termination of translation and lead to the synthesis of truncated proteins without DNA-binding bZIP domains (Fig. 11.1). Given that the CREB promoter contains one CRE site, this alternative splicing may serve to interrupt the autofeedback loop of CREB regulation on its own promoter [25].

In the case of CREM, the situation is even more complex due to:
- the existence of two alternative spliced bZIP domains encoded in this gene provides additional diversity to the capability and specificity of DNA binding and dimerization with other bZIP proteins (Fig. 11.1);
- the alternative use of a translation initiation codon leads to the synthesis of truncated protein; and
- the presence of an alternative, intronic promoter containing several CRE sites drives the expression of a truncated CREM protein, ICER, which exerts strong repressor activity on CRE-dependent transcription. ICER is induced in response to stimulation of the cAMP pathway, but at the same time, represses this pathway, what makes it a good candidate for participating in desensitization mechanisms.

The interaction of these different repressor and activator isoforms of CREM seems to be crucial during spermatogenesis, but their significance in the nervous system is unclear.

11.3
The CREB Binding Protein

11.3.1
Structure and Multifunction

CBP obtained its name, CREB binding protein, because it was originally identified as an interaction partner for CREB [2]. During the following years, many other transcription factors were found to interact physically with CBP or with its homologous p300 [26]. Currently, more than 100 proteins have been found to interact with p300/CBP [8]. Although the primary sequence of P300 and CBP are more than 70% similar, and they have many common interaction partners, these two proteins also appear to have distinct functions and cannot always replace one another [27].

Several structural and functional domains have been identified in CBP (see Fig. 11.2):

- three cysteine/histidine-rich regions (CH1 to CH3) that bind zinc and are involved in protein-protein interactions;
- an histone acetyltransferase (HAT) domain, in the center of the protein;
- a bromodomain (BD) that binds acetylated lysines both in acetylated nucleosome histones and in acetylated transcription factors [28];
- two transactivation domains located in either end of the protein; and finally
- multiple specific interaction domains for different transcription factors such as the KIX domain that mediates CBP interaction with CREB phosphorylated at Ser133, but that do not contain recognizable domains conserved in other proteins.

CBP and p300 carry out a variety of functions related to transcription activation and regulation (Fig. 11.3):

1. Both proteins are co-activators that bridge DNA-binding transcription factors to components of the basal transcription machinery, such as the TATA-box-binding protein (TBP) and the RNApol II complex.
2. The large size of both p300 and CBP (over 2400 amino acids) enables interaction with several proteins at the same time, and thereby to serve as molecular scaffolds that bring a variety of different proteins together to the promoters. For example, the interaction of CBP with MAPKs and the E-Cdk2 complex, promotes not only the phosphorylation of CBP but also the phosphorylation of several CBP-interacting transcription factors [185].
3. p300/CBP are both enzymes that catalyze the transfer of acetyl groups to lysine residues located in the N-terminus of histones – a process that has been associated with enhanced transcription. This enzymatic activity adds a new dimension to the function of CBP as co-activator, since such marking of the chromatin may produce long-term transcriptional effects at specific loci.
4. More recently, CBP was found also to acetylate proteins other than histones. This is the case of the tumor suppressors p53 [29] and pRb [30], diverse transcription factors [31, 32], and components of the RNApol II complex (TFIIE and TFIIF). The functional relevance of this activity in the regulation of the CREB pathway is unknown [33].

A. A bridge between transcription factors that bind to DNA and the RNApol II complex.

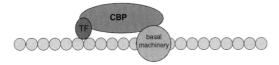

B. A scaffold bringing together different proteins to the promoter.

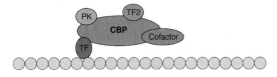

C. Histone acetyltransferase that adds Acetyl groups to histones in nucleosomes.

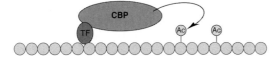

D. Acetyl transferase that adds Acetyl groups to interacting transcription factors.

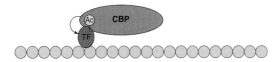

Fig. 11.3 The multiple functions of CBP. The capability of CBP and p300 to co-activate transcription depends on four different activities. TF = transcription factor; PK = protein kinase; Ac = acetyl group. (This figure also appears with the color plates.)

Given the many activities of p300/CBP and their interaction with so many transcription factors, it is not surprising that many critical physiological functions, from cell differentiation to apoptosis, depend on the action of these proteins. Indeed, studies of mouse null mutants in p300 and CBP have demonstrated that at least three out of the four alleles of *cbp* and *p300* must be active during embryogenesis, or the embryos die during early development. Even hemizygotic mutations in either one of these genes have important phenotypic consequences in the development and function of various tissues and organs, including the brain (see Section 11.6.1).

11.4
The CREB Activation Pathway

11.4.1
Post-Translational Regulation of CREB Activity

The phosphorylation of CREB can be triggered by a wide variety of signaling processes, such as an increase of Ca^{2+} through activation of voltage- or ligand-gated channels, an increase in cAMP levels after activation of G-coupled receptors, and an activation of receptor tyrosine kinases by growth factors (Fig. 11.4) [9]. The extent of CREB activation is, therefore, initially affected by proteins, such as cAMP phosphodiesterases or Ca^{2+}-binding proteins, that participate in the regulation of these triggering events. The convergence of multiple intracellular cascades on CREB positions this transcription factor in a ideal situation to integrate different stimuli and regulate neuronal responses. Strikingly, the same stimuli, such as Ca^{2+}-dependent signals or N-methyl-D-aspartate receptor (NMDA-R) signaling, may activate both CREB kinases and CREB phosphatases, leading respectively to the phosphorylation or dephosphorylation of this protein [34]. Furthermore, the final effect of the stimulation depends not only on the nature of the stimulus but also on the cellular context where the stimulation took place. Thus, the synaptic activation of NMDA-R induces CREB phosphorylation and CREB-dependent gene expression, while its extrasynaptic activation shuts off CREB activity by dephosphorylation [35].

The phosphorylation of CREB promotes the recruitment of CBP to the promoter and mediates transcription initiation [2,36–38]. Structural analyses have revealed the molecular details of this interaction and have shown that the phosphorylation of Ser133 of CREB stabilizes its interaction with specific residues in the KIX domain of CBP [22]. Mutations that prevent Ser133 phosphorylation inhibit stimulus-induced CREB activation, while mutations that favor the interaction with CBP lead to CREB-dependent gene expression in the absence of stimulation [39] (see Fig. 11.2). The phosphorylation state of CREB is regulated by the opposing actions of different protein kinases and phosphatases. Both types of enzymes are selectively activated in response to different patterns of neuronal activation. Thus, *in-vivo* studies in the hippocampus have demonstrated that the activation of kinases controls the formation of long-term potentiation, whereas phosphatases are activated during long-term depression [40]. Dozens of different kinases have been reported to phosphorylate CREB *in vitro*, although only a few have been probed to contribute to CREB-dependent transcription regulation *in vivo*. Among them, the PKA, RSK/MAPK and CaM kinase pathways appear specifically to phosphorylate CREB at Ser133 after neuronal stimulation [6, 41]. On the other hand, protein phosphatase 1 and 2A (PP1 and PP2A) appear to be the major CREB phosphatases [42–44]. These enzymes can be activated in response to increased PKA activity as well as Ca^{2+} cascades by the dephosphorylating phosphorylated dopamine and cAMP responsive phosphoprotein (DARP32), which is in turn a substrate of calcineurin [45].

CREB contains additional consensus recognition sites for kinases that suggest a regulation that is more complex than a simple phosphorylation at Ser133 switch.

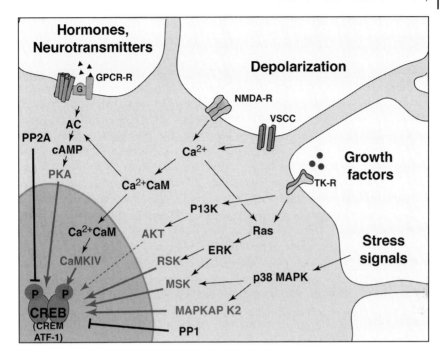

Fig. 11.4 Activation of the CREB signaling pathway. Diverse external stimuli, such as activation of receptors coupled to adenylyl cyclase (AC), such as the G protein-coupled receptors (GPC-R), or opening of Ca^{2+} channels (NMDA-R and VSCC, voltage-sensitive calcium channels), activate protein kinases pathways that converge on CREB phosphorylation at Ser133. Phosphorylation of this residue promotes the recruitment of the co-activator CBP and initiates the transcription of targets genes. However, this is an extremely simplified vision of these regulatory processes. Multiple layers of complexity in the CREB pathway allow the integration of diverse cytoplasmic signals and the divergence of nuclear responses (cartoon adapted from Lonze and Ginty (2002) [9]). (This figure also appears with the color plates.)

The elevation of nuclear Ca^{2+} induces phosphorylation at both Ser 133 and Ser142 by CaM kinases, and this phosphorylation can, in place, mediate a new phosphorylation event at Ser143 by casein kinase II (CKII) [46]. Although previous studies suggested that phosphorylation at Ser142 had a negative effect on CREB-dependent expression by physically disrupting the interaction between CREB and CBP [21], more recent investigations have demonstrated positive regulation by phosphorylation at this site and have suggested that some forms of CREB-dependent gene expression may not require the direct recruitment of CBP by phospho-CREB, but require an alternative, still unknown, mechanism [18, 46].

11.4.2
Regulation of CBP Function

Additional regulatory mechanisms for stimulus-induced CREB-dependent gene expression reside in its cofactor, CBP. CBP itself is a target of several post-transcriptional modifications that modulate its activity. Thus, the transactivation potential of CBP is increased by PKA, CaMKIV and p42/44 MAPKs, while the ability to recruit specific transcription factors selectively increases after PKC phosphorylation [47]. Furthermore, CBP is phosphorylated at Ser301 by CaM kinases [48] and is methylated at Arg residues by the methylase CARM1 [49]. Whereas CBP phosphorylation contributes to CBP-dependent transcription, its methylation appears to block its interaction with CREB and prevent CREB-dependent gene expression. The significance of these regulatory steps in CREB-dependent gene expression *in vivo* is, however, still unclear.

CBP is in limited supply within the cell, and CREB competes with other transcription factor for CBP binding sites. Due to this competition, CREB activity can be indirectly inhibited by the action of transcription factors that are neither functionally nor structurally related to the CREB pathway. This seems to be the case for several viral proteins that sequester CBP from its cellular partners [50].

11.4.3
Other Modulators of the CREB Pathway

In order to appreciate fully the complexity of the regulation of the CREB pathway, it is necessary to better understand other family members and their interactors. The regulated expression and alternative splicing of CREB/CREM adds an additional mechanism for the differential regulation of CREB-dependent gene expression. The expression of transcripts encoding different repressor isoforms of CREM is restricted to specific brain areas [51]. This finding suggests that the presence of CREM antagonists may contribute to determine the region-specific differences observed in CREB-dependent transcription. In particular, the inducible cAMP early repressor (ICER), a group of four proteins produced by an intronic promoter in the CREM gene, appears to play a pivotal role in the regulation of the time-course of expression of various CREB-dependent genes, thereby contributing to CREB function in different processes, from circadian rhythm control to apoptosis. Most of these are related to – but not restricted to – the neuroendocrine system [52]. Other CREB-related protein or even structurally unrelated transcription factors can also contribute to modulate CREB function. For example, ACT, a small LIM protein which is expressed in testis, interacts with CREM and CREB and may activate transcription bypassing the requirements for phosphorylation and CBP interaction [53]. Further studies will be necessary, however, to determine the relevance of these modulatory mechanisms of CREB function in the brain.

11.4.4
CRE-Binding Activity and CREB Downstream Genes

The comparative study of cAMP-responsive promoters and a number of biochemical assays have led to the identification of the consensus palindromic sequence recognized with high affinity by CREB: TGACGTCA. However, the promoters of many cAMP-responsive genes that can also recruit CREB to the promoter often contain only a half-site TGACG sequence.

In-vitro assays [54] and chromatin immunoprecipitation (ChIP) studies *in vivo* [55] have shown that phosphorylation at CREB does not affect its binding to some CRE sites. By contrast, other studies have shown that phosphorylation may enhance the binding affinity of CREB, although unphosphorylated CREB can still bind to the CRE [56, 57].

Recent studies have revealed that some CRE sites are occupied under basal conditions, whereas others are occupied only after cAMP stimulation. Moreover, different cell types express different sets of target genes in response to CREB activation, suggesting that the availability of the promoter for CREB binding is likely different in different cell types. These differences can be due to the specific methylation of some CRE sites that blocks the binding of CREB [58].

Even in a given cell the activation of gene expression by CREB is complex. The electrical or chemical stimulation of a neuron may trigger or enhance the expression of a number of genes which, based on their time-course of induction, can be classified as immediate early genes (IEGs), intermediate, and delayed-response genes. In each of these groups, there are genes that contain CREs in their promoter and are potentially regulated by CREB. However, their time-course and pattern of expression do not correlate precisely with each other and with the kinetics of CREB phosphorylation. This is the case of c-fos and BDNF, two well-known CREB-dependent genes. Whereas c-fos peaks within minutes of stimulation and declines quickly, the induction of BDNF is much slower and reaches a maximum within a few hours, when c-fos has returned to basal levels. Since the transcription of either one of these genes correlates precisely with CREB phosphorylation at Ser133, it is suggested that additional transcription factors or modulators might participate in the induction of these genes.

Earlier studies, both *in vitro* and *in vivo*, demonstrated the participation of CREB in the regulation of the expression of more than 100 genes, while the current widespread application of transcriptional profiling tools, such as high-density expression microarrays, has identified even more potential targets. However, it is still not clear how many of these putative downstream genes are really regulated under physiological conditions [6, 9]. The current list of target genes is heterogeneous, and includes genes with very diverse functions, from transcription and metabolism regulation to cell structure or signaling (Table 11.1). Many CREB targets, such as c-fos, egr1 or C/EBPβ, are themselves transcription factors, the induction of which may trigger a second wave of gene expression. These indirect targets cannot be clearly distinguished using expression arrays, and this limitation has encouraged the use of more sophisticated screening methods.

Table 11.1 Non all-inclusive list of putative CREB target genes categorized into functional groups.

Structural	αA-crystallin; E-cadherin; Fibronectin; ICAM–1; Light neurofilament tracheobronchial mucin; Neurofilament 68 kDa (NF-L); Non-muscle myosin heavy chain.
Cellular metabolism and transport	S-adenosylmethionine decarboxylase; Amino levulinate synthase (ALA-S); Aquaporin–2 (AQP–2); Aryl hydrocarbon receptor; Branched-chain α-keto acid dehydrogenase complex; Carnitine palmitoyl transferase; Cyclooxygenase–2 (COX–2); Cystic fibrosis transmembrane conductance regulator (CFTR); Cytochrome c; Glucose transporter 2 (GLUT2); Glutamine synthetase (GS); Glutathione-S-transferase A3 subunit; Heme oxygenase–1 (HO–1); Hexokinase 2; HMG-CoA synthase; High-mobility-group CoA synthase; Kv3.1 K$^+$- channel; Lactate dehydrogenase; Na$^+$/K$^+$-ATPase α; Neuron-specific enolase (NSE); Ornithine decarboxylase; Phosphoenolpyruvate carboxykinase (PEPCK); Pyruvate carboxylase; Pyruvate dehydrogenase kinase 4; Secretogranin II; Serine dehydratase; Spermine synthase; Superoxide dismutase 2 (SOD2); Type II deiodenase; Ubiquitin-conjugating enzyme; Uncoupling protein–1 (UCP1); Uncoupling protein–2 (UCP2); Uncoupling protein–3 (UCP3).
Signal transduction	14–3–3-ε; Amphiregulin; Bcl-2; BRCA1; Cardiotrophin; Class I MHC; Class II MHC β chain; Cyclin A; Cyclin D1; Cyclin-dependent kinase 5; Cytotoxic T-cell protease 4; DNA polymerase-β; Fibroblast growth factor–6 (FGF–6); Flt–1; GADD34; GEM; Glucose-regulated protein 78; Huntingtin; Immunoglobulinχ 3' enhancer; Inducible nitric oxide synthase (iNOS); Inhibin α; Insulin; Insulin-like growth factor I (IGF-I); Interleukin–2; Interleukin–6; Leptin; Migration inhibitory factor; Mitogen-activated protein kinase kinase phosphatase (MKP–1); Neurofibromatosis 1 (NF–1); Neuronal nitric oxide synthase (nNOS); NoxA; P15INK4b; PDE4D; Proliferating cellular nuclear antigen (PCNA); Proliferin; Prostaglandin synthase–2 (PGS2); Retinoblastoma; Serum and glucocorticoid inducible kinase; Spermatid nuclear transition protein; T-cell receptor-α; Transforming growth factor-β2 (TGF-β2); TrkB; Tumor necrosis factor-α.
Neurotransmission	Acetylcholinesterase (AchE); α1–GABAA receptor; β$_1$-adrenergic receptor; β$_2$-adrenergic receptor; Angiotensinogen; Arylalkylamine N-acetyltransferase (AA-NAT); Brain-derived neurotrophic factor (BDNF); Cardiotrophin–1 (CT–1); Calcitonin gene-related peptide (CGRP); Cholecystokinin (CCK); Chromogranin A; Chromogranin B; Corticotropin-releasing hormone; Dopamine β-hydroxylase (DβH); Follistatin; Galanin; Galanin receptor1 (GalR1); Gastric inhibitory polypeptide receptor (GIPR); Glycoprotein hormone α subunit; Gonadotropin-releasing hormone receptor (GnRHR); Human chorionic gonadotropin-α (hCG-α); Inducible nitric oxide synthase (iNOS); Inhibin A; Murine gastrin-releasing peptide receptor (mGRP-R); Neurotensin/neuromedin N (NT/N); Norepinephrine transporter (NET); Pituitary adenylyl cyclase activating polypeptide (PAC-AP); Preprotachykinin A; Prodynorphin; Proenkephalin; Proglucagon; Prohormone convertase; Secretogranin; Secretogranin II; Somatostatin; Somatostatin receptor (ssr–2); Substance P receptor; Synapsin I; tPA; Tyrosine aminotransferase; Tyrosine hydroxylase; Vasoactive intestinal polypeptide (VIP); Vasopressin (AVP); Vesicular monoamine transporter (VMAT).
Transcription	Activating transcription factor–3 (ATF–3); C/EBP-β; c-fos; c-jun; c-maf; CREB; Egr–1; Epidermal growth factor–1; Fra1; Glucocorticoid receptor; ICER; JunB; JunD; Krox–20; Microphthalmia; mPer1; mPer2; NF-IL6; Nur77; Nurr1; OCA-B; Pit–1; PPAR γ-coactivator 1; STAT3.

Euskirchen and coworkers have recently reported the result of a ChIP-on-chip screening for CREB target genes in chromosome 22, which represents about 1% of the human genome. In this technique, chromatin is immunoprecipitated using a CREB antibody, DNA is amplified by PCR and hybridized to a microarray displaying promoter sequences that, given the current limitations of this technology, only included selected sequences of chromosome 22 thought to be part of promoters. This study revealed 215 binding sites corresponding to 192 loci [59]. Most of these sites did not correspond to classical CREs but rather to shorter variants, and were located in regions not corresponding to known promoters. Only a subset of these candidate genes located in chromosome 22 was affected by forskolin in cultured cells.

Another recent screening for CREB target genes [60], also based on ChIP-on-chip technology but with a genome-wide scope, showed that CREB may occupy at least 4000 promoter sites *in vivo*, depending on the methylation state of the proximal CRE sites. Although the profiles of CREB occupancy were very similar in different human tissues, only a small proportion of CREB target genes were induced in any cell type, most likely due to the differential recruitment of CBP to those promoters. This result confirms that the phosphorylation of CREB on Ser133 is not sufficient to predict transcriptional activation, and the presence of additional CREB regulatory partners is required.

Using a different experimental approach, Impey and colleagues applied serial analysis of chromatin occupancy (SACO), a new technique that combines ChIP and SAGE methodologies, to the identification of all the genes regulated by CREB in rat PC12 cells [61]. In agreement with previous studies, these studies revealed that most (63%) of the CRE sites occupied by CREB in the basal condition identified in the analysis were located in or near transcriptionally active regions. Indeed, many of the identified sequences mapped into the promoter region of previously proposed CREB-regulated genes, although some well-characterized CREB-dependent genes were not identified in this screening, confirming again that at some promoters the binding of CREB is not constitutive. Other interesting findings were the frequent occurrence of functional CRE in bidirectional promoters, and the possible regulation by CREB of the expression of antisense transcripts and miRNAs.

The new availability of the complete mouse and human genome sequences and the development of advanced bioinformatics tools for their analysis have enabled new approaches to be made to this problem. In order to identify new putative target genes, Conkright and colleagues used a hidden Markov model (HMM) focused on known CREB binding sites [55], and confirmed that functional CRE sites are usually located proximal to the transcription start site and that the capability to drive cAMP-responsive transcription is reduced when the CRE site is moved far from the TATA box. Only those promoters with a TATA box proximal to the CRE site exhibited a strong up-regulation in response to forskolin, an activator of adenyl cyclases that increases intracellular levels of cAMP and activates the CREB pathway. Strikingly, when the authors inserted a TATA motif into a TATA-less promoter bearing CRE sites, they found an enhanced activation of the promoter in response to forskolin, showing that external sites contribute greatly to the functionality of CRE sites. The results of this study also confirmed that CREB occupancy of the CRE site does not guarantee transcriptional activation by CREB.

The application of these new, unbiased, genome-wide screening approaches to neurons or specific brain tissues will eventually unravel the CREB regulon and will greatly improve our understanding of how transcription regulation affects brain function.

11.5
Functions of the CREB Activation Pathway in the Nervous System

CREB has been involved in many aspects of nervous system function. In a given neuron, the complex combination of context-specific signals and the differential expression and activation of transcription factors and modulators enables the participation of the CREB pathway in a specific yet large variety of biological processes. Furthermore, the tremendous cellular complexity of the brain enables the same cellular mechanism (e.g., a synaptic plastic change) to underlie very different phenotypic effects. We have outlined above how the nature and intensity of the synaptic stimuli can lead to activation of various signaling pathways and the differential phosphorylation of CREB, CBP and possibly other molecular partners [41,46,62]. In combination with the unique chromatin configuration and molecular environment of different neuronal types these inductive events may determine the activation of different cell type- or tissue-specific patterns of gene expression.

The following sections will discuss first, the contribution of the CREB pathway to the control of intracellular homeostasis and neuronal responses (Fig. 11.5), and second, the participation of CREB directly in some complex functions of the nervous system. The final point for discussion will be how malfunction of the CREB activation pathway leads to important pathological conditions in the nervous system.

11.5.1
Regulation of Cellular Responses by the CREB Pathway

CREB participates in the regulation of neuronal responses to a variety of stimuli, such as changes in intracellular levels of cAMP and calcium, or the binding of neurotrophins or cytokines to their membrane receptors. Although, the activation of the CREB pathway has been involved in a number of different steps along the life of a neuron, from neurogenesis to neuronal death, two major cascades of gene expression have been delineated. The first cascade relates CREB to neuronal survival and protection through the transcriptional control of neurotrophins and anti-apoptotic genes, such as Bcl–2 or BDNF. The second cascade presents CREB as a critical component of the molecular switch that controls the duration of synaptic plasticity changes by regulating the expression of BDNF, tPA, EGR1 and other genes necessary for the formation of new synapses and the strengthening of existing synaptic connections. These two pathways converge on more than just CREB; other molecules, such as ERK or BDNF, appear to play critical roles in both processes, suggesting that the gene expression programs for activity-dependent survival and activity-dependent plasticity may largely overlap. In addition, CREB function has been

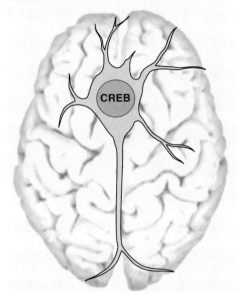

Fig. 11.5 CREB is involved in a variety of processes in the nervous system. Complex brain functions and diverse neuropathologies (in gray) have been related to CREB function. This variety of systemic responses is a consequence of the central role of the CREB pathway in controlling neuronal responses (in orange). Some of these processes are closely related and changes in one of them may influence to others.

also involved in other cellular functions, such as neurogenesis and axonal outgrowth. The participation of CREB in these processes is less well understood in molecular terms, and is likely connected to one of the two major expression cascades discussed above. Thus, although the consequences of CREB activation are diverse, most of them seem to relate to a few specific activation cascades.

11.5.1.1 CREB is Important for Neuronal Survival and Neuroprotection

Strong evidence supports a role for the CREB pathway in neuronal survival. Indeed, CREB-mediated gene expression is necessary and sufficient for the survival of several neuronal subtypes *in vitro* [13,63,64]. Riccio and coworkers screened for candidate survival genes regulated by CREB, and found that bcl–2 was induced in sympathetic neurons and PC12 cells in response to NGF and in cortical neurons in response to BDNF [64]. Bcl–2 belongs to a large family of related proteins that regulate apoptosis, and its promoter contains a consensus CRE site required for CREB-regulated transcription in response to injury and neurotrophin-induced neuronal survival signals [65]. Other pro-survival factors regulated by CREB, such as the neurotrophin BDNF, are also likely contribute to this protective effect.

These *in-vitro* studies have been recently validated *in vivo* using CREB-deficient mice. CREB$^{-/-}$ mutant mice die shortly after birth. Interestingly, dorsal root ganglia (DRG) sensory neurons from these mice showed a complete requirement of CREB activity for survival, whereas neurons of the CNS were not compromised after elimination of CREB [66]. However, when both CREM and CREB are eliminated in the same cell, survival is compromised even in the case of CNS neurons. Thus, CREB$^{-/-}$/CREM$^{-/-}$ double mutants exhibit a marked cell loss due to apoptosis in specific brain structures that results in perinatal lethality [67]. Even when CREB and CREM are eliminated in adult mice, there is a progressive neurodegeneration of specific CNS structures, such as cortex, hippocampus, and striatum [67]. These studies suggest that although CREB and CREM can compensate for each other in neurons of the CNS, CREB cannot be replaced in peripheral sensory neurons. The studies also show that the presence of at least one of these two transcription factors is essential for neuronal survival.

ATF1, the third member of the CREB family, also plays a critical role in cell survival. CREB$^{-/-}$/ATF1$^{-/-}$ double mutants die very early during development, prior to implantation, and even CREB$^{-/-}$/ATF–1$^{+/-}$ mice die at day E9.5, much earlier than CREB$^{-/-}$ single mutant [68].

Recent studies have revealed NMDA-R upstream of CREB as important regulators for both normal transmission and for initiating pathological damage. Indeed, glutamate toxicity is a common source of neuronal damage during stroke. A normal burst of excitatory synaptic transmission results in modulable forms of synaptic plasticity, growth, and survival. An abnormally intense burst of activity may result in excess calcium flux and excitotoxicity. Both processes are mediated by NMDA-R signaling to CREB. Bading and colleagues have solved this dichotomy at the molecular level. These authors found that, in dissociated hippocampal neurons, the activation of extrasynaptic NMDA-R caused a transient phosphorylation of CREB without the corresponding CRE-dependent gene expression and led to neuronal death. In contrast, the activation of synaptic NMDA-R produced a robust phosphorylation of CREB in Ser133, promoted survival, and prevented apoptosis of the dissociated hippocampal neurons [35]. This situation, described in cultured neurons, may have its *in-vivo* counterpart during the refinement of neuronal circuits that takes place in development.

CREB is not only required for neuronal survival, but it may also participate in defensive responses to injury. The activation of CREB under conditions of oxidative stress or hypoxia is a critical component of the cell defensive response, and leads to the expression of genes that prevent neuronal apoptosis [14, 67]. Models of global or focal ischemia models have shown that the hippocampal neurons most resistant to an ischemic challenge are those with the highest level of phosphorylated CREB [69, 70]. In agreement with these results, the expression of a CRE-reporter construct is enhanced after ischemic insult [71]. A variety of studies have demonstrated that the expression of a constitutively active CREB variant or overexpression of CREB protected different types of neurons from apoptotic death, whereas dominant-negative CREB mutants have the opposite effect [72–74].

11.5.1.2 CREB is Required for Axonal Outgrowth and Regeneration

Studies conducted both *in vitro* and *in vivo* have demonstrated a role for CREB in axonal and dendritic outgrowth that seems to be independent of its function in neuronal survival, although the gene expression cascades for these two processes may overlap at some points. It is possible to distinguish at least two different aspects of CREB function on axon outgrowth. First, CREB activation promotes the formation of dendrites and growth of cone bodies in cultures of embryonic neurons or neuroblastoma cells [72, 75], and probably also during development of the nervous system, since deficient CREB mutants display reduced growth of cortical dendrites [76] and defects in axonal projections [66]. Second, several studies have revealed a role for cAMP in preventing the inhibition of axonal regeneration in the CNS [77–79], a process that may have important implications for the recovery after lesions of the adult brain or spinal cord. In a recent set of experiments using recombinant adenovirus and transgenic mice, Filbin and colleagues demonstrated that CREB activation is required to overcome the inhibition of axon growth by myelin and can, by itself, promote axonal regeneration *in vivo* [75]. The enzyme arginase I, which is downstream of CREB, and the synthesis of polyamines seem to be the critical effector molecules that mediate CREB actions on axonal regeneration.

11.5.1.3 CREB has a Role in Neurogenesis and Neuronal Differentiation

CREB regulates the differentiation of a variety of cellular types, from hepatocytes, spermatocytes and cells of the immune system to neurons. In the nervous system, CREB plays an important role in controlling proliferation, differentiation, and survival of newborn neurons. Thus, CREB is activated in Schwann cells during development of the peripheral nervous system [80, 81] and the proliferation of DRG cells seems to be altered in CREB null mice [66].

CREB may also be involved in regulating neurogenesis in the adult hippocampus. The neuronal progenitors residing in the dentate gyrus of adult animals can differentiate to neurons, and are thought to contribute to repairing the brain after injury. The increase in production of new neurons after cerebral ischemic stroke correlates with the activation of CREB-dependent transcription. Moreover, the immature neurons in the dentate gyrus labeled with BrdU also are positive for pCREB [12]. In agreement with this idea, a recent study by Zhu and coworkers showed that the inhibition of the CREB pathway blocked neurogenesis, whilst its enhancement led to an increased number of new neurons [82].

Neurogenesis in the dentate gyrus is regulated by activity, learning, and stress. Indeed, some of the candidate molecules thought to regulate neurogenesis are common to other CREB functions. Thus, CREB is activated through NMDA-R signaling after focal ischemia, and regulates the expression of BDNF that in turn provides an adequate growth environment for the residential precursor cells in dentate gyrus.

11.5.1.4 CREB Activity Contributes to Synaptic Plasticity and the Growth of New Synapses

The term "synaptic plasticity" refers to a variety of activity-dependent processes that result in short-term or long-term changes in synaptic strength. The long-lasting changes in the strength of synaptic connections, such as long-term potentiation (LTP) in the mammalian hippocampus or long-term facilitation (LTF) in *Aplysia* sensory-neurons, are thought to underlie learning and memory storage and to depend on specific alterations in patterns of gene expression [10].

Studies conducted three decades ago in the sea snail *Aplysia* first established the critical role of the cAMP signaling pathway in long-term facilitation, and the long-term strengthening of synaptic connections that takes place during simple forms of learning and memory in this animal [83]. LTF requires new gene expression, and correlates with the formation of new synapses between the presynaptic and post-synaptic neurons [84], suggesting that the new gene products participate in the formation of these new synaptic connections. The first evidence that pointed to CREB as the transcription factor regulating these processes came in 1990 from studies in cultured *Aplysia* sensory neurons in which the intracellular injection of CRE decoy oligonucleotides blocked LTF [85]. Indeed, studies in cultured *Aplysia* neurons have provided the most refined view of the role of CREB in synaptic plasticity (Fig. 11.6).

The neurotransmitter serotonin is liberated by modulatory neurons in *Aplysia* that regulate the strength of the gill-withdrawal reflex in response to a noxious stimulus. Serotonin in turn activates the enzyme adenylyl cyclase (AC) through its interaction with receptors in the membrane of sensory neurons. In dissociated *Aplysia* sensory neurons, a single pulse of serotonin induces a short-term facilitation lasting minutes to hours and mediated by an intracellular increase of cAMP level that releases the catalytic subunit of cAMP-dependent PKA from the regulatory subunits. The liberated catalytic subunits can then phosphorylate substrates in the presynaptic terminals, such as channels and proteins involved in exocytosis, leading to enhanced transmitter availability and release.

Repeated pulses of serotonin cause a persistent increase in the level of cAMP and the recruitment of MAPK activity. Activated PKA and MAPK move to the neuron's nucleus and phosphorylate CREB–1 protein, activating this protein as well as removing the repressive action of CREB–2, an inhibitor isoform of CREB [86]. CREB–1 in turn controls the transcription of several immediate-early response genes, such as ubiquitin hydrolase that contribute to the stabilization of the short-term process [87] and the transcription factor C/EBP that initiates a second wave of gene expression

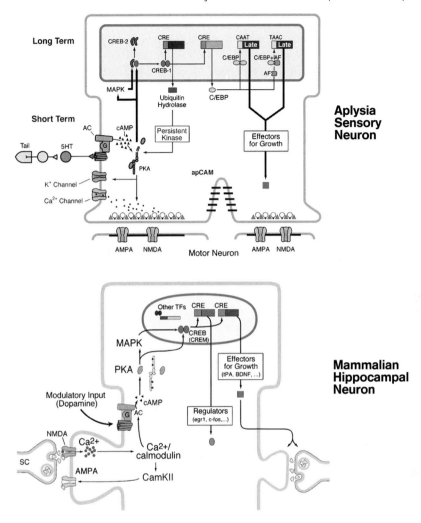

Fig. 11.6 Molecular signaling for short-term and long-term synaptic facilitation in *Aplysia* and mammalian hippocampal neurons. Molecular details are discussed in the text. (This figure also appears with the color plates.)

[88, 89]. The bursts of gene expression triggered by CREB not only stabilize the transient strengthening of synaptic connections, but also lead to the growth of new synapses. In agreement with this model, the injection of phospho-CREB–1 into cultured sensory neurons can by itself initiate long-term synaptic change [90, 91], while the inhibition of ApCREB–2 repressor enhances long-term facilitation [92].

Some of these results reproduced *in vitro* what can also be observed *in vivo* using noxious stimuli, such as electric shocks to the *Aplysia* tail, that trigger the release of serotonin by interneurons that synapse onto sensory neurons.

Most of the upstream signaling cascade leading to CREB activation appears to be conserved through evolution, and many aspects of the role of CREB in synaptic plasticity described in invertebrates have been also observed in the mammalian brain (Fig. 11.6). Although this discussion will focus on the evidence involving CREB in synaptic plasticity in the hippocampus, other brain structure also use CREB for synaptic plasticity changes. CREB function has been related to the regulation of learned fear in the amygdala, motor learning in the cerebellum, and drug abuse in the striatum and diverse regions of the cortex. The activation cascades controlling CREB phosphorylation and the CREB downstream genes mediating the stabilization of the plasticity changes in other brain regions are likely to be similar to those in the hippocampus.

In hippocampal neurons, synaptic release of glutamate triggers the influx of Ca^{2+} through NMDA-R [10] and is thought to increase intracellular levels of cAMP in two ways: (a) through the action of Ca^{2+}-activated adenylyl cyclases [93]; and (b) by the activation of a dopaminergic modulatory system acting through D1/D5 receptors coupled to adenylyl cyclase. When the stimulation reaches a given threshold, the elevated levels of intracellular Ca^{2+} and cAMP activate protein kinases and this leads to the phosphorylation of CREB and the consequent expression of CREB-regulated genes. In agreement with this view, both CREB phosphorylation in CA1 pyramidal neurons and the induction of a CRE-driven *lacZ* reporter construct are triggered by electrical stimuli that induce L-LTP [44,94–96]. The induction of other transcription factors, such as c-fos or EGR–1 [97, 186], and the activation of other kinase cascades, as may be the case for NF-$\varkappa\beta$ [98], also contribute to transcriptional regulation during synaptic plasticity.

A number of genetic studies in mice have confirmed the role of CREB in synaptic plasticity in the mammalian brain (Table 11.2). However, the situation in mammals is more complex than in invertebrates. Silva and colleagues first reported that LTP and long-term memory were defective in mice homozygous for a partial knockout of CREB [99]. However, several groups have had difficulty in replicating these phenotypes. The LTP deficit in this line of α/δ CREB knockout mice has been found to be sensitive to gene dosage and genetic background, indicating that the activity of other genes can compensate for loss of CREB [100]. These discrepancies may be due to compensatory effects between different CRE-binding proteins, since knocking out a specific CREB isoform leads to the overexpression of other CRE-binding proteins, such as CREM and the CREB β isoform, that may compensate the deficiency in CRE-dependent activity and lead to very mild or nonexistent phenotypes [101, 102]. Indeed, a particularly comprehensive analysis using four different strains of CREB-deficient mice, from hypomorphic mutants to neuron-restricted knockouts, failed to demonstrate any deficit in both LTP and LTD in the Schaffer collateral pathway when classical stimulation protocols were used. Although these negative results contrast with the original report by Bourtchuladze et al. [99], they are consistent with other studies of LTP in the hippocampus and amygdala [100, 103, 104, 127]. These results show that although CREB itself often is not sufficient for synaptic plasticity in the mammalian brain due to the compensation by other CRE-binding proteins in CREB-deficient neurons, there is nonetheless strong evidence for a critical role for the CREB pathway in mammalian synaptic plasticity.

Table 11.2 CREB and CBP-deficient mutant mice.

Mouse strain	Phenotype	Reference(s)
CREB$^{-/-}$ (null mutation)	http://www.informatics.jax.org/searches/allele.cgi?7148 Perinatal death Axonal growth defects and degeneration of peripheral neurons	66, 164
CREBα∆ = CREB$^{α\delta-/α\delta-}$ (hypomorphic mutation)	http://www.informatics.jax.org/searches/allele.cgi?3853 Up-regulation of CREBβ and CREM Non clear effects in CRE-driven gene expression Controversial LTP and memory phenotypes Complex addiction phenotype	99–102,126, 127,131,136, 163,165–169
CREBcomp = CREB$^{α\delta-/-}$ (hypomorphic/null mutation)	Normal hippocampal LTP More severe behavioral defects than CREBα∆ mice	100,127
CREB(S142A) (knockin point mutation)	http://www.informatics.jax.org/searches/allele.cgi?7824 Altered circadian rhythms	18
CREB^{CaMKCre7} (postnatal forebrain restricted knockout)	http://www.informatics.jax.org/searches/allele.cgi?7157 Up-regulation of CREM Normal hippocampal LTP and LTD No effect in some hippocampus-dependent tasks	127
CREB^{NesCre} (CNS restricted knockout)	Dwarf phenotype Up-regulation of CREM Normal hippocampal LTP and LTD No effect in some hippocampus-dependent tasks, enhanced tygmotaxis, impaired CTA	127
CREM$^{-/-}$ (null mutation)	http://www.informatics.jax.org/searches/allele.cgi?7154 http://www.informatics.jax.org/searches/allele.cgi?6044 Impaired cardiac function and spermatogenesis Altered emotional and locomotor responses	170–173
CREB^{CaMKCre7}/CREM$^{-/-}$	Progressive neurodegeneration in the hippocampus and dorsolateral striatum	67
CREB^{NesCre}/CREM$^{-/-}$	Perinatal death. Extensive apoptosis of postmitotic neurons	67
ATF$^{-/-}$ (null mutation)	http://www.informatics.jax.org/searches/allele.cgi?21286 No neurological phenotype	68
CREB$^{-/-}$/ATF$^{-/-}$	Early embryonic death (before implantation)	68
CREB$^{-/-}$/ATF$^{+/-}$	Embryonic death (around E9.5)	68
CaMKII-CREB$_{A133}$ (dominant negative transgene)	Normal LTP in amygdala and hippocampus Mild fear conditioning impairment in one out of three lines	103
CaMKII-tTA/tetO-KCREB (inducible dominant negative transgene)	Impaired spatial learning and memory Deficits in some, but not all, forms of LTP	104,105
CaMKII-CREB^{IR} (tamoxifen inducible repressor transgene)	http://www.informatics.jax.org/searches/allele.cgi?32759 Impaired consolidation of fear memories	128
CaMKII-tTA/tetO-VP16CREB (inducible constitutively active transgene)	Lower threshold for L-LTP in hippocampus Enhanced ocular dominance plasticity in visual cortex Reduced inhibition by MAG and myelin	75,106,137
NSE-tTA/tetO-CREB (inducible overexpression of wt protein)	Altered response to cocaine administration	174–176
NSE-tTA/tetO-mCREB (inducible dominant negative transgene)	Anti-depressant-like effect Inhibition of the differentiation and maturation of newborn neurons	176–178

Table 11.2 (continued)

PcP2–CREB (overexpression of wt protein)	Unaltered LTP Impaired habituation to Rotarod	179
CBP$^{-/-}$ (null mutation)	http://www.informatics.jax.org/searches/allele.cgi?3040 Embryonic death (around E11)	180
CBP$^{+/-}$	Skeletal abnormalities Long-term memory and LTP deficits	114,181
CBPΔ$^{-/-}$ (truncated protein)	http://www.informatics.jax.org/searches/allele.cgi?3041 Embryonic death (around E10)	182
CBPΔ$^{+/-}$	Skeletal abnormalities Long-term memory deficit	140,183
P300$^{-/-}$ (null mutation)	http://www.informatics.jax.org/searches/allele.cgi?2245 http://www.informatics.jax.org/searches/allele.cgi?25406 Embryonic death (around E10)	184
CaMKII-tTA/tetO-CBPDN (inducible dominant negative transgene)	Long-term memory deficit	115
CaMKII-CBPΔ1 (dominant negative transgene)	Long-term memory deficit Deficits in some forms of LTP	141

In order to tackle the requirement of the CREB pathway for synaptic plasticity, without the problems associated with the compensation effect of individual CRE-binding proteins, mouse mutants were required that had complete inhibition of CRE-driven gene expression. This was achieved by Pittenger and colleagues, who generated mutant mice expressing KCREB, a dominant-negative form of CREB that prevent its binding to DNA and that can also quench other factors capable of associating with CREB. These authors found that the expression of this inhibitor led to clear deficits in different forms of LTP [104, 105].

The role of CREB in synaptic plasticity has been further explored using the converse approach of examining its sufficiency, instead of its necessity. Using transgenic mice that express a constitutively active CREB variant, VP16–CREB, in CA1 pyramidal neurons, Barco et al. found that enhanced expression of CRE-driven genes favored the formation and stability of LTP [106]. Similarly to what has been described for long-term facilitation in *Aplysia* neurons [91], this study suggested that the products generated after activation of the CREB pathway provide the required support for synaptic strengthening. However, they also showed that the long-term persistence of synaptic strengthening in the mammalian brain requires, in addition to CREB transcription, the concurrence of inductive local events that mark active synapses and stabilize the synaptic changes. Although, the identities of the effector molecules regulated by CREB activity and the molecular mechanisms that underlie the facilitation of LTP are still under investigation, Malenka and colleagues used recombinant Sindbis viruses to confirm that expression of constitutively active CREB favored LTP formation, and suggested that this was the result of enhanced NMDA-R expression and the generation of new, silent synapses [107]. Other CRE-driven genes up-regulated by VP16–CREB, such as BDNF, which contribute to neuronal growth, may also participate in this process [106].

11.5.1.5 CBP, Epigenetics and Long-Term Changes in Neuronal Function

Although, epigenetic mechanisms were widely known to be involved in the formation and long-term storage of cellular information in response to transient environmental signals, the discovery of their putative relevance in adult brain function is relatively recent [86, 108].

During the past few years, a number of studies have explored in detail the chromatin modifications that take place in specific promoters after neuronal stimulation in different systems and paradigms [86,109–111]. Furthermore, several human neurological disorders characterized by severe mental retardation have been directly related to chromatin remodeling. These include Coffin-Lowry syndrome (CLS), X-linked alpha-thalassemia (ETRX), Rett's syndrome (RT) and the Rubinstein-Taybi syndrome (RTS) [112]. In fact, RTS is caused by mutations in the *cbp* gene [113]. The recent characterization of several types of CBP-deficient mice has shown that the HAT activity of CBP is directly involved in late-phase LTP and memory. In turn, these studies have also shown a therapeutic effect in these mutant mice of inhibiting histone deacetylases (HDAC) [114, 115] (see also Section 11.6.2). Moreover, two other recent studies have shown that the induction of LTP and formation of long-term memory in wild-type mice were also enhanced by this family of drugs [33, 116].

All these findings support a critical role for chromatin remodeling in synaptic plasticity and neuronal function. The epigenetic marking of chromatin mediated by CREB activation and CBP function might well underlie the long-term transcriptional effects in specific loci required for long-term modification of synaptic function and have, in consequence, lasting effects on diverse aspect of behavior.

11.5.2
Regulation of Systemic Responses by the CREB Pathway

Given the important cellular functions of CREB activity it is not surprising that this transcription factor had been involved in a variety of systemic responses ranging from memory to addiction and from circadian rhythms to regeneration (see Fig. 11.5). In some cases, the same molecular process seems to be operative in different neuronal cell types and thereby to contribute to different types of brain functions. For example, the role of CREB in memory appears to depend on long-lasting changes in synaptic plasticity in specific hippocampal neurons, whereas its participation in drug addictions depends on plastic changes in neurons of the nucleus accumbens. In other cases, the different functions might really depend on the activation of different downstream genetic programs. For example, CREB is activated in hippocampal neurons under different circumstances, such as after behavioral training or ischemic injury; it is likely that its activation would lead to two different gene programs, the first one required for synaptic plasticity-related to memory, the second participating in neuroprotection.

11.5.2.1 CREB and Memory

Substantial evidence in experimental systems ranging from mollusks to humans indicates that the CREB pathway is a core component of the molecular switch that converts short- to long-term memory. This conclusion is based on independent and parallel evidence that CREB is involved in synaptic plasticity. However, given that synaptic plasticity is widely believed to underlie the formation of long-term memories, it is likely that the same molecular mechanisms that underlie the participation of CREB in synaptic plasticity would also explain the contribution of CREB to learning and memory storage.

As reviewed above, studies in the sea snail *Aplysia* first established the importance of the cAMP and CREB signaling pathway in simple forms of learning and memory in this organism [83, 85]. Parallel findings obtained in other experimental systems soon confirmed the role of CREB as a core component of the molecular switch that converts short- to long-term memory. Thus, *dunce* and *rutabaga*, two of the first memory mutants identified by genetic screenings in *Drosophila* [117], were found to be caused by mutations on genes that participate on the cAMP signaling pathway [118–120]. Genetic manipulation in transgenic flies showed that opposing forms of CREB (activator versus repressor) produce opposite effects on long-term memory (enhancement versus suppression) [121, 122]. These findings in the fly have been recently challenged however by Perazzona et al., 2004, who confirmed the blocking effect in LTM of the CREB repressor, but could not replicate the enhancing effect of the CREB activator.

In the mammalian brain, CREB is phosphorylated [123] and the expression of a CRE-driven *lacZ* reporter construct [124] and diverse CREB downstream genes are induced in CA1 neurons after training in hippocampus-dependent tasks. The correlative evidence, therefore, closely resemble that described for synaptic plasticity in the hippocampus [44,94–96].

Genetic and pharmacological studies have revealed that the activation of the CREB pathway is not just a consequence but rather plays an active role in learning and memory in mammals. A large number of behavioral studies have focused on the analysis of the learning and memory phenotype of CREB mutant mice. Mice homozygous for a deletion of the α and δ isoforms of CREB were originally reported to have a specific deficit in long-term memory, as revealed in several memory tasks [99]. This seminal study represented the first evidence for a direct role of CREB in memory formation in rodents, and it was soon confirmed by experiments in rats, in which the intra-hippocampal infusion of CREB antisense oligos caused deficits in spatial learning [125]. However, further analyses of the CREB hypomorphic mutants showed that the spatial memory defect was sensitive to gene dosage and genetic background [100, 126]. Indeed, the comprehensive analysis of hippocampal function by Balschum and colleagues using four different strains of CREB-deficient mice (including the CREB$^{\alpha\delta-/-}$ hypomorphic mutant) discussed previously in the context of synaptic plasticity, failed to demonstrate any specific deficit in the most classical hippocampus-dependent tasks, including contextual fear conditioning and spatial learning in the water maze [127]. The apparent deficits in the Morris water maze found in some CREB mutants were better explained by an increment in thygmotaxis behavior rather than impaired spatial learning. These discrepancies extend beyond hippocampal function. In the amygdala, where synaptic plasticity processes that are putatively regulated by CREB, have been associated with the formation of fear memories, some studies have shown that CREB-deficient mutants exhibited impaired fear conditioning [99,100,126], whereas others failed to reveal significant deficiencies in this task [103, 127].

These controversial results suggest that either CREB function is dispensable for certain forms of explicit memory or, more likely, that CREB deficiency can be compensated by the action of other CRE-binding factors. As discussed for synaptic plasticity (see Section 11.5.1.4), the existence of different CRE-binding proteins that can compensate each other for the regulation of CRE-driven gene expression together with the parallel role of CREB in neuronal survival complicates the evaluation of the specific role of CREB in learning and memory in mammals.

The application of approaches designed to overcome the obscuring effects of compensation has enabled a more precise examination of the role of CREB on learning and memory. Thus, Pittenger and colleagues showed that the induction of a dominant-negative CREB mutant in the dorsal hippocampus produces spatial memory deficits that could be reversed after turning off the transgene [104]. Silva and coworkers showed that the inducible and transient repression of CREB function specifically blocks the consolidation of long-term fear memories [128]. Josselyn and colleagues have shown, using recombinant herpes viruses, that the acute overexpression of CREB in amygdala facilitated the formation of these types of memories [129], whereas the expression of a dominant-negative CREB mutant inhibited them [130].

CREB has been also involved in other forms of memory, including some hippocampus-independent tasks. Genetic manipulation studies using recombinant viruses or mutant mice have shown that inhibition of CREB leads to deficits in object recognition [104], social transmission of food preferences [131, 132], and conditional taste or odor aversion [127, 133].

In conclusion, the effort of several dozens of research groups during the past ten years has greatly strengthened and refined our understanding of the role of the CREB-dependent transcription in learning and memory, and have also defined the CREB pathway as an attractive target for drugs aimed at improving memory [134]. These studies have also revealed that the role of the CREB pathway in memory in the mammalian brain is more complex than was initially announced, most likely due to the outstanding versatility of function focus of this chapter.

11.5.2.2 CREB and Circadian Rhythms

The discovery of the induction of immediate early genes and the phosphorylation of CREB in neurons of the anterior hypothalamus, and in the suprachiasmatic nucleus (SCN), in a circadian, time-dependent manner supports a role for CREB-dependent gene expression in the re-setting of the mammalian internal clock in response to light. Other CRE-binding proteins work with CREB in synchronization of the circadian clock. Thus, circadian control over ICER expression is observed in the SCN and different neuroendocrine structures such as the pineal and pituitary glands. Regulation of CREB is thought to act on the pineal gland through a genetic loop that involves the daily cycling of CREB and ICER in the promoters of target genes. This daily cycling is thought to be responsible for the rhythmic fluctuations of melatonin, a hormome implicated in the maintenance of the diurnal circadian cycle in synchrony with the day-night cycle. As discussed for other aspects of CREB function, this correlative evidence has been validated by an interesting gene-targeting study [18]. Light and glutamate strongly induce phosphorylation at Ser142 in addition to Ser133. Knockin mice bearing a mutation that blocks this phosphorylation showed light-induced phase shifts of locomotion and an attenuated expression of c-*Fos* and *mPer1*, two genes known to participate in circadian rhythm control in the SCN. These results not only provide genetic evidence supporting a role of CREB in the entrainment of the mammalian clock, but also reveal novel phosphorylation-dependent mechanism of regulation of CREB activity that contributes towards an explanation of its varied responses.

11.5.2.3 CREB Function and Development

Diverse developmental processes in the nervous system have been associated with CREB function. These effects (which are also discussed elsewhere in this chapter) can be divided into two major groups: (a) those due to CREB function in neuronal progenitors and non-neuronal types; and (b) those due to CREB activation in developing and mature neurons.

Brain development does not conclude with the birth of the animal. The outstanding capability of the nervous system for integration and processing of external stimuli rely not only on late developmental processes that refine the neuronal circuits by early experiences, but also on learning in the adult brain. CREB function has been involved in both processes, including different aspects of developmental plasticity, such as ocular dominance in the visual cortex or the formation of anatomical maps in the barrel cortex. Thus, CREB-mediated gene expression is activated by patterns of whisker deprivation in the barrel cortex [135] and following monocular deprivation in the visual cortex [15, 136]. Moreover, CREB seems to play an active role regulating these processes; thus, enhancing the activity of CREB by expressing a constitutively active mutant reverses the decline of plasticity in the visual cortex that takes place during maturation of the juvenile brain [137].

11.6
Dysregulation of CREB Function and Disease in the Nervous System

Since CREB function is involved in so many critical processes in the nervous system, it is not surprising that the consequences of malfunction in its pathway are quite severe. The participation of CREB in signaling cascades that regulate both neuronal survival and plasticity and the connections between these two processes have already been discussed. Some pathological conditions (e.g., drug addiction or memory loss) are more likely caused primarily by failures of synaptic plasticity, whereas other conditions (e.g., neurodegeneration in Huntington disease) are likely due to a defect in neuroprotection. However, given the frequent cross-talking between these two pathways – for example, some neurodegenerative disorders can originate in failures of plasticity, and neuronal loss may have significant effects in synaptic plasticity – it can be difficult to assign a given pathology associated with malfunction of the CREB pathway specifically to either one of them.

11.6.1
CREB and Addiction

Addictive behaviors are generated in response to repeated exposure to drugs of abuse, and can be defined as the loss of control over the use of drugs. Although drugs of abuse are chemically divergent molecules with different activities, their actions converge in a common pathway: the activation of the mesolimbic dopamine system and its forebrain target, particularly the nucleus accumbens. Addiction can be a life-long condition responsible for permanent behavioral abnormalities caused by stable changes in specific brain regions. Recent studies have highlighted the similarities at the cellular and molecular level between learning and memory and addiction. Both processes depend on stimulus-induced, long-lasting changes in neuronal function that correlate at the cellular level with changes in synaptic plasticity at the molecular level with the activation of CREB-dependent gene expression. Therefore, addiction can be considered a pathological manifestation of abnormal synaptic plasticity in specific brain nuclei.

CREB-dependent gene expression seems to be involved in both the acute responses to substance abuse and the development of addiction [16, 17]. The regions of the CNS known to be involved in addiction, such as the locus coeruleus and nucleus accumbens, show significant increases in CREB phosphorylation and CRE-mediated gene expression after exposure to diverse drugs ranging from cocaine to ethanol. Moreover, in CREB mutant mice, a direct role for CREB in addiction has been found, the peculiarities of which vary with the substance abused, the protocol of drug administration, and the brain region involved in the addiction. Overall, these studies suggest that CREB-dependent gene expression may be generally involved in addiction, although its precise role may not be the same under all circumstances.

11.6.2
Mental Retardation

At least two human mental retardation disorders have been directly related to the CREB activation pathway [138]. As mentioned above, mutations in the *cbp* gene lead to a complex autosomal-dominant disease, the Rubinstein-Taybi syndrome, characterized by mental retardation, diverse skeletal abnormalities and a high incidence of neoplasia. The analysis of samples from patients with this syndrome suggested that the loss of the HAT activity of CBP may make an important contribution to this pathology [139]. The recent characterization of four mouse models for Rubinstein-

Taybi syndrome has confirmed this view and demonstrated a direct role of the HAT activity of CBP and the activation of the CREB pathway in this form of mental retardation [114,115,140,141]. In CBP hemizygous mutants, the deficits in some forms of long-term memory correlated with altered chromatin acetylation, and also with a defect in the late phase of hippocampal long-term potentiation (L-LTP) that demonstrated a hippocampal-based component of the disorder [114]. These studies also revealed that CBP may participate in these processes by mechanisms other than co-activation of CREB-dependent gene expression. Specifically, two of these studies revealed a critical role for the HAT activity of CBP and highlighted the importance of epigenetic mechanisms, such as histone acetylation, in neuronal plasticity and memory [114, 115].

Mutations in the gene encoding RSK2, one of the kinases regulating CREB phosphorylation state, also leads to a congenital mental retardation syndrome, the Coffin-Lowry syndrome [138]. An interesting study in human patients correlated their cognitive performance with the cellular capacity to activate RSK2 activity, thus providing evidence for a role for RSK2 and CREB phosphorylation in human learning and memory.

11.6.3
CREB and Age-Related Memory Impairment

The capacity for memory formation declines with age, a process that has been euphemistically termed "benign senescent forgetfulness" or age-associated memory impairment (AAMI). The molecular and cellular events underlying this condition remain unclear, although recent studies have shown that dysregulation of CREB and CBP expression may contribute to the spatial memory deficits of some aged subjects [142, 143]. Interestingly, Bach and colleagues found that rolipram, an inhibitor of a phosphodiesterase that enhances CREB-dependent gene expression, prevented memory decline in old mice [144]. This effect is likely due to the extensively discussed direct role of CREB in synaptic plasticity. However, one intriguing possibility is that enhancing the CREB pathway may prevent memory loss in old animals by reducing neuronal apoptosis or enhancing neurogenesis. These findings encourage the investigation in therapeutic approaches target on the CREB pathway to ameliorate age-related memory deficits [134].

11.6.4
CREB and Neurodegenerative Diseases

The recent finding of the role of CREB-dependent gene expression in neuroprotection and neuronal survival suggests that disruption of this particular genetic pathway may have a critical role in the pathogenesis of some neurological disorders. The putative protective roles of CREB during stroke and spinal cord lesions were discussed previously. In addition, dysfunction in transcriptional regulation seems to play a central role in Huntington disease (HD) and other forms of polyglutamine pathogenesis, and may be involved in early alterations in Alzheimer's disease.

11.6.4.1 Huntington Disease

Dysfunction in transcriptional regulation in the CREB pathway appears to play a central role in pathogenesis in the family of neurodegenerative diseases known as polyglutamine repeat disorders. The most relevant member of this family is HD, a neurodegenerative disorder characterized by motor abnormalities, cognitive dysfunction and other psychiatric symptoms, caused by polyglutamine expansions in the huntingtin protein. At least nine dominantly inherited neurodegenerative diseases are also associated with expansions of a polyglutamine-encoding sequence, including spinocerebellar ataxias (SCAs) [145]. These expansions make the mutant protein toxic to neurons, possibly through abnormal interactions with the polyglutamine tracts of other, normal proteins. HD pathogenesis appears to involve the sequestration of CBP. The C-terminus of CBP, as observed in other transactivation domains, is very rich in glutamine residues that interact with the polyglutamine expansions of mutant huntingtin [146, 147]. As a consequence, CBP may be sequestrated in the cytoplasmic aggregates and depleted from the cell nucleus, resulting in abnormal transcriptional activity and cellular toxicity, although this effect is not observed in all experimental models of HD [148–150]. Conversely, p300 does not contain long poly-Q tracks and does not co-aggregate with the mutant proteins.

Given the critical role of CREB-dependent gene expression in neuroprotection, a severe reduction in functional CBP levels may have severe consequences on neuronal survival and, therefore, play a critical role in the pathogenesis of polyglutamine disorders. Indeed, mice lacking CREB and CREM show progressive neurodegeneration that strikingly resembles that observed during HD [67]. Also consistent with this model, the overexpression of CBP partially rescues the cell death-accompanying expression of mutant huntingtin in neurons [151], whereas the expression of mutant huntingtin leads to a reduction on HAT activity *in vivo* [152]. Moreover, the administration of histone deacetylase (HADC) inhibitors ameliorates poly-Q-dependent neurodegeneration in *Drosophila* and HD mouse models [153–155]. Overall, these data suggest that the reduction on available CBP may lead to neurodegeneration by affecting both the co-activation of CREB-dependent gene expression required for neuronal survival and CBP function on chromatin remodeling.

11.6.4.2 Alzheimer's Disease

Alzheimer's disease (AD) is a progressive neurodegenerative disorder caused by the toxic accumulation of the amyloid β-peptide (Aβ) in neurons and characterized by an initially mild, cognitive impairment followed by severe cortical dysfunctions in later stages. Recent studies in animal models of AD have revealed that the onset of behavioral deficits takes place before the first neuropathological manifestations of the disease, such as the formation of plaques. Shelanski and coworkers have involved the PKA/CREB pathway in the early manifestations of AD, and demonstrated that Aβ treatment of cultured hippocampal neurons leads to the inactivation of PKA and reduced activation of CREB in response to glutamate [156]. These effects were reversed by rolipram or forskolin, compounds that enhance the cAMP-signaling pathway. Moreover, the protective effect of rolipram has been recently confirmed *in vivo* using an Alzheimer mouse model [157]. Although there is no direct evidence that CREB dysfunction contributes to neuronal loss in later stages of AD, the relevance of CREB-dependent gene expression preventing neuronal apoptosis and favoring neuronal survival suggest that enhancement of the PKA/CREB pathway might be beneficial for the treatment of AD.

11.6.5
CREB and Mental Disorders: Depression and other Disorders of Mood

Although our understanding of the molecular bases of depression and other mood disorders is still incomplete, recent studies have provided new insights into the long-term adaptations that underlie the therapeutic effects of treatment with antidepressants, such as 5–hydroxytryptamine (5–HT) and norepinephrine (NE). These studies have revealed a critical role of the cAMP cascade [158]. Both the expression and the activity of CREB are increased by chronic antidepressant treatment [159, 160], suggesting that CRE-driven expression may be one of the targets for these treatments. In agreement with this view, the overexpression of CREB in hippocampus using recombinant herpes virus produced an antidepressant-like effect in behavioral models of depression [161]. However, studies in CREB null mutants and in mice with altered expression of CREB do not support this view [162, 163]. These data suggest that the functional consequences of the regulation of CRE-dependent transcription by antidepressants may be both region-specific and time-dependent.

How does CREB mediate these effects? Among the putative targets, the brain-derived neurotrophic factor (BDNF) is a good candidate for mediating CREB function as anti-depressant. Stress and other environmental insults can induce neuronal atrophy, and the trophic action of BDNF may increase the survival and function of affected neurons. Additional studies are still needed to confirm this model and to understand fully the role of the CREB pathway in mood disorders and antidepressant treatment.

11.7
Conclusions

High-throughput technologies for the analysis of gene expression are revealing the expression programs orchestrated by CREB, and pointing to specific effector molecules that mediate CREB actions in different cell types and physiological contexts. At the same time, technical advances in mouse genetics have allowed one generation of mouse strains with altered CREB function. The anatomical and temporal restriction of the genetic manipulation, combined with the multidisciplinary approach used in the characterization of these mutant mice, have allowed one to address fundamental biological questions related to CREB function unapproachable by previous efforts, such as the molecular nature of memory or addiction. These advances might also, in the near future, enable new therapeutic approaches to tackle nervous diseases that were once thought incurable.

Abbreviations

5–HT	5–hydroxytryptamine
AAMI	age-associated memory impairment
AC	adenylyl cyclase
AD	Alzheimer's disease
BD	bromodomain
BDNF	brain-derived neurotrophic factor
bZIP	basic region/leucine zipper
CBP	CREB-binding protein
ChIP	chromatin immunoprecipitation
CLS	Coffin-Lowry syndrome
CRE	cAMP-responsive-element
CREB	CRE-binding protein
DARP	dopamine and cAMP responsive phosphoprotein
DRG	dorsal root ganglia
ETRX	X-linked alpha-thalassemia
HDAC	histone deacetylase
HAT	histone acetyltransferase
HD	Huntington disease
HMM	hidden Markov model
ICER	inducible cAMP early repressor
IEGs	immediate early genes
KID	kinase-inducible domain
LTF	long-term facilitation
LTP	long-term potentiation
NE	norepinephrine
NMDA-R	N-methyl-D-aspartate receptor
PKA	protein kinase A
PP	protein phosphatase
RT	Rett's syndrome
RTS	Rubinstein-Taybi syndrome
SACO	serial analysis of chromatin occupancy
SCA	spinocerebellar ataxia
SCN	suprachiasmatic nucleus
TBP	TATA-box-binding protein

References

1 M.R. Montminy, L.M. Bilezikjian, *Nature* **1987**, 328, 175–178.
2 J.C. Chrivia, et al., *Nature* **1993**, 365, 855–859.
3 J.F. Habener, C.P. Miller, M. Vallejo, *Vitam. Horm.* **1995**, 51, 1–57.
4 M. Montminy, *Annu Rev Biochem* **1997**, 66, 807–822.
5 D. De Cesare, P. Sassone-Corsi, *Prog. Nucleic Acid Res. Mol. Biol.* **2000**, 64, 343–369.
6 B. Mayr, M. Montminy, *Nat. Rev. Mol. Cell Biol.* **2001**, 2, 599–609.
7 H.M. Chan, N.B. La Thangue, *J. Cell Sci.* **2001**, 114, 2363–2373.
8 R. Janknecht, *Histol. Histopathol.* **2002**, 17, 657–668.
9 B.E. Lonze, D.D. Ginty, *Neuron* **2002**, 35: 605–23.
10 S.J. Martin, P.D. Grimwood, R.G. Morris, *Annu. Rev. Neurosci.* **2000**, 23, 649–711.
11 C. Pittenger, E. Kandel, *C. R. Acad. Sci. III* **1998**, 321, 91–96.
12 S. Nakagawa, et al., *J. Neurosci.* **2002**, 22, 3673–3682.
13 M.R. Walton, I. Dragunow, *Trends Neurosci.* **2000**, 23, 48–53.
14 T.M. Dawson, D.D. Ginty, *Nat. Med.* **2002**, 8, 450–451.
15 T.A. Pham, S. Impey, D.R. Storm, M.P. Stryker, *Neuron* **1999**, 22, 63–72.
16 E.J. Nestler, *Nat. Rev. Neurosci.* **2001**, 2, 119–128.
17 S.C. Pandey, et al., *Alcohol Clin. Exp. Res.* **2005**, 29, 176–184.
18 D. Gau, et al., *Neuron* **2002**, 34, 245–253.
19 E.J. Nestler, et al., *Neuron* **2002**, 34, 13–25.
20 T. Hai, M.G. Hartman, *Gene* **2001**, 273, 1–11.
21 I. Radhakrishnan, et al., *Cell* **1997**, 91, 741–752.
22 D. Parker, et al., *Mol. Cell* **1998**, 2, 353–359.
23 M.A. Schumacher, R.H. Goodman, R.G. Brennan, *J. Biol. Chem.* **2000**, 275, 35242–35247.
24 J.P. Hoeffler, T.E. Meyer, G. Waeber, J.F. Habener, *Mol. Endocrinol.* **1990**, 4, 920–930.
25 T.E. Meyer, G. Waeber, J. Lin, W. Beckmann, J.F. Habener, *Endocrinology* **1993**, 132, 770–780.
26 R. Eckner, et al., *Genes Dev.* **1994**, 8, 869–884.
27 E. Kalkhoven, *Biochem. Pharmacol.* **2004**, 68, 1145–1155.
28 A. Polesskaya, A. Harel-Bellan, *J. Biol. Chem.* **2001**, 276, 44502–44503.
29 W. Gu, R.G. Roeder, *Cell* **1997**, 90, 595–606.
30 H.M. Chan, M. Krstic-Demonacos, L. Smith, C. Demonacos, N.B. La Thangue, *Nat. Cell. Biol.* **2001**, 3, 667–674.
31 J. Boyes, P. Byfield, Y. Nakatani, V. Ogryzko, *Nature* **1998**, 396, 594–598.
32 H.L. Hung, A.Y. Kim, W. Hong, C. Rakowski, G.A. Blobel, *J. Biol. Chem.* **2001**, 276, 10715–10721.
33 S.H. Yeh, C.H. Lin, P.W. Gean, *Mol. Pharmacol.* **2004**, 65, 1286–1292.
34 C. Sala, S. Rudolph-Correia, M. Sheng, *J. Neurosci.* **2000**, 20, 3529–3536.
35 G.E. Hardingham, Y. Fukunaga, H. Bading, *Nat. Neurosci.* **2002**, 5, 405–414.
36 J. Arias, et al., *Nature* **1994**, 370, 226–229.
37 R.P. Kwok, et al., *Nature* **1994**, 370, 223–226.
38 D. Parker, et al., *Mol. Cell. Biol.* **1996**, 16, 694–703.
39 J.R. Cardinaux, et al., *Mol. Cell. Biol.* **2000**, 20, 1546–1552.
40 E.D. Norman, E. Thiels, G. Barrionuevo, E. Klann, *J. Neurochem.* **2000**, 74, 192–198.
41 K. Deisseroth, R.W. Tsien, *Neuron* **2002**, 34, 179–182.
42 M. Hagiwara, et al., *Cell* **1992**, 70, 105–113.
43 B.E. Wadzinski, et al., *Mol. Cell. Biol.* **1993**, 13, 2822–2834.
44 H. Bito, K. Deisseroth, R.W. Tsien, *Cell* **1996**, 87, 1203–1214.
45 P. Svenningsson, et al., *Annu. Rev. Pharmacol. Toxicol.* **2004**, 44, 269–296.
46 J.M. Kornhauser, et al., *Neuron* **2002**, 34, 221–233.
47 K. Zanger, S. Radovick, F.E. Wondisford, *Mol. Cell* **2001**, 7, 551–558.

48 S. Impey, et al., *Neuron* **2002**, 34, 235–244.
49 W. Xu, et al., *Science* **2001**, 294, 2507–2511.
50 M.O. Hottiger, G.J. Nabel, *Trends Microbiol.* **2000**, 8, 560–565.
51 B. Mellstrom, J.R. Naranjo, N.S. Foulkes, M. Lafarga, P. Sassone-Corsi, *Neuron* **1993**, 10, 655–665.
52 B. Mioduszewska, J. Jaworski, L. Kaczmarek, *J. Neurochem.* **2003**, 87, 1313–1320.
53 G.M. Fimia, D. De Cesare, P. Sassone-Corsi, *Nature* **1999**, 398, 165–169.
54 J.P. Richards, H.P. Bachinger, R.H. Goodman, R.G. Brennan, *J. Biol. Chem.* **1996**, 271, 13716–13723.
55 M.D. Conkright, et al., *Mol. Cell* **2003**, 11, 1101–1108.
56 A. Merino, L. Buckbinder, F.H. Mermelstein, D. Reinberg, *J. Biol. Chem.* **1989**, 264, 21266–21276.
57 M. Nichols, et al., *EMBO J.* **1992**, 11, 3337–3346.
58 S.M. Iguchi-Ariga, W. Schaffner, *Genes Dev.* **1989**, 3, 612–619.
59 G. Euskirchen, et al., *Mol. Cell. Biol.* **2004**, 24, 3804–3814.
60 X. Zhang, et al., *Proc. Natl. Acad. Sci. USA* **2005**, 102, 4459–640.
61 S. Impey, et al., *Cell* **2004**, 119, 1041–1054.
62 G.E. Hardingham, F.J. Arnold, H. Bading, *Nat. Neurosci.* **2001**, 4, 261–267.
63 A. Bonni, et al., *Science* **1999**, 286, 1358–1362.
64 A. Riccio, S. Ahn, C.M. Davenport, J.A. Blendy, D.D. Ginty, *Science* **1999**, 286, 2358–2361.
65 R. Meller, et al., *J. Cereb. Blood Flow Metab.* **2005**, 25, 234–246.
66 B.E. Lonze, A. Riccio, S. Cohen, D.D. Ginty, *Neuron* **2002**, 34, 371–385.
67 T. Mantamadiotis, et al., *Nat. Genet.* **2002**, 31, 47–54.
68 S.C. Bleckmann, et al., *Mol. Cell. Biol.* **2002**, 22, 1919–1925.
69 K. Tanaka, et al., *Brain Res.* **1999**, 818, 520–526.
70 M. Walton, et al., *Brain Res. Brain Res. Rev.* **1999**, 29, 137–168.
71 S. Sugiura, et al., *J. Neurosci. Res.* **2004**, 75, 401–407.
72 C.P. Andreatta, P. Nahreini, A.J. Hanson, K.N. Prasad, *J. Neurosci. Res.* **2004**, 78, 570–579.
73 C.P. Glover, et al., *Neuroreport* **2004**, 15, 1171–1175.
74 B. Lee, G.Q. Butcher, K.R. Hoyt, S. Impey, K. Obrietan, *J. Neurosci.* **2005**, 25, 1137–1148.
75 Y. Gao, et al., *Neuron* **2004**, 44, 609–621.
76 L. Redmond, A.H. Kashani, A. Ghosh, *Neuron* **2002**, 34, 999–1010.
77 H. Song, et al., *Science* **1998**, 281, 1515–1518.
78 D. Cai, Y. Shen, M. De Bellard, S. Tang, M.T. Filbin, *Neuron* **1999**, 22, 89–101.
79 J. Qiu, et al., *Neuron* **2002**, 34, 895–903.
80 M.M. Lee, A. Badache, G.H. DeVries, *J. Neurosci. Res.* **1999**, 55, 702–712.
81 B. Stevens, R.D. Fields, *Science* **2000**, 287, 2267–2271.
82 D.Y. Zhu, L. Lau, S.H. Liu, J.S. Wei, Y.M. Lu, *Proc. Natl. Acad. Sci. USA* **2004**, 101, 9453–9457.
83 M. Brunelli, V. Castellucci, E.R. Kandel, *Science* **1976**, 194, 1178–1181.
84 S. Schacher, V.F. Castellucci, E.R. Kandel, *Science* **1988**, 240, 1667–1669.
85 P.K. Dash, B. Hochner, E.R. Kandel, *Nature* **1990**, 345, 718–721.
86 Z. Guan, et al., *Cell* **2002**, 111, 483–493.
87 A.N. Hegde, et al., *Cell* **1997**, 89, 115–126.
88 C.M. Alberini, M. Ghirardi, R. Metz, E.R. Kandel, *Cell* **1994**, 76, 1099–1114.
89 D. Bartsch, et al., *Cell* **2000**, 103, 595–608.
90 D. Bartsch, A. Casadio, K.A. Karl, P. Serodio, E.R. Kandel, *Cell* **1998**, 95, 211–223.
91 A. Casadio, et al., *Cell* **1999**, 99, 221–237.
92 D. Bartsch, et al., *Cell* **1995**, 83, 979–992.
93 S. Poser, D.R. Storm, *Int. J. Dev. Neurosci.* **2001**, 19, 387–394.
94 Y.F. Lu, E.R. Kandel, R.D. Hawkins, *J. Neurosci.* **1999**, 19, 10250–10261.
95 K. Deisseroth, H. Bito, R.W. Tsien, *Neuron* **1996**, 16, 89–101.
96 S. Impey, et al., *Neuron* **1996**, 16, 973–982.

97 W. Tischmeyer, R. Grimm, *Cell. Mol. Life Sci.* **1999**, 55, 564–574.
98 B.C. Albensi, M.P. Mattson, *Synapse* **2000**, 35, 151–159.
99 R. Bourtchuladze, et al., *Cell* **1994**, 79, 59–68.
100 P. Gass, et al., *Learn. Mem.* **1998**, 5, 274–288.
101 E. Hummler, et al., *Proc. Natl. Acad. Sci. USA* **1994**, 91, 5647–5651.
102 J.A. Blendy, K.H. Kaestner, W. Schmid, P. Gass, G. Schutz, *EMBO J.* **1996**, 15, 1098–1106.
103 G. Rammes, et al., *Eur. J. Neurosci.* **2000**, 12, 2534–2546.
104 C. Pittenger, et al., *Neuron* **2002**, 34, 447–462.
105 Y.Y. Huang, C. Pittenger, E.R. Kandel, *Proc. Natl. Acad. Sci. USA* **2004**, 101, 859–864.
106 A. Barco, J.M. Alarcon, E.R. Kandel, *Cell* **2002**, 108, 689–703.
107 H. Marie, W. Morishita, X. Yu, N. Calakos, R.C. Malenka, *Neuron* **2005**, 45, 741–752.
108 J.M. Levenson, J.D. Sweatt, *Nat. Rev. Neurosci.* **2005**, 6, 108–118.
109 Y. Huang, J.J. Doherty, R. Dingledine, *J. Neurosci.* **2002**, 22, 8422–8428.
110 S.W. Park, M.D. Huq, H.H. Loh, L.N. Wei, *J. Neurosci.* **2005**, 25, 3350–3357.
111 A. Mejat, et al., *Nat. Neurosci.* **2005**, 8, 313–321.
112 E.J. Hong, A.E. West, M.E. Greenberg, *Curr. Opin. Neurobiol.* **2005**, 15, 21–28.
113 F. Petrij, et al., *Nature* **1995**, 376, 348–351.
114 J.M. Alarcon, et al., *Neuron* **2004**, 42, 947–959.
115 E. Korzus, M.G. Rosenfeld, M. Mayford, *Neuron* **2004**, 42, 961–972.
116 J.M. Levenson, et al., *J. Biol. Chem.* **2004**, 279, 40545–40559.
117 Y. Dudai, Y.N. Jan, D. Byers, W.G. Quinn, S. Benzer, *Proc. Natl. Acad. Sci. USA* **1976**, 73, 1684–1688.
118 D. Byers, R.L. Davis, J.A. Kiger, Jr., *Nature* **1981**, 289, 79–81.
119 Y. Dudai, A. Uzzan, S. Zvi, *Neurosci. Lett.* **1983**, 42, 207–212.
120 S. Waddell, W.G. Quinn, *Annu. Rev. Neurosci.* **2001**, 24, 1283–1309.
121 J.C. Yin, et al., *Cell* **1994**, 79, 49–58.
122 J.C. Yin, M. Del Vecchio, H. Zhou, T. Tully, *Cell* **1995**, 81, 107–115.
123 S.M. Taubenfeld, K.A. Wiig, M.F. Bear, C.M. Alberini, *Nat. Neurosci.* **1999**, 2, 309–310.
124 S. Impey, et al., *Nat. Neurosci.* **1998**, 1, 595–601.
125 J.F. Guzowski, J.L. McGaugh, *Proc. Natl. Acad. Sci. USA* **1997**, 94, 2693–2698.
126 L. Graves, A. Dalvi, I. Lucki, J.A. Blendy, T. Abel, *Hippocampus* **2002**, 12, 18–26.
127 D. Balschun, et al., *J. Neurosci.* **2003**, 23, 6304–6314.
128 S. Kida, et al., *Nat. Neurosci.* **2002**, 5, 348–355.
129 S.A. Josselyn, et al., *J. Neurosci.* **2001**, 21, 2404–2412.
130 S.A. Josselyn, S. Kida, A.J. Silva, *Neurobiol. Learn. Mem.* **2004**, 82, 159–163.
131 J.H. Kogan, et al., *Curr. Biol.* **1996**, 7, 1–11.
132 J.J. Brightwell, C.A. Smith, R.A. Countryman, R.L. Neve, P.J. Colombo, *Learn. Mem.* **2005**, 12, 12–17.
133 J.J. Zhang, F. Okutani, S. Inoue, H. Kaba, *Neuroscience* **2003**, 117, 707–713.
134 A. Barco, C. Pittenger, E.R. Kandel, *Expert Opin. Ther. Targets* **2003**, 7, 101–114.
135 A.L. Barth, et al., *J. Neurosci.* **2000**, 20, 4206–4216.
136 T.A. Pham, J.L. Rubenstein, A.J. Silva, D.R. Storm, M.P. Stryker, *Neuron* **2001**, 31, 409–420.
137 T.A. Pham, et al., *Learn. Mem.* **2004**, 11, 738–747.
138 E. Trivier, et al., *Nature* **1996**, 384, 567–570.
139 T. Murata, et al., *Hum. Mol. Genet.* **2001**, 10, 1071–1076.
140 R. Bourtchouladze, et al., *Proc. Natl. Acad. Sci. USA* **2003**, 100, 10518–10522.
141 M.A. Wood, et al., *Learn. Mem.* **2005**, 12, 111–119.
142 Y.H. Chung, et al., *Brain Res.* **2002**, 956, 312–318.
143 J.J. Brightwell, M. Gallagher, P.J. Colombo, *Neurobiol. Learn. Mem.* **2004**, 81, 19–26.

144 M.E. Bach, et al., *Proc. Natl. Acad. Sci. USA* **1999**, 96, 5280–5285.
145 C.T. McMurray, *Trends Neurosci.* **2001**, 24, S32–S38.
146 J.S. Steffan, et al., *Proc. Natl. Acad. Sci. USA* **2000**, 97, 6763–6768.
147 A. McCampbell, et al., *Hum. Mol. Genet.* **2000**, 9, 2197–2202.
148 Z.X. Yu, S.H. Li, H.P. Nguyen, X.J. Li, *Hum. Mol. Genet.* **2002**, 11, 905–914.
149 K. Obrietan, K.R. Hoyt, *J. Neurosci.* **2004**, 24, 791–796.
150 D.S. Higgins, K.R. Hoyt, C. Baic, J. Vensel, M. Sulka, *Ann. N. Y. Acad. Sci.* **1999**, 893, 298–300.
151 F.C. Nucifora, Jr., et al., *Science* **2001**, 291, 2423–2428.
152 S. Igarashi, et al., *Neuroreport* **2003**, 14, 565–568.
153 J. S. Steffan, et al., *Nature* **2001**, 413, 739–743.
154 E. Hockly, et al., *Proc. Natl. Acad. Sci. USA* **2003**, 100, 2041–2046.
155 R.J. Ferrante, et al., *J. Neurosci.* **2003**, 23, 9418–9427.
156 O.V. Vitolo, et al., *Proc. Natl. Acad. Sci. USA* **2002**, 99, 13217–13221.
157 B. Gong, et al., *J. Clin. Invest.* **2004**, 114, 1624–1634.
158 A.C. Conti, J.A. Blendy, *Mol. Neurobiol.* **2004**, 30, 143–155.
159 M. Nibuya, E.J. Nestler, R.S. Duman, *J. Neurosci.* **1996**, 16, 2365–2372.
160 J. Thome, et al., *J. Neurosci.* **2000**, 20, 4030–4036.
161 A.C. Chen, Y. Shirayama, K.H. Shin, R.L. Neve, R.S. Duman, *Biol. Psychiatry* **2001**, 49, 753–762.
162 A.M. Pliakas, et al., *J. Neurosci.* **2001**, 21, 7397–7403.
163 A.C. Conti, J.F. Cryan, A. Dalvi, I. Lucki, J.A. Blendy, *J. Neurosci.* **2002**, 22, 3262–3268.
164 D. Rudolph, et al., *Proc. Natl. Acad. Sci. USA* **1998**, 95, 4481–4486.
165 J.A. Blendy, W. Schmid, M. Kiessling, G. Schutz, P. Gass, *Brain Res.* **1995**, 681, 8–14.
166 R. Maldonado, et al., *Science* **1996**, 273, 657–659.
167 Y.H. Cho, K.P. Giese, H. Tanila, A.J. Silva, H. Eichenbaum, *Science* **1998**, 279, 867–869.
168 S.C. Pandey, N. Mittal, A.J. Silva, *Neuroreport* **2000**, 11, 2577–2580.
169 C.L. Walters, M. Godfrey, X. Li, J.A. Blendy, *Brain Res.* **2005**, 1032, 193–199.
170 J.A. Blendy, K.H. Kaestner, G.F. Weinbauer, E. Nieschlag, G. Schutz, *Nature* **1996**, 380, 162–165.
171 F. Nantel, et al., *Nature* **1996**, 380, 159–162.
172 R. Maldonado, C. Smadja, C. Mazzucchelli, P. Sassone-Corsi, *Proc. Natl. Acad. Sci. USA* **1999**, 96, 14094–14099.
173 F.U. Muller, et al., *FASEB J.* **2003**, 17, 103–105.
174 J. Chen, et al., *Mol. Pharmacol.* **1998**, 54, 495–503.
175 N. Sakai, et al., *Mol. Pharmacol.* **2002**, 61, 1453–1464.
176 C. A. McClung, E. J. Nestler, *Nat. Neurosci.* **2003**, 6, 1208–1215.
177 S.S. Newton, et al., *J. Neurosci.* **2002**, 22, 10883–10890.
178 T. Fujioka, A. Fujioka, R.S. Duman, *J. Neurosci.* **2004**, 24, 319–328.
179 C.R. Brodie, et al., *Mol. Cell. Neurosci.* **2004**, 25, 602–611.
180 Y. Tanaka, et al., *Mech. Dev.* **2000**, 95, 133–145.
181 Y. Tanaka, et al., *Proc. Natl. Acad. Sci. USA* **1997**, 94, 10215–10220.
182 Y. Oike, et al., *Blood* **1999**, 93, 2771–2779.
183 Y. Oike, et al., *Hum. Mol. Genet.* **1999**, 8, 387–396.
184 T.P. Yao, et al., *Cell* **1998**, 93, 361–372.
185 N.D. Perkins, et al., *Science* **1997**, 275, 523–7.
186 I. Izguierdo, M. Cammarota, *Science* **2004**, 304, 829–30.
187 B. Perazzona, G. Isabel, T. Preat, R.L. Davis, *J. Neurosci* **2004**, 24, 8823–8.

12
CCAAT Enhancer Binding Proteins in the Nervous System: Their Role in Development, Differentiation, Long-Term Synaptic Plasticity, and Memory

Cristina M. Alberini

Abstract

The CCAAT enhancer binding protein (C/EBP) family of transcription factors is expressed in several cell types, and is implicated in acute phase responses, stress, and terminal differentiation. In the central nervous system (CNS), C/EBPs are expressed in neurons, where they play a role in developmental neurogenesis, neuronal differentiation, and apoptosis. They are also expressed in glia, where they participate in the regulation of energy metabolism and in responses to inflammation. Finally, as part of the gene cascade regulated by cAMP and cAMP response element binding protein (CREB), C/EBP family members play critical roles during long-term synaptic plasticity and memory formation.

12.1
The CCAAT Enhancer Binding Proteins (C/EBPs)

C/EBP is a family of transcription factors that comprise six isoforms defined by distinct genes known as C/EBPα, C/EBPβ, C/EBPγ, C/EBPδ, C/EBPε, and C/EBPζ. Because they were discovered in different laboratories, these genes initially received different names. To avoid confusion, a systematic nomenclature was proposed by Cao et al. [1], in which all members were named C/EBP followed by a Greek letter indicating the chronological order of their discovery (Table 12.1).

The first C/EBP (now known as C/EBPα) was discovered by Landschulz et al. in rat liver nuclei [2] and characterized as a heat-stable DNA-binding protein that bound selectively to the CCAAT motif of several viral promoters, as well as to the core homology domains of many viral enhancers. Later, several homologous proteins that perform a variety of functions in different cell types were identified [3–11].

The C/EBP family belongs to the superfamily of transcription factors known as "b-zip" which, in addition to C/EBP, includes Gcn4, c-Jun, c-Fos and cAMP response element binding protein (CREB) [9, 10]. B-zip indicates the presence of a bipartite structural motif, consisting of a DNA contact region rich in basic amino acids (basic region), and a dimerization interface characterized by a set of amphi-

Table 12.1 Nomenclature of C/EBP genes (modified from [11]).

Gene	Alternative name	Species	References
C/EBPα	C/EBP, c-C/EBP–1	Rat, mouse, human, chicken, bovine, *Xenopus laevis*, *Rana catesbeiana*, zebrafish	1,2,18,87–90
C/EBPβ	NF-IL–6, IL–6-DPB, LAP, CRP2, NF-M, AGP/EBP, ApC/EBP	Rat, mouse, human, chicken, bovine, *Xenopus laevis*, *Aplysia californica*, *Paralichthys olivaveus* (Japanese flounder)	1,3–6,18,70,90–93
C/EBPγ	Ig/EBP–1	Rat, mouse, human, chicken, fish	7,90
C/EBPδ	NF-IL–6β, CRP3, CELF, RcC/EBP–2	Rat, mouse, human, *Rana catesbeiana*, bovine, ovine, zebrafish	1,18,89,90,94–96
C/EBPε	CRP–1	Rat, mouse, human, ovine, fish	18,93,97–100
C/EBPζ	CHOP–10, GADD153	Rat, mouse, human, hamster	8,101,102

pathic alpha-helices in which heptad repeats of leucine are exposed on one side of the helix. This alpha-helix motif intercalates with similar repeats of a dimer partner to produce a domain known as the "leucine zipper", in which the helices are held together by hydrophobic interactions between leucine residues that form a coiled-coil interaction. In addition to this domain, members of the C/EBP family also include an N-terminal transactivating region [10–14].

While the b-zip domain is highly conserved among the members of the C/EBP family (>90%) and the b-zip superfamily, the N-terminal activation domains greatly diverge (<20% sequence identity). C/EBPs – and b-zip proteins in general – function as dimers, and dimerization is required for DNA binding, which is mediated by the basic region. The basic region assumes an α-helical structure when bound to the DNA [15, 16], and its sequence dictates the specificity of binding to the DNA consensus sequence [17]. A model representing a C/EBP homodimer bound to the DNA is shown in Fig. 12.1.

Because the b-zip domain is highly conserved, different C/EBPs are able to form homo- and heterodimers. Moreover, because the trans-activation potential of the various members differs, heterodimerization can produce a variety of complexes that regulate the target gene expression by different mechanisms. These complexes can form different binding structures that target different genes. Alternatively, some isoforms (e.g., C/EBPζ) can act as dominant negative inhibitors. Lastly, these complexes can regulate target gene expression by switching their function from activator to inhibitor. C/EBP can heterodimerize either with other members of the C/EBP family, or with members of different b-zip or other types of transcription factor families, including CREB, Fos, NF-ϰB, Gcn4, and Myc. Heterodimer formation enhances the number of target sequences to which C/EBP can bind in different promoters [18–32].

With the exception of C/EBPζ, the C/EBPs interact with the recognition sequence, identified as the dyad symmetrical repeat RTTGCGYAAY, where R represents A or G

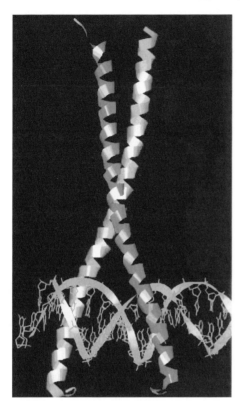

Fig. 12.1 Schematic structural model of the C/EBPβ bound to DNA. Two alpha-helix b-zip domains dimerize and form an inverted Y-shaped structure. The activation domains from each monomer binds to half of the palindromic DNA-binding site. (Reproduced from [11].)

and Y represents C or T. However, the binding site may also include several variations, indicating that C/EBPs are pleiotropic DNA binding factors. C/EBPζ represents an exception because two proline residues in its basic regions disrupt the α-helical structure and its ability to dimerize with other family members, resulting in a heterodimer that cannot bind to the C/EBP binding site in the promoter of target genes [8]. On the other hand, these heterodimers recognize different consensus sequences in the promoters of a subset of genes activated under stress conditions [33]. Thus, C/EBPζ can function either as a dominant negative inhibitor of C/EBP function or as a direct activator of stress genes.

The gene structure of C/EBPs is very simple: C/EBPα, β, γ, and δ are intronless, while C/EBPε and ζ contain two and four exons, respectively [1,3,4–8,11,18]. However, each of the six C/EBP genes can generate the expression of several proteins by alternative use of translation initiation codons in the same mRNA molecule, alternative use of promoters, and differential splicing. Thus, more C/EBP isoforms are expressed than there are encoding sequences [11] (Fig. 12.2).

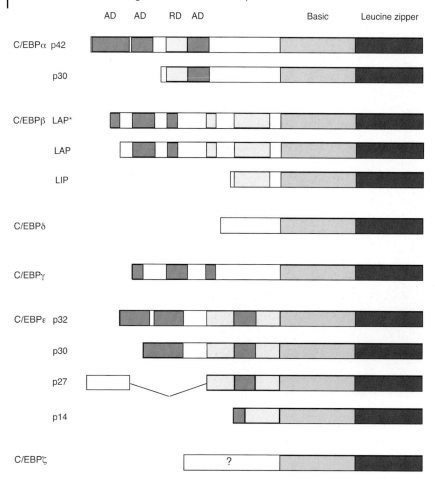

Fig. 12.2 Schematic representation of the C/EBP family members. AD = activation domain; RD = negative regulatory domain. (Modified from Ramji and Foka [11].)

Additionally, post-translational modifications of C/EBP play an important role in directing their function and stability.

C/EBPs are expressed in various tissues, including the CNS, adipose, liver, intestine, lung, adrenal gland, peripheral blood mononuclear cells, and placenta. They play key roles in differentiation, inflammatory response, liver regeneration, metabolism, synaptic plasticity, memory, and in a variety of cellular functions including stress, growth, and apoptosis.

A number of reviews have described in detail the regulation and functional roles of C/EBPs in a variety of tissues [11,13,34–41]. This chapter will provide an overview of the current knowledge about the expression and functions of C/EBP family members in the nervous system. Specifically, the focus will be on the role of C/EBPs in CNS development, differentiation, long-term synaptic plasticity, and memory.

12.2
The Role of C/EBPs in Development and Differentiation

12.2.1
C/EBPs Play a Critical Role in Neurogenesis

C/EBPα, β, δ, and ζ are all expressed in neurons [42–47]. In the following section, some of the most important studies will be discussed that provide evidence for a role of C/EBPs in the CNS and, more specifically, in neuronal development and differentiation.

Development of the cerebral cortex takes place through sequential generation of neurons and glia. In the mouse cortex *in vivo*, neurogenesis occurs during embryonic life, while the majority of gliogenesis takes place postnatally. These differential temporal processes appear to be regulated by specific growth factors, since growth factors can alter the sequence of maturation events. For example, ciliary neurotrophic factor (CNTF) prematurely induces astrocyte formation [48, 49], while fibroblast growth factor 2 (FGF2) is necessary for neurogenesis [50], and platelet-derived growth factor (PDGF) and neurotrophins enhance neurogenesis in the presence of FGF2 [51–53].

According to the recent findings of Menard et al. [54], C/EBPs and mitogen-activated-protein-kinase kinase (MEK) are required for growth factor-regulated cortical neurogenesis. Neural stem cells are multipotent precursors that are capable of self-renewal and give rise to neuronal and glial progenitors. It is not clear how neuronal versus glial fate is determined, but it appears to depend on the combination of both intrinsic cellular mechanisms and environmental signals. In order to identify signaling pathways that promote cortical neurogenesis, Menard et al. [54] used cultured cortical progenitors isolated from embryonic mouse brain at a time when neurogenesis begins. These cells become neurons when cultured in the presence of FGF2, and neurogenesis is enhanced when cultures are supplemented with PDGF [51–53,55–57]. These authors found that neuronal differentiation is mediated by the activation of the extracellular-signal-related kinase (ERK) and C/EBP transcription factors. The application of PDGF to FGF2–treated cultures resulted in an increase in phosphorylation of ERK, a downstream substrate of MEK. The expression of a dominant-negative form of MEK inhibited the expression of neuronal genes even in the presence of FGF2 and PDGF, suggesting that MEK plays a key role in the activation of neuronal-specific pathways.

These investigators also found that cortical progenitors express C/EBPα, β, and δ. When C/EBP was prevented from binding to the C/EBP-DNA consensus sequence by the transfection of an acidic form of C/EBP (A-C/EBP), differentiation of the neurons was suppressed, and the cells were retained in an undifferentiated precursor state. In addition, inhibition of the C/EBPs strongly increased differentiation towards glia in precursors treated with FGF2 and CNTF; this suggested that, in cortical progenitor cells, C/EBP transcriptional activity promotes neurogenesis while it inhibits gliogenesis. The authors discovered that a key mechanism by which C/EBPβ promotes neuronal differentiation during neurogenesis is via phosphorylation of the residue Thr217 by the action of the p90 ribosomal S6 kinase (Rsk). In fact, the expression of a C/EBPβ mutant that mimics phosphorylation at the Thr217 site enhanced the neuronal differentiation of precursor cells. As Rsk is known to be activated by MEK, these results indicate that activation of the MEK-Rsk pathway and C/EBPs (e.g. C/EBPβ) are key events that lead to neurogenesis. Thus, in light of all these results, Menard et al. [54] concluded that the C/EBPs in neuronal precursor cells, as in adipocytes [41], likely act as "differentiation" factors that are responsible for the expression and regulation of specific (in this case, neuronal) genes in response to growth factor signaling.

Further evidence that C/EBPs are involved in the differentiation of neuronal cells has been provided by Cortes-Canteli et al. [43]. These authors found that, similar to hepatocytes [5], lymphocytes [58], and adipocytes [1, 59], C/EBPβ expression promotes differentiation in neuronal cells. Overexpression of C/EBPβ in the N2A neuroblastoma cell line produced a differentiated phenotype with extension of very long neurites, while, conversely, the neurite outgrowth and the differentiated phenotype was inhibited by the expression of the dominant-negative isoform C/EBPζ. These morphological changes were found to be accompanied by an increase in the expression of genes associated with the differentiated state, which appeared to be mediated by C/EBPα. Interestingly, this differentiation required the activation of the phosphatidylinositol 3 kinase (PI3k) signaling pathway, which is known to play an important role in cell growth and survival.

12.2.2
C/EBPs Play a Critical Role in Neuronal Cell Death

Additional findings that confirmed the role of C/EBPs in terminal differentiation also indicated that members of this family of transcription factors, when overexpressed in conditions of growth arrest, are capable of promoting cell death. This is not surprising, as it is known that differentiation is generally accompanied by arrest of the cell cycle. In fact, in N2A neuroblastoma cells that have been deprived of serum for 24 hours, the overexpression of C/EBP β is accompanied by an increase in cell death [43]. This cell death correlates with the activation of p53 and an increase in p21 – two cell-cycle-related proteins known to play a role in apoptosis. Thus, these findings, – like those described above of Menard et al. [54], – indicate that C/EBPs play a critical role in neuronal differentiation. In addition, they suggest a new role for C/EBPs in neurons, namely the activation of apoptosis.

A role of C/EBPs in neuronal apoptosis has also been proposed by Marshall et al. [60], who extended the investigation of the C/EBPs role in differentiation and apoptosis to cerebellar granular cells. These authors found that, in cultured cerebellar granular cells, the cytoplasmic or nuclear localization of C/EBPβ is differentially regulated by calcium influx through either L-type calcium channels or NMDA receptors, respectively. In fact, the L channel-dependent calcium influx correlated with an increase in Ca-calmodulin kinase IV (CaMKIV) and a decrease in C/EBPβ nuclear levels while, conversely, the NMDA-mediated influx correlated with an enhancement of C/EBPβ nuclear levels and resulted in excitotoxic cell death via the activation of the calcium-dependent phosphatase calcineurin. In agreement, suppression of C/EBPs function via expression of a C/EBP dominant negative designed to inhibit the DNA-binding activity of all C/EBPs [61], produced a significant increase in cell survival. Conversely, the overexpression of C/EBPβ triggered a significant increase in cell death. The expression of C/EBPβ appeared to play a critical role in mediating cell death, as its antisense or siRNA-mediated knock-down produced an increase in cell survival. As the C/EBPβ knock-down attenuated NMDA receptor-mediated death, the authors concluded that NMDA receptor stimulation must be involved in the expression, translocation, and/or activation of C/EBPβ. In conclusion, Marshall et al. proposed that C/EBPs, and specifically C/EBPβ, are critically involved in neuronal apoptosis. However, this conclusion is intriguing as, in principle, it is in disagreement with the idea that C/EBPs mediate neurogenesis [54].

How can the opposing roles of C/EBPs in either neurogenesis or neuronal apoptosis be explained? One possible reason might lie in the nature of the differentiation process itself. Since differentiation requires exit from the cell cycle, a differentiated cell likely needs to activate survival mechanisms in order to avoid cell death. Taken together, the results of these studies showed that C/EBPs in neurons, as in many other tissues, act as terminal differentiation factors. Thus, it is reasonable to believe that, under conditions of depletion of growth or survival-promoting factors, the expression of C/EBPs enhances cell death as a result of the induced differentiation state and, perhaps, not because C/EBPs directly regulate the expression of apoptotic genes.

12.2.3
C/EBP Expression in Glia

Glia cells have also been found to express C/EBP family members, and the isoforms better characterized in these cells are C/EBPβ and C/EBPδ. These factors have been found to be expressed in astrocytes and gliomas. The laboratory of Magistretti [62] first described the expression of both isoforms in cortical astrocytes and showed that their level of expression can be regulated in response to the vasoactive intestinal peptide (VIP), VIP-related neuropeptide pituitary adenylate cyclase-activating peptide (PACAP), or noradrenaline (NA). Specifically, C/EBPβ and C/EBPδ were found to be induced by VIP, PACAP, or NA via the cAMP second-messenger pathway, and their expression behaved like that of cAMP-inducible immediate-early genes (IEGs), as indeed the induction of their mRNA occurred in the presence of a protein synthesis inhibitor. Moreover, the analysis of the glycogen metabolism elicited in astrocytes by NA led the authors to suggest that, in these and in other cell types, C/EBPβ and δ regulate the expression of energy metabolism-related enzymes.

The same authors extended the C/EBPβ and δ analysis to other neuromodulators, and reported that these acute-phase proteins are also produced in response to pro-inflammatory cytokines in reactive astrocytes, especially in those surrounding the amyloid plaques of Alzheimer's disease brains. For example, lipopolysaccharides (LPS), interleukin 1b (IL–1b), and tumor necrosis factor α (TNFα) are all able to induce the expression of the C/EBPβ and δ genes in mouse primary astrocytes, and this induction precedes the expression of acute-phase genes. Therefore, it appears that in the brain, as in the liver, C/EBPβ and δ are critical transcription factors involved in inflammation and the regulation of energy metabolism.

As indicated by the results discussed above, the data available thus far seem to emphasize that the cAMP- and MAP-dependent kinase pathways are important for the regulation of the expression of both C/EBPβ and δ in the brain, as well as in other tissues including blood, liver, and adipose. Moreover, they also indicate that both factors often function in the same cells and/or in response to the same activation signals. Cytokines, growth factors, neurotransmitters, and neuromodulators are all capable of mediating the expression of C/EBPβ and C/EBPδ in the CNS. For example, the activation of beta-adrenergic receptors and glucocorticoids seems to act synergistically in regulating the synthesis of nerve growth factor (NGF) in the brain and in C6–2B glioma cells, and the expression of NGF in the brain seems to be under the control of the C/EBPβ, C/EBPδ and CREB transcription factors [63]. These results extended and supported previous findings showing that the NGF-mediated differentiation of pheochromocytoma PC12 cells to neuronal cells is accompanied by the regulation of C/EBPβ expression, and also suggested that C/EBPβ plays an important role in neurotrophin signaling [44].

These data are in many ways very similar to those described below, showing that C/EBPs are regulated by the cAMP-dependent and MAP-dependent pathways during long-term synaptic plasticity responses. Thus, they suggest that common molecular pathways mediate both development/differentiation and long-term synaptic plasticity responses.

Several questions remain to be addressed in the area of CNS or neuronal differentiation and development and the role of C/EBPs:
- What is the role of each C/EBP family member during CNS and neuronal/glia differentiation?
- At which stage of development are they important?
- What are the target genes regulated by C/EBPs in these processes?
- And, finally, which neurotransmitter or growth factor receptors are involved in the C/EBP-dependent responses?

12.3
The Role of C/EBPs in Synaptic Plasticity and Memory

The first demonstration that C/EBPs are expressed in neurons and involved in long-term synaptic plasticity underlying memory formation came from studies in the invertebrate *Aplysia californica*. The simplicity of the *Aplysia* nervous system and the large size of its neurons allowed the establishment of an *in vitro* culture system capable of reproducing the synaptic responses occurring *in vivo* during simple forms of memory. Thus, this system helped pioneer the identification and characterization of molecules and molecular mechanisms underlying memory formation [64–66].

In *Aplysia*, a simple form of memory is the sensitization of the gill and siphon withdrawal reflex. This response occurs when the animal receives a noxious stimulus, such as a tail shock, and therefore learns to enhance its defensive behavior to subsequent neutral stimuli. The defensive response consists of inking and the withdrawal of its siphon and gill, a spout present on the dorsal part of the animal that is used to expel seawater and waste. The duration of gill and siphon withdrawal is used as a measure of the memory. Intensity and duration of this withdrawal is a function of the number of shocks received: a short-term sensitization is induced by a single shock, lasts for a few minutes and depends on post-translational modifications, whereas multiple shocks induce a long-term sensitization that lasts for weeks and requires new protein and RNA synthesis. Short-and long-term memories are mediated by short- or long-term synaptic facilitation of the neurotransmitter release, respectively.

Castellucci et al. [67] and Montarolo et al. [68] found that it is possible to study *in vitro* the changes induced *in vivo* during behavioral sensitization by isolating and culturing the main monosynaptic component of the circuit that contributes to behavioral sensitization. When this sensory-motor neuron circuit is treated with serotonin (a neurotransmitter known to mediate sensitization) it undergoes cellular and synaptic changes similar to those associated with the behavioral response of sensitization. These changes culminate in an increase of the neurotransmitter release (facilitation) at the sensory-motor synapse that can easily be measured and quantified by electrophysiological recording. A single serotonin application to the culture evokes a facilitation that last only a few minutes and requires post-translational modifications, while repeated applications of serotonin result in a long-term facilitation that lasts for days and requires new RNA and protein synthesis. Both short- and long-term facilitation are mediated by the activation of the cAMP-protein kinase A (PKA) signaling pathway. However, only long-term facilitation depends on transcription and translation.

C/EBP was the first transcription factor identified and found to be required in long-term facilitation. This discovery was inspired by the finding of Dash et al. [69] that the activation of cAMP in *Aplysia* cultures leads to the expression of genes that are under the control of cAMP response elements (CRE), and that this CRE-dependent gene expression is required for long-term facilitation. By screening an expression library with C/EBP DNA-binding sequences, Alberini et al. [70] cloned C/EBP from *Aplysia* CNS (ApC/EBP) and found that this factor is expressed in neurons. ApC/EBP in *Aplysia* CNS is present as two splicing isoforms of the same intronless gene. The spliced fragment is a very short sequence in the N-terminal activation domain. Similar to the other C/EBPs, ApC/EBP carries a conserved C-terminal b-zip domain and a very divergent activation domain at the N-terminus. Interestingly, among the mammalian C/EBP isoforms, ApC/EBP appears to be more similar to the C/EBPβ, and, interestingly, they show conserved consensus sequences for the phosphorylation mediated by MAP kinase and CaM Kinase II.

ApC/EBP expression in *Aplysia* neurons is undetectable in the resting condition; however, its transcription and translation are induced in a typical IEG manner following serotonin stimulation. This induction can be mediated by cAMP activation, but it also occurs following injury.

ApC/EBP is an essential transcription factor of long-term facilitation, as its knockdown mediated by antisense DNA or by blocking antibodies results in impaired long-term facilitation, but has no effect on short-term facilitation [70]. The requirement for ApC/EBP during long-term facilitation in cell culture lasts for a relatively long time (more than 9 hours), suggesting that the ApC/EBP-dependent transcriptional phase essential for long-term facilitation is long-lasting. Further studies by Lee et al. [71] confirmed these finding by showing that blocking the expression of ApC/EBP by RNA interference (RNAi) disrupts long-term facilitation without affecting short-term facilitation. Interestingly, these authors also showed that the expression of ApC/EBP in cultured sensory neurons is sufficient for mediating the consolidation of short-term into long-lasting facilitation. In fact, the application of a single pulse of serotonin on sensory neurons overexpressing ApC/EBP produces long-term facilitation, instead of short-term facilitation [71]. Hence, one of the roles of ApC/EBP is probably to mediate structural changes of the synapses, which are believed to represent the storage of information.

Therefore, C/EBP family members are an essential component of the transcriptional phase required for long-term synaptic plasticity in an *in vitro* model of memory formation. In summary, under conditions that induce synaptic plasticity in the CNS, these transcription factors are:
- regulated as IEGs;
- required for a relatively long time, unlike many other regulatory IEGs;
- induced in response to the activation of the cAMP signaling pathway;
- necessary and sufficient for transforming short-term into long-lasting plasticity; and
- induced via cAMP increase and following CREB activation.

These findings indicated that a cascade of gene expression, in which transcription factors regulate the expression of other transcription factors that, in turn, regulate the expression of effector genes, is an essential molecular signature for long-term memory formation [70, 72]. This conclusion is important in that it suggests that the nature of gene-expression required for long-term memory formation is very similar to that of a cell "differentiation" process, where epigenetic mechanisms are responsible for mediating long-lasting changes in gene expression. More recent studies investigating changes in chromatin structure and its post-translational modifications has supported the validity of this hypothesis [73, 74].

Parallel studies that identified several genes important for *Aplysia* long-term facilitation also confirmed the gene cascade hypothesis. CREBs family members were cloned and characterized and found to be regulated by post-translational modification as a consequence of the cAMP activation [74–76]. These post-translational modifications precede the induction of IEGs, among which is ApC/EBP, and both families of b-zip transcription factors are required for long-term facilitation but are dispensable for short-term facilitation. Members of both families are capable of transforming short- into long-term facilitation, suggesting that the activation of this pathway is sufficient to stabilize synaptic plasticity underlying long-term memory formation.

Subsequently, it was shown that mammalian C/EBP isoforms – specifically C/EBPβ and C/EBPδ – are critical for memory formation in mammals, demonstrating that the role of the cAMP-CREB-C/EBP-dependent pathway in memory formation is evolutionarily conserved [46,77–79]. In rat brain, the expression of C/EBPβ and C/EBPδ occurs in the same cell populations in which CREB is phosphorylated at Ser133 (pCREB), a post-translational modification essential for the activation of the CREB-dependent transcriptional activity [46]. Using an inhibitory avoidance (IA) task, in which animals remember to avoid a context previously associated with a foot shock, Taubenfeld et al. [46] found that training leads to a significant increase of Ser133 CREB phosphorylation in the hippocampus, followed by the induction of C/EBP β and C/EBPδ expression. pCREB increases significantly immediately after training, and this enhanced phosphorylation is sustained for at least 20 hours. On the other hand, C/EBPβ and C/EBPδ expression are significantly augmented several hours after training, specifically between 6 and 9 hours; this increase is sustained for at least 28 hours and returns to control levels by 48 hours after training. The training-induced increase of pCREB, C/EBPβ and C/EBPδ is likely regulated by modulatory neurotransmitter systems because it requires an intact fornix [80], though the nature of the modulation(s) involved is still unclear. Notably, the C/EBPβ isoform plays an essential role in the hippocampus during IA memory consolidation. Indeed, if the hippocampal induction of C/EBPβ expression is blocked by injections of antisense DNA into the hippocampi of trained rats, IA retention is completely disrupted. In agreement with its profile of induction, the requirement for hippocampal C/EBPβ during IA memory consolidation is transient, but lasts for more than a day, indicating that the transcription phase regulated by C/EBPβ and mediating memory formation is activated for a relatively long time [77].

Hence, as we have pointed out above, C/EBP and CREB family members participate in a fundamental, evolutionarily conserved cascade of events required for the consolidation of new memories [81, 82] (Fig. 12.3).

Similar to the findings obtained in *Aplysia* cultures, C/EBPβ expression regulation in the hippocampus during mammalian memory formation is under the control of CREB and can be evoked by the Ca^{2+} and cAMP-induced intracellular pathways. In fact, blocking CRE-dependent gene expression before contextual fear conditioning training completely inhibits the increase of C/EBPβ in hippocampi of mice [83].

C/EBPδ also plays a critical role in memory formation, but apparently by acting in an opposite direction compared to that of C/EBPβ, implying that distinct C/EBP isoforms exert different roles in memory formation. In mice, the knockout of C/EBPδ results in a selective enhancement of contextual fear conditioning, but has no effect on water maze memory [78]. Similarly, an enhancement in memory and long-term plasticity can also be induced when a broad dominant-negative inhibitor of C/EBP (EGFP-AZIP), which preferentially interacts with several inhibiting isoforms of both C/EBP and of ATF4 (a distant member of the C/EBP family of transcription factors), is selectively expressed in the forebrain [79]. This suggests that the relief of C/EBP or C/EBP-like-mediated inhibitory mechanisms lowers the threshold for hippocampal-dependent synaptic long-term potentiation (LTP) and memory storage in mice. On the other hand, it should be kept in mind that because of the

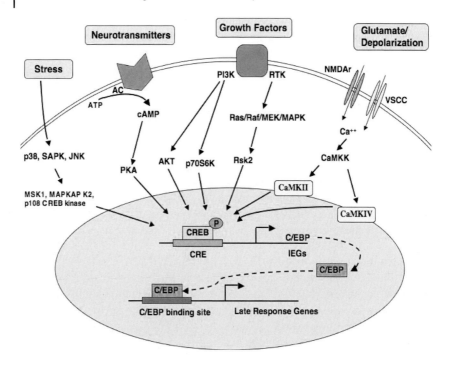

Fig. 12.3 Schematic representation of the CREB-C/EBP pathway activated during long-term memory.

spatial and temporal limitations of the knockout approach and the unspecific targeting of general C/EBP inhibitors, it is possible that the enhancement of memory is a result of compensatory effects or the targeting of other C/EBP inhibitor isoforms.

The cAMP and Ca^{2+}-dependent signal transduction pathways have a prominent role in the regulation of expression of C/EBPs in neurons and neuronal long-term synaptic plasticity. The first example of this regulation was observed, as described above, in the pheochromocytoma PC12 cell line, where C/EBPβ has been found to promote neuronal differentiation. In these cells, C/EBPβ is expressed in the cytoplasm and translocates to the nucleus upon cAMP stimulation [84]. Although the contribution of the translocation of C/EBPs has not yet been investigated, interestingly, C/EBPs have been found to be expressed in the cytoplasm of Aplysia and rat hippocampal neurons, suggesting that translocation may be a mechanism for functionally linking the distal (synaptic) compartment to nuclear functions, in particular, gene expression. However, it is clear that, in hippocampal neurons, the overall expression and binding activity of C/EBPs are increased by the activation of the cAMP and Ca^{2+} [45].

Another level of regulation of the C/EBP function in neurons is through the activity of kinases, among which MAP kinase, CaM Kinase II and IV, and protein kinase C play a critical role [85]. These kinase-dependent regulations participate in both the activation of C/EBPs as transcription factors as well as the control of their degradation. For example, ApC/EBP is degraded through the ubiquitin-proteasome pathway, but its phosphorylation via MAPK prevents its proteolysis [85]. Thus, it seems that phosphorylation of ApC/EBP by MAPK synergistically acts on two levels to ensure that C/EBP-dependent gene expression is sufficiently prolonged during the consolidation phase. As pointed out above, it is interesting to note that the same MAPK or its related kinases are also critical for the activation of C/EBPβ in developmental neurogenesis [54], axonal injury [86], and following terminal differentiation [43, 60].

Since C/EBPs are transcription factors, it is important to determine which target genes they regulate. This question intrigues the field of molecular neuroscience as well as other fields in which C/EBPs play important roles. Advancing knowledge in transcription regulation suggests that the specificity of target gene expression depends on the convergence of different signaling pathways, which act in concert over space and time to dictate the pattern of the expressed target genes. This knowledge will hopefully be available in the near future and should clarify why C/EBPs – factors discovered in the induction of acute-phase response in liver – are also fundamental for long-term memory formation.

Abbreviations

CaMK	Ca^{2+}-calmodulin-dependent protein kinase
CNS	central nervous system
CNTF	ciliary neurotrophic factor
CRE	cAMP response element
CREB	cAMP response element binding protein
EBP	enhancer binding protein
ERK	extracellular-signal-related kinase
FGF	fibroblast growth factor
IA	inhibitory avoidance
IEG	immediate-early gene
IL	interleukin
LPS	lipopolysaccharide
LTP	long-term potentiation
MEK	mitogen-activated-protein-kinase kinase
NA	noradrenaline
NGF	nerve growth factor
NMDA	N-methyl-D-aspartate
PACAP	pituitary adenylate cyclase-activating peptide
PDGF	platelet-derived growth factor
PI3k	phosphatidylinositol 3 kinase
PKA	protein kinase A

RNAi   RNA interference
Rsk   ribosomal S6 kinase
TNF   tumor necrosis factor
VIP   vasoactive intestinal peptide

References

1. Cao, Z., Umek, R.M., McKnight, S.L., *Genes Dev.* **1991**, 5, 1538–1552.
2. Landschulz, W.H., Johnson, P.F., Adashi, E.Y., Graves, B.J., McKnight, S.L., *Genes Dev.* **1988**, 2, 786–800.
3. Akira, S., Isshiki, H., Sugita, T., Tanabe, O., Kinoshita, S., Nishio, Y., Nakajima, T., Hirano, T., Kishimoto, T., *EMBO J.* **1990**, 9, 1897–1906.
4. Chang, C.J., Chen, T.T., Lei, H.Y., Chen, D.S., Lee, S.C., *Mol. Cell Biol.* **1990**, 10, 6642–6653.
5. Descombes, P., Chojkier, M., Lichtsteiner, S., Falvey, E., Schibler, U., *Genes Dev.* **1990**, 4, 1541–1551.
6. Poli, V., Mancini, F.P., Cortese, R., *Cell* **1990**, 63, 643–653.
7. Roman, C., Platero, J.S., Shuman, J., Calame, K., *Genes Dev.* **1990**, 4, 1404–1415.
8. Ron, D., Habener, J.F., *Genes Dev.* **1992**, 6, 439–453.
9. Takiguchi, M., *Int. J. Exp. Pathol.* **1998**, 79, 369–391.
10. Niu, X., Renshaw-Gegg, L., Miller, L., Guiltinan, M.J., *Plant Mol. Biol.* **1999**, 41, 1–13.
11. Ramji, D.P., Foka, P., *Biochem. J.* **2002**, 365, 561–575.
12. Metallo, S.J., Schepartz, A., *Chem. Biol.* **1994**, 1, 143–151.
13. Schrem, H., Klempnauer, J., Borlak, J., *Pharmacol. Rev.* **2004**, 56, 291–330.
14. Agre, P., Johnson, P.F., McKnight, S.L., *Science* **1989**, 246, 922–926.
15. Vinson, C.R., Sigler, P.B., McKnight, S.L., *Science* **1989**, 246, 911–916.
16. Hurst, H.C., *Protein Profile* **1995**, 2, 101–168.
17. Johnson, P.F., *Mol. Cell. Biol.* **1993**, 13, 6919–6930.
18. Williams, S.C., Cantwell, C.A., Johnson, P.F., *Genes Dev.* **1991**, 5, 1553–1567.
19. Bannister, A.J., Cook, A., Kouzarides, T., *Oncogene* **1991**, 6, 1243–1250.
20. Parkin, S.E., Baer, M., Copeland, T.D., Schwartz, R.C., Johnson, P.F., *J. Biol. Chem.* **2002**, 277, 23563–23572.
21. Podust, L.M., Krezel, A.M., Kim, Y., *J. Biol. Chem.* **2001**, 276, 505–513.
22. Pan, Z., Hetherington, C.J., Zhang, D.E., *J. Biol. Chem.* **1999**, 274, 23242–23248.
23. Sok, J., Wang, X.Z., Batchvarova, N., Kuroda, M., Harding, H., Ron, D., *Mol. Cell. Biol.* **1999**, 19, 495–504.
24. Kinugawa, K., Shimizu, T., Yao, A., Kohmoto, O., Serizawa, T., Takahashi, T., *Circ. Res.* **1997**, 81, 911–921.
25. Omori, S.A., Smale, S., O'Shea-Greenfield, A., Wall, R., *J. Immunol.* **1997**, 159, 1800–1808.
26. Shuman, J.D., Cheong, J., Coligan, J.E., *J. Biol. Chem.* **1997**, 272, 12793–12800.
27. Chen, B.P., Wolfgang, C.D., Hai, T., *Mol. Cell. Biol.* **1996**, 16, 1157–1168.
28. Batchvarova, N., Wang, X.Z., Ron, D., *EMBO J.* **1995**, 14, 4654–4661.
29. Tsukada, J., Saito, K., Waterman, W.R., Webb, A.C., Auron, P.E., *Mol. Cell. Biol.* **1994**, 14, 7285–7297.
30. Kinoshita, S., Akira, S., Kishimoto, T., *Proc. Natl. Acad. Sci. USA* **1992**, 89, 1473–1476.
31. Brandt-Rauf, P.W., Pincus, M.R., Chen, J.M., Lee, G., *J. Protein Chem.* **1989**, 8, 679–688.
32. Ross, H.L., Nonnemacher, M.R., Hogan, T.H., Quiterio, S.J., Henderson, A., McAllister, J.J., Krebs, F.C., Wigdahl, B., *J. Virol.* **2001**, 75, 1842–1856.
33. Ubeda, M., Wang, X.Z., Zinszner, H., Wu, I., Habener, J.F., Ron, D., *Mol. Cell. Biol.* **1996**, 16, 1479–1489.

34 Nerlov, C., *Nat. Rev. Cancer* **2004**, 4, 394–400.
35 Buck, M., Chojkier, M., *Hepatology* **2003**, 31, 731–738.
36 Wilson, H.L., Roesler, W.J., *Mol. Cell. Endocrinol.* **2002**, 188, 15–20.
37 McKnight, S.L., *Cell* **2001**, 107, 259–261.
38 Roesler, W.J., *Annu. Rev. Nutr.* **2001**, 21, 141–165.
39 Lekstrom-Himes, J.A., *Stem Cells* **2001**, 19, 125–133.
40 Rosen, E.D., Walkey, C.J., Puigserver, P., Spiegelman, B.M., *Genes Dev.* **2000**, 14, 1293–1307.
41 Lane, M.D., Tang, Q.Q., Jiang, M.S., *Biochem. Biophys. Res. Commun.* **1999**, 266, 677–683.
42 Lekstrom-Himes, J., Xanthopoulos, K.G., *J. Biol. Chem.* **1998**, 273, 28545–28548.
43 Cortes-Canteli, M., Pignatelli, M., Santos, A., Perez-Castillo, A., *J. Biol. Chem.* **2002**, 277, 5460–5467.
44 Sterneck, E., Johnson, P.F., *J. Neurochem.* **1998**, 70, 2424–2433.
45 Yukawa, K., Tanaka, T., Tsuji, S., Akira, S., *J. Biol. Chem.* **1998**, 273, 31345–31351.
46 Taubenfeld, S.M., Wiig, K.A., Monti, B., Dolan, B., Pollonini, G., Alberini, C.M., *J. Neurosci.* **2001**, 21, 84–91.
47 Tajiri, S., Oyadomari, S., Yano, S., Morioka, M., Gotoh, T., Hamada, J.I., Ushio, Y., Mori, M., *Cell Death Differ.* **2004**, 11, 403–415.
48 Bonni, A., Sun, Y., Nadal-Vicens, M., Bhatt, A., Frank, D.A., Rozovsky, I., Stahl, N., Yancopoulos, G.D., Greenberg, M.E., *Science* **1997**, 278, 477–483.
49 Nakashima, K., Yanagisawa, M., Arakawa, H., Kimura, N., Hisatsune, T., Kawabata, M., Miyazono, K., Taga, T., *Science* **1999**, 284, 479–482.
50 Raballo, R., Rhee, J., Lyn-Cook, R., Leckman, J.F., Schwartz, M.L., Vaccarino, F.M., *J. Neurosci.* **2000**, 20, 5012–5023.
51 Williams, B.P., Park, J.K., Alberta, J.A., Muhlebach, S.G., Hwang, G.Y., Roberts, T.M., Stiles, C.D., *Neuron* **1997**, 18, 553–562.
52 Park, J.K., Williams, B.P., Alberta, J.A., Stiles, C.D., *J. Neurosci.* **1999**, 19, 10383–10389.
53 Ghosh, A., Greenberg, M.E., *Neuron* **1995**, 15, 89–103.
54 Menard, C., Hein, P., Paquin, A., Savelson, A., Yang, X.M., Lederfein, D., Barnabe-Heider, F., Mir, A.A., Sterneck, E., Peterson, A.C., Johnson, P.F., Vinson, C., Miller, F.D., *Neuron* **2002**, 36, 597–610.
55 Slack, R.S., El-Bizri, H., Wong, J., Belliveau, D.J., Miller, F.D., *J. Cell Biol.* **1998**, 140, 1497–1509
56 Gloster, A., El-Bizri, H., Bamji, S.X., Rogers, D., Miller, F.D., *J. Comp. Neurol.* **1999**, 405, 45–60.
57 Toma, J.G., El-Bizri, H., Barnabe-Heider, F., Aloyz, R., Miller, F.D., *J. Neurosci.* **2000**, 20, 7648–7656.
58 Chen, X., Liu, W., Ambrosino, C., Ruocco, M.R., Poli, V., Romani, L., Quinto, I., Barbieri, S., Holmes, K.L., Venuta, S., Scala, G., *Blood* **1997**, 90, 156–164.
59 Yeh, W.C., Cao, Z., Classon, M., McKnight, S.L., *Genes Dev.* **1995**, 9, 168–181.
60 Marshall, J., Dolan, B.M., Garcia, E.P., Sathe, S., Tang, X., Mao, Z., Blair, L.A., *Neuron* **2003**, 39, 625–639.
61 Olive, M., Williams, S.C., Dezan, C., Johnson, P.F., Vinson, C., *J. Biol. Chem.* **1996**, 271, 2040–2047.
62 Cardinaux, J.R., Magistretti, P.J., *J. Neurosci.* **1996**, 16, 919–929.
63 Colangelo, A.M., Mallei, A., Johnson, P.F. Mocchetti, I., *Brain Res. Mol. Brain Res.* **2004**, 124, 97–104.
64 Kandel, S., Jessel, T., Schwartz, J. *Principles of Neural Science.* McGraw-Hill, Fourth edition, **2000**.
65 Bailey, C.H., Bartsch, D., Kandel, E.R., *Proc. Natl. Acad. Sci. USA* **1996**, 93, 13445–13452.
66 Alberini, C.M., Ghirardi, M., Huang, Y.Y., Nguyen, P.V., Kandel, E.R., *Ann. N. Y. Acad. Sci.* **1995**, 758, 261–286.
67 Castellucci, V.F., Frost, W.N., Goelet, P., Montarolo, P.G., Schacher, S., Morgan, J.A., Blumenfeld, H., Kandel, E.R., *J. Physiol.* **1986**, 81, 349–357.
68 Montarolo, P.G., Goelet, P., Castellucci, V. F., Morgan, J., Kandel, E.R.,

Schacher, S., *Science* **1986**, 234, 1249–1254.
69 Dash, P.K., Hochner, B., Kandel, E.R., *J. Neurosci.* **1990**, 7, 718–721.
70 Alberini, C.M., Ghirardi, M., Metz, R., Kandel, E.R., *Cell* **1994**, 76, 1099–1114.
71 Lee, J.A., Kim, H.K., Kim, K.H., Han, J.H., Lee, Y.S., Lim, C.S., Chang, D.J., Kubo, T., Kaang, B.K., *Learn. Mem.* **2001**, 8, 220–226.
72 Goelet, P., Castellucci, V.F., Schacher, S., Kandel, E.R., *Nature* **1986**, 322, 419–422.
73 Levenson, J.M., Sweatt, J.D., *Nat. Rev. Neurosci.* **2005**, 6, 108–118.
74 Guan, Z., Giustetto, M., Lomvardas, S., Kim, J.H., Miniaci, M.C., Schwartz, J.H., Thanos, D., Kandel, E.R., *Cell* **2002**, 111, 483–493.
75 Bartsch, D., Ghirardi, M., Skehel, P., Karl, K.A., Herder, S.P., Chen, M., Bailey C.H., Kandel, E.R., *Cell* **1995**, 83, 979–992.
76 Bartsch, D., Casadio, A., Karl, K.A., Serodio, P., Kandel, E.R., *Cell* **1998**, 95, 211–223.
77 Taubenfeld, S.M., Milekic, M., Monti, B., Alberini, C.M., *Nat. Neurosci.* **2001**, 4, 813–818.
78 Sterneck, E., Paylor, R., Jackson-Lewis, V., Libbey, M., Przedborski, S., Tessarollo, L., Crawley, J.N., Johnson, P.F., *Proc. Natl. Acad. Sci. USA* **1998**, 95, 10908–10913.
79 Chen, A., Muzzio, I.A., Malleret, G., Bartch, D., Verbitsky, M., Pavlidis, P., Yonan, A.L. Vronskaya, S., Grody, M.B., Cepeda, J., Gilliam, T.C., Kandel, E.R., *Neuron* **2003**, 39, 655–669.
80 Taubenfeld, S.M., Wiig, K.A., Bear, M.F., Alberini, C.M., *Nat. Neurosci.* **1999**, 2, 309–310.
81 Alberini, C.M., *J. Exp. Biol.* **1999**, 202, 2887–2891.
82 Carew, T.J., Sutton, M.A., *Nat. Neurosci.* **2001**, 4, 769–771.
83 Athos, J., Impey, S., Pineda, V.V., Chen, X., Storm, D.R., *Nat. Neurosci.* **2002**, 5, 1119–1120.
84 Metz, R., Ziff, E., *Genes Dev.* **1991**, 5, 1754–1766.
85 Yamamoto, N., Hedge, A.N., Chain, D.G., Schwartz, J.H., *J. Neurochem.* **1999**, 73, 2415–2423.
86 Sung, Y.J., Povelones, M., Ambron, R.T., *J. Neurobiol.* **2001**, 47, 67–79.
87 Antonson, P., Xanthopoulos, K.G., *Biochem. Biophys. Res. Commun.* **1995**, 215, 106–113.
88 Calkhoven, C.F., Muller, C., Leutz, A., *Genes Dev.* **2000**, 14, 1920–1932.
89 Chen, Y., Hu, H., Atkinson, B.G., *Dev. Genet.* **1994**, 15, 366–377.
90 Lyons, S.E., Shue, B.C., Lei, L., Oates, A.C., Zon, L.I., Liu, P.P., *Gene* **2001**, 281, 43–51.
91 Katz, S., Kowenz-Leutz, E., Muller, C., Meese, K., Ness, S.A., Leutz, A., *EMBO J.* **1993**, 12, 1321–1332.
92 Kousteni, S., Kockar, F.T., Sweeney, G.E., Ramji, D.P., *Mech. Dev.* **1998**, 77, 143–148.
93 Tucker, C.S., Hirono, I., Aoki, T., *Dev. Comp. Immunol.* **2002**, 26, 271–282.
94 Kinoshita, S., Akira, S., Kishimoto, T., *Proc. Natl. Acad. Sci. USA* **1992**, 89, 1473–1476.
95 Kageyama, R., Sasai, Y., Nakanishi, S., *J. Biol. Chem.* **1991**, 266, 15525–15531.
96 Davies, G.E., Sabatakos, G., Cryer, A., Ramji, D.P., *Biochem. Biophys. Res. Commun.* **2000**, 271, 346–352.
97 Antonson, P., Stellan, B., Yamanaka, R., Xanthopoulos, K.G., *Genomics* **1996**, 35, 30–38.
98 Chumakov, A.M., Grillier, I., Chumakova, E., Chih, D., Slater, J., Koeffler, H.P., *Mol. Cell. Biol.* **1997**, 17, 1375–1386.
99 Yamanaka, R., Kim, G.D., Radomska, H.S., Lekstrom-Himes, J., Smith, L.T., Antonson, P., Tenen, D.G., Xanthopoulos, K.G., *Proc. Natl. Acad. Sci. USA* **1997**, 94, 6462–6467.
100 Sabatakos, G., Kousteni, S., Cryer, A., Ramji, D.P., *J. Anim. Sci.* **1998**, 76, 2953–2954.
101 Luethy, J.D., Fargnoli, J., Park, J.S., Fornace, A.J. Jr., Holbrook, N.J., *J. Biol. Chem.* **1990**, 265, 16521–16526.
102 Park, J.S., Luethy, J.D., Wang, M.G., Fargnoli, J., Fornace, A.J. Jr., McBride, O.W., Holbrook, N.J., *Gene* **1992**, 116, 259–267.

13
The Role of c-Jun in Brain Function

Gennadij Raivich and Axel Behrens

Abstract

The transcription factor AP–1 consists of a variety of dimers composed of members of the Jun and Fos families of proteins. However, it is the up-regulation of c-jun that is a particularly common event in the developing, adult, and also injured nervous system that serves as a model of transcriptional control of brain function. In view of the long list of excellent in-depth overviews on different members of the Jun family and associated molecules, the primary aim of this chapter is to focus on c-jun specifically and to discuss evidence for the involvement of this transcription factor in ischemia and stroke, in seizures, during learning and memory, or following axonal injury and during successful regeneration. Functional studies employing *in-vivo* strategies using gene deletion, targeted expression of dominant-negative isoforms and pharmacological inhibitors all suggest a bipotential role of c-jun, in mediating neurodegeneration and cell death, as well as in plasticity and repair. Phosphorylation of c-jun, and the activation of its upstream kinases, is required in many (but not in all) forms of these events, with only a partial overlap of the jun-, JNK- or JNKK(n)-dependent functions. Moreover, a better understanding of the non-overlapping roles could considerably increase the potential of pharmacological agents to improve neurological outcome following trauma, neonatal encephalopathy and stroke, and neurodegenerative disease.

13.1
Introduction

Up-regulation of the c-Jun protein, the principal component of the AP–1 transcription factor, is a common event in the developing, adult, and injured nervous system. Initially noted as the counterpart of c-Fos in the brain transcription factor response to the convulsants pentylenetetrazole and picrotoxin (Morgan et al., 1987; Saffen et al., 1988), over the past 17 years increased *c-jun* gene expression, protein and phosphorylation has been noted under a variety of conditions. These include neuronal differentiation and survival during normal embryonic and postnatal development

Transcription Factors. Edited by Gerald Thiel
Copyright © 2006 WILEY-VCH Verlag GmbH & Co. KGaA, Weinheim
ISBN 3-527-31285-4

(Mellström et al., 1991; Kockel et al., 1997) and in transplantation (Dragunow et al., 1991). C-Jun expression is strongly increased in seizures (Morgan and Curran, 1998; Gall et al., 1990; Gass et al., 1993), addiction (Hayward et al., 1990; Freeman et al., 2001), pain (Wisden et al., 1990; Herdegen et al., 1991a; Naranjo et al., 1991) and, more selectively in some (but not other) cases of long-term potentiation and/or depression and memory formation (Abraham et al., 1993; Tischmeyer et al., 1994). Last, but not least, enhanced activity of c-Jun is a common and critical event in cerebral ischemia and stroke (Kindy et al., 1991; Wessel et al., 1991), axotomy and other forms of trauma (Herdegen et al., 1991b,c; Jenkins and Hunt, 1991), as well as in the post-traumatic repair (Chaisuksunt et al., 2003; Raivich et al., 2004).

In most cases, this enhanced expression of *c-jun* gene, protein and function is not a solitary event, but can be accompanied by an induction of transcription factors that are related to c-Jun (e.g., Jun-B, Jun-D), the Fos family members (Fos, FosB, Fra–1/2), or the ATF-family members (for a review, see Herdegen and Leah, 1998), allowing the formation of functionally different heterodimers. Moreover, differences in the regulation of these complementary molecules after inducing events – whether it is the hypoxic ischemic insult, axotomy, long-term potentiating paradigms, addictive or antipsychotic substances – could also play a role in the time-dependent expression of its multiple downstream targets. However, in view of the long list of excellent in-depth overviews on the different members of the c-Jun family and associated molecules (Morgan and Curran, 1988; Dragunow et al., 1989; Nestler, 1993; Herdegen and Leah, 1998; Gelderblom et al., 2004), the primary focus of this chapter is to focus on c-Jun specifically and to review the evidence for the direct involvement of this transcription factor in the developing, adult, and injured nervous system.

13.2
C-Jun Phosphorylation and Upstream Signaling

A key regulatory component in the post-translational function of c-Jun is the serine and threonine phosphorylation of the N-terminal transactivational domain of this transcription factor. N-terminal phosphorylation does correlate with the growth rate of fibroblasts (Castellazi et al., 1991) and the transcriptional activation of AP–1 promoters (Radler-Pohl et al., 1993). However, it has only moderate effects on the transactivation by c-Jun homodimers (Baker et al., 1992), and does not affect dimerization or DNA-binding (Binetruy et al., 1991; Smeal et al., 1992). This point that phosphorylation of c-Jun only incurs moderate effects is clearly supported by the much more severe phenotype of global *c-jun* deletion that is embryonic lethal (Hillberg et al., 1993), compared with the much milder phenotype of the junAA mice where the two N-terminal Ser63 and Ser73 photoacceptor sites are replaced with alanine residues (Behrens et al., 1999). However, phosphorylation of c-Jun does modulate c-Jun's intrinsic affinity for promoter sites (Hirai et al., 1990). Both the constitutive and induced phosphorylation of N-terminal sites is specific to c-Jun, because in JunB and JunD it is impeded by amino acids on either the docking or the phosphorylation sites (Franklin et al., 1992; Karin, 1995).

13.2.1
Mitogen-Activated/Stress-Activated Protein Kinase (MAPK/SAPK) Level

In vitro, phosphorylation of Ser63 and Ser73 following TNF, anisomycin and alkylating agents is catalyzed by the family of c-Jun N-terminal kinases JNK1, JNK2, and JNK3 that are also known as stress-activated protein kinases (SAPKs). However, differential splicing gives rise to a total of ten different variants known so far (Casanova et al., 1996; Gelderblom et al., 2004). Together with ERK and p38, JNKs belong to the group of mitogen-activated protein kinases (MAPKs). JNKs and p38s are involved in the regulation of growth arrest, apoptosis, and proliferation induced by stress signals (e.g., UV irradiation, heat- or cold-shock, osmotic stress), cytokines (e.g., interleukin (IL)–1 or tumor necrosis factor-α (TNFα)) (Ichijo, 1999), as well as by G protein-coupled receptor agonists such as thrombin (Collins et al., 1996).

13.2.2
MAP Kinase Kinase (MEK/MKK) and MAP Kinase Kinase Kinase (MEKK) Level

Stress-activated JNKs are regulated by two specific MAPK kinases or MKKs: MKK4 (also known as SEK1 and JNKK1), and MKK7 (also known as SEK2 and JNKK2); MKK4 and MKK7 phosphorylate and activate JNKs (Hagemann and Blank, 2001). One layer above, MKK4 and 7 are themselves activated by the MEK kinases (MEKK) MEKK1, 2, 3, and 4 for MKK4 and MEKK1–3 for MKK7. In addition, MKK7 can be activated by the mixed lineage kinase MLK3 (Hehner et al., 2000). For the other three MAPK components, ERK1 and ERK2 are both activated by MEK1 and 2; the more recently discovered ERK5 by MEK5 (Zhou et al., 1995), and the p38 is activated directly by MKK3 and MKK6 (Hagemann and Blank, 2001). Most of the direct target specificity appears to have been lost by this stage, since all or a selection of MEKK1–4 can also activate MEK1–2, the activators of ERK1/2, MKK3 and 6, the activators of p38, and MEK5, the activator of ERK5 (for a review, see (Hagemann and Blank, 2001).

13.2.3
Scaffolding Proteins

The recently discovered scaffold proteins provide a means to recover specificity of the top-bottom signaling by bringing together specific and consecutive members of the signaling cascade. The JNK Interacting Protein 1, abbreviated as JIP1, provides a scaffold for the MLK3/MKK7/JNK module (Yasuda et al., 1999), and JSAP1, an alternatively spliced isoform of JIP3, for MEKK1, MKK4 (SEK) and JNK3 (Akechi et al., 2001). Other scaffolding proteins such as beta-arrestin 2 are specific for JNK3, bringing it together with its upstream kinases – the apoptosis signal-regulating kinase 1 (ASK1) and MKK4 (McDonald et al., 2000). MEKK2 may also serve a dual function of an activator and scaffold, binding JNK1 and MKK7, and directly activating MKK7 (Cheng et al., 2000).

Many of these signaling components are particularly strongly expressed in the brain, spinal cord and peripheral neurons, underscoring their importance for the regulation of JNK activity, and indirectly that of c-Jun, throughout the nervous system (English et al., 1995; Casanova et al., 1996; Gupta et al., 1996; Kim et al., 1999; Lee et al., 1999; Zhang et al., 2003a,b; Nateri et al., 2004). Importantly, the scaffolding proteins, JIP1, JIP2 and JSAP1 also provide the main adaptor molecules for the cargo of the kinesin motors mediating the microtubule-associated axonal transport between neuronal cell bodies and neurite terminals (Verhey et al., 2001), as well as that of the MEKK/MLK, MKK, JNK and ATF2 components (Lindwall and Kanje, 2005).

13.2.3.1 Multimodal Effects of Deletion

Despite the potential streamlining effects of the scaffolding proteins there is considerable variation in the effects of interfering with different levels of the upstream signaling cascade terminating with c-Jun during neural development. Thus, complete excision of neural *c-jun*, or the exchange of N-terminal Ser63 and Ser73 residues with alanines does not result in a visible change in the overall CNS morphology or gross numeric differences in selected neuronal populations (Behrens et al., 1999; Raivich et al., 2004).

The deletion of JNK1 leads to a disruption in the formation of the anterior commissure (Chang et al., 2003), but with only minor additional effects in mutants lacking JNK1, JNK2 or JNK3 alone, or combinations of JNK1 and JNK3 or JNK2 and JNK3 (Gelderblom et al., 2004). The deletion of JSAP-1 causes an axon guidance defect of the telencephalic commissures (Kelkar et al., 2003), with a partial rescue following transgenic expression of JIP1 in embryonic glial cells (Ha et al., 2005). Excision of exons 3 through to exon 8 of the 12 exon-containing JIP1 gene causes very early embryonic lethality, before the formation of the nervous system (Thompson et al., 2001), and simply underscores the non-redundancy of the JIP scaffolding and motor adaptor molecule for many different, and early, embryogenic processes. A selective deletion of just the JIP1 exon 3, that contains the JNK binding domain (JBD) is compatible with life, but interferes with stress-induced induction of JNK and with excitotoxic hippocampal neuron cell death (Whitmarsh et al., 2001). Mice with a more recent deletion of exon 2 and upstream part of exon 3 (Delta JIP1–ex2/u3), as reported by Im et al. (2003), are also viable and show a phenotype similar to the Delta JIP1–ex3 mutants (Whitmarsh et al., 2001).

Mutant mice with a combined deletion of JNK1 and JNK2 develop exancephaly, with precocious degeneration in the forebrain (Kuan et al., 1999; Sabapathy et al., 1999). Effects on neuroepithelial cell death are complex, with an early reduction of apoptotic PCD in the lateral hindbrain at embryonic day 9, followed by an increase at day 10 in the forebrain and hindbrain areas. Loss of upstream activators of JNK, MKK4, and MEKK4 show a simple increase in apoptosis (Ganiatsas et al., 1998; Nishina et al., 1999; Chi et al., 2005), suggesting supplementary targets of upstream activity, above that of JNK. Interestingly, deletion of only MEKK4, but not of the more downstream MKK4, interferes with neural tube closure (Chi et al., 2005).

The multimodality of the effects described for the JNKs and their upstream regulator enzymes in the mammalian nervous system is probably due to their involvement in many other cellular transduction events. For example, the three JNKs are also involved in the direct regulation of other nuclear transcription factors such as ATF2 and ELK1 and the mitochondria-associated Bc2 and BH3–only proteins (Gelderblom et al., 2004). In line with the latter function, JNK2 translocates to the mitochondria following stimulation with the oxidative stressor 6–hydroxy-dopamine, followed by release of cytochrome c and cell death. These actions are prevented by a transfection with dominant negative JNK2 (Eminel et al., 2004).

JNK1 also plays an important role in the maintenance of neuronal microtubules and cytoskeletal integrity. MAP2 and MAP1B polypeptides are hypophosphorylated in Jnk1$^{-/-}$ brains, resulting in a compromised ability to bind microtubules and promote their assembly (Chang et al., 2003). In the case of MKK4 and MEKK4, these signaling enzymes can also lead to the activation of p38 (Hagemann and Blank, 2001).

In contrast to mammals, the deletion of *jun* in a simpler organism such as *Drosophila melanogaster* shows the same phenotype as that of its upstream regulators during the phenomenon termed "dorsal closure" (Kockel et al., 1997; Riesgo-Escovar and Hafen, 1997). In this morphogenetic process, ectodermal cells of the lateral epithelium stretch in a coordinated fashion to internalize the amnioserosa cells and close the embryo dorsally, a process that relies on the AP–1 dependent, local synthesis of decapentaplegic (*dpp*), the *Drosophila* homolog of transforming growth factor beta. Embryos lacking jun activity fail to show localized synthesis of *dpp* and display a profound dorsal closure phenotype (Kockel et al., 1997).

The same phenotype is also observed in animals with mutated basket (*bsk*), the *Drosophila* homolog of JNK1, and hemipterous (*hep*), the homolog of MKK7 (Riesgo-Escovar and Hafen, 1997), indicating a simple signaling cascade from *hep* to *bsk* to *jun* to *dpp* (Hou et al., 1997). Here, the maintenance of the same phenotype across several key signaling layers in Drosophila, and its loss in the mouse embryogenesis may reflect the growing complexity of the *jun* upstream signaling in the vertebrate nervous system.

13.3
Development

In the rat, c-jun is widely expressed during embryogenesis, starting with neurulation (Benett et al., 1997). High neuroepithelial levels were also observed for upstream kinases JNK2 alpha 1, MKK4 and MEKK4 (Ishikawa et al., 1997; Chi et al., 2005). Following neural tube closure and the ensuing formation of the marginal zone, high levels of nuclear phosphorylated jun are expressed on neuroblasts migrating out of the neuroepithelial layer in the forebrain and hindbrain regions. In the developing hindbrain, c-jun expression depended on the transcription factor MafB, but not on KROX20 transcription factor (Mechta-Grigoriou et al., 2003). Jun and MafB also form a heterodimer, participate in the regulation of HoxB3, and may play a role in the hindbrain patterning (Mechta-Grigoriou et al., 2003) At somewhat later stage, in the late embryonic nervous system of the rodent (E15–E19 in mouse, E15–E21 in rat), particularly high levels are present on sympathetic, sensory and motor neurons.

Postnatally, a comparative analysis in rat central nervous system showed the highest levels at day 15 (P15) and thereafter declining to the low adult levels (Wilkinson et al., 1989; Mellström et al., 1991). Postnatal levels of c-Jun are greatest in the visual areas, but c-Jun is also present in the olfactory epithelium, ventricular layers (especially of the telencephalon), and in restricted sites in the mid- and hindbrain, including cranial and spinal visceromotor and somatic motor neurons, which display a basal level of c-Jun expression in the adult (Wilkinson et al., 1989; Herdegen et al., 1991c; Kalla et al., 2001). Despite the particularly high levels of c-Jun in the cat visual cortex, this expression was not affected by visual exposure, unlike that of many other immediate-early genes such as *egr1*, *c-fos*, and *junB* (Rosen et al., 1992), suggesting that this part of the cortical *c-jun* expression profile is not activity-driven.

What, therefore, is the function of *c-jun* during embryonic and postnatal development? *In vitro*, NGF withdrawal leads to an accumulation of P-Jun. Overexpression of c-Jun induces cell death in PC12 cells, and dominant-negative Jun (DN-jun) prevents this form of neuronal cell death (Ham et al., 1995). Moreover, coexpression of DN-jun with the JNK-blocking JBD domain of JIP1 shows involvement of the same pathway in the JNK and jun-mediated death (Eilers et al., 2001; Harding et al., 2001). Inhibition with CEP–1347 of mixed lineage kinases, the upstream inducers of p38 and JNK, also promotes the survival of embryonic sensory, sympathetic and motor neurons, as well as the PC12 cells (Borasio et al., 1998; Maroney et al., 1999). The results of more recent studies also showed that neurite outgrowth in PC12 cells depends on EGR/jun complex (Levkovitz and Baraban, 2002).

The *in-vivo* data are more complex. Confirming the situation *in vitro*, all apoptotic-looking, postnatal sympathetic neurons expressed *c-jun*, suggesting that *c-jun* is involved with the commitment to die (Messina et al., 1996). Phosphorylation of c-Jun during embryonic programmed cell death (PCD) in motor neurons showed direct correlation with experimental procedures that would later increase PCD, such as limb removal (Sun et al., 2005). Procedures increasing survival (neuromuscular activity blockade) reduced the number of the phospho-Jun-positive (PJ+) motor neurons. Finally, PCD-rescue by deletion of the more downstream proapoptotic gene *Bax*, causes a transient increase in the number of the PJ+ motor neurons, indicating that activation of c-Jun signaling is potentially necessary, but not sufficient for the induction of motor neuron PCD (Sun et al., 2005).

Nevertheless, deletion of *c-fos* and *c-jun*, the primary AP1 components, does not enhance the normal cell death in the developing CNS (Roffler-Tarlov et al., 1996). C-Jun is not needed for developmental cell death or axonal outgrowth in the embryonic retina (Herzog et al., 1999; Grimm et al., 2001). Moreover, neural excision of the floxed *c-jun* gene, with neuroepithelial expression of the cre recombinase using the nestin promoter (nestin::cre), did not result in an overt change in brain morphology. Interestingly, an absence of neural jun did cause any slight increase in the number of facial motor neurons, but was apparently without any effect on other motor nuclei or on sensory ganglia (Raivich et al., 2004).

Unlike the situation *in vitro*, this may reflect a partially redundant role of c-jun during late mouse neural development that is reminiscent of the D-jun function in eye and wing development in *Drosophila*. There, jun participates in a separate signaling pathway that is comprised of Ras, Raf, and *Rolled*, an ERK kinase (MEK1/2) homolog that can also phosphorylate jun. In contrast to the strict requirement for Jun in dorsal closure, its role in the eye is redundant but can be uncovered by mutations in other signaling components (Weber et al., 2000; Kockel et al., 2001). Similar redundancy may also serve as a second line of defense in the formation of the vertebrate CNS and contribute to the precision of neuronal differentiation and the establishment of synaptic connections.

13.4
Novelty, Learning and Memory, and Addiction

There is considerable evidence pointing to enhanced expression of c-jun in the physiological function of the adult brain, during conditions associated with novelty, short- and long-term memory formation and pain-causing or rewarding behavior (Wisden et al., 1990; Herdegen et al., 1991a; Abraham et al., 1993; Papa et al., 1993). These regulatory patterns may implicate c-jun in contributing to synaptic plasticity. For example, inhibition of c-jun function using overexpression of a dominant-negative isoform in *Drosophila*, has been shown to inhibit synaptic plasticity, by a reduction in synaptic strength as well as synaptic bouton number, via CREB as well as ADF1 and FasII mechanisms in the larval neuromuscular junction (Sanyal et al., 2002).

13.4.1
Novelty and Pain

First-time exposure to spatial novelty stimuli in mazes induced extensive c-Fos- and c-Jun immunoreactivity in the reticular formation, the caudate-putamen complex, the hippocampus (granular and pyramidal neurons), the cerebellum (granular neurons), and all layers of somatosensory cortex. Maximal c-jun expression was observed at 2–6 hours after the event, with a strong reduction in evoked expression with habituation, in repeatedly exposed animals (Papa et al., 1993). Spinal cord up-regulation of this known to lead to a strong up-regulation of immediate-early genes, including the transcription factors c-fos, c-jun, and krox–24 (Wisden et al., 1990; Herdegen et al., 1991a). This up-regulation has been postulated to convert short-term stimulations into long-lasting responses in dorsal-horn neurons, and can be blocked by a pretreatment with strong analgesics such as morphine (Presley et al., 1990; Tolle et al., 1994; Giorgi et al., 1997).

The learning paradigms involving food grain picking and passive avoidance in newly hatched chicks cause an almost four-fold increase in c-jun mRNA and protein levels in the chick forebrain, particularly in the lateral intermediate ventral hyperstriatum (Anokhin and Rose, 1991) and lateral paraolfactory nucleus (Freeman and

Rose, 1999). In mice, memory processing training in an appetitive bar-pressing task and apamin also induces a spatially selective induction of c-jun, along with c-fos, in the hippocampal CA1, CA3, and dentate gyrus regions (Heurteaux et al., 1993a). Interestingly, continuous training with trace eyeblink conditioning for 24 hours or longer periods can actually induce a down-regulation in c-jun levels (Donahue et al., 2002).

13.4.2
Learning

Application of antisense oligonucleotides for c-jun, but not those for junB, has been shown to inhibit sequence-specific learning following aversive stimulus using the foot shock-motivated brightness discrimination (Tischmeyer et al., 1994). However, complete deletion of the c-jun gene using its neural excision with nestin::cre did not interfere with spatial learning in the Morris water maze or with fear conditioning paradigms (Raivich et al., 2004).

There is also only an imperfect correlation between hippocampal long-term potentiation (LTP) and jun expression, with the latter detected in some (Cole et al., 1990; Worley et al., 1993) but not in other studies (Wisden et al., 1990; Abraham et al., 1992). This imperfect correlation also holds true for the N-terminal jun kinases. Mice deficient for JNK2 show impaired hippocampal LTP (Chen et al., 2005). Interestingly, the pharmacological inhibition of hippocampal JNK with SP600125 enhances short-term memory but appears to block the long-term memory formation and retrieval of an inhibitory avoidance task (Bevilaqua et al., 2003).

In addition to these direct effects, the activation of c-Jun-N-terminal kinase is also critical in mediating the blocking effects of amyloid beta (Costello and Herron, 2004) or the lipopolysaccharide-induced cytokines on hippocampal LTP (Curran et al., 2003; Barry et al., 2005). It is possible that these stimulatory or neurotoxin-mediated, blocking effects are produced by different JNK isoforms acting on distinct and non-overlapping downstream targets. Similar disparity is known from studies in stroke and ischemia models, where deletion of JNK2 or JNK3 confers protection, and deletion of JNK1 actually enhances the forebrain tissue loss (Kuan et al., 2003; Brecht et al., 2005).

13.4.3
Addiction

Strong up-regulation of c-Jun during initial exposure to addictive stimuli such as cocaine or its withdrawal following long-term habituation, was observed in locus coeruleus (Hayward et al., 1990) and nucleus accumbens (Hope et al., 1992), but not in striatum, which shows a targeted expression of other, related IEG/transcription factors such as c-Fos and JunB (Moratalla et al., 1993; Couceyro et al., 1994). Interestingly, transgenic, doxycycline-regulated expression of the dominant-negative form of c-Jun (Delta c-Jun) in the striatum and certain other brain regions of adult mice decreases their development of cocaine-induced conditioned place preference, suggesting reduced sensitivity to the rewarding effects of cocaine (Peakman et al., 2003). These behavioral effects were accompanied by a block in the ability of chronically administered cocaine to induce three known targets for AP–1 in the nucleus accumbens: the AMPA glutamate receptor subunit GluR2; the cyclin-dependent protein kinase Cdk5; and the transcription factor nuclear factor-kappaB (NF-𝜘B). However, inhibition of the AP–1 function had no effect on cocaine-induced locomotor activity or sensitization, indicating the dissociation between the rewarding and locomotor effects (Peakman et al., 2003).

13.5
Seizures and Excitotoxicity

The wave of excitation during generalized seizures, whether through electroshock, trauma or chemical excitotoxins, in most cases flows from the dentate gyrus to CA1 and then to CA3/4 and cerebral cortex (Gass et al., 1993). First increases in mRNA encoding inducible transcription factors are detectable in the dentate gyrus, with the strongest appearance of junB, c-fos and krox–24 occurring within 10–15 minutes (Lanaud et al., 1993; Yount et al., 1994). They are then enhanced in the hippocampal CA1 and CA3 areas, and in the cortex (White and Gall, 1987; Cole et al., 1990). Protein increases are considerably slower, probably due the inhibition in peri- and post-ictal protein synthesis. The amount of mRNA declines to basal levels within 1–2 hours in the dentate gyrus and 2–4 hours in the CA1/3 areas. Compared with the other induced transcription factors, the up-regulation of c-Jun is on the whole more moderate, but also more prolonged, for 24–72 hours (Sonnenberg et al., 1989; Cole et al., 1990; Kaminska et al., 1994). Chronic electroconvulsive seizures may down-regulate the expression of inducible transcription factors such as Fos and Jun (Winston et al., 1990; Brecht et al., 1999). Nevertheless, constitutively epileptic E1 animals show higher cortical and hippocampal AP1 transcription factor levels than their non-epileptogenic ddY parents (Yoneda et al., 1993).

There is considerable evidence pointing to a functional role of jun, its JNK-mediated phosphorylation and associated scaffolding proteins in seizure activity and excitotoxic effects. Neuronal injury is known to cause increased c-jun N-terminal kinase activity coupled with N-terminal phosphorylation of c-Jun (Herdegen et al., 1998; Schauwecker, 2000). Transgenic deletion of JNK3 protects against kainic acid-mediated excitotoxicity, reducing seizure activity and preventing the associated apoptosis of hippocampal neurons (Yang et al., 1997; Brecht et al., 2005). Similar protective effects were also observed in the junAA mice, where the phosphorylatable Ser63 and Ser73 residues are replaced with alanines (Behrens et al., 1999). In line with the critical JNK activation which permits function of the scaffolding proteins such as JIP1, deletion of the JNK binding domain containing exon3 of JIP1 interferes with the kainic acid-mediated JNK activation and promotes hippocampal neuron survival (Whitmarsh et al., 2001). A broader excision of JIP1 exons, beginning with exon 3 and including exon 8, is not compatible with even early embryogenesis in the homozygous mice (Thompson et al., 2001). However, the hemizygous Delta ex3/8JIP1 animals with a 50% reduced gene dosage of the normal JIP1 protein show a surprising increase in the activation of JNK and in the hippocampal vulnerability to systemic application of kainic acid. This points to protective and JNK-down-regulating effects of JIP1, that are located outside of the JBD domain in exon 3 (Magara et al., 2003).

13.6
Ischemia, Stroke, and Brain Trauma

13.6.1
Biochemical Regulation

As in seizures and excitotoxic injuries, brain ischemia, elicited through: (1) the occlusion of a local terminal artery (e.g., medial cerebral artery); (2) a combination of unilateral carotid occlusion followed by hypoxia; (3) a four-vessel occlusion in rat; or (4) a bicarotid occlusion in gerbil, will cause a very rapid up-regulation of inducible transcription factors, such as fos, c-jun, junB and NGFI-A/egr1, as well as other transcription factors and immediate-early genes (Kindy et al., 1991; Woodburn et al., 1993; Dragunow et al., 1994; Herdegen and Leah, 1998). Similar induction of transcription factors, including that of c-Jun, is also observed in traumatic brain injury (Raghupathi et al., 1995; Kobori et al., 2002), and in early postnatal hypoxic-ischemic (HI) brain (Munell et al., 1994), a commonly used model of human neonatal encephalopathy following pre- or peri-natal hypoxia (Vannucci and Vannucci, 2005).

On the mechanistic level, a minimal hybridization signal for either immediate-early gene is detected in animals perfused with fixative immediately following ischemia, suggesting that cellular energy levels may have to be restored to a certain level before efficient *de-novo* mRNA synthesis can occur (Wessel et al., 1991). Moderate HI injury following unilateral carotid occlusion and hypoxia induced c-Jun in the damaged regions (Dragunow et al., 1993). However, this extended into non-damaged regions on the unoccluded side following severe HI associated with necrotic cell death. In the damaged forebrain regions, c-Jun (and to a lesser extent c-Fos/FRAs) showed a prolonged expression in neurons undergoing delayed, but not necrotic, cell death; this suggested that they may be involved in the biochemical cascade that causes selective delayed neuronal death (Dragunow et al., 1994). This late neuronal cell death is normally associated with the appearance of nuclear DNA breaks using *in-situ* TUNEL staining and the regular DNA laddering on gel electrophoresis (Dragunow et al., 1994; Macaya et al., 1998), the latter being the classical tell-tale sign of the apoptotic form of PCD.

13.6.2
Role of Jun

At present, the evidence tying the up-regulation of c-jun to the post-ischemic, PCD is still indirect. Application of ATP-sensitive K-channel openers prevents the expression of ischemia-induced transcription factors as well as the hippocampal neuron cell death (Heurteaux et al., 1993b). Similar effects were also observed using antagonists of the NMDA-type glutamate receptor antagonists dizocilpine and MK–801 (Woodburn et al., 1993; Collaco-Moraes et al., 1994), as well as that of dexamethasone in perinatal HI injury (Macaya et al., 1998). On the other hand, some neuroprotective agents such the kappa opiate receptor agonist enadoline or post-ischemic brain cooling did not inhibit the increased levels of c-jun, suggesting that either the induction per se is not detrimental, or that enadoline and brain cooling act at different, further downstream levels (Woodburn et al., 1993; Akaji et al., 2003). Interestingly, short-term exposure to ischemia, a procedure known to induce an ischemia-tolerant state, is associated with a strong and persistent induction of c-jun, but not other transcription factors such as fos, fosB, junB, junD, or KROX–24 (Sommer et al., 1995).

13.6.3
Functional Role of JNK Cascade

In contrast to c-jun, there is considerable evidence for the involvement of N-terminal jun kinase. Activity-wise, analysis of the mitogen-activated protein kinases in post-ischemic brain reveals a strong and rapid increase in the levels of activated, phosphorylated JNK as well as ERK1/2, but not that of the p38 stress kinase (Herdegen et al., 1998; Otani et al., 2002). This is accompanied by a rapid activation of MKK4 and its upstream kinase ASK1 within 10 minutes of reperfusion following global ischemia (Zhang et al., 2003a), but also showing a second maximum at three days, that is associated with a peak in neuronal cell death. MKK7 and MLK3 are already induced directly at the resumption of reperfusion, and remain activated during the entire tested period of three days (Pan et al., 2005; but see also Zhang et al., 2003b). In the context of scaffolding proteins, one study has reported an increase of beta-arrestin–1 following a 90–minute HI insult in 12–day postnatal rats (Lombardi et al., 2004). The JIP1 protein also appears constitutively present, but there is a massive increase in its association with activated MKK7, within 30 minutes of reperfusion following global ischemia (Li et al., 2005).

As with c-jun, there are several indirect lines of evidence pointing to the involvement of the JNK cascade in mediating hypoxic and/or ischemic brain damage *in vivo*. Thus, protective hypothermia and preconditioning ischemia enhance the phosphorylation and activation of the mitogen-activated protein kinases ERK1 and 2, but not that of JNK (Gu et al., 2000; Hicks et al., 2000); the preconditioning effect appears mediated by inhibition of MLK3 via NMDA receptor-mediated Akt1 activation (Yin et al., 2005). A similar effect via glutamate receptors, MLK3 and JNK3 is also observed with oligonucleotide-mediated inhibition of PSD95 (Pei et al., 2004). Moreover, the protective effects of FK506 (Martin-Villalba et al., 1999), Ca-permeable AMPA receptor antagonists (Zhang et al., 2003a) or the MLK3 inhibitor K252a (Pan et al., 2005) are all associated with decreased phosphorylation of c-jun, JNK and the components of the upstream cascade.

13.6.4
Direct Evidence

Information on the critical role of the JNK family comes from the data on the *in-vivo* application of a global JNK inhibitor (Borsello et al., 2003; Hirt et al., 2004), the deletion of the JIP1 scaffolding protein (Im et al., 2003), and selective deletion of JNK1 and JNK3 (Kuan et al., 2003; Brecht et al., 2005). In the case of D-JNKI–1, a cell penetrating JNK-inhibitor which interferes by blocking the JNK binding domain, the systemic or the local, intracerebroventricular application of D-JNKI–1 strongly inhibited the post-ischemic neuronal cell death following medial cerebral artery occlusion (MCAO). This protective effect was consistent, present after transient or permanent MCAO, in postnatal day 14 as well as in adult rats, and was still observed when the peptide was given as long as 12 hours after the insult, indicating considerable therapeutic time window (Borsello et al., 2003; Hirt et al., 2004; Gao et al., 2005). Similar neuroprotective effects were also shown with the ATP-binding site competitive JNK inhibitor AS601245 (Carboni et al., 2005) and the anthrapyrazolone blocker SP600125 (Gao et al. 2005).

Experiments using genetically modified mice also show a more than 80% reduction in infarct size following transient MCAO in mice lacking JIP1 exon2/3, compared with their littermate controls (Im et al., 2003). This study underscores the importance of JIP1 in the post-ischemic induction of the JNK activity and its pro-degenerative effects following ischemic brain injury. However, it is unclear which of the JNK isoforms is mediating this damaging effect. Studies using genetic deletions of JNK1, JNK2, and JNK3 have come to somewhat differing results. Transgenic deletion of JNK3, but not JNK2, leads to a two-fold increase in the number of animals surviving without lesion at five days after HI insult following unilateral carotid occlusion and exposure to 7.5% oxygen (Kuan et al., 2003).

In the second JNK deletion study (Brecht et al., 2005), the absence of JNK1 in animals with a permanent occlusion of the medial cerebral artery caused a significant, two-fold increase in cortical tissue loss. The absence of JNK3 had no effect. The absence of JNK2 caused a slight but not significant reduction in tissue loss, and combination of JNK2 and JNK3 also caused a moderate but not significant loss, as with JNK2 alone (Brecht et al., 2005). As an ad hoc explanation, both studies (Kuan et al., 2003; Brecht et al., 2005) used different experimental models and outcome measures. However, unlike the JIP1 study (Im et al., 2003), it is important to note that the latter experiments also used non-littermate controls. This procedure can introduce an additional measure of variability in modifying the experimental outcome, even if performed on apparently the same background (Werner et al., 2001).

13.7
Axotomy

13.7.1
Regulation

Disruption or disconnection of the main axonal process is known to cause a strong increase in the neuronal c-jun expression (Herdegen et al., 1991c; Jenkins and Hunt, 1991). Although the levels of both mRNA and protein are augmented, the increase is considerably slower than that observed following ischemia or seizure activity, in the range of 12 to 24 hours. In line with the axonal transport hypothesis, the slower increase may be mediated by the interruption in retrograde fast axonal transport that is needed to maintain neurons (Leah et al., 1991, 1993; Makwana and Raivich, 2005), due to the time needed for the signal to travel between the lesion site and the neuronal cell body. The application of neurotrophic factors (NGF, NT4, GDNF, FGF2), all of which are normally transported retrogradely from the innervation target sites to the neuronal cell body, has been shown to inhibit the up-regulation of c-Jun in axotomized neurons (Hughes et al., 1997; Blottner and Herdegen, 1998; Vaudano et al., 2001). A similar up-regulation of c-Jun is also known to occur in neurodegenerative diseases such as amyotrophic lateral sclerosis affecting spinal and cranial motor neurons (Jaarsma et al., 1996), and which are associated with a disturbance in retrograde axonal transport (Breuer et al., 1987; Kieran et al., 2005). Interestingly, there is increasing evidence that the retrograde transport of activated N-terminal jun kinases, their upstream activators and scaffolding and transport proteins, may in itself play a key role in transmitting the injury signal from the lesion site to the neuronal cell body (Cavalli et al., 2005; Lindwall and Kanje, 2005).

Up-regulation of c-jun, and other inducible transcription factors coexpressed after injury appears to play a vital role in transforming a normal, differentiated neuron performing its physiological function into an injured cell capable of mounting a successful regenerative response (Herdegen and Leah, 1998; Makwana and Raivich, 2005). The increase in c-jun is present in all neuronal groups that project through peripheral nerves and are capable of successful regeneration, including sensory neurons, sympathetic neurons, motor neurons, and visceral motor neurons innervating autonomic ganglia (Rutherfurd et al., 1992; Koistinaho et al., 1993; Herdegen et al., 1993).

Neurons in the CNS that do not project through peripheral nerves, do show an up-regulation of c-jun, but mostly only transiently and/or following axotomy very near to the neuronal cell body (Dragunow, 1992; Mason et al., 2003; Schmitt et al., 2003). Several lines of evidence point to a tight correlation between the ability to regenerate and the post-traumatic expression of c-jun. Ola or WLDs mutant mice, that show a four-week delay in anterograde or Wallerian degeneration in injured peripheral nerves, also show a much more transient expression of c-jun in injured sensory and motor neurons (Gold et al., 1994). The transplantation of peripheral nerves, which serves as a regeneration-stimulating substrate to many central neurons, is associated with a strong up-regulation of c-jun axotomized retinal ganglion cell neurons (Hull and Bahr, 1994; Robinson, 1995), as well as those in the striatum, thalamus, substantia nigra, brainstem, and cerebellum (Anderson et al., 1998). Moreover, antibody-mediated inactivation of central neurite growth inhibitory proteins is also associated with an up-regulation of c-jun in the axotomized Purkinje cell neurons (Zagrebelsky et al., 1998). At present, it is unclear if the local inflammatory response plays a role. Local inflammation following intraganglionic *Clostridium diphtheriae* injection caused a strong induction in previously uninjured neurons (Lu and Richardson, 1995); on the other hand, the almost complete inhibition of local inflammation in macrophage colony-stimulating factor (MCSF)-deficient animals did not affect normal c-jun induction in motor neurons following facial nerve cutting (Kalla et al., 2001).

13.7.2
Functional Role: Only Partial Overlap with Jun and JNK

In-vivo analysis of c-jun points to a bipotential or a double role of c-jun, in its ability to enhance cell death as well as successful axonal regeneration (Herdegen et al., 1997; Dragunow et al., 2000). As found in *in-vitro* investigations (Ham et al., 1995; Eilers et al., 2001; Harding et al., 2001), functional *in-vivo* and *in-situ* studies concur that enhanced expression of c-jun induces cell death. Striatal injection of an adenovirus vector dominant-negative form of c-jun delayed the death of dopaminergic neurons after transection of the medial forebrain bundle (Crocker et al., 2001). Similar inhibition was also observed in axotomized retinal ganglion cell neurons injected with short interfering RNA (siRNA) against c-jun (Lingor et al., 2005). Purkinje cell promoter (L7) -driven overexpression of c-jun also enhanced their cell death in organotypic cell cultures (Carulli et al., 2002). These results were confirmed in a recent study where removal of c-jun activity by cutting out the floxed *jun* gene in the CNS using nestin-driven cre recombinase completely abolished neuronal cell death following facial axotomy (Raivich et al., 2004).

In addition to cell death, an absence of c-jun following nerve transection also interfered with the cell body response to axonal injury. C-Jun-deficient motor neurons were atrophic, and failed to activate neighboring microglia, to recruit blood-borne lymphocytes, or to show perineuronal axonal sprouting. Compared with controls that did not express cre, the neural jun-deficient mice displayed a four-fold decrease in the speed of regeneration, the reinnervation target muscle and functional recovery. Expression of CD44, galanin, and $\alpha_7\beta_1$ integrin – molecules known to be involved in regeneration and to carry AP–1 responsive elements in their promoter regions – was greatly impaired, suggesting a mechanism for c-Jun-mediated axonal growth (Raivich et al., 2004). Taken together, these results identify c-Jun as a central regulator of axonal regeneration and the overall neuronal cell body response in the injured CNS.

Although axotomy activates the JNK system (Herdegen et al., 1998; Masui et al., 2002), only part of the effects appears to be mediated through c-jun. Thus, replacement of the two N-terminal serine phosphoacceptor sites (Ser63, Ser73) with alanine inhibited PCD in retinal ganglion cell neurons after optic nerve cutting (Yoshida et al., 2002), but had no effect on adult facial motor neurons (Brecht et al., 2005; G. Raivich and A. Behrens, unpublished results). Interestingly, the deletion of JNK3 protects motor neurons and dorsal root ganglia sensory neurons against neonatal axotomy-induced death, but through mechanisms that are independent of c-jun phosphorylation (Keramaris et al., 2005). On the whole, these data reinforce the notion of only a partial overlap between the jun, and the JNK-mediated mechanisms (Gelderblom et al., 2004), pointing to a considerable, independent role of these injury-activated signaling kinases.

13.8
Conclusions

The strong neuronal expression of c-jun is a consistent feature of gene expression during embryogenesis, following injury as well as during other forms of stimulation in the adult organism. c-Jun is intensely up-regulated in ischemia and stroke, in seizures, during learning and memory, or following axonal injury and during successful regeneration. Functional studies employing *in-vivo* strategies using gene deletion, targeted expression of dominant-negative isoforms and pharmacological inhibitors all appear to confirm the bipotential role of c-jun, in mediating neurodegeneration and cell death, as well as in plasticity and repair. The phosphorylation of c-jun, and the activation of its upstream kinases, is required in many – but not in all – forms of these events, pointing to a complex picture, with only a partial overlap of the jun-, JNK-, or JNKK(n)-dependent functions. Here, a better understanding of the non-overlapping roles could considerably increase the potential of pharmacological agents to improve neurological outcome following trauma, neonatal encephalopathy and stroke, or neurodegenerative disease.

Abbreviations

ASK1	apoptosis signal-regulating kinase 1
HI	hypoxic-ischemic
JBD	JNK binding domain
JIP1	JNK interacting protein
LTP	long-term potentiation
MAPK	mitogen-activated protein kinase
MCAO	medial cerebral artery occlusion
MCSF	macrophage colony-stimulating factor
MEK/MKK	MAP kinase kinase
MEKK	MAP kinase kinase kinase
PCD	programmed cell death
SAPK	stress-activated protein kinase
siRNA	short interfering RNA

References

Abraham WC, Mason SE, Demmer J, Williams JM, Richardson CL, Tate WP, Lawlor PA, Dragunow M (1993) Correlations between immediate early gene induction and the persistence of long-term potentiation. *Neuroscience* 56: 717–727.

Akaji K, Suga S, Fujino T, Mayanagi K, Inamasu J, Horiguchi T, Sato S, Kawase T (2003) Effect of intra-ischemic hypothermia on the expression of c-Fos and c-Jun, and DNA binding activity of AP–1 after focal cerebral ischemia in rat brain. *Brain Res.* 975: 149–157.

Akechi M, Ito M, Uemura K, Takamatsu N, Yamashita S, Uchiyama K, Yoshioka K, Shiba T (2001) Expression of JNK cascade scaffold protein JSAP1 in the mouse nervous system. *Neurosci. Res.* 39: 391–400.

Anderson PN, Campbell G, Zhang Y, Lieberman AR (1998) Cellular and molecular correlates of the regeneration of adult mammalian CNS axons into peripheral nerve grafts. *Prog. Brain Res.* 117: 211–232.

Anokhin KV, Rose SP (1991) Learning-induced increase of immediate early gene messenger RNA in the chick forebrain. *Eur. J. Neurosci.* 3: 162–167.

Baker SJ, Kerppola TK, Luk D, Vandenberg MT, Marshak DR, Curran T, Abate C (1992) Jun is phosphorylated by several protein kinases at the same sites that are modified in serum-stimulated fibroblasts. *Mol. Cell. Biol.* 12: 4694–4705.

Barry CE, Nolan Y, Clarke RM, Lynch A, Lynch MA (2005) Activation of c-Jun-N-terminal kinase is critical in mediating lipopolysaccharide-induced changes in the rat hippocampus. *J. Neurochem.* 93: 221–231.

Behrens A, Sibilia M, Wagner EF (1999) Amino-terminal phosphorylation of c-Jun regulates stress-induced apoptosis and cellular proliferation. *Nat. Genet.* 21: 326–329.

Bennett GD, Lau F, Calvin JA, Finnell RH (1997) Phenytoin-induced teratogenesis: a molecular basis for the observed developmental delay during neurulation. *Epilepsia* 38: 415–423.

Bevilaqua LR, Kerr DS, Medina JH, Izquierdo I Cammarota M (2003) Inhibition of hippocampal Jun N-terminal kinase enhances short-term memory but blocks long-term memory formation and retrieval of an inhibitory avoidance task. *Eur. J. Neurosci.* 17: 897–902.

Binetruy B, Smeal T, Karin M (1991) Ha-Ras augments c-jun activity and stimulates phosphorylation of its activation domain. *Nature* 351: 122–127.

Blottner D, Herdegen T (1998) Neuroprotective fibroblast growth factor type–2 down-regulates the c-Jun transcription factor in axotomized sympathetic preganglionic neurons of adult rat. *Neuroscience* 82: 283–292.

Borasio GD, Horstmann S, Anneser JM, Neff NT, Glicksman MA. (1998) CEP-1347/KT7515, a JNK pathway inhibitor, supports the in vitro survival of chick embryonic neurons. *Neuroreport* 9: 1435–1439.

Borsello T, Clarke PG, Hirt L, Vercelli A, Repici M, Schorderet DF, Bogousslavsky J, Bonny C (2003) A peptide inhibitor of c-Jun N-terminal kinase protects against excitotoxicity and cerebral ischemia. *Nat. Med.* 9: 1180–1186.

Brecht S, Simler S, Vergnes M, Mielke K, Marescaux C, Herdegen T (1999) Repetitive electroconvulsive seizures induce activity of c-Jun N-terminal kinase and compartment-specific desensitization of c-Jun phosphorylation in the rat brain. *Mol. Brain Res.* 68: 101–108.

Brecht S, Kirchhof R, Chromik A, Willesen M, Nicolaus T, Raivich G, Wessig J, Waetzig V, Goetz M, Claussen M, Pearse D, Kuan CY, Vaudano E, Behrens A, Wagner E, Flavell RA, Davis RJ, Herdegen T (2005) Specific pathophysiological functions of JNK isoforms in the brain. *Eur. J. Neurosci.* 21: 363–377.

Breuer AC, Lynn MP, Atkinson MB, Chou SM, Wilbourn AJ, Marks KE, Culver JE, Fleegler EJ (1987) Fast axonal transport in amyotrophic lateral sclerosis: an intra-axonal organelle traffic analysis. *Neurology* 37: 738–748.

Carboni S, Antonsson B, Gaillard P, Gotteland JP, Gillon JY, Vitte PA (2005) Control of death receptor and mitochondrial-dependent apoptosis by c-Jun N-terminal kinase in hippocampal CA1 neurones following global transient ischaemia. *J. Neurochem.* 92: 1054–1060.

Carulli D, Buffo A, Botta C, Altruda F, Strata P (2002) Regenerative and survival capabilities of Purkinje cells overexpressing c-Jun. *Eur. J. Neurosci.* 16: 105–118.

Casanova E, Garate C, Ovalle S, Calvo P, Chinchetru MA (1996) Identification of four splice variants of the mouse stress-activated protein kinase JNK/SAPK alpha-isoform. *Neuroreport* 7: 1320–1324.

Castellazzi M, Spyrou G, La Vista N, Dangy J-P, Piu F, Yaniv M, and Mand Brun G (1991) Overexpression of c-jun, junB, or junD affects cell growth differently. *Proc. Natl. Acad. Sci. USA* 88: 8890–8894.

Cavalli V, Kujala P, Klumperman J, Goldstein LS (2005) Sunday Driver links axonal transport to damage signaling. *J. Cell Biol.* 168: 775–787.

Chang L, Jones Y, Ellisman MH, Goldstein LS, Karin M (2003) JNK1 is required for maintenance of neuronal microtubules and controls phosphorylation of microtubule-associated proteins. *Dev. Cell* 4: 521–533.

Chaisuksunt V, Campbell G, Zhang Y, Schachner M, Lieberman AR, Anderson PN (2003) Expression of regeneration-related molecules in injured and regenerating striatal and nigral neurons. *J. Neurocytol.* 32: 161–183.

Chen JT, Lu DH, Chia CP, Ruan DY, Sabapathy K, Xiao ZC (2005) Impaired long-term potentiation in c-Jun N-terminal kinase 2–deficient mice. *J. Neurochem.* 93: 463–473.

Cheng J, Yang J, Xia Y, Karin M, Su B (2000) Synergistic interaction of MEK kinase 2, c-Jun N-terminal kinase (JNK) kinase 2, and JNK1 results in efficient and specific JNK1 activation. *Mol. Cell. Biol.* 20: 2334–2342.

Chi H, Sarkisian MR, Rakic P, Flavell RA (2005) Loss of mitogen-activated protein kinase kinase kinase 4 (MEKK4) results in enhanced apoptosis and defective neural tube development. *Proc. Natl. Acad. Sci. USA* 102: 3846–3851.

Cole AJ, Abu-Shakra S, Saffen DW, Baraban JM, Worley PF (1990) Rapid rise in transcription factor mRNAs in rat brain after electroshock-induced seizures. *J. Neurochem.* 55: 1920–1927.

Collaco-Moraes Y, Aspey BS, de Belleroche JS, Harrison MJ (1994) Focal ischemia causes an extensive induction of immediate early genes that are sensitive to MK-801. *Stroke* 25: 1855–1860.

Collins LR, Minden A, Karin M, Brown JH (1996) Galpha12 stimulates c-Jun NH_2-terminal kinase through the small G proteins Ras and Rac. *J. Biol. Chem.* 271:17349–17353.

Costello DA, Herron CE (2004) The role of c-Jun N-terminal kinase in the Abeta-mediated impairment of LTP and regulation of synaptic transmission in the hippocampus. *Neuropharmacology* 46: 655–662.

Couceyro P, Pollock KM, Drews K, Douglass J (1994) Cocaine differentially regulates activator protein–1 mRNA levels and DNA-binding complexes in the rat striatum and cerebellum. *Mol. Pharmacol.* 46: 667–676.

Crocker SJ, Lamba WR, Smith PD, Callaghan SM, Slack RS, Anisman H, Park DS (2001) c-Jun mediates axotomy-induced dopamine neuron death in vivo. *Proc. Natl. Acad. Sci. USA* 98: 13385–13390.

Curran BP, Murray HJ, O'Connor JJ (2003) A role for c-Jun N-terminal kinase in the inhibition of long-term potentiation by interleukin–1beta and long-term depression in the rat dentate gyrus in vitro. *Neuroscience* 118: 347–357.

Donahue CP, Jensen RV, Ochiishi T, Eisenstein I, Zhao M, Shors T, Kosik KS (2002) Transcriptional profiling reveals regulated genes in the hippocampus during memory formation. *Hippocampus* 12: 821–833.

Dragunow M (1992) Axotomized medial septal-diagonal band neurons express Jun-like immunoreactivity. *Mol. Brain Res.* 15: 141–144.

Dragunow M, Currie RW, Faull RL, Robertson HA, Jansen K (1989) Immediate-early genes, kindling and long-term potentiation. *Neurosci. Biobehav. Rev.* 13: 301–313.

Dragunow M, Faull RL, Waldvogel HJ, Williams MN, Leah J (1991) Elevated expression of jun and fos-related proteins in transplanted striatal neurons. *Brain Res.* 558: 321–324.

Dragunow M, Young D, Hughes P, MacGibbon G, Lawlor P, Singleton K, Sirimanne E, Beilharz E, Gluckman P (1993) Is c-Jun involved in nerve cell death following status epilepticus and hypoxic-ischaemic brain injury? *Mol. Brain Res.* 18: 347–352.

Dragunow M, Beilharz E, Sirimanne E, Lawlor P, Williams C, Bravo R, Gluckman P (1994) Immediate-early gene protein expression in neurons undergoing delayed death, but not necrosis, following hypoxic-ischaemic injury to the young rat brain. *Mol. Brain Res.* 25: 19–33.

Dragunow M, Xu R, Walton M, Woodgate A, Lawlor P, MacGibbon GA, Young D, Gibbons H, Lipski J, Muravlev A, Pearson A, During M (2000) c-Jun promotes neurite outgrowth and survival in PC12 cells. *Mol. Brain Res.* 83: 20–33.

Eilers A, Whitfield J, Shah B, Spadoni C, Desmond H, Ham J (2001) Direct inhibition of c-Jun N-terminal kinase in sympathetic neurones prevents c-jun promoter activation and NGF withdrawal-induced death. *J. Neurochem.* 76: 1439–1454.

Eminel S, Klettner A, Roemer L, Herdegen T, Waetzig V (2004) JNK2 translocates to the mitochondria and mediates cytochrome c release in PC12 cells in response to 6–hydroxydopamine. *J. Biol. Chem.* 279: 55385–55392.

English JM, Vanderbilt CA, Xu S, Marcus S, Cobb MH (1995) Isolation of MEK5 and differential expression of alternatively spliced forms. *J. Biol. Chem.* 270: 28897–28902.

Franklin CC, Sanchez V, Wagner F, Woodgett JR, Kraft AS (1992) Phorbol ester-induced amino-terminal phosphorylation of human jun but not junB regulates transcriptional activation. *Proc. Natl. Acad. Sci. USA* 8: 7247–7251.

Freeman FM, Rose SP (1999) Expression of Fos and Jun proteins following passive avoidance training in the day-old chick. *Learn. Mem.* 6: 389–397

Freeman WM, Nader MA, Nader SH, Robertson DJ, Gioia L, Mitchell SM, Daunais JB, Porrino LJ, Friedman DP, Vrana KE (2001) Chronic cocaine-mediated changes in non-human primate nucleus accumbens gene expression. *J. Neurochem.* 77: 542–549

Gall C, Lauterborn J, Isackson P, White J (1990) Seizures, neuropeptide regulation, and mRNA expression in the hippocampus. *Prog. Brain Res.* 83: 371–390

Ganiatsas S, Kwee L, Fujiwara Y, Perkins A, Ikeda T, Labow MA, Zon LI (1998) SEK1 deficiency reveals mitogen-activated protein kinase cascade crossregulation and leads to abnormal hepatogenesis. *Proc. Natl. Acad. Sci. USA* 95: 6881–6886

Gao Y, Signore AP, Yin W, Cao G, Yin XM, Sun F, Luo Y, Graham SH, Chen J (2005) Neuroprotection against focal ischemic brain injury by inhibition of c-Jun N-terminal kinase and attenuation of the mitochondrial apoptosis-signaling pathway. *J. Cereb. Blood Flow Metab.* 25: 694–712

Gass P, Herdegen T, Bravo R, Kiessling M (1993) Spatiotemporal induction of immediate early genes in the rat brain after limbic seizures: effects of NMDA receptor antagonist MK–801. *Eur. J. Neurosci.* 5: 933–943

Gelderblom M, Eminel S, Herdegen T, Waetzig V (2004) c-Jun N-terminal kinases (JNKs) and the cytoskeleton—functions beyond neurodegeneration. *Int. J. Dev. Neurosci.* 22: 559–564

Giorgi L, Fanfani E, Paternostro E, Trovati F, Falchetti A, Novelli GP (1997) Immediate-early genes expression in spinal cord as related to acute noxious stimulus. *Int. J. Clin. Pharmacol. Res.* 17: 59–61

Gold BG, Austin DR, Storm-Dickerson T (1994) Multiple signals underlie the axotomy-induced up-regulation of c-jun in adult sensory neurons. *Neurosci. Lett.* 176: 123–127

Grimm C, Wenzel A, Behrens A, Hafezi F, Wagner EF, Reme CE (2001) AP–1 mediated retinal photoreceptor apoptosis is independent of N-terminal phosphorylation of c-Jun. *Cell Death Differ.* 8: 859–867.

Gu Z, Jiang Q, Zhang G, Cui Z, Zhu Z (2000) Diphosphorylation of extracellular signal-regulated kinases and c-Jun N-terminal protein kinases in brain ischemic tolerance in rat. *Brain Res.* 860: 157–160.

Gupta S, Barrett T, Whitmarsh AJ, Cavanagh J, Sluss HK, Derijard B, Davis RJ (1996) Selective interaction of JNK protein kinase isoforms with transcription factors. *EMBO J.* 15: 2760–2770.

Ha HY, Cho IH, Lee KW, Lee KW, Song JY, Kim KS, Yu YM, Lee JK, Song JS, Yang SD, Shin HS, Han PL (2005) The axon guidance defect of the telencephalic commissures of the JSAP1–deficient brain was partially rescued by the transgenic expression of JIP1. *Dev. Biol.* 277: 184–199.

Hagemann C, Blank JL (2001) The ups and downs of MEK kinase interactions. *Cell Signal.* 13: 863–875.

Ham J, Babij C, Whitfield J, Pfarr CM, Lallemand D, Yaniv M, Rubin LL (1995) A c-Jun dominant negative mutant protects sympathetic neurons against programmed cell death. *Neuron* 14: 927–939.

Harding TC, Xue L, Bienemann A, Haywood D, Dickens M, Tolkovsky AM, Uney JB (2001) Inhibition of JNK by overexpression of the JNL binding domain of JIP–1 prevents apoptosis in sympathetic neurons. *J. Biol. Chem.* 276: 4531–4534.

Hayward MD, Duman RS, Nestler EJ (1990) Induction of the c-fos proto-oncogene during opiate withdrawal in the locus coeruleus and other regions of rat brain. *Brain Res.* 525: 256–266.

Hehner SP, Hofmann TG, Dienz O, Droge W, Schmitz ML (2000) Tyrosine-phosphorylated Vav1 as a point of integration for T-cell receptor- and CD28–mediated activation of JNK, p38, and interleukin–2 transcription. *J. Biol. Chem.* 275: 18160–18171.

Herdegen T, Leah JD (1998) Inducible and constitutive transcription factors in the mammalian nervous system: control of gene expression by Jun, Fos and Krox, and CREB/ATF proteins. *Brain Res. Rev.* 28: 370–490.

Herdegen T, Tolle TR, Bravo R, Zieglgansberger W, Zimmermann M (1991a) Sequential expression of junB, junD and fosB proteins in rat spinal neurons: cascade of transcriptional operations during nociception. *Neurosci. Lett.* 129: 221–224.

Herdegen T, Leah JD, Manisali A, Bravo R, Zimmermann M (1991b) C-jun-like immunoreactivity in the CNS of the adult rat: basal and transynaptically induced expression of an immediate-early gene. *Neuroscience* 41(2–3): 643–654.

Herdegen T, Kummer W, Fiallos CE, Leah J, Bravo R (1991c) Expression of c-jun, junB and junD proteins in rat nervous system following transection of vagus nerve and cervical sympathetic trunk. *Neuroscience* 45(2): 413–422.

Herdegen T, Fiallos-Estrada CE, Bravo R, Zimmermann M (1993) Colocalisation and covariation of c-jun transcription factor with galanin in primary afferent neurons and with CGRP in spinal motoneurons following transection of rat sciatic nerve. *Mol. Brain Res.* 17: 147–154.

Herdegen T, Skene P, Bahr M (1997) The c-Jun transcription factor – bipotential mediator of neuronal death, survival and regeneration. *Trends Neurosci.* 20: 227–231.

Herdegen T, Claret FX, Kallunki T, Martin-Villalba A, Winter C, Hunter T, Karin M (1998) Lasting N-terminal phosphorylation of c-Jun and activation of c-Jun N-terminal kinases after neuronal injury. *J. Neurosci.* 18: 5124–5135.

Herzog KH, Chen SC, Morgan JI (1999) c-jun Is dispensable for developmental cell death and axogenesis in the retina. *J. Neurosci.* 19: 4349–4359.

Heurteaux C, Messier C, Destrade C, Lazdunski M (1993a) Memory processing and apamin induce immediate-early gene expression in mouse brain. *Mol. Brain Res.* 18: 17–22.

Heurteaux C, Bertaina V, Widmann C, Lazdunski M (1993b) K^+ channel openers prevent global ischemia-induced expression of c-fos, c-jun, heat shock protein, and amyloid beta-protein precursor genes and neuronal death in rat hippocampus. *Proc. Natl. Acad. Sci. USA* 90: 9431–9435.

Hicks SD, Parmele KT, DeFranco DB, Klann E, Callaway CW (2000) Hypothermia differentially increases extracellular signal-regulated kinase and stress-activated protein kinase/c-Jun terminal kinase activation in the hippocampus during reperfusion after asphyxial cardiac arrest. *Neuroscience* 98: 677–685.

Hilberg F, Aguzzi A, Howells N, Wagner EF (1993) c-jun is essential for normal mouse development and hepatogenesis. *Nature* 365: 179–181.

Hirai S, Bourachot B, Yaniv M (1990) Both jun and fos contribute to transcription activation by the heterodimer. *Oncogene* 5: 39–46.

Hirt L, Badaut J, Thevenet J, Granziera C, Regli L, Maurer F, Bonny C, Bogousslavsky J (2004) D-JNKI1, a cell-penetrating c-Jun-N-terminal kinase inhibitor, protects against cell death in severe cerebral ischemia. *Stroke* 35(7): 1738–1743. e-pub 2004 June 3.

Hope B, Kosofsky B, Hyman SE, Nestler EJ (1992) Regulation of immediate early gene expression and AP–1 binding in the rat nucleus accumbens by chronic cocaine. *Proc. Natl. Acad. Sci. USA* 89: 5764–5768.

Hou XS, Goldstein ES, Perrimon N (1997) *Drosophila* Jun relays the Jun amino-terminal kinase signal transduction pathway to the Decapentaplegic signal transduction pathway in regulating epithelial cell sheet movement. *Genes Dev.* 11: 1728–1737.

Hughes PE, Alexi T, Hefti F, Knusel B (1997) Axotomized septal cholinergic neurons rescued by nerve growth factor or neurotrophin–4/5 fail to express the inducible transcription factor c-Jun. *Neuroscience* 78: 1037–1049.

Hull M, Bahr M (1994) Regulation of immediate-early gene expression in rat retinal ganglion cells after axotomy and during regeneration through a peripheral nerve graft. *J. Neurobiol.* 25: 92–105.

Ichijo H (1999) From receptors to stress-activated MAP kinases. *Oncogene* 18: 6087–6093.

Im JY, Lee KW, Kim MH, Lee SH, Ha HY, Cho IH, Kim D, Yu MS, Kim JB, Lee JK, Kim YJ, Youn BW, Yang SD, Shin HS, Han PL (2003) Repression of phospho-JNK and infarct volume in ischemic brain of JIP1–deficient mice. *J. Neurosci. Res.* 74: 326–332.

Ishikawa T, Nakada-Moriya Y, Ando C, Tanda N, Nishida S, Minatogawa Y, Nohno T (1997) Expression of the JNK2–alpha1 gene in the developing chick brain. *Biochem. Biophys. Res. Commun.* 234: 489–492.

Jaarsma D, Holstege JC, Troost D, Davis M, Kennis J, Haasdijk ED, de Jong VJ (1996) Induction of c-Jun immunoreactivity in spinal cord and brainstem neurons in a transgenic mouse model for amyotrophic lateral sclerosis. *Neurosci. Lett.* 219: 179–182.

Jenkins R, Hunt SP (1991) Long-term increase in the levels of c-jun mRNA and jun protein-like immunoreactivity in motor and sensory neurons following axon damage. *Neurosci. Lett.* 129: 107–110.

Kalla R, Liu Z, Xu S, Koppius A, Imai Y, Kloss CU, Kohsaka S, Geschwendtner A, Moller JC, Werner A, Raivich G (2001) Microglia and the early phase of immune surveillance in the axotomized facial motor nucleus: impaired microglial activation and lymphocyte recruitment but no effect on neuronal survival or axonal regeneration in macrophage-colony stimulating factor-deficient mice. *J. Comp. Neurol.* 436: 182–201.

Kaminska B, Filipkowski RK, Zurkowska G, Lason W, Przewlocki R, Kaczmarek L (1994) Dynamic changes in the composition of the AP–1 transcription factor DNA-binding activity in rat brain following kainate-induced seizures and cell death. *Eur. J. Neurosci.* 6: 1558–1566.

Karin M (1995) The regulation of AP–1 activity by mitogen-activated protein kinases. *J. Biol. Chem.* 270: 16483–16486.

Kelkar N, Delmotte MH, Weston CR, Barrett T, Sheppard BJ, Flavell RA, Davis RJ (2003) Morphogenesis of the telencephalic commissure requires scaffold protein JNK-interacting protein 3 (JIP3). *Proc. Natl. Acad. Sci. USA* 100: 9843–9848.

Keramaris E, Vanderluit JL, Bahadori M, Mousavi K, Davis RJ, Flavell R, Slack RS, Park DS (2005) c-Jun N-terminal kinase 3 deficiency protects neurons from axotomy-induced death in vivo through mechanisms independent of c-Jun phosphorylation. *J. Biol. Chem.* 280: 1132–1141.

Kieran D, Hafezparast M, Bohnert S, Dick JR, Martin J, Schiavo G, Fisher EM, Greensmith L (2005) A mutation in dynein rescues axonal transport defects and extends the life span of ALS mice. *J. Cell Biol.* 169: 561–567.

Kim IJ, Lee KW, Park BY, Lee JK, Park J, Choi IY, Eom SJ, Chang TS, Kim MJ, Yeom YI, Chang SK, Lee YD, Choi EJ, Han PL (1999) Molecular cloning of multiple splicing variants of JIP–1 preferentially expressed in brain. *J. Neurochem.* 72: 1335–1343.

Kindy MS, Carney JP, Dempsey RJ, Carney JM (1991) Ischemic induction of protooncogene expression in gerbil brain. *J. Mol. Neurosci.* 2: 217–228.

Kobori N, Clifton GL, Dash P (2002) Altered expression of novel genes in the cerebral cortex following experimental brain injury. *Mol. Brain Res.* 104: 148–158.

Kockel L, Zeitlinger J, Staszewski LM, Mlodzik M, Bohmann D (1997) Jun in *Drosophila* development: redundant and nonredundant functions and regulation by two MAPK signal transduction pathways. *Genes Dev.* 11: 1748–1758.

Kockel L, Homsy JS, Bohmann D (2001) *Drosophila* AP–1: Lessons from an invertebrate. *Oncogene* 20: 2347–2364.

Koistinaho J, Pelto-Huikko M, Sagar SM, Dagerlind A, Roivainen R, Hokfelt T (1993) Injury-induced long-term expression of immediate early genes in the rat superior cervical ganglion. *Neuroreport* 4(1): 37–40.

Kuan CY, Yang DD, Samanta Roy DR, Davis RJ, Rakic P, Flavell RA (1999) The Jnk1 and Jnk2 protein kinases are required for regional specific apoptosis during early brain development. *Neuron* 22: 667–676.

Kuan CY, Whitmarsh AJ, Yang DD, Liao G, Schloemer AJ, Dong C, Bao J, Banasiak KJ, Haddad GG, Flavell RA, Davis RJ, Rakic P (2003) A critical role of neural-specific JNK3 for ischemic apoptosis. *Proc. Natl. Acad. Sci. USA* 100: 15184–15189.

Lanaud P, Maggio R, Gale K, Grayson DR (1993) Temporal and spatial patterns of expression of c-fos, zif/268, c-jun and jun-B mRNAs in rat brain following seizures evoked focally from the deep prepiriform cortex. *Exp. Neurol.* 119: 20–31.

Leah JD, Herdegen T, Bravo R (1991) Selective expression of Jun proteins following axotomy and axonal transport block in peripheral nerves in the rat: evidence for a role in the regeneration process. *Brain Res.* 566(1–2): 198–207.

Leah JD, Herdegen T, Murashov A, Dragunow M, Bravo R (1993) Expression of immediate early gene proteins following axotomy and inhibition of axonal transport in the rat central nervous system. *Neuroscience* 57: 53–66.

Lee JK, Hwang WS, Lee YD, Han PL (1999) Dynamic expression of SEK1 suggests multiple roles of the gene during embryogenesis and in adult brain of mice. *Mol. Brain Res.* 66: 133–140.

Levkovitz Y, Baraban JM (2002) A dominant negative Egr inhibitor blocks nerve growth factor-induced neurite outgrowth by suppressing c-Jun activation: role of an Egr/c-Jun complex. *J. Neurosci.* 22: 3845–3854.

Li CH, Wang RM, Zhang QG, Zhang GY (2005) Activated mitogen-activated protein kinase kinase 7 redistributes to the cytosol and binds to Jun N-terminal kinase-interacting protein 1 involving oxidative stress during early reperfusion in rat hippocampal CA1 region. *J. Neurochem.* 93: 290–298.

Lindwall C, Kanje M (2005) Retrograde axonal transport of JNK signaling molecules influence injury induced nuclear changes in p-c-Jun and ATF3 in adult rat sensory neurons. *Mol. Cell. Neurosci.* 29: 269–282.

Lingor P, Koeberle P, Kugler S, Bahr M (2005) Down-regulation of apoptosis mediators by RNAi inhibits axotomy-induced retinal ganglion cell death in vivo. *Brain* 128: 550–558.

Lombardi MS, van den Tweel E, Kavelaars A, Groenendaal F, van Bel F, Heijnen CJ (2004) Hypoxia/ischemia modulates G protein-coupled receptor kinase 2 and beta-arrestin–1 levels in the neonatal rat brain. *Stroke* 35: 981–986.

Lu X, Richardson PM (1995) Changes in neuronal mRNAs induced by a local inflammatory reaction. *J. Neurosci. Res.* 41(1): 8–14.

Macaya A, Munell F, Ferrer I, de Torres C, Reventos J (1998) Cell death and associated c-jun induction in perinatal hypoxia-ischemia. Effect of the neuroprotective drug dexamethasone. *Mol. Brain Res.* 56: 29–37.

Magara F, Haefliger JA, Thompson N, Riederer B, Welker E, Nicod P, Waeber G (2003) Increased vulnerability to kainic acid-induced epileptic seizures in mice underexpressing the scaffold protein Islet-Brain 1/JIP–1. *Eur. J. Neurosci.* 17: 2602–2610.

Makwana M, Raivich G (2005) Molecular mechanisms in successful peripheral regeneration. *FEBS J.* 272: 2628–2638.

Maroney AC, Finn JP, Bozyczko-Coyne D, O'Kane TM, Neff NT, Tolkovsky AM, Park

DS, Yan CY, Troy CM, Greene LA (1999) CEP-1347 (KT7515), an inhibitor of JNK activation, rescues sympathetic neurons and neuronally differentiated PC12 cells from death evoked by three distinct insults. *J. Neurochem.* 73: 1901–1912.

Martin-Villalba A, Herr I, Jeremias I, Hahne M, Brandt R, Vogel J, Schenkel J, Herdegen T, Debatin KM (1999) CD95 ligand (Fas-L/APO–1L) and tumor necrosis factor-related apoptosis-inducing ligand mediate ischemia-induced apoptosis in neurons. *J. Neurosci.* 19: 3809–3817.

Mason MR, Lieberman AR, Anderson PN (2003) Corticospinal neurons up-regulate a range of growth-associated genes following intracortical, but not spinal, axotomy. *Eur. J. Neurosci.* 18: 789–802.

Masui K, Yamada E, Shimokawara T, Mishima K, Enomoto Y, Nakajima H, Yoshikawa T, Sakaki T, Ichijima K (2002) Expression of c-Jun N-terminal kinases after axotomy in the dorsal motor nucleus of the vagus nerve and the hypoglossal nucleus. *Acta Neuropathol.* 104: 123–129.

McDonald PH, Chow CW, Miller WE, Laporte SA, Field ME, Lin FT, Davis RJ, Lefkowitz RJ (2000) Beta-arrestin 2: a receptor-regulated MAPK scaffold for the activation of JNK3. *Science* 290: 1574–1577.

Mechta-Grigoriou F, Giudicelli F, Pujades C, Charnay P, Yaniv M (2003) c-jun regulation and function in the developing hindbrain. *Dev. Biol.* 258: 419–431.

Mellström B, Achaval M, Montero D, Naranjo JR, Sassone-Corsi P (1991) Differential expression of the jun family members in rat brain. *Oncogene* 6: 1959–1964.

Messina A, Jaworowski A, Bell C (1996) Detection of jun but not fos protein during developmental cell death in sympathetic neurons. *J. Comp. Neurol.* 372: 544–550.

Moratalla R, Vickers EA, Robertson HA, Cochran BH, Graybiel AM (1993) Coordinate expression of c-fos and junB is induced in the rat striatum by cocaine. *J. Neurosci.* 13: 423–433.

Morgan JI, Curran T (1988) Calcium as a modulator of the immediate-early gene cascade in neurons. *Cell Calcium* 9: 303–311.

Morgan JI, Cohen DR, Hempstead JL, Curran T (1987) Mapping patterns of c-fos expression in the central nervous system after seizure. *Science* 237: 192–197.

Munell F, Burke RE, Bandele A, Gubits RM (1994) Localization of c-fos, c-jun, and hsp70 mRNA expression in brain after neonatal hypoxia-ischemia. *Dev. Brain Res.* 77: 111–121.

Naranjo JR, Mellstrom B, Achaval M, Sassone-Corsi P (1991) Molecular pathways of pain: Fos/Jun-mediated activation of a noncanonical AP-1 site in the prodynorphin gene. *Neuron* 6: 607–617.

Nateri AS, Riera-Sans L, Da Costa C, Behrens A (2004) The ubiquitin ligase SCFFbw7 antagonizes apoptotic JNK signaling. *Science* 303: 1374–1378.

Nestler EJ (1993) Cellular responses to chronic treatment with drugs of abuse. *Crit. Rev. Neurobiol.* 7: 23–39.

Nishina H, Vaz C, Billia P, Nghiem M, Sasaki T, De la Pompa JL, Furlonger K, Paige C, Hui C, Fischer KD, Kishimoto H, Iwatsubo T, Katada T, Woodgett JR, Penninger JM (1999) Defective liver formation and liver cell apoptosis in mice lacking the stress signaling kinase SEK1/MKK4. *Development* 126: 505–516.

Otani N, Nawashiro H, Fukui S, Nomura N, Yano A, Miyazawa T, Shima K (2002) Differential activation of mitogen-activated protein kinase pathways after traumatic brain injury in the rat hippocampus. *J. Cereb. Blood Flow Metab.* 22: 327–334.

Pan J, Zhang QG, Zhang GY (2005) The neuroprotective effects of K252a through inhibiting MLK3/MKK7/JNK3 signaling pathway on ischemic brain injury in rat hippocampal CA1 region. *Neuroscience* 131: 147–159.

Papa M, Pellicano MP, Welzl H, Sadile AG (1993) Distributed changes in c-Fos and c-Jun immunoreactivity in the rat brain associated with arousal and habituation to novelty. *Brain Res. Bull.* 32: 509–515.

Peakman MC, Colby C, Perrotti LI, Tekumalla P, Carle T, Ulery P, Chao J, Duman C, Steffen C, Monteggia L, Allen MR, Stock JL, Duman RS, McNeish JD, Barrot M, Self DW, Nestler EJ, Schaeffer E (2003) Inducible, brain region-specific expression of a dominant negative mutant of c-Jun in transgenic mice decreases sensitivity to cocaine. *Brain Res.* 970: 73–86.

Pei DS, Sun YF, Guan QH, Hao ZB, Xu TL, Zhang GY (2004) Postsynaptic density protein 95 antisense oligodeoxynucleotides inhibits the activation of MLK3 and JNK3 via the GluR6.PSD–95.MLK3 signaling module after transient cerebral ischemia in rat hippocampus. *Neurosci. Lett.* 367: 71–75.

Presley RW, Menetrey D, Levine JD, Basbaum AI (1990) Systemic morphine suppresses noxious stimulus-evoked Fos protein-like immunoreactivity in the rat spinal cord. *J. Neurosci.* 10: 323–335.

Radler-Pohl A, Sachsenmaier Ch, Gebel S, Auer H-P, Bruder JT, Rapp U, Angel P, Rahmsdorf HJ, Herrlich P (1993) UV-induced activation of AP–1 involves obligatory extranuclear steps including Raf–1 kinase. *EMBO J.* 12: 1005–1021.

Raghupathi R, McIntosh TK, Smith DH (1995) Cellular responses to experimental brain injury. *Brain Pathol.* 5: 437–442.

Raivich G, Bohatschek M, Da Costa C, Iwata O, Galiano M, Hristova M, Nateri AS, Makwana M, Riera-Sans L, Wolfer DP, Lipp HP, Aguzzi A, Wagner EF, Behrens A (2004) The AP–1 transcription factor c-Jun is required for efficient axonal regeneration. *Neuron* 43: 57–67.

Riesgo-Escovar JR, Hafen E (1997) *Drosophila* Jun kinase regulates expression of decapentaplegic via the ETS-domain protein Aop and the AP–1 transcription factor Djun during dorsal closure. *Genes Dev.* 11: 1717–1727.

Robinson GA (1995) Axotomy-induced regulation of c-Jun expression in regenerating rat retinal ganglion cells. *Mol. Brain Res.* 30: 61–69.

Roffler-Tarlov S, Brown JJ, Tarlov E, Stolarov J, Chapman DL, Alexiou M, Papaioannou VE (1996) Programmed cell death in the absence of c-Fos and c-Jun. *Development* 122: 1–9.

Rosen KM, McCormack MA, Villa-Komaroff L, Mower GD (1992) Brief visual experience induces immediate early gene expression in the cat visual cortex. *Proc. Natl. Acad. Sci. USA* 89: 5437–5441.

Rutherfurd SD, Louis WJ, Gundlach AL (1992) Induction of c-jun expression in vagal motoneurones following axotomy. *Neuroreport* 3: 465–468.

Sabapathy K, Jochum W, Hochedlinger K, Chang L, Karin M, Wagner EF (1999) Defective neural tube morphogenesis and altered apoptosis in the absence of both JNK1 and JNK2. *Mech. Dev.* 89: 115–124.

Saffen DW, Cole AJ, Worley PF, Christy BA, Ryder K, Baraban JM (1988) Convulsant-induced increase in transcription factor messenger RNAs in rat brain. *Proc. Natl. Acad. Sci. USA* 85: 7795–7799.

Sanyal S, Sandstrom DJ, Hoeffer CA, Ramaswami M (2002) AP–1 functions upstream of CREB to control synaptic plasticity in *Drosophila*. *Nature* 416: 870–874.

Schauwecker PE (2000) Seizure-induced neuronal death is associated with induction of c-Jun N-terminal kinase and is dependent on genetic background. *Brain Res.* 884: 116–128.

Schmitt AB, Breuer S, Polat L, Pech K, Kakulas B, Love S, Martin D, Schoenen J, Noth J, Brook GA. (2003) Retrograde reactions of Clarke's nucleus neurons after human spinal cord injury. *Ann. Neurol.* 54: 534–539.

Smeal T, Binetruy B, Mercola D, Grover-Bardwick A, Heidecker G, Rapp UR, M Karin (1992) Oncoprotein-mediated signalling cascade stimulates c-jun activity by phosphorylation of serines 63 and 73. *Mol. Cell. Biol.* 12: 3507–3513.

Sommer C, Gass P, Kiessling M (1995) Selective c-jun expression in CA1 neurons of the gerbil hippocampus during and after acquisition of an ischemia-tolerant state. *Brain Pathol.* 5: 135–144.

Sonnenberg JL, Macgregor-Leon PF, Curran T, Morgan JI (1989) Dynamic alterations occur in the levels and composition of transcription factor AP–1 complexes after seizure. *Neuron* 3(3): 359–365.

Sun W, Gould TW, Newbern J, Milligan C, Choi SY, Kim H, Oppenheim RW (2005) Phosphorylation of c-Jun in avian and mammalian motoneurons in vivo during programmed cell death: an early reversible event in the apoptotic cascade. *J. Neurosci.* 25: 5595–5603.

Thompson NA, Haefliger JA, Senn A, Tawadros T, Magara F, Ledermann B, Nicod P, Waeber G (2001) Islet-brain1/JNK interacting protein–1 is required for early embryogenesis in mice. *J. Biol. Chem.* 276: 27745–27748.

Tischmeyer W, Grimm R, Schicknick H, Brysch W, Schlingensiepen KH (1994) Sequence-specific impairment of learning by c-jun antisense oligonucleotides. *Neuroreport* 5: 1501–1504.

Tolle TR, Herdegen T, Schadrack J, Bravo R, Zimmermann M, Zieglgansberger W (1994) Application of morphine prior to noxious stimulation differentially modulates expression of Fos, Jun and Krox–24 proteins in rat spinal cord neurons. *Neuroscience* 58: 305–321.

Vannucci RC, Vannucci SJ (2005) Perinatal hypoxic-ischemic brain damage: evolution of an animal model. *Dev. Neurosci.* 27: 81–86.

Vaudano E, Rosenblad C, Bjorklund A (2001) Injury induced c-Jun expression and phosphorylation in the dopaminergic nigral neurons of the rat: correlation with neuronal death and modulation by glial-cell-line-derived neurotrophic factor. *Eur. J. Neurosci.* 13: 1–14.

Verhey KJ, Meyer D, Deehan R, Blenis J, Schnapp BJ, Rapoport TA, Margolis B (2001) Cargo of kinesin identified as JIP scaffolding proteins and associated signaling molecules. *J. Cell Biol.* 152: 959–970.

Weber U, Paricio N, Mlodzik M (2000) Jun mediates Frizzled-induced R3/R4 cell fate distinction and planar polarity determination in the *Drosophila* eye. *Development* 127: 3619–3629.

Werner A, Martin S, Gutierrez-Ramos JC, Raivich G (2001) Leukocyte recruitment and neuroglial activation during facial nerve regeneration in ICAM–1–deficient mice: effects of breeding strategy. *Cell Tissue Res.* 305: 25–41.

Wessel TC, Joh TH, Volpe BT (1991) In situ hybridization analysis of c-fos and c-jun expression in the rat brain following transient forebrain ischemia. *Brain Res.* 567: 231–240.

White JD, Gall CM (1987) Differential regulation of neuropeptide and proto-oncogene mRNA content in the hippocampus following recurrent seizures. *Brain Res.* 427: 21–29.

Whitmarsh AJ, Kuan CY, Kennedy NJ, Kelkar N, Haydar TF, Mordes JP, Appel M, Rossini AA, Jones SN, Flavell RA, Rakic P, Davis RJ (2001) Requirement of the JIP1 scaffold protein for stress-induced JNK activation. *Genes Dev.* 15: 2421–2432.

Wilkinson DG, Bhatt S, Ryseck RP, Bravo R (1989) Tissue-specific expression of c-jun and junB during organogenesis in the mouse. *Development* 106: 465–471.

Winston SM, Hayward MD, Nestler EJ, Duman RS (1990) Chronic electroconvulsive seizures down-regulate expression of the immediate-early genes c-fos and c-jun in rat cerebral cortex. *J. Neurochem.* 54: 1920–1925.

Wisden W, Errington ML, Williams S, Dunnett SB, Waters C, Hitchcock D, Evan G, Bliss TV, Hunt SP (1990) Differential expression of immediate early genes in the hippocampus and spinal cord. *Neuron* 4: 603–614.

Woodburn VL, Hayward NJ, Poat JA, Woodruff GN, Hughes J (1993) The effect of dizocilpine and enadoline on immediate early gene expression in the gerbil global ischaemia model. *Neuropharmacology* 32: 1047–1059.

Worley F, Bhat RV, Baraban JM, Erickson CA, McNaughton BL, Barnes CA (1993) Thresholds for synaptic activation of transcription factors in hippocampus: correlation with long-term enhancement. *J. Neurosci.* 13: 4776–4786.

Yang DD, Kuan CY, Whitmarsh AJ, Rincon M, Zheng TS, Davis RJ, Rakic P, Flavell RA (1997) Absence of excitotoxicity-induced apoptosis in the hippocampus of mice lacking the Jnk3 gene. *Nature* 389: 865–870.

Yasuda J, Whitmarsh AJ, Cavanagh J, Sharma M, Davis RJ (1999) The JIP group of mitogen-activated protein kinase scaffold proteins. *Mol. Cell. Biol.* 19: 7245–7254.

Yin XH, Zhang QG, Miao B, Zhang GY (2005) Neuroprotective effects of preconditioning ischaemia on ischaemic brain injury through inhibition of mixed-lineage kinase 3 via NMDA receptor-mediated Akt1 activation. *J. Neurochem.* 93: 1021–1029.

Yoneda Y, Ogita K, Kabutoz H, Mori A (1993) Selectively high expression of the transcription factor AP1 in telencephalic structures of epileptic E1 mice. *Neurosci. Lett.* 161: 161–164.

Yoshida K, Behrens A, Le-Niculescu H, Wagner EF, Harada T, Imaki J, Ohno S, Karin M (2002) Amino-terminal phosphorylation of c-Jun regulates apoptosis in the retinal ganglion cells by optic nerve transection. *Invest. Ophthalmol. Vis. Sci.* 43: 1631–1635.

Yount GL, Ponsalle P, White JD (1994) Pentylenetetrazole-induced seizures stimulate transcription of early and late response genes. Mol. *Brain Res.* 21: 219–224.

Zagrebelsky M, Buffo A, Skerra A, Schwab ME, Strata P, Rossi F (1998) Retrograde regulation of growth-associated gene expression in adult rat Purkinje cells by myelin-associated neurite growth inhibitory proteins. *J. Neurosci.* 18: 7912–7929.

Zhang Q, Zhang G, Meng F, Tian H (2003a) Biphasic activation of apoptosis signal-regulating kinase 1–stress-activated protein kinase 1–c-Jun N-terminal protein kinase pathway is selectively mediated by Ca2+-permeable alpha-amino–3–hydroxy–5–methyl–4–isoxazolepropionate receptors involving oxidative stress following brain ischemia in rat hippocampus. *Neurosci. Lett.* 337: 51–55.

Zhang Q, Tian H, Fu X, Zhang G (2003b) Delayed activation and regulation of MKK7 in hippocampal CA1 region following global cerebral ischemia in rats. *Life Sci.* 74: 37–45.

Zhou G, Bao ZQ, Dixon JE (1995) Components of a new human protein kinase signal transduction pathway. *J. Biol. Chem.* 270: 12665–12669.

14
Expression, Function, and Regulation of Transcription Factor MEF2 in Neurons

Zixu Mao and Xuemin Wang

Abstract

Myocyte enhancer factor 2 (MEF2), a transcription factor originally identified as playing a critical role in muscle differentiation, has been shown to play diverse roles in an increasing number of non-muscle cells. The results of recent studies have suggested that MEF2 is highly expressed in neurons and is critically involved in the regulation of several important neuronal functions. Studies on MEF2 in neurons have broadened the cellular processes which are controlled by MEF2 and revealed novel regulatory mechanisms by which MEF2 is modulated. In this chapter, the findings of MEF2 in neurons including its expression, function, and regulation by several signal transduction pathways are discussed, and some of the recent progress in identifying the mechanisms through which MEF2 activity is fine-tuned in cells is summarized.

14.1
Introduction

Transcriptional regulation underlies the basis of diverse neuronal functions, including differentiation, maturation, survival, and plasticity. The diverse ability that transcription factor myocyte enhancer factor 2 (MEF2) displays in receiving and integrating signals from several important regulatory pathways suggests that it may play a dynamic role in neurons. Indeed, the results of recent studies have provided a growing body of evidence to implicate MEF2 in the regulation of these fundamental processes in neurons. In this chapter, our current understanding of MEF2, its biochemical properties, expression, function, and regulation in neurons is described.

14.2
The MEF2 Family of Transcription Factors

14.2.1
MEF2 Genes and Transcripts

Early studies of muscle differentiation revealed the presence of a DNA-binding activity that specifically recognizes an A/T DNA sequence present in the regulatory region of many muscle-specific genes (for a review, see Black and Olson, 1998). Subsequent cloning experiments identified MEF2 as the factor that binds to the A/T-rich sequence. Four vertebrate MEF2s – MEF2A, B, C, and D – have been identified (Pollock and Treisman, 1991; Yu et al., 1992; Breitbart et al., 1993; McDermott et al., 1993; Martin et al., 1994). Using unique MEF2 cDNA sequence on the DNA of human-rodent hybrid clones, the location of MEF2A to D has been mapped to human chromosomal regions 15q26, 19p12, 5q14, and 1q12–q23, respectively (Hobson et al., 1995). Together, these findings verify the existence of at least four distinct loci for members of this gene family.

All four vertebrate mef2 gene transcripts are alternatively spliced among coding region exons to give rise to splicing variants (Fig. 14.1). MEF2A, C, and D have highly similar gene structure and alternative splicing patterns in their coding exons. MEF2A and D mRNAs have four potential distinct coding regions. For MEF2C, there are eight potential variants because of the existence of an alternative cryptic splice acceptor within exon 9 that gives rise to γ variants. Recent findings have begun to shed light on the potential functions of these alternatively spliced domains (see below).

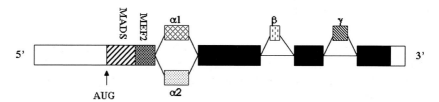

Fig. 14.1 Structure of MEF2 gene product. α, β, and γ indicate alternative splicing exons (γ is MEF2C-specific).

14.2.2
Structure of MEF2 Proteins

The four MEF2 proteins, A-D, share a highly homologous N-terminal region (over 95% similarity) that spans the first 86 amino acid residues. This is followed by a divergent large C-terminal transcriptional activation domain (TAD) (Fig. 14.2). The N-terminal region is further divided into two subdomains (Black and Olson, 1998). Amino acid residues 1 to 56 comprise the so-called MADS domain due to its sequence homology to transcription factors MCM1, Agamous, Deficiens, and Serum response factor. The MADS domain is a highly conserved structure motif that regulates homeotic fate, growth, and differentiation of many organisms (Yun and Wold, 1996; Ng and Yanofsky, 2001). The domain is required for mediating MEF2 hetero- or homo-dimerization, the binding of MEF2 to specific DNA sequence, and the interaction between MEF2 and other transcription factors and regulators (McKinsey et al., 2002). The next 30 or so amino acid residues share sequence homology among the four MEF2 isoforms themselves, thus acquiring the name of the MEF2 domain. Similar to MADS, MEF2 domain participates in mediating high-affinity DNA binding and MEF2 dimer formation (McDermott et al., 1993; Molkentin et al., 1996a). Together, these two subdomains constitute the entire structural requirement for specific DNA binding by MEF2.

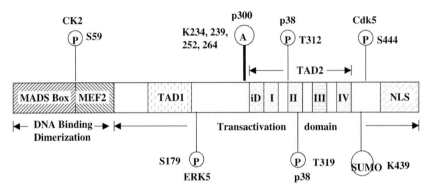

Fig. 14.2 Schematic structure of MEF2 protein. TAD = transactivation domain; iD = inhibitory domain; NLS = nuclear localization signal; P = phosphorylation site; A = acetylation site; SUMO = sumoylation site.

Several groups have reported the crystal structure of N-terminal MEF2 (Huang et al., 2000; Santelli and Richmond, 2000). Studies of MEF2A amino acid 2–78 bound to its consensus DNA sequence show that, in contrast to SRF and MCM1, the MADS domain in MEF2 binds to relatively unbent DNA (Santelli and Richmond, 2000). The binding specificity is achieved through selecting a narrow minor groove and making specific contacts with two base pairs per half-site. The MEF2 domain provides the primary contacting basis for MEF2–MEF2 dimer formation and interaction with other regulators. Indeed, the crystal structure of the MADS-MEF2 domain of human MEF2B bound to DNA and a motif of its transcriptional repressor Cabin1 reveals that Cabin1 binds to a hydrophobic groove on the MEF2 domain (Han et al., 2003).

The larger C-terminal part of the MEF2 molecule is the region required for MEF2-mediated transcription transactivation. Although the overall sequence homology in this region is low among the four MEF2 isoforms, it is interspersed with several short stretches of segments with some degree of homology and containing conserved key residues (Molkentin et al., 1996a; Yu, 1996). Several subdomain structures in TAD have been identified which appear able to activate transcription independently (Molkentin et al., 1996a; Yu, 1996). At least one of these, TAD2, is conserved among MEF2 isoforms, and can be further delineated into several conserved regions. How precisely these subdomains regulate MEF2 transcription activity is not entirely clear. However, they may very well contain sites of post-translational modification which are important for mediating interaction between MEF2 and co-regulatory molecules. Recent studies have demonstrated clearly that transcription activation property of MEF2 is highly regulated (Han and Molkentin, 2000; McKinsey et al., 2002). Novel sites of regulation which reside outside the previously defined transactivation subdomains are being discovered, suggesting that the structure-function relationship of the C-terminus of MEF2 is far more complicated than our current diagram depicts.

A bipartite nuclear localization sequence (NLS) is present near the very end of the C-terminal region. Its presence and pattern of distribution seem to be conserved among MEF2 isoforms (Borghi et al., 2001). Deletion studies have suggested that MEF2 NLS is functionally required to properly target MEF2 and its interacting protein HDAC4 to the nucleus. Although MEF2 is a nuclear transcription factor, reports have been made that it may not reside constitutively in the nucleus, or it may shuttle between the nuclear and cytoplasmic compartments under certain conditions (see below).

14.2.3
Specific Interaction Between MEF2 and DNA

MADS proteins recognize a canonical A/T-rich cis-element. Analysis of the regulatory region of MEF2 target genes in muscle allows the identification of the core motif of MEF2 binding sequence as CT(A/t)(a/t)AAATAG (Yu et al., 1992). Interestingly, *in-vitro* studies using a pool of degenerate oligonucleotides suggest that although MEF2s present in the brain lysates bind to this same core motif, they seem to have additional sequence constraints outside this core motif that are not observed in sequences selected with muscle extracts (Andres et al., 1995). Since very few MEF2 targets in neurons have been identified, it remains to be determined whether MEF2 may indeed prefer a slightly different binding sequence in neurons than in muscle cells.

Point mutation and deletion analysis demonstrates conclusively that MEF2 and MADS domains mediate dimerization and DNA binding (Nurrish and Treisman, 1995; Molkentin et al., 1996a). Such studies indicate that the N-terminal half of the MADS domain and entire MEF2 domain are mainly responsible for DNA binding. The C-terminal half of MADS domain contributes primarily to the control of dimerization. This conclusion is further refined by the crystal structure analysis showing that protein-DNA contacts are confined to amino acids 1 to 36 (Santelli and Richmond, 2000). The structure difference between the MEF2 domain and the equivalent domain in other MADS box proteins such as SRF and MCM1 accounts for the absence of cross-reactivity in DNA binding among them. Interestingly, mutation studies reveal the presence of amino acid residues in MEF2 domain that are not required for DNA binding but are essential for site-specific transcription, suggesting a functional interdependence between MEF2 domain and TAD (Molkentin et al., 1996a).

14.3
Expression of mef2 in Neurons

Early tissue survey studies following the initial cloning of MEF2s confirmed that the transcripts of MEF2 isoforms are expressed at high levels in muscle. However, what is also evident from these earlier experiments is that MEF2 transcripts and proteins are not restricted to muscle but are expressed at different levels in non-muscle tissues, particularly in brain (McDermott et al., 1993). In fact, shortly after its initial identification in muscle, MEF2C was also independently cloned from brain (Leifer et al., 1993), underscoring its presence in the central nervous system (CNS). Indeed, all MEF2 isoforms are expressed at variable levels in distinct, but overlapping, patterns in the CNS during embryogenesis, as well as in the adult.

14.3.1
Expression of mef2 Transcripts in the Central Nervous System

mef2C is the first mef2 isoform the expression of which in the CNS was investigated in some detail. Northern blot studies showed that mef2C mRNA is expressed at variable levels in various parts of rat and human brain, including the cerebral cortex, cerebellum, and basal ganglia (Leifer et al., 1993). More detailed *in-situ* hybridization analysis revealed that mef2C transcript shows a striking laminar distribution in the cerebral cortex with consistently stronger signals in the outer layers of the neocortex than in the infragranular layers and striatal neurons. This general pattern of highly variable and region-specific expression of mef2C transcript in the CNS is also seen in other mef2 isoforms and in quail, mouse, and *Drosophila* (Ikeshima et al., 1995; Lyons et al., 1995; Schulz et al., 1996; Xue et al., 2000). Studies of the four *mef2* gene transcripts in the mouse CNS perhaps offer the most comprehensive illustration of the dynamic nature of their expression. In mouse brain, each of the four *mef2* genes is expressed in the front cortex, midbrain, thalamus, hippocampus, and hindbrain by 13.5 days pre-coitus (dpc). The temporal and spatial patterns of each mef2 gene evolve during embryogenesis, which appears to follow the gradients of neuronal maturation and correlates with withdrawal of neurons from the cell cycle (Schulz et al., 1996). Together, studies of *mef2* gene expression suggest a role of this subfamily of MADS proteins in neuronal differentiation. The expression of *mef2* genes in the CNS is summarized in Table 14.1.

Table 14.1 MEF2 gene expression in mouse brain

	Adult				E14.5			
	2A	2B	2C	2D	2A	2B	2C	2D
Cerebral cortex	+	+++	+++	++	−	++	+++	++
Cerebellum	+	−	+	++	−	−	+	+
Hippocampus	+++	+/−	++	+++	−	+	+++	+
Thalamus	+++	−	++	++	++	+	+	+
Midbrain	−	++	+++	++	+	−	++	++

E14.5, embryo day 14.5

14.3.2
Expression of MEF2 Proteins in Neurons

Although the pattern of mef2 transcripts in the CNS is consistent with neuronal expression, direct proof of this point came from a combination of immunohistochemical analysis of brain tissues and immunocytochemical analysis of cultured neurons. MEF2C immunoreactivity in brain slice shows a laminate pattern with stronger staining in the outer cortical layers that is consistent with the observed pattern of its transcript (Leifer et al., 1994). Immunocytochemistry studies of primary culture of cortical and cerebellar granule neurons showed co-localization of MEF2 and neuronal marker TuJ–1, demonstrating clearly the neuronal nature of the MEF2–positive cells (Mao et al., 1999). Consistent with this observation, MEF2C does not co-localize with glia marker GFAP. More importantly, anti-MEF2C antibody does not stain proliferating neuronal precursors identified in culture by BrDU, suggesting that MEF2C-positive neurons are postmitotic. Consistent with this, a high level of MEF2C protein is expressed at embryonic day (E) 17 to 21 in the cortical plate where postmitotic neurons further differentiate and mature. In contrast, there is little MEF2C expression in the ventricular zone where the dividing neuronal precursors reside. The exact point at which postmitotic neurons start to express MEF2C is still unclear. Close examination of E17 to E21 rat brain seems to suggest that some neurons may start to express MEF2C soon after they leave the ventricular zone and begin to migrate along the radial glial cells. The level of MEF2C begins to decline in the cortical region sometime after birth and is reduced to a low level by 2 weeks after birth.

In the cerebellum, MEF2A and D proteins appear to be the dominant species, although MEF2C is also expressed (Leifer et al., 1994; Mao et al., 1999). There, MEF2A and D are detectable at the inner half region of the external granule layer of P6 rat brain, where the granule neuron precursors have stopped dividing and are preparing to migrate through the Purkinje layer into the internal granule layer. MEF2A and D are expressed at high levels by the internal granule neurons (Mao et al., 1999). Interestingly, Purkinje neurons also express high levels of MEF2s.

Taken together, the results of these studies suggest that the MEF2 proteins largely follow the patterns of their transcripts in the CNS and are robustly expressed in postmitotic neurons. However, to date, there have not been any comprehensive immunohistochemical studies that systemically examine the expression of each MEF2 protein in the CNS.

14.4
Function of MEF2 in Neurons

The robust expression of MEF2 transcripts and proteins in neurons correlates well with the strong MEF2 activity determined by either DNA binding or MEF2–dependent reporter gene assays, suggesting that MEF2s in neurons are functional.

14.4.1
The Role of MEF2 in Neuronal Differentiation

Several lines of evidence suggest that MEF2 may regulate neuronal differentiation. First, the expression patterns of *mef2* genes correlate closely with neuronal maturation during development of the CNS (see above). Second, the expression of MEF2 proteins increases significantly following exit of cell cycles in developing neurons and remains at high levels during early stages of development (Mao et al., 1999). Its level in neurons decreases after the early stage of development. Third, the increase in MEF2 levels coincides with an increase in the expression of neuronal markers such as TuJ1. Fourth, MEF2 protein interacts physically with the neurogenic basic helix-loop-helix (bHLH) transcription factor MASH–1, a protein implicated in the development of neurons, to synergistically regulate target gene expression (Black et al., 1996; Mao and Nadal-Ginard, 1996). Fifth, in chemical-induced neuronal differentiation of P19 cells, the expression of specific isoforms of MEF2s correlates with the onset of neuronal differentiation (Mao and Nadal-Ginard, 1996). Furthermore, forced expression of MEF2C induces the expression of MASH1, neuN, as well as neurofilament protein NF68 in aggregated P19 cells, and facilitates their differentiation into neurons (Skerjanc and Wilton, 2000). However, these data remain largely correlative. More direct evidence to link MEF2 to mechanisms of neuronal differentiation is still lacking.

14.4.2
The Role of MEF2 in Neuronal Survival

Searching the function for MEF2 has uncovered an unexpected role of MEF2 in neuronal survival. The first clue that MEF2 regulates the survival of neurons came from experiments where several dominant-negative mutants of MEF2 were overexpressed in primary cortical neurons cultured from E17 rat brain to block the function of endogenous MEF2 (Mao et al., 1999). The inhibition of MEF2 induced neuronal apoptosis. Inhibition of MEF2 by the same approach in E14 cortical neurons during neurogenesis and differentiation induced neuronal death with a kinetic that parallels the expression of endogenous MEF2. The role of MEF2 in neuronal survival was tested in cerebellar granule neurons cultured in the presence of membrane-depolarizing concentration of potassium chloride (25–30 mM), a well-characterized model which is used widely to mimic neuronal activity-dependent survival. Similarly, inhibition of MEF2 also blocked neuronal activity-induced survival of primary cerebellar granule neurons. Conversely, enhancing MEF2 function by over-expression of a constitutively active form of MEF2 rescued neurons from survival factor-withdrawal induced death. Together, these results suggest that MEF2 is both necessary and sufficient for the survival of different types of neurons.

The key observation that MEF2 is an important regulator of neuronal survival was subsequently re-confirmed by several studies using other neuronal survival paradigms. For example, reducing MEF2 activity during retinoic acid induced-neuronal differentiation of P19 cells also resulted in neuronal apoptosis (Okamoto et al., 2000). Similarly, MEF2 is required for BDNF-mediated survival of E17 cortical (Liu et al., 2003) and newly generated cerebellar granule neurons (Shalizi et al., 2003). The concept that MEF2 regulates cellular survival has also been extended to other types of cells. For example, the deletion of big mitogen-activated protein kinase 1 (BMK1; also known as ERK5) led to apoptosis of endothelial cells due to the loss of positive regulation of MEF2C by ERK5. The enhancement of MEF2 protected endothelial cells from apoptosis (Hayashi et al., 2004).

While MEF2 prevents neuronal death in several of the survival paradigms tested, it was interesting to note that its activity is not always required for the survival of neurons. The first example of this was an observation made when a dominant-negative MEF2 was expressed in E14 primary cortical neurons (Mao et al., 1999). In this experimental paradigm, neuronal apoptosis was much delayed compared to the E17 model. This delay coincided with a window period when neuronal precursors express little or no endogenous MEF2, suggesting that MEF2 is not required for the survival of neuronal precursors or neurons at a very early stage of development. Consistent with this, MEF2 was shown to protect E17 cortical neurons against BDNF withdrawal-induced death, though it does not seem to offer protection for P0 (postnatal day 0) cortical neurons (Liu et al., 2003). Similarly, MEF2 activity is required for BDNF-enhanced survival of newly generated (cultured for 3 days *in vitro* from P6 rat) but not older (8 days *in vitro*) cerebellar granule neurons (Shalizi et al., 2003). These latter two studies exemplify that MEF2 may not mediate the survival of more mature neurons. Collectively, these findings suggest that there may be a critical window during the early stages of neuronal development when MEF2 function is required for neuronal survival. Whether there are other window periods or conditions under which MEF2 activity is required for survival remains to be determined. Taken together, these findings provide additional support that the survival effect of MEF2 observed is specific and most likely reflects a genuine function of endogenous MEF2.

How exactly MEF2 promotes neuronal survival is not entirely clear. One possibility is that MEF2 promotes survival of neurons by regulating the expression of survival-related genes. One recent study supported this simple and straightforward hypothesis. For example, in response to BDNF, MEF2 has been shown to participate in the regulation of neurotrophin–3 (nt3) gene expression (Shalizi et al., 2003). NT3 is a well-characterized neurotrophic factor that is known to play an important role in supporting neuronal survival. This provides an example that MEF2 directly enhances the expression of pro-survival genes. However, the close temporal correlation of neuronal differentiation and the narrow window when MEF2 functions to promote survival raises the possibility that the roles of MEF2 in these two processes may be intricately related. It is possible that MEF2's role in promoting differentiation may contribute, in part, to neuronal survival. Promoting differentiation by MEF2 in a timely and orderly manner may help guard neurons against improper triggering of death mechanisms. On the other hand, it is also likely that activating adequate survival mechanisms by MEF2 may be a prerequisite for proper neuronal differentiation. Further experiments are needed to delineate these different possibilities.

14.4.3
Regulatory Targets of MEF2 in Neurons

In contrast to the wealth of information available on the regulatory targets of MEF2 in muscle, relatively little is known about the identities of the MEF2–controlled genes in neurons. However, some published studies have suggested that MEF2 participates in the regulation of neuronal-specific or enriched genes. Using the neuronal differentiation of P19 cells as a model, it has been shown that forced expression of MEF2C in P19 cells aggregated with dimethyl sulfoxide facilitates neurogenesis (Skerjanc and Wilton, 2000). This correlates with increased expression of several neuronal proteins including neuronal-specific transcription factor MASH1 and neurofilament protein NF68, suggesting that MEF2C may regulate their expression during neuronal differentiation. However, it remains to be determined whether MEF2C promotes the expression of these genes through direct activation of their promoters.

Two potential direct targets of MEF2 have been reported in neurons. The first example is the regulation of NR1 subunit of N-methyl-D-aspartate (NMDA) subtype of glutamate receptor, a protein which is critically involved in neuronal development, plasticity, and cell death (Krainc et al., 1998). Analysis of the promoter of NR1 reveals the presence of a MEF2 DNA binding site in its regulatory region. MEF2 cooperates with transcription factor SP1 to synergistically activate the NR1 promoter. However, the role of MEF2 may be relatively limited in NR1 regulation since disruption of its DNA binding site only moderately reduces NR1 promoter activity in neurons. Nevertheless, this study highlights the versatility of MEF2 in controlling gene expression in conjunction with other factors. This theme is echoed by a later study on the regulation of the *nt3* gene by MEF2 (Shalizi et al., 2003). There, MEF2 is proposed to cooperate with transcription factor CREB at the NT–3 promoter to regulate its expression in response to BDNF stimulation.

In addition to tissue-specific genes, MEF2 has also been shown to regulate the expression of genes expressed more broadly. For example, promoter analysis demonstrates that MEF2 mediates the expression of the *c-jun* gene in several cell types, including HeLa cells, Jurkat T cells, and monocytic cells (Han and Prywes, 1995; Ornatsky and McDermott, 1996; Coso et al., 1997). However, whether and under what condition MEF2 may regulate the expression of c-*jun* in neuronal cells remains to be determined. Undoubtedly, the control of many of the important regulatory targets by MEF2 may be cellular context and stimulus-specific. The importance of this is underscored by the finding that in T cells, MEF2 promotes apoptosis by up-regulating a pro-apoptotic gene *nur77* (Youn et al., 1999). Clearly, target selection by MEF2 in itself is a point of regulation. With the development and improvement of microarray, chromatin immunoprecipitation, and bioinformatic approaches, one should expect to see rapid advance in identifying direct MEF2 targets in neurons.

14.5
Regulation of MEF2 in Neurons

One of the most exciting areas in MEF2 research is the advance in our understanding of how MEF2 activity is regulated in cells. Although many of the basic regulatory mechanisms were initially revealed in non-neuronal systems, some of them have been verified to regulate MEF2 in neurons. In addition, regulatory mechanisms that are revealed in and may be (to some degree) unique to neurons have also begun to emerge. Collectively, these studies demonstrate convincingly that multiple signaling pathways converge on MEF2. This leads to timely and precise regulation of MEF2 activity in response to a diverse array of stimuli. The regulation of MEF2 function is quite complex, and is coordinated at multiple levels. This may involve steps that control mRNA alternative splicing, translation, transactivation domain activity, DNA binding, subcellular localization, and stability, which are modulated through MEF2 phosphorylation, acetylation, sumoylation, and interaction with other cofactors.

14.5.1
Regulation of MEF2 Transactivation Potential

Structurally, the transactivation domain located at the C' portion of MEF2 constitutes the majority of the mass of the MEF2 molecule. Therefore, it is probably not surprising that this domain has also received the most attention. It is quite evident that the activity of TAD is highly regulated. Regulation of TAD may involve Ca^{2+} signal, p38 MAPK signaling pathway, ERK5 pathway, or cyclin-dependent kinase 5 (Cdk5), depending on the stimuli and paradigms used.

A yeast two-hybrid screening with p38 MAPK as bait first revealed that p38 MAPK directly phosphorylates MEF2C (Han et al., 1997; Han and Molkentin, 2000). Later studies confirmed that each of the four p38MAPK isoforms (α, β, γ, and δ) can phosphorylate MEF2 at multiple sites. For example, MEF2C is phosphorylated at Thr293, Thr300, and Ser387, and MEF2A at Thr312 and 319, respectively (Zhao et al., 1999). Phosphorylation of MEF2A and C by p38 MAPK enhances their transcriptional activation activity in a MEF2–dependent reporter gene assay. The efficient target of MEF2A and C by p38 is mediated by a conserved docking domain present between amino acids 266 to 282 in MEF2A, and between amino acids 249 to 264 in MEF2C (Yang et al., 1999). Interestingly, MEF2D lacks this p38 docking domain in its sequence, providing an explanation as to why MEF2D is not a p38 target. However, the new member of MAPK family, ERK5, directly phosphorylates multiple isoforms of MEF2 including MEF2D (Yang et al., 1998; Marinissen et al., 1999; Kato et al., 2000). Similar to the regulation by p38 MAPK, phosphorylation by ERK5 also increases the transcriptional potential of MEF2.

The relevance of p38 MAPK-mediated regulation of MEF2 in neurons was first confirmed in a study in which the role of this pathway was tested in an activity-dependent survival model (Mao et al., 1999). Neuronal activity-induced survival of cerebellar granule neurons was blocked by the inhibition of p38 MAPK pathway via a dominant negative p38 MAPK. Conversely, neuronal death following neuronal activity withdrawal was attenuated by constitutive activation of MKK6–p38MAPK–MEF2 pathway. More importantly, p38 MAPK-mediated survival requires phosphorylation and activation of MEF2. Overexpression of a MEF2 mutant that cannot be phosphorylated by p38 blocked membrane depolarization- and p38 MAPK-induced neuronal survival. This study established the role of p38 MAPK-MEF2 in neuronal survival, a conclusion later confirmed in the neurogenesis model of P19 cells (Okamoto et al., 2000).

The regulation of MEF2 by ERK5 was first shown in Chinese hamster ovary (CHO) cells. For MEF2C, ERK5 phosphorylates Ser387 to activate its transcriptional activity (Yang et al., 1998; Marinissen et al., 1999; Kato et al., 2000). In contrast to p38 MAPK, ERK5 also interacts with and directly phosphorylates MEF2D in addition to MEF2A and C. However, there are conflicting reports on whether phosphorylation of MEF2D by ERK5 regulates its activity. One study showed that phosphorylation of MEF2D at Ser179 by ERK5 is required for its enhanced transcriptional activity in response to EGF stimulation in transfected HeLa cells. However, other studies indicated that phosphorylation of the same site cannot account for ERK5–mediated enhancement of MEF2D transcription potential in response to PMA/ionomycin treatment in DO11.10 hybridomas. Additional transfection studies performed in COS cells demonstrated that ERK5 specifically up-regulates the activity of MEF2A and C, but not of MEF2D. The reason for these differences is not clear, although it is possible that the regulation of MEF2D by ERK5 may be stimulus or cellular context-dependent.

Two separate studies have addressed the issue of whether ERK5 regulates MEF2 in neurons. Using a BDNF-mediated survival model, it was shown that BDNF protects E17 cortical neurons against trophic withdrawal-induced death in an ERK5–dependent mechanism. The transfection of a dominant-negative ERK5 mutant blocked BDNF-induced survival. Overexpression of a constitutively active form of MEF2, MEF2C-VP16, attenuated the neuronal death induced by the dominant-negative ERK5. Blocking MEF2 function also attenuated BDNF-induced, ERK5–dependent survival. Together, these data demonstrate that ERK5–mediated survival effect of BDNF requires the activity of its downstream target MEF2. This conclusion is supported by the findings of a second study that specifically investigated the role of the

ERK5–MEF2 pathway in newly generated cerebellar granule neurons. Similar to the findings made in cortical neurons, disruption of the ERK5–MEF2 signaling pathway blocked BDNF-induced survival of young cerebellar granule neurons.

The results of the above studies have demonstrated the positive regulation of MEF2 by phosphorylation. The clue that phosphorylation may also negatively regulate MEF2 came from a study where MEF2A protein was observed to migrate at a slower rate in cerebellar granule neurons upon withdrawal of neuronal activity support (Mao and Wiedmann, 1999; Li et al., 2001). The slow migration of MEF2A is due to hyperphosphorylation, which correlates with reduced MEF2 activity, suggesting that hyperphosphorylation inhibits its function. Subsequent studies identified Cdk5 as a negative regulator of MEF2 in neurons (Gong et al., 2003). Neurotoxins activate nuclear Cdk5 kinase activity. Activated Cdk5 directly phosphorylates MEF2 at a conserved serine residue present in the C' transactivation domain (Ser 408 and Ser444 for MEF2A and D, respectively). Phosphorylation of MEF2 by Cdk5 inhibits its transcriptional potential and induces neuronal apoptosis. Overexpression of a MEF2 mutant that is resistant to Cdk5–mediated phosphorylation rescues MEF2–dependent gene transcription activity and prevents neurons from oxidative stress- or excitotoxin glutamate-induced death. This study establishes phosphorylation and inhibition of MEF2–dependent survival function by Cdk5 as a critical mechanism that mediates toxin-induced neuronal death. Interestingly, unlike neuronal activity withdrawal, the activation of Cdk5 by neurotoxins does not seem to alter the pattern of MEF2 migration, suggesting that the hyperphosphorylation of MEF2 observed following neuronal activity withdrawal may be due to the activities of yet-to-be-identified additional negative regulators. Although the identity of these kinases remains to be revealed, the phosphatase that counters the inhibitory effect of these kinases has been identified. In cerebellar granule neurons, the calcium-sensitive protein phosphatase calcineurin seems to be required to maintain MEF2 in a hypophosphorylated and active state. Inhibition of calcineurin alone is sufficient to recapitulate neuronal activity withdrawal-induced hyperphosphorylation and inhibition of MEF2, suggesting that calcineurin either directly or indirectly de-phosphorylates MEF2 in neurons.

In addition to phosphorylation, recent studies have shown that MEF2 transcriptional activity is also regulated by other means of modification. For example, acetylase p300 interacts with MEF2 and enhances MEF2 function by acetylating lysine residues at its C-terminal transactivation domain (Ma et al., 2005). Mutation of these lysines affects MEF2 transcriptional activity and its synergistic effect with another transcription factor myogenine. Overexpression of these mutants blocks myogenic differentiation. Similarly, the C-terminal domain of MEF2C and D has been shown recently to be modified by SUMO2 and SUMO3 (Gregoire and Yang, 2005). Sumoylation inhibits MEF2 transcriptional activity. The SUMO protease SENP3 reverses this inhibitory effect and augments the myogenic activity of MEF2. Together, these findings encourage an examination of the role of other modifications of MEF2 in neurons.

14.5.2
Regulation of MEF2 DNA Binding

Biochemical studies have identified two phosphorylation sites in the N terminus of MEF2. Casein kinase II phosphorylates the conserved Ser59 in the MEF2 domain to increase MEF2 DNA binding (Molkentin et al., 1996b). However, phosphorylation of this site appears to be constitutive instead of inducible. Protein kinase A (PKA) phosphorylates the conserved Thr20 residue present in the MADS domain *in vitro* (Wang et al., 2005). A recombinant MEF2 fragment phosphorylated at Thr20 by PKA *in vitro* shows enhanced activity by *in-vitro* DNA-binding assay. Although it is not clear whether Thr20 is phosphorylated *in vivo*, activation and inhibition of the cAMP-PKA signaling pathway correlates well with the increase and decrease in MEF2 DNA binding activity, respectively, consistent with the notion that cAMP-PKA pathway functions to regulate MEF2 in neurons.

In addition to phosphorylation, MEF2 DNA-binding activity is also regulated through interaction with other proteins. For example, the repressor protein Ki–1/57 binds to the N terminus of MEF2C to inhibit its DNA binding. MEF2 is acetylated by transcription co-activator p300 *in vitro* and in cells. Although the acetylation sites are mapped to the transactivation domain of MEF2, acetylation of these conserved sites enhances not only its transcriptional activity but also its DNA binding potential (Ma et al., 2005). These sites are functionally important, since overexpression of non-acetylatable MEF2 mutants inhibited myogenic differentiation. These studies underscore the complexity of intra-molecular/inter-domain regulation of MEF2 function.

14.5.3
Regulation of MEF2 Stability

Recent studies have demonstrated that the control of MEF2 stability represents a key mechanism of regulation in neurons. Neuronal activity withdrawal or neurotoxic stress induces a clear and gradual decline of MEF2 protein levels in cerebellar granule neurons, which accompanies neuronal apoptosis (Tang et al., 2005). This appears to involve in part a caspase-dependent degradation step *in vivo*, since the inhibition of caspase protects MEF2 from degradation (Li et al., 2001). Further studies showed that co-incubation of MEF2 generated *in vitro* with several caspases results in specific cleavage of various MEF2 isoforms. However, in contrast to the efficient degradation in neurons, the cleavage of MEF2 generated *in vitro* seems to be highly inefficient, generating very limited degradation products (Okamoto et al., 2002). The results of a recent study provide an explanation for this. It appears that for MEF2 to be cleaved efficiently by caspase–3, it must be phosphorylated by Cdk5 (Tang et al., 2005). One model that emerged from this study was that excitotoxin activates Cdk5, which leads to Cdk5–dependent phosphorylation of MEF2A and D. This phosphorylation facilitates caspase-mediated degradation of MEF2 (Fig. 14.3). Blocking Cdk5 by either a dominant-negative Cdk5 mutant or its pharmacological inhibitor attenuated MEF2A and D degradation. A MEF2 in which the Cdk5 phosphorylation site was mutated became highly resistant to caspase-induced degradation, both *in vitro* and in neurons, and was more efficient than wild-type MEF2 in rescuing neurons from neurotoxin- or Cdk5–induced apoptosis. Consistently, MEF2C isoform which is not phosphorylated by Cdk5 in granule neurons in response to excitotoxicity was not cleaved. These findings suggest that Cdk5 and caspases are coordinated to regulate MEF2 stability, thereby controlling the viability of cells upon neurotoxic insult. The means by which phosphorylation by Cdk5 makes MEF2 a better caspase substrate are still unclear, although it is conceivable that such phosphorylation leads to a change in MEF2 conformation that allows better access to the cleavage sites by caspases.

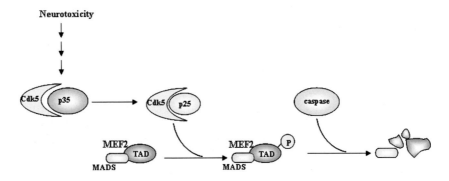

Fig. 14.3 Model of neurotoxin-induced phosphorylation and degradation of MEF2.

Phosphorylation-dependent regulation of MEF2 stability is certainly not limited to neurons. Biochemical studies have shown that phosphorylation of MEF2 at sites distinct from the Cdk5 site by yet-to-be-identified kinase(s) also regulates MEF2 stability in non-neuronal cells. For example, Ser255 of MEF2A becomes phosphorylated when p38MAPK activity is enhanced. Mutation of Ser255 to aspartic acid destabilizes MEF2A and leads to its degradation in COS7 cells (Cox et al., 2003). It is not known if, and under what conditions, Ser255 is phosphorylated in neurons. Neither is it clear whether Ser255 phosphorylation-induced degradation involves caspase. However, this study serves as a reminder that regulation of MEF2 stability may be a far more delicate process.

14.5.4
Regulation of MEF2 Subcellular Localization

MEF2 proteins normally reside in the nuclear compartment in most cells examined, although its proper targeting to the nucleus requires the nuclear localization signal present in the C terminus of MEF2 (Yu, 1996). Interestingly, it was reported that in embryonal rhabdomyosarcoma cells MEF2 is present in the cytoplasm. In these cells, the withdrawal of serum induces nuclear translocation of MEF2, providing a unique example of cytoplasm to nuclear shuttling of MEF2 (Chen et al., 2001). One study examined the effect of transforming growth factor β (TGF-β) on the differentiation of myogenic cells and reported that, when these cells are grown at high density, TGF-β induces a nucleus to cytoplasm translocation of MEF2 (De Angelis et al., 1998). However, this translocation has not been observed by others under similar experimental conditions (Quinn et al., 2001). To add another layer of complexity, interaction of MEF2 with its co-regulators can alter its subnuclear localization. For example, the co-expression of a steroid receptor cofactor GRIP–1 was found to recruit MEF2 into a punctate subnuclear structure, which is disrupted by cyclin D-Cdk4 (Lazaro et al., 2002). Taken together, the results of these studies point to cytoplasm-nucleus shuttling and subnuclear localization as a potentially important mechanism to regulate MEF2 function. However, the role that this mechanism might play in neurons remains to be determined.

14.5.5
Regulation of MEF2 by Alternative Splicing

Early studies of different splicing variants of MEF2C in brain and muscle showed that exon β is preferentially expressed in brain, but not in muscle (McDermott et al., 1993), providing the first example of tissue-specific alternative splicing. Further evidence to support this conclusion came from studies that showed exon α2 of MEF2D to be specifically expressed in differentiated, but not in undifferentiated, muscle tissue or myogenic cells. More recent studies showed that MEF2s with exon β seem to be more potent in activating MEF2 responsive reporter gene expression than MEF2s without exon β, suggesting that alternative splicing variants may process distinct regulatory capabilities *in vivo* (Zhu et al., 2005). As discussed above, Cdk5 is a critical regulator that directly phosphorylates MEF2A, C, and D. Coincidentally, the conserved putative Cdk5 phosphorylation site in MEF2C resides in the alternative splicing exon γ, a domain unique to MEF2C (Gong et al., 2003). This raises the possibility that, depending upon whether exon γ is present, MEF2C may respond to Cdk5 regulation differently in neurons.

14.5.6
Regulation of MEF2 by Interaction with Co-Regulators

An important aspect of MEF2 regulation is achieved through its interaction with a diverse array of cofactors. These include specific transcription factors, general transcriptional activators and repressors, and adaptor/chaperon proteins.

Studies from many laboratories have shown that MEF2 can interact with many different transcriptional factors to synergistically regulate target gene expression in a variety of cellular models. Almost exclusively, this interaction is mediated through the MADS-MEF2 domain of MEF2. Rather than a complete summary of all the transcription factors that have been shown to interact with MEF2, two examples of MEF2 interaction with neuronal specific and with broadly expressed transcription factors are provided to illustrate the versatility of MEF2 in neurons. MEF2 interacts with the bHLH transcription factor MASH1 to regulate gene expression. This interaction is mediated through the MADS-MEF2 domain of MEF2 and bHLH region of MASH1 (Black et al., 1996; Mao and Nadal-Ginard, 1996). This interaction allows MEF2 and MASH1 to activate gene expression through their respective DNA-binding sites in a cooperative and synergistic manner. Interestingly, MEF2 can also cooperate with MASH1 through protein interaction alone without having to bind to DNA, thereby providing an operation mode that potentially could significantly expand its regulatory targets. To underscore the importance of cooperativity, MEF2 can also cooperate with the broadly expressed transcription factors SP1 and CREB to regulate the promoters of the neuronal gene NMDA receptor subtype NR1 and trophic factor NT3, respectively (Krainc et al., 1998; Shalizi et al., 2003).

Transcription repressor histone deacetylases (HDACs) silence gene expression through deacetylating the N-terminal tails of core histones, which causes ensuing chromatin condensation (McKinsey et al., 2001). HDACs are classified into three groups (I, II, and III) based on their homology to distinct yeast HDACs. Members of class II HDACs (HDACs 4, 5, 7, and 9) interact with MEF2 through its MADS-MEF2 domain. This interaction brings HDACs close to MEF2 target genes and represses their expression. A truncated form of HDAC 9, termed MEF2–interacting transcription repressor (MITR), lacks intrinsic HDAC activity. It binds MEF2 and inhibits its gene transcription by recruiting additional co-repressors HDACs or CtBP (Zhang et al., 2001). The interaction between HDACs and MEF2 is regulated by calcium/calmodulin-dependent protein kinase (CaMK). CaMK IV and I directly phosphorylate two conserved serines at the N-terminus of class II HDACs. This creates docking sites for intracellular chaperone protein 14–3–3. The binding of 14–3–3 to HDACs disrupts its interaction with MEF2, resulting in the nuclear export of HDACs and releasing MEF2 from repression (McKinsey et al., 2000). In addition to its chaperone role, 14–3–3 can also associate with MEF2D directly and enhance its activity. The role of the HDAC-mediated regulation of MEF2 has just begun to be explored in neurons. In cerebellar granule neurons, the withdrawal of neuronal activity induces a rapid cytoplasm-to-nuclear translocation of HDAC5 (Linseman et al., 2003). Blocking CaMKII expression mimics this effect. This is accompanied by loss of MEF2 activity and neuronal viability, suggesting that HDAC plays a role in silencing MEF2–dependent gene expression in activity withdrawal-induced death.

14.5.7
Regulation of MEF2 by Calcium Signaling

Many of the regulatory mechanisms described above are sensitive to calcium signaling, linking calcium-dependent pathways to MEF2–mediated gene response (McKinsey et al., 2002). Calcium-dependent regulation of MEF2s is briefly summarized below, with some of their functional relevance in neurons remaining to be demonstrated.

CaMK, in response to a calcium signal, regulates MEF2 by at least two mechanisms. First, CaMKIV and I phosphorylate and remove the repressive effect of HDACs on MEF2. Second, CaMKIV may also directly phosphorylate MEF2D and enhance its function. However, the specific sites of regulation in MEF2D have not been identified and there is also no evidence that CaMKIV directly regulates other MEF2 isoforms.

Calcineurin, a calcium/calmodulin-sensitive protein phosphatase, signals to MEF2 in response to neuronal activity. Membrane depolarization promotes neuronal survival by a calcineurin-dependent mechanism that involves the dephosphorylation of MEF2A and D, thus keeping them in a hypophosphorylated and active state (Mao and Wiedmann, 1999; Li et al., 2001). Calcineurin-dependent regulation of MEF2 appears to be responsible for a hypertrophic response in functionally overloaded and electrically stimulated mouse muscle cells. Calcineurin activates MEF2 by facilitating NFAT nuclear translocation, which associates with MEF2 and recruits coactivator p300 (Youn et al., 2000). Whether calcineurin can directly dephosphorylate MEF2 remains to be definitively determined.

Calmodulin regulates MEF2 by several mechanisms. It associates with and disrupts the MEF2–HDAC complex (Youn et al., 1999); facilitates CaMK-dependent phosphorylation of HDACs; and binds to Cabin1, a transcriptional repressor, and possibly prevents Cabin1 from associating with and inhibiting MEF2. Interestingly, Cabin1 also inhibits MEF2 through several mechanisms: (a) by recruiting class I HDACs to MEF2; (b) by attenuating calcineurin function; and (c) by preventing the association between MEF2 and ERK5.

MAP kinases are sensitive to calcium signals. In cerebellar granule neurons, membrane depolarization activates voltage-sensitive calcium channels and causes calcium influx. This induces a p38 MAPK-dependent activation of MEF2 (Mao et al., 1999). Blocking either p38 or MEF2 inhibits calcium-mediated survival. A calcium signal has also been shown to stimulate the interaction between ERK5 and MEF2 in T cells (Kasler et al., 2000). In neurons, MEF2 function is required for ERK5-mediated neuronal survival.

14.6
Future Studies

It is clear that MEF2 is a dynamic molecule, and is a key target of many signaling pathways. Its pattern of expression in brain raises many interesting functional questions that remain to be answered. Definitive proof as to if and how MEF2 may be involved in neuronal differentiation is still lacking. The mechanisms by which MEF2 supports neuronal survival need to be fully explored, and the target genes of MEF2 in neurons remain to be identified. Different roles of various MEF2 isoforms and splicing variants in neurons also require further study. Some of the key regulatory mechanisms initially identified in other cell types should be verified in neurons, and an understanding these in developing and mature neurons should allow us to assess the function of MEF2 under both physiological and pathological conditions.

Acknowledgments

The authors thank Dr. John Marshall for his critical reading of the manuscript. These studies were supported by grants to Z.M. (NIH RO1 HD39446, AG023695, and NS048254).

Abbreviations

bHLH	basic helix-loop-helix
CaMK	calcium/calmodulin-dependent protein kinase
CHO	Chinese hamster ovary
CNS	central nervous system
dpc	days pre-coitus
HDAC	transcription repressor histone deacetylase
MEF2	myocyte enhancer factor 2
MITR	MEF2–interacting transcription repressor
NLS	nuclear localization sequence
NMDA	N-methyl-D-aspartate
nt3	neurotrophin–3
PKA	protein kinase A
TGF-β	transforming growth factor β

References

Andres, V., Cervera, M., Mahdavi, V. (1995). Determination of the consensus binding site for MEF2 expressed in muscle and brain reveals tissue-specific sequence constraints. *J. Biol. Chem.* 270, 23246–23249.

Black, B.L., Olson, E.N. (1998). Transcriptional control of muscle development by myocyte enhancer factor–2 (MEF2) proteins. *Annu. Rev. Cell. Dev. Biol.* 14, 167–196.

Black, B.L., Ligon, K.L., Zhang, Y., Olson, E.N. (1996). Cooperative transcriptional activation by the neurogenic basic helix-loop-helix protein MASH1 and members of the myocyte enhancer factor–2 (MEF2) family. *J. Biol. Chem.* 271, 26659–26663.

Borghi, S., Molinari, S., Razzini, G., Parise, F., Battini, R., Ferrari, S. (2001). The nuclear localization domain of the MEF2 family of transcription factors shows member-specific features and mediates the nuclear import of histone deacetylase 4. *J. Cell Sci.* 114, 4477–4483.

Breitbart, R.E., Liang, C.S., Smoot, L.B., Laheru, D.A., Mahdavi, V., Nadal-Ginard, B. (1993). A fourth human MEF2 transcription factor, hMEF2D, is an early marker of the myogenic lineage. *Development* 118, 1095–1106.

Chen, S.L., Wang, S.C., Hosking, B., Muscat, G.E. (2001). Subcellular localization of the steroid receptor coactivators (SRCs) and MEF2 in muscle and rhabdomyosarcoma cells. *Mol. Endocrinol.* 15, 783–796.

Coso, O.A., Montaner, S., Fromm, C., Lacal, J.C., Prywes, R., Teramoto, H., Gutkind, J.S. (1997). Signaling from G protein-coupled receptors to the c-jun promoter involves the MEF2 transcription factor. Evidence for a novel c-jun amino-terminal kinase-independent pathway. *J. Biol. Chem.* 272, 20691–20697.

Cox, D.M., Du, M., Marback, M., Yang, E.C., Chan, J., Siu, K.W., McDermott, J.C. (2003). Phosphorylation motifs regulating the stability and function of myocyte enhancer factor 2A. *J. Biol. Chem.* 278, 15297–15303.

De Angelis, L., Borghi, S., Melchionna, R., Berghella, L., Baccarani-Contri, M., Parise, F., Ferrari, S., Cossu, G. (1998). Inhibition of myogenesis by transforming growth factor beta is density-dependent and related to the translocation of transcription factor MEF2 to the cytoplasm. *Proc. Natl. Acad. Sci. USA* 95, 12358–12363.

Gong, X., Tang, X., Wiedmann, M., Wang, X., Peng, J., Zheng, D., Blair, L.A., Marshall, J., Mao, Z. (2003). Cdk5–mediated inhibition of the protective effects of transcription factor MEF2 in neurotoxicity-induced apoptosis. *Neuron* 38, 33–46.

Gregoire, S., Yang, X.J. (2005). Association with class IIa histone deacetylases upregulates the sumoylation of MEF2 transcription factors. *Mol. Cell. Biol.* 25, 2273–2287.

Han, A., Pan, F., Stroud, J.C., Youn, H.D., Liu, J.O., Chen, L. (2003). Sequence-specific recruitment of transcriptional co-repressor Cabin1 by myocyte enhancer factor–2. *Nature* 422, 730–734.

Han, J., Molkentin, J.D. (2000). Regulation of MEF2 by p38 MAPK and its implication in cardiomyocyte biology. *Trends Cardiovasc. Med.* 10, 19–22.

Han, J., Jiang, Y., Li, Z., Kravchenko, V.V., Ulevitch, R.J. (1997). Activation of the transcription factor MEF2C by the MAP kinase p38 in inflammation. *Nature* 386, 296–299.

Han, T.H., Prywes, R. (1995). Regulatory role of MEF2D in serum induction of the c-jun promoter. *Mol. Cell. Biol.* 15, 2907–2915.

Hayashi, M., Kim, S.W., Imanaka-Yoshida, K., Yoshida, T., Abel, E.D., Eliceiri, B., Yang, Y., Ulevitch, R.J., Lee, J.D. (2004). Targeted deletion of BMK1/ERK5 in adult mice perturbs vascular integrity and leads to endothelial failure. *J. Clin. Invest.* 113, 1138–1148.

Hobson, G.M., Krahe, R., Garcia, E., Siciliano, M.J., Funanage, V.L. (1995). Regional chromosomal assignments for four members of the MADS domain transcription enhancer factor 2 (MEF2) gene family to human chromosomes 15q26, 19p12, 5q14, and 1q12–q23. *Genomics* 29, 704–711.

Huang, K., Louis, J.M., Donaldson, L., Lim, F.L., Sharrocks, A.D., Clore, G.M. (2000). Solution structure of the MEF2A-DNA complex: structural basis for the modulation of DNA bending and specificity by MADS-box transcription factors. *EMBO J.* 19, 2615–2628.

Ikeshima, H., Imai, S., Shimoda, K., Hata, J., Takano, T. (1995). Expression of a MADS box gene, MEF2D, in neurons of the mouse central nervous system: implication of its binary function in myogenic and neurogenic cell lineages. *Neurosci. Lett.* 200, 117–120.

Kasler, H.G., Victoria, J., Duramad, O., Winoto, A. (2000). ERK5 is a novel type of mitogen-activated protein kinase containing a transcriptional activation domain. *Mol. Cell. Biol.* 20, 8382–8389.

Kato, Y., Zhao, M., Morikawa, A., Sugiyama, T., Chakravortty, D., Koide, N., Yoshida, T., Tapping, R.I., Yang, Y., Yokochi, T., Lee, J. D. (2000). Big mitogen-activated kinase regulates multiple members of the MEF2 protein family. *J. Biol. Chem.* 275, 18534–18540.

Krainc, D., Bai, G., Okamoto, S., Carles, M., Kusiak, J.W., Brent, R.N., Lipton, S.A. (1998). Synergistic activation of the N-methyl-D-aspartate receptor subunit 1 promoter by myocyte enhancer factor 2C and Sp1. *J. Biol. Chem.* 273, 26218–26224.

Lazaro, J.B., Bailey, P.J., Lassar, A.B. (2002). Cyclin D-cdk4 activity modulates the subnuclear localization and interaction of MEF2 with SRC-family coactivators during skeletal muscle differentiation. *Genes Dev.* 16, 1792–1805.

Leifer, D., Golden, J., Kowall, N.W. (1994). Myocyte-specific enhancer binding factor 2C expression in human brain development. *Neuroscience* 63, 1067–1079.

Leifer, D., Krainc, D., Yu, Y.T., McDermott, J., Breitbart, R.E., Heng, J., Neve, R.L., Kosofsky, B., Nadal-Ginard, B., Lipton, S.A. (1993). MEF2C, a MADS/MEF2–family transcription factor expressed in a laminar distribution in cerebral cortex. *Proc. Natl.*

Acad. Sci. USA 90, 1546–1550.

Li, M., Linseman, D.A., Allen, M.P., Meintzer, M.K., Wang, X., Laessig, T., Wierman, M.E., Heidenreich, K.A. (2001). Myocyte enhancer factor 2A and 2D undergo phosphorylation and caspase-mediated degradation during apoptosis of rat cerebellar granule neurons. *J. Neurosci.* 21, 6544–6552.

Linseman, D.A., Bartley, C.M., Le, S.S., Laessig, T.A., Bouchard, R.J., Meintzer, M.K., Li, M., Heidenreich, K.A. (2003). Inactivation of the myocyte enhancer factor–2 repressor histone deacetylase–5 by endogenous Ca(2+)//calmodulin-dependent kinase II promotes depolarization-mediated cerebellar granule neuron survival. *J. Biol. Chem.* 278, 41472–41481.

Liu, L., Cavanaugh, J.E., Wang, Y., Sakagami, H., Mao, Z., Xia, Z. (2003). ERK5 activation of MEF2–mediated gene expression plays a critical role in BDNF-promoted survival of developing but not mature cortical neurons. *Proc. Natl. Acad. Sci. USA* 100, 8532–8537.

Lyons, G.E., Micales, B.K., Schwarz, J., Martin, J.F., Olson, E.N. (1995). Expression of mef2 genes in the mouse central nervous system suggests a role in neuronal maturation. *J. Neurosci.* 15, 5727–5738.

Ma, K., Chan, J.K., Zhu, G., Wu, Z. (2005). Myocyte enhancer factor 2 acetylation by p300 enhances its DNA binding activity, transcriptional activity, and myogenic differentiation. *Mol. Cell. Biol.* 25, 3575–3582.

Mao, Z., Nadal-Ginard, B. (1996). Functional and physical interactions between mammalian achaete-scute homolog 1 and myocyte enhancer factor 2A. *J. Biol. Chem.* 271, 14371–14375.

Mao, Z., Wiedmann, M. (1999). Calcineurin enhances MEF2 DNA binding activity in calcium-dependent survival of cerebellar granule neurons. *J. Biol. Chem.* 274, 31102–31107.

Mao, Z., Bonni, A., Xia, F., Nadal-Vicens, M., Greenberg, M.E. (1999). Neuronal activity-dependent cell survival mediated by transcription factor MEF2. *Science* 286, 785–790.

Marinissen, M.J., Chiariello, M., Pallante, M., Gutkind, J.S. (1999). A network of mitogen-activated protein kinases links G protein-coupled receptors to the c-jun promoter: a role for c-Jun NH_2-terminal kinase, p38s, and extracellular signal-regulated kinase 5. *Mol. Cell. Biol.* 19, 4289–4301.

Martin, J.F., Miano, J.M., Hustad, C.M., Copeland, N.G., Jenkins, N.A., Olson, E.N. (1994). A Mef2 gene that generates a muscle-specific isoform via alternative mRNA splicing. *Mol. Cell. Biol.* 14, 1647–1656.

McDermott, J.C., Cardoso, M.C., Yu, Y.T., Andres, V., Leifer, D., Krainc, D., Lipton, S.A., Nadal-Ginard, B. (1993). hMEF2C gene encodes skeletal muscle- and brain-specific transcription factors. *Mol. Cell. Biol.* 13, 2564–2577.

McKinsey, T.A., Zhang, C.L., Lu, J., Olson, E.N. (2000). Signal-dependent nuclear export of a histone deacetylase regulates muscle differentiation. *Nature* 408, 106–111.

McKinsey, T.A., Zhang, C.L., Olson, E.N. (2001). Control of muscle development by dueling HATs and HDACs. *Curr. Opin. Genet. Dev.* 11, 497–504.

McKinsey, T.A., Zhang, C.L., Olson, E.N. (2002). MEF2: a calcium-dependent regulator of cell division, differentiation and death. *Trends Biochem. Sci.* 27, 40–47.

Molkentin, J.D., Black, B.L., Martin, J.F., Olson, E.N. (1996a). Mutational analysis of the DNA binding, dimerization, and transcriptional activation domains of MEF2C. *Mol. Cell. Biol.* 16, 2627–2636.

Molkentin, J.D., Li, L., Olson, E.N. (1996b). Phosphorylation of the MADS-Box transcription factor MEF2C enhances its DNA binding activity. *J. Biol. Chem.* 271, 17199–17204.

Ng, M., Yanofsky, M.F. (2001). Function and evolution of the plant MADS-box gene family. *Nat. Rev. Genet.* 2, 186–195.

Nurrish, S.J., Treisman, R. (1995). DNA binding specificity determinants in MADS-box transcription factors. *Mol. Cell. Biol.* 15, 4076–4085.

Okamoto, S., Krainc, D., Sherman, K., Lipton, S.A. (2000). Antiapoptotic role of the p38 mitogen-activated protein kinase-myocyte enhancer factor 2 transcription factor pathway during neuronal differentiation. *Proc. Natl. Acad. Sci. USA* 97, 7561–7566.

Okamoto, S., Li, Z., Ju, C., Scholzke, M.N., Mathews, E., Cui, J., Salvesen, G.S., Bossy-Wetzel, E., Lipton, S.A. (2002). Dominant-interfering forms of MEF2 generated by caspase cleavage contribute to NMDA-induced neuronal apoptosis. *Proc. Natl. Acad. Sci. USA* 99, 3974–3979.

Ornatsky, O.I., McDermott, J.C. (1996). MEF2 protein expression, DNA binding specificity and complex composition, and transcriptional activity in muscle and non-muscle cells. *J. Biol. Chem.* 271, 24927–24933.

Pollock, R., Treisman, R. (1991). Human SRF-related proteins: DNA-binding properties and potential regulatory targets. *Genes Dev.* 5, 2327–2341.

Quinn, Z.A., Yang, C.C., Wrana, J.L., McDermott, J.C. (2001). Smad proteins function as co-modulators for MEF2 transcriptional regulatory proteins. *Nucleic Acids Res.* 29, 732–742.

Santelli, E., Richmond, T.J. (2000). Crystal structure of MEF2A core bound to DNA at 1.5 A resolution. *J. Mol. Biol.* 297, 437–449.

Schulz, R.A., Chromey, C., Lu, M.F., Zhao, B., Olson, E.N. (1996). Expression of the D-MEF2 transcription in the *Drosophila* brain suggests a role in neuronal cell differentiation. *Oncogene* 12, 1827–1831.

Shalizi, A., Lehtinen, M., Gaudilliere, B., Donovan, N., Han, J., Konishi, Y., Bonni, A. (2003). Characterization of a neurotrophin signaling mechanism that mediates neuron survival in a temporally specific pattern. *J. Neurosci.* 23, 7326–7336.

Skerjanc, I.S., Wilton, S. (2000). Myocyte enhancer factor 2C upregulates MASH–1 expression and induces neurogenesis in P19 cells. *FEBS Lett.* 472, 53–56.

Tang, X., Wang, X., Gong, X., Tong, M., Park, D., Xia, Z., Mao, Z. (2005). Cyclin-dependent kinase 5 mediates neurotoxin-induced degradation of the transcription factor myocyte enhancer factor 2. *J. Neurosci.* 25, 4823–4834.

Wang, X., Tang, X., Li, M., Marshall, J., Mao, Z. (2005). Regulation of neuroprotective activity of myocyte-enhancer Factor 2 by cAMP-protein kinase A signaling pathway in neuronal survival. *J. Biol. Chem.* 280, 16705–16713.

Xue, Z.G., Xue, J.X., Roncier, B., Chamagne, A.M., Portier, M.M. (2000). Isolation of quail qMEF2D gene and its expression pattern in the developing central nervous system. *Biochim. Biophys. Acta* 1492, 543–547.

Yang, C.C., Ornatsky, O.I., McDermott, J.C., Cruz, T.F., Prody, C.A. (1998). Interaction of myocyte enhancer factor 2 (MEF2) with a mitogen-activated protein kinase, ERK5/BMK1. *Nucleic Acids Res.* 26, 4771–4777.

Yang, S.H., Galanis, A., Sharrocks, A.D. (1999). Targeting of p38 mitogen-activated protein kinases to MEF2 transcription factors. *Mol. Cell. Biol.* 19, 4028–4038.

Youn, H.D., Chatila, T.A., Liu, J.O. (2000). Integration of calcineurin and MEF2 signals by the coactivator p300 during T-cell apoptosis. *EMBO J.* 19, 4323–4331.

Youn, H.D., Sun, L., Prywes, R., Liu, J.O. (1999). Apoptosis of T cells mediated by Ca^{2+}-induced release of the transcription factor MEF2. *Science* 286, 790–793.

Yu, Y.T. (1996). Distinct domains of myocyte enhancer binding factor–2A determining nuclear localization and cell type-specific transcriptional activity. *J. Biol. Chem.* 271, 24675–24683.

Yu, Y.T., Breitbart, R.E., Smoot, L.B., Lee, Y., Mahdavi, V., Nadal-Ginard, B. (1992). Human myocyte-specific enhancer factor 2 comprises a group of tissue-restricted MADS box transcription factors. *Genes Dev.* 6, 1783–1798.

Yun, K., Wold, B. (1996). Skeletal muscle determination and differentiation: story of a core regulatory network and its context. *Curr. Opin. Cell Biol.* 8, 877–889.

Zhang, C.L., McKinsey, T.A., Olson, E.N. (2001). The transcriptional corepressor MITR is a signal-responsive inhibitor of myogenesis. *Proc. Natl. Acad. Sci. USA* 98, 7354–7359.

Zhao, M., New, L., Kravchenko, V.V., Kato, Y., Gram, H., di Padova, F., Olson, E.N., Ulevitch, R.J., Han, J. (1999). Regulation of the MEF2 family of transcription factors by p38. *Mol. Cell. Biol.* 19, 21–30.

Zhu, B., Ramachandran, B., Gulick, T. (2005). Alternative pre-mRNA splicing governs expression of a conserved acidic transactivation domain in MEF2 factors of striated muscle and brain. *J. Biol. Chem* 280, 28749–28760.

15
RORα: An Orphan that Staggers the Mind

Peter M. Gent and Bruce A. Hamilton

Abstract

RORα is a nuclear receptor family member closely related to the retinoic acid receptors. A spontaneous null mutation of the mouse *Rora* gene in the *staggerer* mouse blocks the differentiation of cerebellar Purkinje cells, resulting in a hypoplastic cerebellum and congenital ataxia. Recent studies have identified several direct transcriptional targets of RORα in developing Purkinje cells, including a mitogenic signal to granule cell precursors, *Sonic hedgehog*, and genes required for postsynaptic reception of granule parallel fiber inputs, including metabotropic glutamatergic and calcium second messenger components. RORα is also implicated in transcriptional control of the circadian clock in suprachiasmatic nucleus and for physiological functions in several peripheral tissues, including regulation of circulating apolipoproteins. The ligand binding domain of RORα has been found in complex with cholesterol and cholesterol derivatives, but the extent to which such binding represents a conventional ligand interaction is not clear.

15.1
Introduction

Retinoic acid-related Orphan Receptor alpha (RORα, also known as RZRα or NR1F1) is an orphan member of the nuclear receptor superfamily, one of three paralogous genes in mammals (RORα, RORβ, and RORγ) closely related to the retinoic acid receptors. Of the three ROR genes, RORα is the most conserved across currently sequenced vertebrates, and is presently the most extensively studied, owing both to pioneering studies of Giguere and colleagues and to a well-characterized classical mutation in mice, *staggerer*, which provided 30 years of phenotypic data prior to its identification with RORα. RORα has a highly complex expression pattern in many tissues, but is most thoroughly studied through its critical role in the differentiation of cerebellar Purkinje neurons. RORα is also thought to play a role in circadian behavior and is required in peripheral tissues, the endocrine effectors of which may secondarily impact the nervous system, particularly with regard to lipid

---

Transcription Factors. Edited by Gerald Thiel
Copyright © 2006 WILEY-VCH Verlag GmbH & Co. KGaA, Weinheim
ISBN 3-527-31285-4

metabolism and the production of apolipoproteins. Analyses of RORα physiological function must therefore be made with care in order to distinguish direct effects of RORα activities from indirect effects, and from effects whose proximity to RORα is not yet established. In this chapter, the biochemical properties of RORα, the phenotypes of both classical and targeted mutations, and current understanding of the organization of its target genes are summarized and discussed.

15.2
Identification and Biochemical Properties of RORα

15.2.1
Identification

RORα was identified independently by Giguere and Becker-Andre and their colleagues in homology screens for novel RAR and RXR-related genes [1, 2]. RORα proteins contain a C4 zinc finger DNA-binding domain (DBD) and a HOLI domain and AF–2 motif typical of nuclear hormone ligand-binding domains (LBDs). Initial studies identified homology between the DNA-binding domains and, to a lesser extent, the ligand-binding domains of RORα and RARα, Rev-Erbα, and the *Drosophila* nuclear receptor DHR3 [2]. Two additional RAR-related orphan receptors, RORβ and RORγ, have been characterized [3, 4], as well as invertebrate homologues DHR3 in *Drosophila* [5, 6] and CHR3 in *Caenorhabditis elegans* [7]. The genomic location of RORα was mapped to chromosome 15 in humans and chromosome 9 in mice in 1995 [8], and identified as the gene mutated in the classical *staggerer* mutation in 1996 [9].

15.2.2
Isoforms

Four isoforms of RORα have been identified in humans, but only two are known in mice (Fig. 15.1). Each of the four isoforms has a different N-terminal domain created by alternate promoter usage and exon splicing, but share a DNA-binding domain, hinge region, and ligand-binding domain [2]. Both murine isoforms, RORα1 and α4, are expressed within the cerebellum [10]. The extent to which the isoforms are functionally differentiated *in vivo* is not yet clear, but *in-vitro* studies by Giguere identified different DNA binding preferences for each isoform and characterized the role of interaction of the isoform-specific N-terminal domains with the DNA-binding domain to provide binding specificity [2, 11].

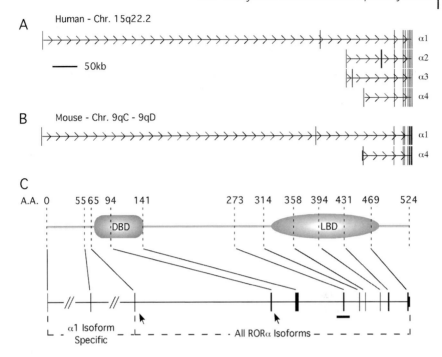

Fig. 15.1 Structure of RORα gene and protein. (A) The human genome expresses four RORα isoforms, covering 732 kb, that differ in their first and second exons, encoding proteins that differ in amino-terminal extensions before the canonical DNA binding domain. (B) The mouse genome expresses two RORα isoforms, homologous to human α1 and α4, covering 735 kb. Displays are based on public genome assemblies at http://genome.ucsc.edu. (C) Domain structure of human RORα1 protein predicted by SMART (http://smart.embl-heidelberg.de) indicates the composition by exon. Arrows point to locations of insertions in gene-targeted alleles [57, 58]; the thick line denotes *staggerer* deletion [9]. (This figure also appears with the color plates.)

15.2.3
RORα Binding and Response Elements

RORα binds as a monomer to a subset of canonical RGGTCA half-sites typical of its group of nuclear receptors, with efficiencies that depend primarily on 5' sequence context. Unlike many other nuclear receptors in its family, RORα does not require dimerization with itself or other receptors for binding or activity [2, 12]. Mutagenesis studies showed that the replacement of four residues in the DBD of RORα to the residues found in the highly homologous Rev-erbα DBD allow mutant RORα to form a homodimer, providing a structural basis for this distinction [13]. For native RORα isoforms, bound half-sites are sufficient for transactivation in transfection assays [2]. An RORα response element (RORE) was characterized by Giguere et al. using a PCR-based strategy and electrophoretic mobility shift assays (EMSAs) [2].

These experiments showed that RORα binds as a monomer to a single RGGTCA consensus half-site motif, but prefers half-sites preceded by a six-base AT-rich region (together called the extended half-site). Isoform-specific interactions between the variable N-terminal domains and the DBD may alter preference for DNA binding. Thus, RORα1 and RORα2 – the two isoforms most extensively characterized [2,11,12] – show different consensus sequences for extended half-site preference. The RORα1 consensus site, which is of interest here due to its role in the cerebellum, was identified as DWWWNWAGGTCA, where D is an A, T, or G and W is an A or T [2].

15.2.4
Crosstalk Between Factors

The high degree of conservation between the DNA-binding domains of RORα and related nuclear receptors raises the possibility of crosstalk among nuclear receptor signaling pathways at RORα response elements [14]. RORα1, Rev-erbα and Rev-erbβ bind to the same response element *in vitro*, and overexpression of Rev-erb α/β in a luciferase reporter assay is sufficient to repress transcription at an RORE activated by RORα [14]. As discussed below, opposing activities of Rev-erbα and RORα have subsequently been proposed to play a role in regulating circadian rhythm as a functional output of such crosstalk [15–18]. In another example, RORα-dependent activation of a γF-Crystallin promoter driving luciferase in P19 cells is repressed by a RAR/RXR heterodimer in the absence of retinoic acid [19]. Similarly, RORα crosstalk with PPARα/RXR has also been reported in activation of transcription at a peroxisome proliferator response element (PPRE) upstream of the enoyl-CoA hydratase/3–hydroxyacyl-CoA dehydrogenase gene [20].

RORα also intersects with nuclear receptor-mediated thyroid hormone (TH) signaling in the cerebellum. Messer showed that exogenous TH is capable of stimulating precocious cell division by granule cell precursors in the external germinal zone of early postnatal wild-type, but not *staggerer* (RORα-deficient) cerebellum [21]. TH receptors are also reported to activate the transcription of Purkinje cell protein 2 (*Pcp2*), both *in vitro* and *in vivo* [22, 23]; however, this activation is absolutely dependent on RORα [9, 24]. Additional overexpression and *in-vitro* experiments with additional TH response elements further indicate some level of crosstalk between RORα and TH signaling [25, 26]. Chin and colleagues have shown effects of TH on RORα expression levels *in vivo*, and suggested this as at least one level of interaction between these pathways [27]. A similar relationship between DHR3, a homologue of RORα and ecdysone receptor signaling has been reported in *Drosophila* [6].

15.2.5
Ligands or Cofactors?

RORα is somewhat unusual among characterized nuclear receptor family members in that it appears to be constitutively active and does not require exogenous ligand to function in mammalian or yeast cells [2, 28]. Surprisingly, a crystal structure of

RORα LBD overexpressed in insect SF9 cells showed that cholesterol bound in the ligand-binding domain of RORα. Mass spectrometry, mutagenesis, and cholesterol depletion experiments support the idea that cholesterol functions as an important cofactor for RORα [29]. Exchange experiments showed that cholesterol derivatives can also bind to RORα, whilst X-ray crystallography, luciferase reporter assays, and mutagenesis experiments suggest that cholesterol sulfate stimulates RORα activity better than cholesterol [29–31]. It seems unlikely that cholesterol acts as a regulatory ligand in the classical sense as its ubiquity might make regulating its interaction with RORα difficult, and mice engineered to lack cholesterol do not show RORα-related phenotypes [32]. It is worth noting, however, that RORα has independently been reported to regulate the expression of several apolipoproteins [33–37]. Thus, it is enticing to consider that having some level of sensitivity to cholesterol (or its derivatives) could facilitate RORα acting as a sensor in this pathway.

15.2.6
Co-activators

RORα contains a C-terminal Activation Function 2 (AF–2) helix in the C terminus of the ligand-binding domain. This creates a binding pocket that interacts with highly conserved LXXLL motifs and is required for cofactor recruitment and transactivation [11,28,38–41]. Several nuclear receptor co-activators have been reported to interact with RORα *in vitro* and in cell culture experiments (Table 15.1). These provide a useful guide for likely interactions *in vivo*, but further studies are needed to decode which interactions occur on endogenous regulatory sequences *in vivo*, under what conditions, and which are physiologically important.

In studies conducted in our laboratory by Gold et al., chromatin immunoprecipitation (ChIP) was used to identify several co-activators present *in vivo* on RORα target promoters within developing mouse cerebella [24]. Co-activators p300, Tip60, GRIP1, SRC1, and β-catenin were each found on at least one tested promoter by ChIP from control but not from *staggerer* mice. As the latter lack RORα, this indicates that RORα is required for the recruitment of these co-activators to those sites on endogenous promoters that are RORα-responsive *in vivo*. β-catenin also co-immunoprecipitated with RORα from native extracts of cerebellum. Whether RORα recruits these factors directly or through intermediary factors *in vivo* has not been shown, but protein interactions *in vitro* suggest that direct contact *in vivo* is likely. Interestingly, the findings showed that RORα mediates recruitment of different sets of cofactors to different promoters (see Table 15.1). Functional requirements for several of the co-activators that co-localize to the *Pcp2* promoter were further demonstrated in a cell culture model. CV-1 cells were co-injected with a *Pcp2*–lacZ reporter and an RORα expression plasmid with or without blocking antibodies against individual cofactors. As predicted by ChIP data, blocking antibodies to RORα-dependent cofactors β-catenin, SRC–1, and Tip60, but not CBP or p/CIP, eliminated RORα-induced lacZ expression driven by the *Pcp2* promoter.

Table 15.1 Co-activators associated with by RORα.

RORα co-activators	Cerebellar promoters	Other promoters	Method	Reference(s)
SRC1	Pcp2		ChIP	24
			GST, TH	40,92
β-Catenin	Pcp2		ChIP	24
	Pcp4		ChIP	24
	Shh		ChIP	24
Tip60	Pcp2		ChIP	24
	Pcp4		ChIP	24
	Slc1a6		ChIP	24
p300	Shh		ChIP	24
	Slc1a6		ChIP	24
		CPT1	Luc	82
		Caveolin-3	Luc	82
			GST, TH	41,81
GRIP1/SRC2	Slc1a6		ChIP	24
		Caveolin-3	Luc	82
		Reverbα	Luc	92
			GST, MS, TH	28,30,40,41
PGC1		CPT1	Luc	82
p/CIP/SRC3			GST	40
TRIP11			TH	28
TIF1			TH	28
PBP/DRIP205			GST, TH	28,41

Abbreviations: ChIP = chromatin immunoprecipitation; GST = GST pull down; Luc = co-transfection with Luciferase Reporter; MS = mass spectrometry; TH = two- hybrid.

15.2.7
Co-repressors

The AF–2 domain is also responsible for recruiting the co-repressor *hairless* (Hr), which can repress transcriptional activity by protecting RORα from proteosomal degradation, thus blocking the exchange of cofactors and recycling of the activator complex [40, 42]. Notably, Hr interaction with RORα allows concurrent interaction of the same domain with co-activators [40]. This is unlikely to play a role in the classical cerebellar phenotypes of *staggerer* mice as Hr is not co-expressed with RORα in Purkinje cells [43], but may mediate important effects in other sites of RORα expression. *In-vitro* data indicate that, under certain conditions, RORα can interact with N-CoR and SMRT, although it is unclear whether this holds true under physiological conditions [38]. It remains to be seen whether other co-repressors may act in concert with RORα independently of Rev-erb to shape gene expression in the cerebellum.

15.2.8
Activation and Regulation of RORα Expression

Comparatively little is known about which factors initiate RORα expression; however, the nexus between factors that modulate later RORα expression and some recently described targets of RORα are provocative. Factors that turn on RORα expression in postmitotic Purkinje cells (by embryonic day (E)12.5 in mice [24]) have not yet been identified, but physiological regulation of RORα expression by both hypoxia and calcium has been reported. As discussed below, RORα regulates a number of calcium-handling genes during neuronal development. Under hypoxic conditions, neuronal calcium concentrations show both short-term and long-term effects on cellular physiology [44–46]. HIF1α stimulates RORα expression under hypoxic conditions in cell culture [47–49], and HIF1α itself is activated in part by changes in cytosolic calcium [50, 51]. Intriguingly, RORα activity has also been reported to be calcium-sensitive [39]. Mice with a loss of calcium/calmodulin-dependent protein kinase IV (CaMKIV) show *staggerer*-like phenotypes, and CaMKIV is a potent activator of RORα activity in cell culture. CaMKIV does not directly phosphorylate RORα, but the addition of peptides containing AF–2 interacting LXXLL motifs inhibited the CaMKIV stimulatory effect on RORα, suggesting that the kinase may act indirectly on cofactors [39].

These observations may provide testable hypotheses for extending the physiological logic of RORα-mediated gene expression patterns. For example, if HIF1α modulates RORα expression to induce the same calcium-responding genes reported in cerebellum development [24], this might provide a simple mechanism for coupling a developmental gene expression module to later physiological processes and provide a further rationale for the genome-wide architecture of response to this transcription factor. Calcium-dependent activation of RORα could likewise provide an important homeostatic feedback mechanism by increasing the expression of calcium buffers such as Calbindin and other calcium-handling gene products. Further studies to identify both transcriptional and post-transcriptional control of RORα activity are clearly needed, and data relating to the concentration-dependent effects of RORα would be desirable to place the importance of RORα regulation in a physiological context.

15.2.9
RORα Expression in the Nervous System

In-situ hybridization detects RORα expression first in developing cerebellum and later in other brain and peripheral sites. RORα expression is first detectable in newly postmitotic Purkinje cells by E12.5, and continues through adulthood (Fig. 15.2). This is among the earliest markers for Purkinje cells. Later, RORα is expressed in interneurons of the molecular layer (basket and stellate cells), which derive from the same ventricular germinal zone. RORα is next apparent in the developing thalamus and later in the olfactory bulb, suprachiasmatic nuclei, and other areas of the central nervous system, including some regions of the cortex [9,24,52–57].

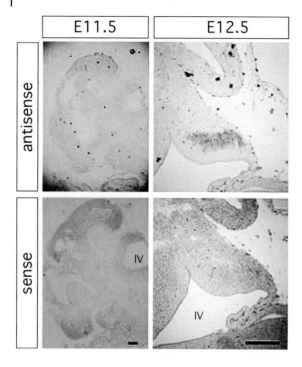

Fig. 15.2 *In-situ* hybridization of RORα to sectioned embryos shows no apparent expression at E11.5, but substantial expression in newly postmitotic Purkinje cells at E12.5. Commercially obtained 10 μm sections processed as described in [9]. Scale bars = 200 μm.

15.3
Role of RORα in the Developing Cerebellum

The role of RORα *in vivo* was initially characterized in the spontaneous mutant *staggerer* (*sg*) in 1962 by Sidman et al., and has been extensively characterized by multiple groups since. Positional cloning of *staggerer* some 34 years later identified the mutation as a genomic deletion that removes a single exon encoding part of the hinge between the DBD and LBD of RORα, causing a frameshift and premature termination [9]. Three other groups subsequently confirmed this finding independently [10,57,58]. The identification of *staggerer* has allowed three decades of phenotypic data to be re-interpreted in terms of RORα function.

Homozygous *staggerer* mice display a severe, nonprogressive congenital ataxia and cerebellar hypoplasia [59]. Cerebella of mice lacking RORα are characterized by a thin molecular layer, a disordered Purkinje cell layer, and a near-absence of granule cells (Fig. 15.3). Experiments using wild-type ↔ *staggerer* chimeric mice showed that *staggerer* is intrinsic to Purkinje cells, as genetically mutant granule cells can be rescued by wild-type Purkinje cells *in vivo*, but mutant Purkinje cells always show the typical immature, stunted morphology [60–62]. *Staggerer* Purkinje cells appear blocked in differentiation according to morphological, biochemical, and synaptic characteristics. They retain embryonic cell surface markers including lectin-binding properties [63, 64], and apparent failure to switch N-CAM from the E to A isoform [65]. *Staggerer* Purkinje cells fail to form mature synaptic arrangements and do not develop the dendritic arbor and tertiary spiny branchlets characteristic of mature nonmutant Purkinje cells [66, 67]. In particular, *staggerer* mice fail to prune supernumerary climbing fibers from the inferior olive that innervate Purkinje cells early in development and that in normal mice are pruned after Purkinje cells receive parallel fibers from the granule cells [68, 69]. Granule cells, which are reduced in number to begin with, are unable to form synapses on Purkinje cells, despite being able to synapse with stellate and basket cells [66, 67]. Thus, *staggerer* Purkinje cells are competent to receive innervation from climbing fibers, but not subsequent innervation from parallel fibers, and fail to undergo subsequent refinements to form mature synaptic arrangements.

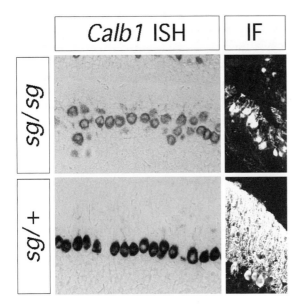

Fig. 15.3 Purkinje cell defects in *staggerer* cerebellum revealed by Calbindin D28 (*Calb1*) staining. *In-situ* hybridization to antisense *Calb1* probe illustrates ectopic positioning of surviving Purkinje cells in *staggerer* compared with a heterozygote control. Of note is the reduced *Calb1* expression in *staggerer* mice. Confocal microscopy of Purkinje cells labeled with an anti-Calbindin monoclonal antibody illustrates the extreme poverty of dendritic arbor in the mutant.

Given the Purkinje cell-intrinsic nature of the *staggerer* mutation, defects in cerebellar granule cells of the mutants should be viewed as indicators of defective RORα-dependent Purkinje-granule (or possibly molecular layer-granule) cell signaling. These effects begin early and affect several steps in cerebellum development. Granule precursors in the external germinal zone (EGL) are unable to respond to exogenous thyroid hormone, which induces proliferation in wild-type mice [21]. This indicates that normally RORα in Purkinje cells is required to retransmit an endogenous thyroid hormone signal to granule cells. Gene expression and slice culture experiments suggest that this mitogenic signal is most likely *sonic hedgehog* [24]. Perhaps as a consequence of this impaired signaling, the EGL persists longer in *staggerer* than wild-type mice [59,67,70] and fewer granule cells are produced. From this combination of impaired granule cell genesis and synaptic incompetence, Purkinje and granule neurons fail to establish mutual trophic support. Granule cells in *staggerer* cerebella degenerate during postnatal development, and by P30 most Purkinje cells (and subsequently the inferior olivary cells) are also lost [66,67,71]. Interestingly, heterozygous mice, despite having apparently normal cerebellar development, exhibit accelerated Purkinje cell atrophy and loss with age, indicating a continued role for RORα in cerebellar maintenance during normal aging [72]. Conditional alleles of RORα would be useful in confirming a pure maintenance, rather than latent developmental, origin and Purkinje cell autonomy for these age-related phenotypes.

These observations in *staggerer* and knockout mice indicate that RORα promotes Purkinje cell differentiation and coordinates the differentiation of other cell types by activating pathways necessary for signaling between Purkinje cells and their afferents. These RORα-dependent intercellular signaling pathways are necessary for numerical matching between communicating cell types during development. Observations in heterozygous mice indicate that RORα is also required for maintenance of Purkinje cell states.

15.4
Roles of RORα in Other Tissues

15.4.1
Suprachiasmatic Nuclei

The results of recent investigations have indicated that RORα may play a role in generating circadian rhythm in the suprachiasmatic nuclei (SCN), a key brain center for regulating circadian behavior [73]. RORα mRNA levels oscillate in a circadian fashion within the SCN, but not in other tissues [15,17,18]. *Staggerer* mice show altered free-running activity period lengths in constant dark after 12–hour light/12–hour dark cycle entrainment, and heterozygous mice appear to re-entrain to alternate light cycles faster than wild-type controls [17, 18]. Opposing activities of RORα and Rev-erbα appear to regulate expression of Bmal1 in the SCN [16–18]. Bmal1 heterodimerizes with Clock as part of the established transcriptional control of the SCN circadian pacemaker [74, 75]. Overexpression of RORα1 or RORα4 can up-regulate Bmal1 more than 10–fold, while dominant-negative RORα1 suppresses Bmal1 oscillations in a luciferase reporter assay in NIH3T3 cells [17, 18]. Endogenous RORα can bind to a *Bmal1*–RORE oligonucleotide probe, and mutagenesis of *Bmal1* promoter shows that two ROREs are required for *in-vitro* interactions between RORα and *Bmal1* and for transcriptional oscillation in cell culture. Hogenesch and colleagues provide evidence that RORα levels are dysregulated in *Clock* mutants and propose a model in which Clock/Bmal1 activates RORα and Rev-erbα to form a transcriptional feedback loop in SCN [17]. Whether this circadian role for RORα interacts directly with its roles in regulating other sets of target genes (see below) remains an interesting question.

15.4.2
Peripheral Tissues

Several reports implicate RORα in lipid homeostasis outside the nervous system. *Staggerer* mice fed a high-cholesterol and high-fat diet develop severe atherosclerosis compared to nonmutant controls, possibly due to decreased ApoA-I levels in *staggerer* [33, 37]. RORα is required for regulation of triglyceride levels through transcriptional activation of ApoA-V and ApoC-III, which show potential ROREs that bind RORα *in vitro* [34–36]. A role for RORα in suppressing the inflammatory response [76–79] may also contribute to the development of atherosclerosis in *staggerer* mice.

Additional RORα-dependent activities in bone, skeletal muscle, immune system and vascular function have been reported. *Staggerer* mice show reduced bone mineral content, and RORα is up-regulated in mesenchymal stem cells during osteogenic differentiation. Expression of RORα in cell culture activates the promoters of mouse bone sialoprotein and represses osteocalcin [80]. RORα may also act in muscle development. Expression of caveolin–3 and muscle carnitine palmitoyl-transferase–1 (CPT1) can be manipulated in opposite directions by transfection of either putative dominant-negative or full-length RORα in cultured muscle cells [81, 82]. RORα is expressed in vascular cells, and a role for RORα in vascular function has been reported based on observations of smooth muscle dysfunction and increased ischemia-induced angiogenesis in *staggerer* mice [47,83,84]. Additional characterization of the role of RORα outside the cerebellum will no doubt be highly interesting as the important aspects of these RORα functions are more finely elucidated.

15.5
In-Vivo Identification of RORα Targets

15.5.1
Genetic Program Controlled by RORα in the Cerebellum

The strongest evidence for direct transcriptional activation of specific target genes by RORα *in vivo* comes from studies in the developing cerebellum. To define the genetic circuit regulated by RORα during Purkinje cell differentiation, Gold et al. initiated a systematic search for transcriptional targets of RORα in developing cerebella [24]. Commercial high-density oligonucleotide arrays were used to profile RNA expression levels in cerebellum every two days from E15.5 through P4 in both *staggerer* and sex-matched wild-type littermates. A variety of statistical analyses, including standard ANOVA tests, detected statistically significant changes between *staggerer* and wild-type mice at several developmental stages in this perinatal window (Fig. 15.4). Focusing analysis on the earliest gene expression changes primarily identified genes required in cell proliferation and intracellular calcium signaling as downregulated in the absence of RORα. Integrating this genomic data with developmental timing allowed Gold et al. to identify a logic to this set of candidate target genes. As illustrated in Fig. 15.5, RORα stimulates expression of *sonic hedgehog* (*Shh*), a potent mitogen for proliferating granule cell precursors, and simultaneously stimulates the expression of genes involved in calcium second messenger handling downstream of metabotropic neurotransmitter receptor pathways in Purkinje cells, including genes for the glutamate transporter EAAT4 (*Slc1a6*), its cytoplasmic anchor (*Spnb3*), Calbindin (*Calb1*), the IP3–gated intracellular calcium channel (*Itpr1*), its interaction partner (*Cals1*), a calcium channel-interacting GoLoco protein (*Pcp2*), and a calmodulin kinase partner (*Pcp4*). RORα is also required for the expression of a second mitogenic signal, kit ligand (Kitl, or Steel factor). It is unclear what function this signal may have, as cerebellar phenotypes have not been reported in *Kitl* mutant mice, but the cognate receptor is expressed on adjacent cells [85].

15.5.2
Direct or Indirect Targets?

To test whether these genes are direct targets of RORα, Gold et al. analyzed several promoters with ChIP experiments. Promoters for five out of five genes tested, but not control promoters, were precipitated from the cerebella of normal mice by antibodies to RORα. Importantly, ChIP experiments from littermate *staggerer* animals did not precipitate these promoters with RORα, confirming both RORα specificity of the assay and RORα-dependence of several co-activators, notably TIP60, p300 and β-catenin, at some of these sites. Interestingly, each promoter showed unique sets of co-activator recruitment (see Table 15.1). *Itpr1* is unique in that none of the tested cofactors was found at the RORα bound site. Direct action of RORα on genes for both outgoing mitogenic signals to afferent cells and signal transduction machinery required to receive incoming signals from afferents cells led Gold et al. to propose a reciprocal signaling circuit as a logic for the early targets of RORα in the developing cerebellum.

15.5.3
Developmental Signaling Genes

RORα activity on the *Shh* promoter is of particular interest. Shh is a potent mitogen/morphogen, the expression of which in multiple tissues is tightly regulated. Mechanisms that stimulate *Shh* expression in discrete elements of its normal pattern are not fully understood, and its regulatory factors for its expression in Purkinje cells had not been previously reported. The secretion of Shh from Purkinje cells is responsible for providing a mitogenic signal to granule cell precursors in the external germinal zone. ChIP experiments showed RORα binding at both a proximal

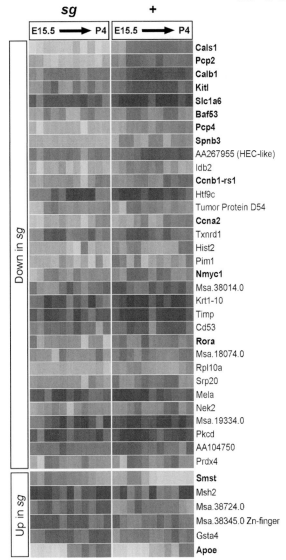

Fig. 15.4 RORα-dependent gene expression in developing cerebellum. Microarray data from Gold et al. [24] are shown here in an alternate view. Whole-cerebellum RNA from paired *staggerer* (sg) and nonmutant littermates of several developmental stages was analyzed on GeneChip arrays (Affymetrix). Data are normalized across all hybridizations for each gene. Red indicates a relative increase and green a relative decrease in expression compared to the mean of all measurements. (See Gold et al. [24] for details of data handling and rank ordering of RORα-responsive genes.) (This figure also appears with the color plates.)

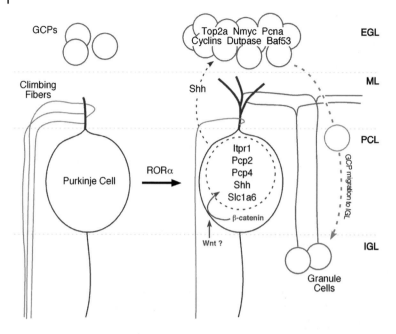

Fig. 15.5 RORα coordinates reciprocal signaling between Purkinje cells and afferent neurons. Shh signaling from Purkinje cells stimulates proliferation of granule cell precursors (GPCs) in the external granule cell layers. Granule cells migrate through the molecular layer (ML) and Purkinje cell layer (PCL) to the internal granule cell layer (IGL). At the same time, RORα activates genes necessary for receiving innervation from granule cells and reduction of supernumerary synapses from climbing fibers. (This figure also appears with the color plates.)

promoter (probe ~1 kb upstream of the start site) and a distal enhancer (~15 kb upstream). Interestingly, RORα is required for the recruitment of identical sets of co-activators detected at each of these two sites (p300 and β-catenin). Recruitment of β-catenin by RORα to both sites on the *Shh* locus is especially intriguing in light of the possibility that Wnt1 signaling by granule progenitors in the rhombic lip or possibly granule-dependent Wnt3 expression by Purkinje cells [86, 87] could create either feedback or relay pathways through the canonical Wnt/β-catenin pathway (see Fig. 15.5).

15.5.4
Calcium Signaling and Synaptic Function Genes

Characterization of the *staggerer* mouse shows that RORα has a cell-autonomous role in preparing Purkinje cells to receive input from afferent granule neurons. Both gene expression and ChIP data indicate genes required for both metabotropic postsynaptic function and downstream calcium signaling are up-regulated by RORα.

ChIP on the promoters of *Itpr1*, *Pcp2*, *Pcp4*, and *Slc1a6* showed direct interaction with RORα. Calcium regulator genes *Calb1* and *Cals1* have not yet been tested for direct interaction with RORα, nor the metabotropic glutamate receptor *Grm1* and the cytoskeletal anchor *Spnb3*. The role of RORα in the activation of calcium homeostasis genes is especially interesting in light of the reported ability of CaMKIV to stimulate RORα activity [39], and the possibility of a regulatory feedback loop through calcium signaling. In order to explain the coordinate regulation of *Shh* and postsynaptic signaling genes by RORα, Gold et al. propose that RORα coordinates expression of interacting synaptic function and calcium signaling gene sets that work together to allow differentiation of Purkinje cells such that they become competent to receive innervation from afferent neurons (see Fig. 15.5).

15.6
Implication of RORα in SCA1 Disorder

One of the more intriguing observations made by Gold et al. is a substantial overlap between genes down-regulated in spinocerebellar ataxia type 1 (SCA1) transgenic mice [88] and those down-regulated in *staggerer*, raising the possibility that polyglutamine repeat expansions in Ataxin–1 may interfere with transcriptional activation by RORα [24]. Subsequent profiling experiments with the SCA1 model have increased the extent of overlap between these two conditions [89]. It is unclear at this point whether this effect is due to direct interaction between RORα and ataxin–1. An alternative possibility is that key cofactors, or possibly RORα regulators, are sequestered by aggregates of the polyglutamine repeat expanded ataxin–1 protein [88,90,91], removing them from the available pool needed to maintain RORα activity. While more work is needed to clarify the mechanistic connection, current data provide an intriguing nexus between the developmental pathways regulated by RORα and the pathogenesis of mature onset disease in SCA1. Whether this relates directly to maintenance functions of RORα also remains to be seen.

15.7
Summary

The orphan nuclear receptor RORα plays an increasingly clear role in the biology of several key systems. How these different functions might relate to each other, if at all, will require further characterization of the non-cerebellar genetic networks regulated by RORα. RORα coordinates sets of genes required for differentiation of Purkinje cells and their signaling to other differentiating cell types in the developing cerebellum. By regulating genes involved in developmental signaling and postsynaptic function, RORα is able to stimulate proliferation of granule cell precursors, allow Purkinje cells to become competent to receive synapses from those same granule cells, and allow activity-dependent synaptic arrangements to progress. Outside of the cerebellum, RORα is implicated in many functions including circadian rhythm, bone development, inhibition of inflammatory response, and lipid metabolism. Future studies of the role and function of RORα will continue to illuminate the role of this important receptor in these and perhaps other functions, as well as providing additional details into both how RORα itself is regulated and the mechanisms by which it activates its many downstream targets.

Abbreviations

CaMKIV	calcium/calmodulin-dependent protein kinase IV
ChIP	chromatin immunoprecipitation
CPT1	carnitine palmitoyltransferase–1
DBD	DNA-binding domain
EGL	external germinal zone
EMSA	electrophoretic mobility shift assay
LBD	ligand binding domain
PPRE	peroxisome proliferator response element
RORα	Retinoic acid-related Orphan Receptor alpha
RORE	RORα response element
SCA1	spinocerebellar ataxia type 1
SCN	suprachiasmatic nuclei
TH	thyroid hormone

References

1 Becker-Andre M, Andre E, DeLamarter JF. *Biochem. Biophys. Res. Commun.*, **1993**, 194, 1371–1379.
2 Giguere V, Tini M, Flock G, Ong E, Evans RM, Otulakowski G. *Genes Dev.*, **1994**, 8, 538–553.
3 Carlberg C, Hooft van Huijsduijnen R, Staple JK, DeLamarter JF, Becker-Andre M. *Mol. Endocrinol.*, **1994**, 8, 757–770.
4 Hirose T, Smith RJ, Jetten AM. *Biochem. Biophys. Res. Commun.*, **1994**, 205, 1976–1983.
5 Koelle MR, Segraves WA, Hogness DS. *Proc. Natl. Acad. Sci. USA*, **1992**, 89, 6167–6171.
6 King-Jones K, Thummel CS. *Nat. Rev. Genet.*, **2005**, 6, 311–323.
7 Kostrouchova M, Krause M, Kostrouch Z, Rall JE. *Development*, **1998**, 125, 1617–1626.
8 Giguere V, Beatty B, Squire J, Copeland NG, Jenkins NA. *Genomics*, **1995**, 28, 596–598.
9 Hamilton BA, Frankel WN, Kerrebrock AW, Hawkins TL, FitzHugh W, Kusumi K, Russell LB, Mueller KL, van Berkel V, Birren BW, et al. *Nature*, **1996**, 379, 736–739.
10 Matysiak-Scholze U, Nehls M. *Genomics*, **1997**, 43, 78–84.
11 McBroom LD, Flock G, Giguere V. *Mol. Cell. Biol.*, **1995**, 15, 796–808.
12 Giguere V, McBroom LD, Flock G. *Mol. Cell. Biol.*, **1995**, 15, 2517–2526.
13 Moraitis AN, Giguere V. *Mol. Endocrinol.*, **1999**, 13, 431–439.
14 Forman BM, Chen J, Blumberg B, Kliewer SA, Henshaw R, Ong ES, Evans RM. *Mol. Endocrinol.*, **1994**, 8, 1253–1261.
15 Ueda HR, Chen W, Adachi A, Wakamatsu H, Hayashi S, Takasugi T, Nagano M, Nakahama K, Suzuki Y,

Sugano S, et al. *Nature*, **2002**, 418, 534–539.

16 Nakajima Y, Ikeda M, Kimura T, Honma S, Ohmiya Y, Honma K. *FEBS Lett.*, **2004**, 565, 122–126.

17 Sato TK, Panda S, Miraglia LJ, Reyes TM, Rudic RD, McNamara P, Naik KA, FitzGerald GA, Kay SA, Hogenesch JB. *Neuron*, **2004**, 43, 527–537.

18 Akashi M, Takumi T. *Nat. Struct. Mol. Biol.* **2005** 12, 441–448.

19 Tini M, Fraser RA, Giguere V. *J. Biol. Chem.*, **1995**, 270, 20156–20161.

20 Winrow CJ, Capone JP, Rachubinski RA. *J. Biol. Chem.*, **1998**, 273, 31442–31448.

21 Messer A. *J. Neurochem.*, **1988**, 51, 888–891.

22 Strait KA, Zou L, Oppenheimer JH. *Mol. Endocrinol.*, **1992**, 6, 1874–1880.

23 Zou L, Hagen SG, Strait KA, Oppenheimer JH. *J. Biol. Chem.*, **1994**, 269, 13346–13352.

24 Gold DA, Baek SH, Schork NJ, Rose DW, Larsen DD, Sachs BD, Rosenfeld MG, Hamilton BA. *Neuron*, **2003**, 40, 1119–1131.

25 Koibuchi N, Liu Y, Fukuda H, Takeshita A, Yen PM, Chin WW. *Endocrinology*, **1999**, 140, 1356–1364.

26 Kuno-Murata M, Koibuchi N, Fukuda H, Murata M, Chin WW. *Endocrinology*, **2000**, 141, 2275–2278.

27 Koibuchi N, Chin WW. *Endocrinology*, **1998**, 139, 2335–2341.

28 Atkins GB, Hu X, Guenther MG, Rachez C, Freedman LP, Lazar MA. *Mol. Endocrinol.*, **1999**, 13, 1550–1557.

29 Kallen JA, Schlaeppi JM, Bitsch F, Geisse S, Geiser M, Delhon I, Fournier B. *Structure (Camb).*, **2002**, 10, 1697–1707.

30 Bitsch F, Aichholz R, Kallen J, Geisse S, Fournier B, Schlaeppi JM. *Anal. Biochem.*, **2003**, 323, 139–149.

31 Kallen J, Schlaeppi JM, Bitsch F, Delhon I, Fournier B. *J. Biol. Chem.*, **2004**, 279, 14033–14038.

32 Wechsler A, Brafman A, Shafir M, Heverin M, Gottlieb H, Damari G, Gozlan-Kelner S, Spivak I, Moshkin O, Fridman E, et al. *Science*, **2003**, 302, 2087.

33 Mamontova A, Seguret-Mace S, Esposito B, Chaniale C, Bouly M, Delhaye-Bouchaud N, Luc G, Staels B, Duverger N, Mariani J, et al. *Circulation*, **1998**, 98, 2738–2743.

34 Genoux A, Dehondt H, Helleboid-Chapman A, Duhem C, Hum DW, Martin G, Pennacchio LA, Staels B, Fruchart-Najib J, Fruchart JC. *Arterioscler. Thromb. Vasc. Biol.*, **2005** 25, 1186–1192.

35 Lind U, Nilsson T, McPheat J, Stromstedt PE, Bamberg K, Balendran C, Kang D. *Biochem. Biophys. Res. Commun.*, **2005**, 330, 233–241.

36 Raspe E, Duez H, Gervois P, Fievet C, Fruchart JC, Besnard S, Mariani J, Tedgui A, Staels B. *J. Biol. Chem.*, **2001**, 276, 2865–2871.

37 Vu-Dac N, Gervois P, Grotzinger T, De Vos P, Schoonjans K, Fruchart JC, Auwerx J, Mariani J, Tedgui A, Staels B. *J. Biol. Chem.*, **1997**, 272, 22401–22404.

38 Harding HP, Atkins GB, Jaffe AB, Seo WJ, Lazar MA. *Mol. Endocrinol.*, **1997**, 11, 1737–1746.

39 Kane CD, Means AR. *EMBO J.*, **2000**, 19, 691–701.

40 Moraitis AN, Giguere V, Thompson CC. *Mol. Cell. Biol.*, **2002**, 22, 6831–6841.

41 Harris JM, Lau P, Chen SL, Muscat GE. *Mol. Endocrinol.*, **2002**, 16, 998–1012.

42 Moraitis AN, Giguere V. *J. Biol. Chem.*, **2003**, 278, 52511–52518.

43 Potter GB, Zarach JM, Sisk JM, Thompson CC. *Mol. Endocrinol.*, **2002**, 16, 2547–2560.

44 Yao H, Haddad GG. *Cell Calcium*, **2004**, 36, 247–255.

45 White BC, Wiegenstein JG, Winegar CD. *JAMA*, **1984**, 251, 1586–1590.

46 Nicholson C, Bruggencate GT, Steinberg R, Stockle H. *Proc. Natl. Acad. Sci. USA*, **1977**, 74, 1287–1290.

47 Besnard S, Heymes C, Merval R, Rodriguez M, Galizzi JP, Boutin JA, Mariani J, Tedgui A. *FEBS Lett.*, **2002**, 511, 36–40.

48 Chauvet C, Bois-Joyeux B, Danan JL. *Biochem. J.*, **2002**, 364, 449–456.

49 Chauvet C, Bois-Joyeux B, Berra E, Pouyssegur J, Danan JL. *Biochem. J.*, **2004**, 384, 79–85.
50 Berchner-Pfannschmidt U, Petrat F, Doege K, Trinidad B, Freitag P, Metzen E, de Groot H, Fandrey J. *J. Biol. Chem.*, **2004**, 279, 44976–44986.
51 Yuan G, Nanduri J, Bhasker CR, Semenza GL, Prabhakar NR. *J. Biol. Chem.*, **2005**, 280, 4321–4328.
52 Ino H. *J. Histochem. Cytochem.*, **2004**, 52, 311–323.
53 Nakagawa S, Watanabe M, Inoue Y. *Neurosci. Res.*, **1997**, 28, 177–184.
54 Nakagawa Y, O'Leary DD. *Dev. Neurosci.*, **2003**, 25, 234–244.
55 Matsui T, Sashihara S, Oh Y, Waxman SG. *Brain Res. Mol. Brain Res.*, **1995**, 33, 217–226.
56 Sashihara S, Felts PA, Waxman SG, Matsui T. *Brain Res. Mol. Brain Res.*, **1996**, 42, 109–117.
57 Steinmayr M, Andre E, Conquet F, Rondi-Reig L, Delhaye-Bouchaud N, Auclair N, Daniel H, Crepel F, Mariani J, Sotelo C, et al. *Proc. Natl. Acad. Sci. USA*, **1998**, 95, 3960–3965.
58 Dussault I, Fawcett D, Matthyssen A, Bader JA, Giguere V. *Mech. Dev.*, **1998**, 70, 147–153.
59 Sidman RL, Lane PW, Dickie MM. *Science*, **1962**, 137, 610–612.
60 Herrup K, Mullen RJ. *Brain Res.*, **1979**, 178, 443–457.
61 Herrup K, Mullen RJ. *Brain Res.*, **1981**, 227, 475–485.
62 Herrup K. *Brain Res.*, **1983**, 313, 267–274.
63 Hatten ME, Messer A. *Nature*, **1978**, 276, 504–506.
64 Trenkner E. *Nature*, **1979**, 277, 566–567.
65 Edelman GM, Chuong CM. *Proc. Natl. Acad. Sci. USA*, **1982**, 79, 7036–7040.
66 Sotelo C, Changeux JP. *Brain Res.*, **1974**, 67, 519–526.
67 Landis DM, Sidman RL. *J. Comp. Neurol.*, **1978**, 179, 831–863.
68 Mariani J, Changeux JP. *J. Neurobiol.*, **1980**, 11, 41–50.
69 Crepel F, Delhaye-Bouchaud N, Guastavino JM, Sampaio I. *Nature*, **1980**, 283, 483–484.
70 Yoon CH. *Neurology*, **1972**, 22, 743–754.
71 Herrup K, Mullen RJ. *Brain Res.*, **1979**, 172, 1–12.
72 Zanjani HS, Mariani J, Delhaye-Bouchaud N, Herrup K. *Brain Res. Dev. Brain Res.*, **1992**, 67, 153–160.
73 Ralph MR, Foster RG, Davis FC, Menaker M. *Science*, **1990**, 247, 975–978.
74 Gekakis N, Staknis D, Nguyen HB, Davis FC, Wilsbacher LD, King DP, Takahashi JS, Weitz CJ. *Science*, **1998**, 280, 1564–1569.
75 Jin X, Shearman LP, Weaver DR, Zylka MJ, de Vries GJ, Reppert SM. *Cell*, **1999**, 96, 57–68.
76 Delerive P, Monte D, Dubois G, Trottein F, Fruchart-Najib J, Mariani J, Fruchart JC, Staels B. *EMBO Rep.*, **2001**, 2, 42–48.
77 Dzhagalov I, Giguere V, He YW. *J. Immunol.*, **2004**, 173, 2952–2959.
78 Migita H, Satozawa N, Lin JH, Morser J, Kawai K. *FEBS Lett.*, **2004**, 557, 269–274.
79 Migita H, Morser J. *Arterioscler. Thromb. Vasc. Biol.*, **2005**, 25, 710–716.
80 Meyer T, Kneissel M, Mariani J, Fournier B. *Proc. Natl. Acad. Sci. USA*, **2000**, 97, 9197–9202.
81 Lau P, Bailey P, Dowhan DH, Muscat GE. *Nucleic Acids Res.*, **1999**, 27, 411–420.
82 Lau P, Nixon SJ, Parton RG, Muscat GE. *J. Biol. Chem.*, **2004**, 279, 36828–36840.
83 Besnard S, Silvestre JS, Duriez M, Bakouche J, Lemaigre-Dubreuil Y, Mariani J, Levy BI, Tedgui A. *Circ. Res.*, **2001**, 89, 1209–1215.
84 Besnard S, Bakouche J, Lemaigre-Dubreuil Y, Mariani J, Tedgui A, Henrion D. *Circ. Res.*, **2002**, 90, 820–825.
85 Zhang SC, Fedoroff S. *J. Neurosci. Res.*, **1997**, 47, 1–15.
86 Salinas PC, Fletcher C, Copeland NG, Jenkins NA, Nusse R. *Development*, **1994**, 120, 1277–1286.
87 Shimamura K, Hirano S, McMahon AP, Takeichi M. *Development*, **1994**, 120, 2225–2234.
88 Lin X, Antalffy B, Kang D, Orr HT,

Zoghbi HY. *Nat. Neurosci.*, **2000**, 3, 157–163.
89 Serra HG, Byam CE, Lande JD, Tousey SK, Zoghbi HY, Orr HT. *Hum. Mol. Genet.*, **2004**, 13, 2535–2543.
90 Kazantsev A, Preisinger E, Dranovsky A, Goldgaber D, Housman D. *Proc. Natl. Acad. Sci. USA*, **1999**, 96, 11404–11409.
91 Perez MK, Paulson HL, Pendse SJ, Saionz SJ, Bonini NM, Pittman RN. *J. Cell Biol.*, **1998**, 143, 1457–1470.
92 Delerive P, Chin WW, Suen CS. *J. Biol. Chem.*, **2002**, 277, 35013–35018.

16
The Role of NF-κB in Brain Function

Barbara Kaltschmidt, Ilja Mikenberg, Darius Widera, and Christian Kaltschmidt

Abstract

Nuclear factor kappa B (NF-κB) is an inducible transcription factor which is detected in neurons and glia. NF-κB is composed of three subunits: two DNA-binding and one inhibitory subunit. Activation of NF-κB takes place in the cytoplasm and results in degradation of the inhibitory subunit, thus enabling nuclear import of the DNA-binding subunits. Within the nucleus many target genes could be activated. A physiological function of NF-κB was shown for innate immune response and a pathophysiological function for cancer. For brain function, recent genetic models have identified a novel role for NF-κB in neuroprotection against various neurotoxins. Furthermore, genetic evidence for a role of NF-κB in learning and memory is now emerging. In this chapter our current understanding of neuronal NF-κB in response to synaptic transmission and potential physiological or pathophysiological modulators of NF-κB activity in the brain will be summarized. Synaptic NF-κB activated by glutamate and Ca^{2+} will be discussed in the context of retrograde signaling. The controversial role of NF-κB in neurodegenerative diseases will be discussed in the light of a model explaining the physiological amount of NF-κB activation.

16.1
Introduction

Nuclear factor kappa B (NF-κB) was discovered in David Baltimore's laboratory as an inducible transcription factor in lymphocytes (Nabel and Baltimore, 1987). NF-κB is involved in many biological processes such as inflammation and innate immunity, development, apoptosis and anti-apoptosis. Recent evidence also suggests an involvement in neuronal plasticity.

16.2
The NF-κB/Rel Family of Transcription Factors

The NF-ϰB/Rel family contains five members of DNA-binding proteins: p50, p52, p65 (RelA), c-Rel, and RelB (Baeuerle and Baltimore, 1996) (Fig. 16.1). A common denominator of all family members is an N-terminal 300 amino acids-long Rel-homology domain (RHD). The RHD contains essential sequence motifs for specific DNA-binding, homologous and heterologous dimerization, regulation of nuclear import and for the interaction with IϰB proteins. The transcription factor NFAT (nuclear factor of activated T-cells) also contains a RHD, and is therefore sometimes also regarded as a member of the NF-ϰB/Rel family (Nolan and Baltimore, 1992). Rel-Proteins are transcription factors binding to specific DNA-sequences (ϰB elements with the consensus sequence: 5'-GGGACTTTCC3'). The ϰB element was detected first in the enhancer of the ϰB element-light chain of immunoglobolins in B-cells (Sen and Baltimore, 1986). Inducible NF-ϰB resides in the cytoplasm in a latent DNA-binding form. There are five DNA-binding isoforms: RelB, c-Rel, p65 (RelA), p50, and p52, the latter two being proteolytic products of larger precursors. Surprisingly, p50 and p52 are the products of specific protein degradation within the proteasome. In some proteins, the proteasome only degrades specific parts of the protein and leaves the rest intact. This process, termed the regulated ubiquitin proteasome-dependent pathway (RUP), was recently reviewed (Rape and Jentsch, 2004). The C-terminal part of the NF-ϰB p50 precursor p105 is degraded by the proteasome in RUP-dependent manner. Degradation occurs either co- or post-translationally. Mechanistically, a protection of degradation via dimerization has been suggested. The dimerization of NF-ϰB can occur between two nascent polypeptide chains, emerging from successive ribosomes on the mRNA (Lin et al., 1998). This complex method of processing might add further ways of regulation, especially in neurons where proteasome activity seems to be regulated by synaptic activity (Ehlers, 2003). Within the cell, NF-ϰB is composed of two identical or non-identical subunits, although interestingly not all possible subunit compositions have been detected within cells. NF-ϰB subunits may be classified into two groups based on the presence of a transactivation domain (TAD): Class I contains p50 and p52 without a TAD, whilst Class II includes Rel-B, c-Rel, and p65 (RelA), which contain TADs that are capable of activating transcription without the help of other NF-ϰB subunits. Homo- and hetero-dimers of Class I subunits might act as repressors of transcription. The p50 and p65 heterodimers are detected most frequently. The domain structure and size of NF-ϰB subunits are depicted in Fig. 16.1.

16.2.1
The IκB Proteins: Inhibitors of NF-κB

DNA-binding occurs as a dimer, whereas the latent non-DNA-binding NF-ϰB complex contains an inhibitory subunit called IϰB. Later, more IϰB subunits were described (see Fig. 16.1.). Well-characterized inhibitory subunits include IϰB-α, IϰB-β, IϰB-γ(p105), IϰB-δ(p100), IϰB-ε, and IϰB-ζ (Whiteside and Israel, 1997; Yamazaki et

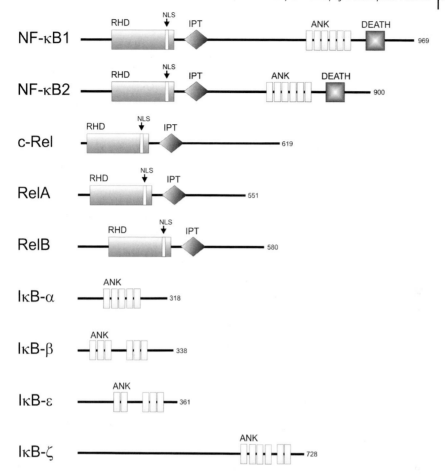

Fig. 16.1 Domain structure of NF-κB subunits. Domain motifs were assigned using EMBL SMART (http://smart.embl-heidelberg.de). Protein size is given on the right as amino acids. RHD = Rel homology domain; NLS = nuclear localization signal; IPT = Ig-like, plexins, transcription factors; ANK = ankyrin repeats; DEATH = DEATH domain, found in proteins involved in cell death (apoptosis).

al., 2001). Bcl–3 and the precursors of p50 and p52 (p105 or p100) could also act as inhibitory proteins (Karin and Ben-Neriah, 2000). Two IκB proteins, Cactus and Relish, were also present in *Drosophila*.

All IκB proteins contain several ankyrin repeats, which constitutes the ankyrin repeat domain (ANK; see Fig. 16.1). One ankyrin repeat contains 33 amino acids, and was initially detected within the protein Ankyrin, which has the function of a linker protein connecting membrane proteins with the cytoskeleton. ANKs are detected in many proteins such as membrane channel proteins, enzymes, toxins, signal cascade proteins, and transcription factors. The alpha-helical stack of ANK repeats provides a specific protein interaction interface (see Fig. 16.2, alpha-helical stacks at the right). The ANK of IκB proteins is responsible for the interaction with the RHD of NF-κB/Rel proteins. Interaction of ANK with the RHD mediates the interaction with the nuclear localization signal (NLS) in its alpha-helical conformation (Jacobs and Harrison, 1998) (see Fig. 16.2.). This interferes with the interaction of the NLS and the nuclear import machinery and keeps the NF-κB/IκB complex within the cytoplasm. After degradation of IκB, the NLS forms a random coil structure which is recognized by importin-α. Interestingly, this structure structural switch of the NLS sequence forms the basis for an activity-specific antibody (Kaltschmidt et al., 1995a).

Only IκB-α, IκB-β, and IκB-ε contain the domain essential for stimulus-dependent degradation of the N-terminal regulatory region. Expression of all IκB proteins, with respect to IκB-β , is directed by NF-κB (Karin and Ben-Neriah, 2000). IκB-α and IκB-β associates preferentially with dimers containing either c-Rel or p65.

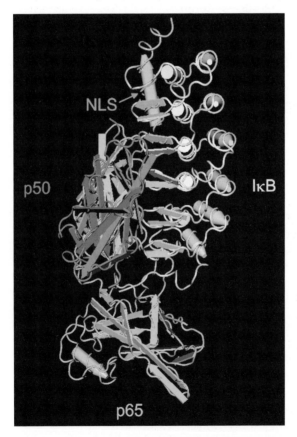

Fig. 16.2 Three-dimensional model of a co-crystal containing IκB, p50, and p65. The alpha-helical conformation of the p65 NLS (marked by an arrow) is due to an interaction with IκB-α. After degradation of IκB-α, the NLS loses its alpha-helical conformation and can be recognized by the nuclear import machinery via interaction with importin-α. Drawn after pbd: molecule 1IKN (Huxford et al., 1998). (This figure also appears with the color plates.)

16.3
Canonical NF-κB Activation

The canonical pathway of NF-κB activation via tumor necrosis factor (TNF) (Bonizzi and Karin, 2004; Schmitz et al., 2004) will be presented here to summarize part of the general knowledge on activation mechanisms (Fig. 16.3).

Within the nervous system, TNF (a 17–kDa protein) is able to bind to TNF receptors (TNF-Rs) expressed on both the glia and on neurons (Bruce et al., 1996). Expression of the TNF-α gene is subject to autoregulation via activated NF-κB (Collart et al., 1990), with two different receptors – p55 (TNF-R1) and 75 (TNF-R2) – having been identified. The p55 receptor is thought to be the major NF-κB activating TNF-R (Kolesnick and Golde, 1994). Initially, it was shown that TNF-mediated signaling via NF-κB could protect neurons against excitotoxic stress and against neurotoxic amyloid β (Cheng et al., 1994; Barger et al., 1995; Kaltschmidt et al., 1999a). In respect of this, it is noteworthy that the NF-κB and CREB signaling pathways are the major neuroprotective pathways identified as being protective against Alzheimer's disease (for a detailed discussion, see Mattson, 2004). A reduction of this protective NF-κB activation within Alzheimer patients' brains around late plaque stages might be one of the reasons for increased neurodegeneration (Kaltschmidt et al., 1999a). Presumably, part of the neuroprotective pathway mediated by NF-κB is due to the inhibition of the caspase-mediated apoptotic pathway (see Fig. 16.3).

Central to NF-κB activation seems to be the IκB kinase complex (IKK), which catalyzes the signal-dependent phosphorylation of the NF-κB inhibiting IκB. Thus, the IKK complex initiates NF-κB activation via phosphorylation of IκB, which is a signal for degradation. An amazing wealth of information has accumulated describing how receptor activation might be connected to IKK activation. There are themes of ubiquitination after activation of the trimeric TNF receptor. Binding of soluble or cell-bound TNF leads to activation of the latent trimerized TNF receptors. These receptors share an intracellular so-called death domain (DD) with several other TNF receptors such as TRAIL receptors or DR3, DR6 and various other receptors with non-TNF ligands such as the CD 95 (Apo/Fas) receptor or the p75 low-affinity nerve growth factor receptor. TNF-RI is unique in its composition of intracellular interaction proteins (Wajant et al.). Genetic ablation of TNF-R1 (p55) exacerbates traumatic brain injury and correlates with a reduced NF-κB activation (Bruce et al., 1996). Recent data have suggested that the TNF-R2 is responsible for a persistent NF-κB activation and neuroprotection (Yang et al., 2002; Marchetti et al., 2004).

Trimerization of the non-signaling TNF-R complex is mediated by a N-terminal pre- ligand assembly domain. TNF binding activates the pre-assembled receptor trimer via release of silencer of death domains (SODD) from the intracellular death domains (for a discussion, see Henkler and Wajant, 2004). However, debate persists as to the physiological role of SODD. The trimeric death domains appears to function as an assembly platform for further intracellular interactors such as the adapter protein, TRADD. TRADD, In turn, appears to enable a bifurcation in either the apoptotic pathway, leading to caspase activation, or in the anti-apoptotic NF-κB-dependent pathway, which involves the transcription of genes encoding survival fac-

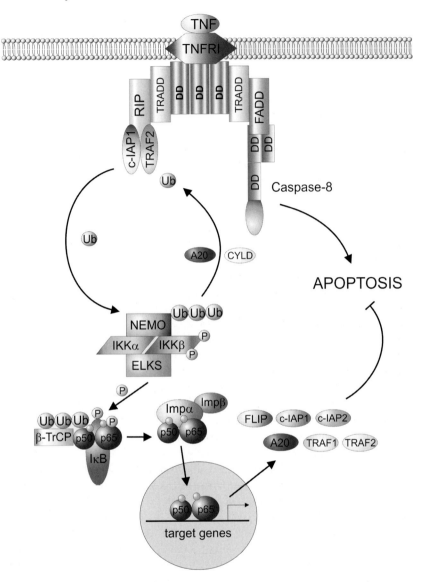

Fig. 16.3 Canonical pathway of NF-κB activation by tumor necrosis factor (TNF). The anti-apoptosis (NF-κB) and caspase-mediated apoptosis pathway is shown. Activation of the TNF receptor by ligand binding is transmitted to the IKK complex, which phosphorylates IκB family inhibitory molecules (see text for details). This targets IkB for degradation within the 26S proteasome, freeing nuclear localization signals on the DNA-binding p65/p50 subunits. After nuclear import, target gene transcription is initiated. Not all signaling components depicted in the canonical pathway have been investigated in the nervous system, but appear to be present in neurons and glia. Proteins are depicted as icons which illustrate a functional category (receptor, enzyme etc.) as suggested by the Alliance for Signalling convention (www.signaling-gateway.org). Ub = ubiquitination; P = phosphorylation. (This figure also appears with the color plates.)

tors. Apoptosis or anti-apoptosis involve TRADD in the context of different signaling complexes (Micheau and Tschopp, 2003). The initial receptor-bound complex (see Fig. 16.3) might contain TRADD, c-IAP1/2, RIP, and TRAF2 (Danial and Korsmeyer, 2004). This provides a signaling scaffold for the activation of NF-κB. An apoptotic pathway could be triggered by FADD, which might interact with RIP and TRADD in a non-receptor-bound cytoplasmic complex. With its N-terminal domain, TRADD could facilitate the interactions with TRAF1 and TRAF2. Anti-apoptotic proteins are targeted to the receptor complex (e.g., cIAP–1 and cIAP–2), and this may be the reason for the relatively good protection of activated TNF-R1 signaling against apoptosis. Several studies with murine neuronal cultures have used human TNF-α, which has been shown to activate the TNF-R1 (Cheng et al., 1994; Carlson et al., 1999).

The IκB kinase complex might provide a common denominator of many NF-κB activating stimuli culminating on NF-κB activation within the nervous system such as TNF, LPS, IL–1, NGF or glutamate-mediated signaling (for a review, see O'Neill and Kaltschmidt, 1997).

Crucial to NF-κB activation is the phosphorylation of IκB. This is most commonly due to interaction of activated IκB-kinase complex (IKK) with IκB. The IKK complex is composed of two catalytic subunits (IKK-α and IKK-β), a receptor-targeting/oligomerization subunit NEMO, and a recently discovered IκB-α targeting subunit ELKS (Ducut Sigala et al., 2004). Activation of the IKK complex might be regulated by multiple mechanisms (Schmitz et al., 2004). A classical phosphorylation of an activation loop has been reported for the main IκB phosphorylating kinase IKK-β at Ser177 and Ser181 (Delhase et al., 1999). This might be either due to autophosphorylation or due to upstream kinases. Surprisingly, genetic evidence for an involvement in the TNF pathway could be provided only for the upstream kinase MEKK3 (Yang et al., 2001). Other activation mechanisms dependent on oligomerization of the IKK complex have also been described (Tegethoff et al., 2003; Agou et al., 2004). NEMO-mediated recruitment of the IKK complex to the T-cell receptor complex has been identified as an activation mechanism (Weil et al., 2003). Also surprisingly, the membrane localization of NEMO could activate the IKK complex (Weil et al., 2003). NEMO, the regulatory subunit of the IKK complex, is inducibly ubiquitinated. The receptor-bound complex containing TRAF2 and TRAF5 is able to recruit the IKK complex to the membrane-bound TNF receptor, and this leads to activation via IKK oligomerization (Chen et al., 2002). IκB is recruited to the IKK complex via an interaction with ELKS (Ducut Sigala et al., 2004). The activated IKK complex may then catalyze phosphorylation of Ser32 and Ser36 on the IκB-α molecule.

As summarized above, activation of the IKK complex seems to rely on multimerization. Ubiquitination of NEMO might be an essential prerequisite of the activation process, since deubiquitinating enzymes such as CYLD (Kovalenko et al., 2003; Trompouki et al., 2003) are essential to deactivate the IKK complex. These ubiquitination pathways seem to be independent of the proteasomal degradation, but will modulate the oligomeric state. Other deubiquitination and ubiquitination activity is found to be encoded in the protein A20, which could target RIP for degradation (Heyninck and Beyaert, 2005).

Ubiquitin is a small (86 kDa) protein used to tag proteins either for degradation or for signaling (multimerization). Ubiquitin is conjugated to the amino groups of lysine residues on target proteins by a cascade of enzymes called E1, E2, and E3 (Ciechanover and Schwartz, 2004). A SCF (Skp–1/Cul/F box) -type multisubunit E3 ubiquitin ligase holoenzyme contains the phospho-IκB-specific acceptor subunit bTrCF, and is responsible for IκB poly-ubiquitination (Yaron et al., 1998). Interestingly, one of the frequently used pharmacological inhibitors of NF-κB activation, pyrrolidone dithiocarbamate (PDTC), acts as an inhibitor of the IκB ubiquitin-ligase complex (Hayakawa et al., 2003). The signal for the ubiquitin-ligase appears to be the phosphorylation of Ser32 and Ser36 on IκB-α.

Within the nervous system, inducible NF-κB is most frequently composed of two DNA-binding subunits (e.g., p50 or p65) that are either constitutively active or form a complex with the inhibitory subunit IκB-α (Bakalkin et al., 1993; Kaltschmidt et al., 1993, 1994, 1995a, 1995b, 1997; Rattner et al., 1993; Guerrini et al., 1995). There are reports of other κB-binding activities such as brain-specific transcription factor (BETA), which is specifically detected in gray matter extracts (Körner et al., 1989), developing brain factors (DBFs), which were reported to be highly enriched in the developing cortex (Cauley and Verma, 1994), and neuronal κB binding factor (NKBF) with different target sequence requirements (Moerman et al., 1999). To date, these other κB binding factors have not been assigned to specific genes, nor could they be tested directly in reporter gene assays. On the other hand, it seems to be accepted that there is an additional level of complexity, added by overlapping mutually exclusive or synergistically acting binding sites for other transcription factors. In glia and neurons, a sustained NF-κB activation for up to 72 hours was observed, though novel data suggest that this might be due to a differential use of IκB isoforms α and β. For glial cells, it has been reported that sustained NF-κB activity induced via interleukin (IL)–1 was still present even after the level of IκB-α protein returned that found pre-stimulation. In contrast, IκB-α protein levels remained low, suggesting that IκB-β is the negative regulator for sustained NF-κB activation (Bourke et al., 2000). In contrast to this, a biphasic response which is repressed by IκB-α was reported for TNF-stimulated neural cells (Kemler and Fontana, 1999). IκB proteins are essential regulators of nuclear import, which can interact with the NLS to induce an alpha-helical conformation (Huxford et al., 1998; Jacobs and Harrison, 1998). Interestingly, IκB-α only interacts with the NLS of p65, whereas IκB-β interacts with both the NLS of p50 and p65 (Malek et al., 1998). This conformation cannot be recognized by the nuclear import receptor importin alpha (Jacobs and Harrison, 1998), and in contrast the IκB free nuclear localization signal assumes a random coil conformation, which is the basis for the interaction with importin (see Fig. 16.2). Recently, it was shown that the NLS of p50 and p65 were recognized mainly by importin α3 (Fagerlund et al., 2005). A fresh view on import/export suggests an additional complexity as the trimeric p50/p65 IκB-α complex is shuttled between nucleus and cytoplasm (Malek et al., 1998). However, the nucleocytoplasmic shuttling of the trimeric complex of p50, p65 and IκB-α could not be observed (Fagerlund et al., 2005). In addition to inhibition by IκB, there seems to be a novel mechanism to terminate NF-κB signaling, namely promoter-specific degradation of p65 via nuclear proteasomes (Saccani et al., 2004). Whereas much knowledge on the mechanisms of TNF-mediated signaling has been acquired from non-neuronal cells, a systematic analysis of the TNF pathways within the nervous system has not yet been carried out.

In a microarray study using U373 human glioblastoma cells, it could be shown that many genes responded to TNF ($>$880 from 7500 tested) with a more than two-fold induction rate (Schwamborn et al., 2003). In this study, several novel TNF-responsive genes (about 60% of the genes regulated by a factor $\geq$3) were detected. A comparison of the TNF-induced gene expression profiles of U373, with profiles from 3T3 and HeLa cells revealed a striking cell-type specificity. Several of the TNF-induced genes were repressed by the inhibitor of IKK, PDTC, and these might therefore constitute novel target genes. It is not easy to discriminate between a role for NF-$\varkappa$B in glia or neurons *in vivo*, although cell type-specific knockout models might help to clarify this issue. A remaining problem might be the penetrance of cell type-specific knockouts, as this might not result in a deletion in all of the targeted cell types. Therefore, neuron-specific expression of transdominant negative I$\varkappa$B was used to analyze the role of neuronal NF-$\varkappa$B *in vivo* (Fridmacher et al., 2003).

16.3.1
Activators of NF-κB

Over many years, a wide range of activators of NF-$\varkappa$B have been identified (Table 16.1). Some of these are specific to the nervous system, such as the neurotransmitter glutamate which acts as an NF-$\varkappa$B activator via the main ionotropic glutamate receptors, or the neurotrophins, which display an amazing specificity. NGF activates NF-$\varkappa$B via the p75 receptor (Carter et al., 1996), whereas other cytokines (e.g., NT–3 or NT4/5) do not activate. In microglia, all neurotrophins activate NF-$\varkappa$B (Nakajima et al., 1998). Taken together, there is clearly much complexity in the system, as one molecule such as TNF can either activate or repress NF-$\varkappa$B in neurons (Kaltschmidt et al., 1999a), and there are also cell type-specific effects.

Table 16.1 Examples of molecules which activate NF-$\varkappa$B in the nervous system.

Molecule	Cell type	Reference(s)
Glutamate	Neurons	Guerrini et al. (1995)
Kainate	Neurons	Kaltschmidt et al. (1995b)
NMDA	Neurons	Lipsky et al. (2001)
TNF	Neuroblastoma	Drew et al. (1993)
TNF	Neurons	Barger et al. (1995), Kaltschmidt et al. (1999b)
TNF	Astrocytes	Sparacio et al. (1992)
TNF	Microglia	Lee et al. (2000)
IL–1	Glia cells	O'Neill and Kaltschmidt (1997)
IL–1	Neurons	Grilli et al. (1996)

Table 16.1 (continued)

sAPP	Neurons	Barger and Mattson (1996)
sAPP	Microglia	Barger and Harmon (1997)
Aβ	Neurons	Behl et al. (1994), Kaltschmidt et al. (1997)
Aβ	Astrocytes	Kaltschmidt et al. (1997)
Aβ/IFN γ	Microglia	Bonaiuto et al. (1997)
ATP	Microglia	Ferrari et al. (1997)
ATP/IL–1	Astrocytes	John et al. (2001)
Adenosine	Neurons	Basheer et al. (2001)
LPS	Microglia	Bauer et al. (1997)
LPS	Astrocytes	Pistritto et al. (1999)
EPO	Neurons	Digicaylioglu and Lipton (2001)
Bradykinin	Neurons	Schwaninger et al. (1999)
SDF–1	Astrocytes	Han et al. (2001)
EGF	Astrocytes	Zelenaia et al. (2000)
VEGF	Neuroblastoma	Jin et al. (2000)
NGF	Neurons	Wood (1995)
NGF	Schwann C.	Carter et al. (1996)
NGF	Oligodendrocyte	Yoon et al. (1998)
CNTF	PNS neurons	Middleton et al. (2000)
LIF	PNS neurons	Middleton et al. (2000)
IL–6	PNS neurons	Middleton et al. (2000)
CT–1	PNS neurons	Middleton et al. (2000)
BDNF	Microglia	Nakajima et al. (1998)
NT–3	Microglia	Nakajima et al. (1998)
NT4/5	Microglia	Nakajima et al. (1998)
IGF–1	Neuroblastoma	Heck et al. (1999)
ADNF	Neuroblastoma	Glazner et al. (2000)
PEDF	Neurons	Yabe et al. (2001)
H_2O_2	Neurons	Kaltschmidt et al. (1999b)
H_2O_2	Oligodendrocyte	Vollgraf et al. (1999)
Gangliosides	Neurons	Pyo et al. (1999)
Glutaredoxin	Neurons	Daily et al. (2001)
Focal cerebral ischemia	Neurons	Schneider et al. (1999)
RAGE-dependent diabetic neuropathy	Neurons	Bierhaus et al. (2004)
Non-Aβ amyloid	Neurons	Tanaka et al. (2002)
Amitriptyline	Neurons	Bartholoma et al. (2002)
Desipramine	Neurons	Bartholoma et al. (2002)
Fluoxetine	Neurons	Bartholoma et al. (2002)
IL–1	Neurons	Pizzi et al. (2002)
Lack of prion protein	Neurons	Brown et al. (2002)
Selenium deficiency	Neurons	Savaskan et al. (2003)
Intracellular calcium	Neurons	Lilienbaum and Israel (2003)
AMPA	Neurons	de Erausquin et al. (2003)
BDNF	Neurons	Burke and Bothwell (2003)
NGF	Neurons	Burke and Bothwell (2003)
NT–3	Neurons	Burke and Bothwell (2003)
NT–4/5	Neurons	Burke and Bothwell (2003)
Pre-myelination	Schwann cells	Nickols et al. (2003)

Table 16.1 (continued)

Dehydroepiandrosterone (DHEA)	PC–12	Charalampopoulos et al. (2004)
Allopregnanolone (Allo)	PC–12	Charalampopoulos et al. (2004)
Sleep deprivation	Neurons	Brandt et al. (2004)
FAIM	Neurons	Sole et al. (2004)
TGF-β1	Neurons	Zhu et al. (2004)

NMDA = N-methyl-D-aspartate; TNF = tumor necrosis factor; IL–1 = interleukin–1; sAPP = secreted beta-amyloid precursor; Aβ = beta-amyloid; IFN γ = interferon γ; LPS = lipopolysaccharide; EPO = erythropoietin; SDF–1 = stromal-derived cell factor–1 alpha; EGF = epidermal growth factor; AMPA = alpha-amino–3–hydroxy–5–methyl–4–isoxazolepropionic acid.

16.3.2
Repressors of NF-κB

Several anti-inflammatory cytokines known from the immune system (e.g., TGFβ or IL–10) also inhibit NF-κB in the nervous system (Table 16.2). Mechanistically, it is less clear how these molecules act in the nervous system, but one possibility might be the induction of IκB transcription (Arsura et al., 1996). Some molecules are already known to be activators, but appear to act in higher concentrations as repressors. Interestingly, the lipid peroxidation product 4–hydroxy–2,3–nonenal inhibits both constitutive and inducible NF-κB activity (Camandola et al., 2000).

Table 16.2 Examples of molecules which repress NF-κB in the nervous system.

Molecule	Cell type	Reference(s)
IL–4	Astrocytes	Pousset et al. (2000)
IL–10	Astrocytes	Pousset et al. (2000)
IL–10	Neurons	Bachis et al. (2001)
IL–10	Microglia	Ehrlich et al. (1998)
TGFβ	Microglia	Hu et al. (1999)
TGFβ	Neurons	Kaltschmidt and Kaltschmidt (2001)
TNF/H_2O_2	Neurons	Ginis et al. (2000)
TNF	Neurons	Kaltschmidt et al. (1999b)
NO	Neurons	Togashi et al. (1997)
NO	Microglia	Colasanti and Persichini (2000)
Hydroxy-nonenal	Neurons	Camandola et al. (2000)
Glucocorticoids	Neurons	Braun et al. (2000)
Aβ	Neurons	Kaltschmidt et al. (1997)
Melatonin	Neurons	Lezoualc'h et al. (1998)
Corticotropin- releasing-hormone	Neurons	Lezoualc'h et al. (2000)

Table 16.2 (continued)

Aspirin	Neurons	Grilli et al. (1996)
Triflusal	Glia	Acarin et al. (1998)
LY341122	Neurons	Stephenson et al. (2000)
Ganglioside	Astrocytes	Massa (1993)
PDTC	Neurons	Kaltschmidt et al. (1995b)
PDTC	Microglia	Bauer et al. (1997)
Vitamin E	Neurons	Behl (2000)
Dexanabinol (cannabinoid)	Neurons	Juttler et al. (2004)
Hypericin	Neurons	Kaltschmidt et al. (2002)
Selegiline	Neuroblastoma	Sharma et al. (2003)
Methylpyridinium (MPP(+))–	Neuroblastoma	Halvorsen et al. (2002)
Silymarin	Microglia	Wang et al. (2002)
Mutant preseniline–1	Neurons	Kassed et al. (2003)
Caffeic acid phenethyl ester (CAPE)	Neurons	Amodio et al. (2003)
Selenite	Neurons	Rossler et al. (2004)
BAY 11 7082	Neurons	Gutierrez et al. (2005)

16.3.3
Synaptic NF-κB

It has been proposed by Routtenberg (Routtenberg, 2000) that synaptic plasticity leads to information storage as the result of a synaptic dialogue. The first step would be glutamate release from the presynaptic site, followed by a modification of the presynaptic release process. This process is thought to require a retrograde messenger, which travels along the axon to switch on gene expression, in order to replenish the presynaptic protein supply. We, and others, consider that NF-κB might be crucially involved in this important process of synaptic plasticity, and the evidence for this which has been accumulated to date is reviewed in the following section (Fig. 16.4).

The ability of NF-κB to transmit information from active synapses to the nucleus is supported by several studies demonstrating the presence of NF-κB in synapses (Kaltschmidt et al., 1993; Meberg et al., 1996; Meffert et al., 2003). Synaptosomes contain presynaptic proteins, that are sealed and stabilized by the postsynaptic density. Inducible forms of NF-κB have been found in synaptosomal preparations (Kaltschmidt et al., 1993). Low-salt extracts prepared from synaptosomes contain NF-κB proteins, such as p50 and p65, together with IκB-α. Synaptophysin co-fractionates with NF-κB proteins during purification, whilst co-localization of synaptophysin and NF-κB proteins has also been detected in rat cerebral cortex (Kaltschmidt et al., 1993). In cortical extracts, NF-κB DNA-binding can be activated with the detergent desoxycholate (DOC), resulting in two specific DNA-binding complexes with different sensitivities for DOC. Supershifting and inhibition with recombinant IκB-α

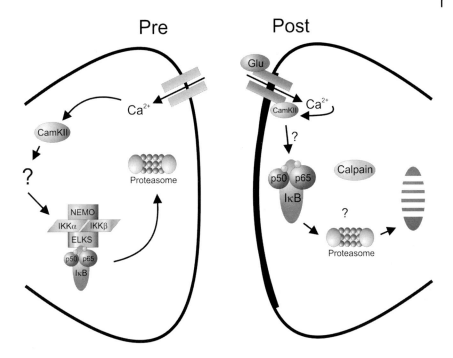

Fig. 16.4 Potential activation pathways of NF-κB activation machinery within the synapse. Presynapse (left) and postsynapse (right) might be discriminated due to different calcium channels. Both, presynapse and postsynapse seem to contain the necessary components. The importance of presynaptic and postsynaptic activation mechanisms requires clarification.

showed a bona fide DNA-binding complex that includes the p65 and p50 subunit. Similar complexes were detected using hippocampal synaptosomal preparations (Meberg et al., 1996; Meffert et al., 2003). In addition, Meberg and coworkers reported a robust increase in p65 mRNA after long-term potentiation, *in vivo*. It is possible that this is part of a feed-forward mechanism leading to increased DNA-binding to κB elements during long-term potentiation. Recently, an important influence of NF-κB on long-term suppression of synaptic transmission was also reported (Albensi and Mattson, 2000). Purkinje cell synapses were analyzed using light microscopy and en-passant synapses were found to contain NF-κB (Guerrini et al., 1995). Using electron microscopy, NF-κB- and IκB-α-like immunoreactivities within dendrites, including dendritic spines and postsynaptic densities, of neurons in the hippocampus and the cerebral cortex (Suzuki et al., 1997) were reported. With the help of an activity-specific anti-p65 antibody, it was possible to detect activated NF-κB in granule cell dendrites within the fascia dentata of rat hippocampus (Kaltschmidt et al., 2000a). NF-κB is activated in neurons by glutamate and depolarization (Guerrini et al., 1995; Kaltschmidt et al., 1995b). In *Drosophila melanogaster*, the NF-κB homologue Dorsal co-localizes with the IκB homologue Cactus within the nervous system. Both proteins are detected at high levels in postsynaptic sites of glutamatergic neuromuscular junctions (Cantera et al., 1999). Thus, NF-κB is utilized as a retrograde messenger in both presynaptic and postsynaptic compartments (see Fig. 16.4). Memory consolidation in crab also involves the activation of NF-κB-like activity (Freudenthal et al., 1998). Activated NF-κB was detected in *Aplysia* axons (Povelones et al., 1997) and in rat (Sulejczak and Skup, 2000). In rat, traumatic brain injury first causes an activation of axonal NF-κB, whilst at a later stage activated NF-κB is detected in neuronal nuclei. Activation was detectable for up to one year after brain injury (Nonaka et al., 1999). As an injury signal, sensed by synapses, Aβ from diffuse plaques is able to activate synaptic NF-κB (Ferrer et al., 1998). This might explain the activation of NF-κB in neurons around diffuse plaques, which is lost in neurons around later plaque stages (Kaltschmidt et al., 1999a). Recently, the transport of NF-κB in living neurons was analyzed (Wellmann et al., 2001; Meffert et al., 2003). To enable an analysis of the translocation of GFP-tagged p65 in living hippocampal neurons, a GFP tag was fused to the p65 subunit of NF-κB, and it was then confirmed that this fusion protein (GFP-p65) retained its functionality as a transcription factor. GFP-p65 was present in the nuclei of neurons, but after overexpression together with IκB-α the distribution of the protein changed from nuclear to neuritic (in dendrites and axons). A return of GFP-p65 from a neuritic to a nuclear distribution was observed in glutamate-stimulated hippocampal neurons, and Meffert and colleagues reported a similar result (Meffert et al., 2003). Glutamate-induced movement of GFP-p65 was detected in hippocampal neurons. When using small quantities of GFP-p65 expression vectors, endogenous IκB was able to keep the fusion protein in a neuritic/cytoplasmic localization, although glutamate agonists were able to overcome the localization to neurites and activate a nuclear localization. Interestingly, a recent study suggested a crucial involvement of NF-κB activity in BDNF-induced neuritic arborization (Gutierrez et al., 2005). Synaptic localization was crucially dependent on p65 (Meffert et al., 2003). In KO p65 animals, which are viable when crossed to a TNF-RI background, no synaptic NF-κB activity was detected (Meffert et al., 2003). Interestingly, the redistribution of GFP-p65 was dependent on a functional NLS (Wellmann et al., 2001). This NLS-dependent retrograde transport has already been described for *Aplysia* axons (Schmied and Ambron, 1997). Recently, it was reported that axonal injury led to increased retrograde transport of NLS peptides, this transport being mediated by an importin α/β complex interacting with the retrograde motor protein dynein (Hanz and Fainzilber, 2004). NF-κB activation by the neurotransmitter glutamate was identified in cerebellar granule cells (Kaltschmidt et al., 1995b; Guerrini et al., 1995, 1997; Lilienbaum and Israel, 2003). Constitutive activity of NF-κB was initially identified within neurons from the hippocampus and the cerebral cortex, using EMSA and immunofluorescence with an antibody specific for the activated p65 and with reporter gene assays (Kaltschmidt et al., 1993, 1994; Schmidt-Ullrich et al., 1996; Bhakar et al., 2002; Fridmacher et al., 2003). It was suggested that constitutive NF-κB activity is the result of synaptic activity (Kaltschmidt et al., 1994, 1995b; O'Neill and Kaltschmidt, 1997). Basal constitutive NF-κB activity in neurons could be repressed by specific inhibitors of action potential generation, glutamate receptors and L-type calcium channels (Meffert et al., 2003). The blockade of constitutive active NF-κB was most effective through inhibition of *N*-methyl-D-aspartate (NMDA) receptors using APV. Blockade of L-type calcium channels with nimodipine was also effective. This suggests that an extracellular influx of Ca^{2+} either through NMDA receptors or via L-type Ca^{2+} channels could activate NF-κB. Potential presynaptic and postsynaptic activation mechanisms for NF-κB are depicted in Fig. 16.2, and there is some evidence for a presynaptic NF-κB activation machinery. The proteasome and ubiquitination enzymes could be detected in the presynapse (Speese et al., 2003), and presynaptic mechanisms might involve the localized action of the IKK complex and voltage-gated Ca^{2+} channels, activated during action potential propagation. Calcium/calmodulin-dependent protein kinase II (CaMKII) is present in the presynapse, awaiting calcium ion-mediated activation (Ninan and Arancio, 2004), whilst also within the postsynapse a potential NF-κB activation machinery is in place. Activation by NMDA suggests a postsynaptic signaling where at least some isoforms of CamK (mainly isoform 2) have been reported to be associated with the NMDA receptor. Autophosphorylation of αCaMKII at Thr286 by Ca^{2+} influx through NMDA receptors (Ouyang et al., 1997) switches the kinase into a calcium/calmodulin-independent active status. It has been shown that CamKII can activate NF-κB in neurons (Lilienbaum and Israel, 2003; Meffert et al., 2003). The means by which the activated CamK activates NF-κB has not yet been identified, although a contribution of the proteasomal degradation machine seems likely. Indeed, activation of the postsynaptic proteasomal degradation has been shown to occur after the induction of neuronal activity (Ehlers, 2003). The role of calpain remains conflicting, as it appears to be active in IκB degradation in cerebellar granule cells (Scholzke et al., 2003), but not in matured hippocampal cultures (Meffert et al., 2003).

Activation of NF-κB by glutamate in the cerebellum and constitutive activity within the basal forebrain neurons could be also detected in mice containing NF-κB reporter genes (Schmidt-Ullrich et al., 1996; Guerrini et al., 1997; Bhakar et al., 2002). There are spurious reports on the failure to activate NF-κB in neurons by glutamate (Moerman et al., 1999). A potential reason for these negative results might be the inverted U-shaped activation curve of NF-κB (see Kaltschmidt et al., 1999a) or the sensitivity of the EMSA method, which might be only used to detect activated NF-κB in brain neurons, when isolated nuclei were prepared (Kaltschmidt et al., 1994). Another reason for these conflicting results might be different culture conditions.

It is concluded that NF-κB is capable of being a signal transducer, transmitting information from, for example active synapses to the nucleus, in addition to its well-known role as a transcription factor. In this way, NF-κB might be involved in translating short-term signals from distant sites in neurites into long-term changes in gene expression, and this may have a key role in plasticity, development, and survival. Indeed, it has been shown that $p65^{-/-}$ on a $TNF-RI^{-/-}$ background results in a severe learning deficit (Meffert et al., 2003). There are two other mouse models where NF-κB was repressed via tetracycline-regulated expression of transdominant negative IκB: (1) a model with CamKII promoter-driven expression of tTA in basal forebrain neurons (Fridmacher et al., 2003); and (2) a model with prion promoter-driven tTA expression in neurons and glia (work of Warner Greene and coworkers, see Ben-Neriah and Schmitz, 2004). Both have a modulation of learning and memory. For an extensive discussion see the recent comprehensive review by (Meffert and Baltimore, 2005). In accordance with Meffert and coworkers, we have found that repression of NF-κB by IκB in neurons resulted in behavioral deficits and a reduction in LTP and LTD induction. These effects could be correlated with a strongly reduced CREB phosphorylation (B. Kaltschmidt, A. Israel, C. Kaltschmidt, S. Memet and coworkers, submitted). On the other hand, IκB expression driven by the prion promoter-expressed tTA resulted in enhanced learning in older animals (Meffert and Baltimore, 2005). These might be two sides of the same coin. On the one side, a repression of neuroinflammation in elder age via inhibition of pathological NF-κB hyperactivation, might enhance learning, whereas on the other side neuronal NF-κB at physiological levels is needed for learning.

Initial reports of the neuroprotective role of TNF (Cheng et al., 1994) were followed by suggestions of a neuroprotective role for NF-κB. DNA decoy with NF-κB binding sites competed in cultured neurons with NF-κB activity induced by TNF and abolished neuroprotection (Barger et al., 1995; Mattson et al., 1997). Similarly, the survival of adult sensory neurons is dependent on TNF-mediated NF-κB activation (Fernyhough et al., 2005). In cerebellar granule cells, which have a low basal NF-κB activity (Kaltschmidt et al., 1995b), an inverted U-shaped dose-response for TNF-mediated NF-κB activation could be detected. An inverted U-shaped protection curve was also reported when TNF-mediated protection was analyzed against NMDA excitotoxicity (Carlson et al., 1999). A low dose of Aβ was able to activate NF-κB and to protect against a high cytotoxic dose of Aβ. This led to the discovery of an essential role for NF-κB in preconditioning (Kaltschmidt et al., 1999b; Blondeau et al., 2001; Ravati et al., 2001). Preconditioning describes an old observation worded by Paracelsus as "Alle Ding' sind Gift und nichts ohn' Gift; allein die Dosis macht, das ein Ding' kein Gift ist." (all things are poisons, only the dose makes the poison). In this line, a toxin in a low dose could activate a cellular response program, which later on protects against a high dose of toxin. It is considered that the mechanisms might be similar to a process described by David Baltimore as intracellular immunization (Baltimore, 1988) against virus infection. The preconditioning effect of NF-κB might be completely abolished by overexpression of transdominant IκB-α. On the other hand, there are toxic stimuli such as staurosporine where repression of NF-κB activation was protective (Kaltschmidt et al., 2002). This might be a general concept, which was first described in non-neuronal cells: the nature of the apoptotic stimulus dictates the pro- or anti-apoptotic action of NF-κB (Kaltschmidt et al., 2000b). Genetic evidence suggests that constitutive NF-κB activity is essential for neuronal survival (Bhakar et al., 2002). Similarly, pharmacological repression of the IKK complex (Aleyasin et al., 2004; B. Kaltschmidt et al., unpublished results) results in neuronal death. Transgenic overexpression of transdominant IκB-α sensitizes neurons against excitotoxic lesions (Fridmacher et al., 2003), whilst NF-κB activation protects neurons against Aβ toxicity (Barger et al., 1995; Kaltschmidt et al., 1999a). This might be a neuroprotective mechanism which is perturbed during Alzheimer's disease (Mattson, 2004). On the other hand, there exist several diseases where NF-κB hyperactivation is disease-promoting; examples include ischemia

(Schneider et al., 1999), Parkinson's disease (Hunot et al., 1997), and conditions where NF-κB-dependent p53 transcription mediates neuronal death (Aleyasin et al., 2004). The situation of p53–mediated death seems in some paradigms also to rely on a repression NF-κB activity (Culmsee et al., 2003).

To solve this dilemma, an optimal activation hypothesis is proposed similar to the NMDA receptor activation (Hardingham and Bading, 2003). A too-low activation of NF-κB in neurons is disastrous, as is a too-high activation (hyperactivation). The optimal basal constitutive level in neurons is maintained by synaptic activity which activates NF-κB and is repressed by one of its target genes IκB or CYLD in an autoregulatory fashion. A potential protective role of NF-κB in neurons should be discriminated from a potential degenerative role in glia.

Acknowledgments

These studies were supported in part by grants from the Deutsche Forschungsgemeinschaft (DFG), the Volkswagen-Stiftung and the Land NRW and the Universität Witten/Herdecke.

Abbreviations

BETA	brain-specific transcription factor
DBF	developing brain factor
DD	death domain
DOC	desoxycholate
IKK	IκB kinase complex
NFAT	nuclear factor of activated T-cells
NKBF	neuronal κB binding factor
NLS	nuclear localization signal
NMDA	N-methyl-D-aspartate
PDTC	pyrrolidone dithiocarbamate
RHD	Rel-homology domain
RUP	regulated ubiquitin proteasome-dependent pathway
SODD	silencer of death domain
TAD	transactivation domain
TNF	tumor necrosis factor
TNF-R	TNF receptor

References

Acarin L, Gonzalez B, Castellano B. **1998**. Stat3 and NFkappaB glial expression after excitotoxic damage to the postnatal brain. *Neuroreport* 9: 2869–2873.

Agou F, Courtois G, Chiaravalli J, Baleux F, Coic YM, Traincard F, Israel A, Veron M. **2004**. Inhibition of NF-kappa B activation by peptides targeting NF-kappa B essential modulator (nemo) oligomerization. *J. Biol. Chem.* 279: 54248–54257.

Albensi BC, Mattson MP. **2000**. Evidence for the involvement of TNF and NF-kappaB in hippocampal synaptic plasticity. *Synapse* 35: 151–159.

Aleyasin H, Cregan SP, Iyirhiaro G, O'Hare MJ, Callaghan SM, Slack RS and Park DS. **2004**. Nuclear factor-(kappa)B modulates the p53 response in neurons exposed to DNA damage. *J. Neurosci.* 24: 2963–2973.

Amodio R, De Ruvo C, Sacchetti A, Di Santo A, Martelli N, Di Matteo V, Lorenzet R, Poggi A, Rotilio D, Cacchio M, Esposito E. **2003**. Caffeic acid phenethyl ester blocks apoptosis induced by low potassium in cerebellar granule cells. *Int. J. Dev. Neurosci.* 21: 379–389.

Arsura M, Wu M, Sonenshein GE. **1996**. TGF beta 1 inhibits NF-kappa B/Rel activity inducing apoptosis of B cells: transcriptional activation of I kappa B alpha. *Immunity* 5: 31–40.

Bachis A, Colangelo AM, Vicini S, Doe PP, De Bernardi MA, Brooker G, Mocchetti I. **2001**. Interleukin–10 prevents glutamate-mediated cerebellar granule cell death by blocking caspase–3–like activity. *J. Neurosci.* 21: 3104–3112.

Baeuerle PA, Baltimore D. **1996**. NF-kappa B: ten years after. *Cell* 87: 13–20.

Bakalkin G, Yakovleva T, Terenius L. **1993**. NF-kappa B-like factors in the murine brain. Developmentally-regulated and tissue-specific expression. *Brain Res. Mol. Brain Res.* 20: 137–146.

Baltimore D. **1988**. Intracellular immunization. *Nature* 335: 395–396.

Barger SW, Harmon AD. **1997**. Microglial activation by Alzheimer amyloid precursor protein and modulation by apolipoprotein E. *Nature* 388: 878–881.

Barger SW, Mattson MP. **1996**. Induction of neuroprotective kappa B-dependent transcription by secreted forms of the Alzheimer's beta-amyloid precursor. *Mol. Brain Res.* 40: 116–126.

Barger SW, Horster D, Furukawa K, Goodman Y, Krieglstein J, Mattson MP. **1995**. Tumor necrosis factors alpha and beta protect neurons against amyloid beta-peptide toxicity: evidence for involvement of a kappa B-binding factor and attenuation of peroxide and Ca^{2+} accumulation. *Proc. Natl. Acad. Sci. USA* 92: 9328–9332.

Bartholoma P, Erlandsson N, Kaufmann K, Rossler OG, Baumann B, Wirth T, Giehl KM, Thiel G. **2002**. Neuronal cell death induced by antidepressants: lack of correlation with Egr–1, NF-kappa B and extracellular signal-regulated protein kinase activation. *Biochem. Pharmacol.* 63: 1507–1516.

Basheer R, Rainnie DG, Porkka-Heiskanen T, Ramesh V, McCarley RW. **2001**. Adenosine, prolonged wakefulness, and A1–activated NF-kappaB DNA binding in the basal forebrain of the rat. *Neuroscience* 104: 731–739.

Bauer MK, Lieb K, Schulze-Osthoff K, Berger M, Gebicke-Haerter PJ, Bauer J, Fiebich BL. **1997**. Expression and regulation of cyclooxygenase–2 in rat microglia. *Eur. J. Biochem.* 243: 726–731.

Behl C. **2000**. Vitamin E protects neurons against oxidative cell death in vitro more effectively than 17–beta estradiol and induces the activity of the transcription factor NF-kappaB. *J. Neural Transm.* 107: 393–407.

Behl C, Davis JB, Lesley R, Schubert D. **1994**. Hydrogen peroxide mediates amyloid beta protein toxicity. *Cell* 77: 817–827.

Ben-Neriah Y, Schmitz ML. **2004**. Of mice and men. *EMBO Rep.* 5: 668–673.

Bhakar AL, Tannis LL, Zeindler C, Russo MP, Jobin C, Park DS, MacPherson S, Barker PA. **2002**. Constitutive nuclear factor-kappa B activity is required for central neuron survival. *J. Neurosci.* 22: 8466–8475.

Bierhaus A, Haslbeck KM, Humpert PM, Liliensiek B, Dehmer T, Morcos M, Sayed AA, Andrassy M, Schiekofer S, Schneider JG, Schulz JB, Heuss D, Neundorfer B,

Dierl S, Huber J, Tritschler H, Schmidt AM, Schwaninger M, Haering HU, Schleicher E, Kasper M, Stern DM, Arnold B, Nawroth PP. **2004**. Loss of pain perception in diabetes is dependent on a receptor of the immunoglobulin superfamily. *J. Clin. Invest.* 114: 1741–1751.

Blondeau N, Widmann C, Lazdunski M, Heurteaux C. **2001**. Activation of the nuclear factor-kappaB is a key event in brain tolerance. *J. Neurosci.* 21: 4668–4677.

Bonaiuto C, McDonald PP, Rossi F, Cassatella MA. **1997**. Activation of nuclear factor-kappa B by beta-amyloid peptides and interferon-gamma in murine microglia. *J. Neuroimmunol.* 77: 51–56.

Bonizzi G, Karin M. **2004**. The two NF-kappaB activation pathways and their role in innate and adaptive immunity. *Trends Immunol.* 25: 280–288.

Bourke E, Kennedy EJ, Moynagh PN. **2000**. Loss of I-kappa B-beta is associated with prolonged NF-kappa B activity in human glial cells. *J. Biol. Chem.* 275: 39996–40002.

Brandt JA, Churchill L, Rehman A, Ellis G, Memet S, Israel A, Krueger JM. **2004**. Sleep deprivation increases the activation of nuclear factor kappa B in lateral hypothalamic cells. *Brain Res.* 1004: 91–97.

Braun S, Liebetrau W, Berning B, Behl C. **2000**. Dexamethasone-enhanced sensitivity of mouse hippocampal HT22 cells for oxidative stress is associated with the suppression of nuclear factor-kappaB. *Neurosci. Lett.* 295: 101–104.

Brown DR, Nicholas RS, Canevari L. **2002**. Lack of prion protein expression results in a neuronal phenotype sensitive to stress. *J. Neurosci. Res.* 67: 211–224.

Bruce AJ, Boling W, Kindy MS, Peschon J, Kraemer PJ, Carpenter MK, Holtsberg FW, Mattson MP. **1996**. Altered neuronal and microglial responses to excitotoxic and ischemic brain injury in mice lacking TNF receptors. *Nat. Med.* 2: 788–794.

Burke MA, Bothwell M. **2003**. p75 neurotrophin receptor mediates neurotrophin activation of NF-kappa B and induction of iNOS expression in P19 neurons. *J. Neurobiol.* 55: 191–203.

Camandola S, Poli G, Mattson MP. **2000**. The lipid peroxidation product 4-hydroxy-2,3-nonenal inhibits constitutive and inducible activity of nuclear factor kappa B in neurons. *Brain Res. Mol. Brain Res.* 85: 53–60.

Cantera R, Roos E, Engstrom Y. **1999**. Dif and cactus are colocalized in the larval nervous system of *Drosophila melanogaster*. *J. Neurobiol.* 38: 16–26.

Carlson NG, Wieggel WA, Chen J, Bacchi A, Rogers SW, Gahring LC. **1999**. Inflammatory cytokines IL-1 alpha, IL-1 beta, IL-6, and TNF-alpha impart neuroprotection to an excitotoxin through distinct pathways. *J. Immunol.* 163: 3963–3968.

Carter BD, Kaltschmidt C, Kaltschmidt B, Offenhauser N, Bohm-Matthaei R, Baeuerle PA, Barde YA. **1996**. Selective activation of NF-κB by nerve growth factor through the neurotrophin receptor p75. *Science* 272: 542–545.

Cauley K, Verma IM. **1994**. Kappa B enhancer-binding complexes that do not contain NF-kappa B are developmentally regulated in mammalian brain. *Proc. Natl. Acad. Sci. USA* 91: 390–394.

Charalampopoulos I, Tsatsanis C, Dermitzaki E, Alexaki VI, Castanas E, Margioris AN, Gravanis A. **2004**. Dehydroepiandrosterone and allopregnanolone protect sympathoadrenal medulla cells against apoptosis via antiapoptotic Bcl–2 proteins. *Proc. Natl. Acad. Sci. USA* 101: 8209–8214.

Chen G, Cao P, Goeddel DV. **2002**. TNF-induced recruitment and activation of the IKK complex require Cdc37 and Hsp90. *Mol. Cell* 9: 401–410.

Cheng B, Christakos S, Mattson MP. **1994**. Tumor necrosis factors protect neurons against metabolic-excitotoxic insults and promote maintenance of calcium homeostasis. *Neuron* 12: 139–153.

Ciechanover A, Schwartz AL. **2004**. The ubiquitin system: pathogenesis of human diseases and drug targeting. *Biochim. Biophys. Acta* 1695: 3–17.

Colasanti M, Persichini T. **2000**. Nitric oxide: an inhibitor of NF-kappaB/Rel system in glial cells. *Brain Res. Bull.* 52: 155–161.

Collart MA, Baeuerle P, Vassalli P. **1990**. Regulation of tumor necrosis factor alpha transcription in macrophages: involvement of four kappa B-like motifs and of constitutive and inducible forms of NF-kappa B. *Mol. Cell. Biol.* 10: 1498–1506.

Culmsee C, Siewe J, Junker V, Retiounskaia M, Schwarz S, Camandola S, El-Metainy S, Behnke H, Mattson MP, Krieglstein J. **2003**. Reciprocal inhibition of p53 and nuclear factor-kappaB transcriptional activities determines cell survival or death in neurons. *J. Neurosci.* 23: 8586–8595.

Daily D, Vlamis-Gardikas A, Offen D, Mittelman L, Melamed E, Holmgren A, Barzilai A. **2001**. Glutaredoxin protects cerebellar granule neurons from dopamine-induced apoptosis by dual activation of the ras-phosphoinositide 3–kinase and jun n-terminal kinase pathways. *J. Biol. Chem.* 276: 21618–21626.

Danial NN, Korsmeyer SJ. **2004**. Cell death: critical control points. *Cell* 116: 205–219.

de Erausquin GA, Hyrc K, Dorsey DA, Mamah D, Dokucu M, Masco DH, Walton T, Dikranian K, Soriano M, Garcia Verdugo JM, Goldberg MP, Dugan LL. **2003**. Nuclear translocation of nuclear transcription factor-kappa B by alpha-amino–3–hydroxy–5–methyl–4–isoxazolepropionic acid receptors leads to transcription of p53 and cell death in dopaminergic neurons. *Mol. Pharmacol.* 63: 784–790.

Delhase M, Hayakawa M, Chen Y, Karin M. **1999**. Positive and negative regulation of IkappaB kinase activity through IKKbeta subunit phosphorylation. *Science* 284: 309–313.

Digicaylioglu M, Lipton SA. **2001**. Erythropoietin-mediated neuroprotection involves cross-talk between Jak2 and NF-kappaB signalling cascades. *Nature* 412: 641–647.

Drew PD, Lonergan M, Goldstein ME, Lampson LA, Ozato K, McFarlin DE. **1993**. Regulation of MHC class I and beta 2–microglobulin gene expression in human neuronal cells. Factor binding to conserved cis-acting regulatory sequences correlates with expression of the genes. *J. Immunol.* 150: 3300–3310.

Ducut Sigala JL, Bottero V, Young DB, Shevchenko A, Mercurio F, Verma IM. **2004**. Activation of transcription factor NF-kappaB requires ELKS, an IkappaB kinase regulatory subunit. *Science* 304: 1963–1967.

Ehlers MD. **2003**. Activity level controls postsynaptic composition and signaling via the ubiquitin-proteasome system. *Nat. Neurosci.* 6: 231–242.

Ehrlich LC, Hu S, Peterson PK, Chao CC. **1998**. IL–10 down-regulates human microglial IL–8 by inhibition of NF-kappaB activation. *Neuroreport* 9: 1723–1726.

Fagerlund R, Kinnunen L, Kohler M, Julkunen I, Melen K. **2005**. NF-kappa B is transported into the nucleus by importin alpha 3 and importin alpha 4. *J. Biol. Chem.* 280: 15942–51.

Fernyhough P, Smith DR, Schapansky J, Van Der Ploeg R, Gardiner NJ, Tweed CW, Kontos A, Freeman L, Purves-Tyson TD, Glazner GW. **2005**. Activation of nuclear factor-κB via endogenous tumor necrosis factor α regulates survival of axotomized adult sensory neurons. *J. Neurosci.* 25: 1682–1690.

Ferrari D, Wesselborg S, Bauer M, Schulze OK. **1997**. Extracellular ATP activates transcription factor NF-kappaB through the P2Z purinoreceptor by selectively targeting NF-kappaB p65. *J. Cell Biol.* 139: 1635–1643.

Ferrer I, Marti E, Lopez E, Tortosa A. **1998**. NF-kB immunoreactivity is observed in association with beta A4 diffuse plaques in patients with Alzheimer's disease. *Neuropathol. Appl. Neurobiol.* 24: 271–277.

Freudenthal R, Locatelli F, Hermitte G, Maldonado H, Lafourcade C, Delorenzi A, Romano A. **1998**. Kappa-B like DNA-binding activity is enhanced after spaced training that induces long-term memory in the crab Chasmagnathus. *Neurosci. Lett.* 242: 143–146.

Fridmacher V, Kaltschmidt B, Goudeau B, Ndiaye D, Rossi FM, Pfeiffer J, Kaltschmidt C, Israel A, Memet S. **2003**. Forebrain-specific neuronal inhibition of nuclear factor-kappaB activity leads to loss of neuroprotection. *J. Neurosci.* 23: 9403–9408.

Ginis I, Hallenbeck JM, Liu J, Spatz M, Jaiswal R, Shohami E. **2000**. Tumor necrosis factor and reactive oxygen species cooperative cytotoxicity is mediated via inhibition of NF-kappaB. *Mol. Med.* 6: 1028–1041.

Glazner GW, Camandola S, Mattson MP. **2000**. Nuclear factor-kappaB mediates the cell survival-promoting action of activity-dependent neurotrophic factor peptide–9.

J. Neurochem. 75: 101–108.

Grilli M, Goffi F, Memo M, Spano P. **1996**. Interleukin–1beta and glutamate activate the NF-kappaB/Rel binding site from the regulatory region of the amyloid precursor protein gene in primary neuronal cultures. *J. Biol. Chem.* 271: 15002–15007.

Guerrini L, Blasi F, Denis DS. **1995**. Synaptic activation of NF-κB by glutamate in cerebellar granule neurons in vitro. *Proc. Natl. Acad. Sci. USA* 92: 9077–9081.

Guerrini L, Molteni A, Wirth T, Kistler B, Blasi F. **1997**. Glutamate-dependent activation of NF-kappaB during mouse cerebellum development. *J. Neurosci.* 17: 6057–6063.

Gutierrez H, Hale VA, Dolcet X, Davies A. **2005**. NF-κB signalling regulates the growth of neural processes in the developing PNS and CNS. *Development* 132: 1713–1726.

Halvorsen EM, Dennis J, Keeney P, Sturgill TW, Tuttle JB, Bennett JB, Jr. **2002**. Methylpyridinium (MPP(+))– and nerve growth factor-induced changes in pro- and anti-apoptotic signaling pathways in SH-SY5Y neuroblastoma cells. *Brain Res.* 952: 98–110.

Han Y, He T, Huang DR, Pardo CA, Ransohoff RM. **2001**. TNF-alpha mediates SDF–1 alpha-induced NF-kappa B activation and cytotoxic effects in primary astrocytes. *J. Clin. Invest.* 108: 425–435.

Hanz S, Fainzilber M. **2004**. Integration of retrograde axonal and nuclear transport mechanisms in neurons: implications for therapeutics. *Neuroscientist* 10: 404–408.

Hardingham GE, Bading H. **2003**. The Yin and Yang of NMDA receptor signalling. *Trends Neurosci.* 26: 81–89.

Hayakawa M, Miyashita H, Sakamoto I, Kitagawa M, Tanaka H, Yasuda H, Karin M, Kikugawa K. **2003**. Evidence that reactive oxygen species do not mediate NF-kappaB activation. *EMBO J.* 22: 3356–3366.

Heck S, Lezoualc'h F, Engert S, Behl C. **1999**. Insulin-like growth factor–1–mediated neuroprotection against oxidative stress is associated with activation of nuclear factor kappaB. *J. Biol. Chem.* 274: 9828–9835.

Heyninck K, Beyaert R. **2005**. A20 inhibits NF-kappaB activation by dual ubiquitin-editing functions. *Trends Biochem. Sci.* 30: 1–4.

Hu S, Chao CC, Ehrlich LC, Sheng WS, Sutton RL, Rockswold GL, Peterson PK. **1999**. Inhibition of microglial cell RANTES production by IL–10 and TGF-beta. *J. Leukoc. Biol.* 65: 815–821.

Hunot S, Brugg B, Ricard D, Michel PP, Muriel MP, Ruberg M, Faucheux BA, Agid Y, Hirsch EC. **1997**. Nuclear translocation of NF-kappaB is increased in dopaminergic neurons of patients with parkinson disease. *Proc. Natl. Acad. Sci. USA* 94: 7531–7536.

Huxford T, Huang DB, Malek S, Ghosh G. **1998**. The crystal structure of the IkappaBalpha/NF-kappaB complex reveals mechanisms of NF-kappaB inactivation. *Cell* 95: 759–770.

Jacobs MD, Harrison SC. **1998**. Structure of an IkappaBalpha/NF-kappaB complex. *Cell* 95: 749–758.

Jin KL, Mao XO, Greenberg DA. **2000**. Vascular endothelial growth factor rescues HN33 neural cells from death induced by serum withdrawal. *J. Mol. Neurosci.* 14: 197–203.

John GR, Simpson JE, Woodroofe MN, Lee SC, Brosnan CF. **2001**. Extracellular nucleotides differentially regulate interleukin–1beta signaling in primary human astrocytes: implications for inflammatory gene expression. *J. Neurosci.* 21: 4134–4142.

Juttler E, Potrovita I, Tarabin V, Prinz S, Dong-Si T, Fink G, Schwaninger M. **2004**. The cannabinoid dexanabinol is an inhibitor of the nuclear factor-kappa B (NF-kappa B). *Neuropharmacology* 47: 580–592.

Kaltschmidt B, Kaltschmidt C. **2001**. DNA array analysis of the developing rat cerebellum: transforming growth factor-beta2 inhibits constitutively activated NF-kappaB in granule neurons. *Mech. Dev.* 101: 11–19.

Kaltschmidt C, Kaltschmidt B, Baeuerle PA. **1993**. Brain synapses contain inducible forms of the transcription factor NF-κB. *Mech. Dev.* 43: 135–147.

Kaltschmidt C, Kaltschmidt B, Neumann H, Wekerle H, Baeuerle PA. **1994**. Constitutive NF-κB activity in neurons. *Mol. Cell. Biol.* 14: 3981–3992.

Kaltschmidt C, Kaltschmidt B, Henkel T, Stockinger H, Baeuerle PA. **1995a**. Selective recognition of the activated form of transcription factor NF-kappa B by a monoclonal antibody. *Biol. Chem. Hoppe Seyler* 376: 9–16.

Kaltschmidt C, Kaltschmidt B, Baeuerle PA. **1995b**. Stimulation of ionotropic glutamate receptors activates transcription factor NF-𝜘B in primary neurons. *Proc. Natl. Acad. Sci. USA* 92: 9618–9622.

Kaltschmidt B, Uherek M, Volk B, Baeuerle PA, Kaltschmidt C. **1997**. Transcription factor NF-𝜘B is activated in primary neurons by amyloid β peptides and in neurons surrounding early plaques from patients with Alzheimer disease. *Proc. Natl. Acad. Sci. USA* 94: 2642–2647.

Kaltschmidt B, Uherek M, Wellmann H, Volk B, Kaltschmidt C. **1999a**. Inhibition of NF-𝜘B potentiates amyloid-β-mediated neuronal apoptosis. *Proc. Natl. Acad. Sci. USA* 96: 9409–9414.

Kaltschmidt B, Uherek M, Wellmann H, Volk B, Kaltschmidt C. **1999b**. Inhibition of NF-kappaB potentiates amyloid beta-mediated neuronal apoptosis. *Proc. Natl. Acad. Sci. USA* 96: 9409–9414.

Kaltschmidt B, Deller T, Frotscher M, Kaltschmidt C. **2000a**. Ultrastructural localization of activated NF-kappaB in granule cells of the rat fascia dentata. *Neuroreport* 11: 839–844.

Kaltschmidt B, Kaltschmidt C, Hofmann TG, Hehner SP, Droge W, Schmitz ML. **2000b**. The pro- or anti-apoptotic function of NF-kappaB is determined by the nature of the apoptotic stimulus. *Eur. J. Biochem.* 267: 3828–3835.

Kaltschmidt B, Heinrich M, Kaltschmidt C. **2002**. Stimulus-dependent activation of NF-kappaB specifies apoptosis or neuroprotection in cerebellar granule cells. *Neuromol. Med.* 2: 299–309.

Karin M, Ben-Neriah Y. **2000**. Phosphorylation meets ubiquitination: the control of NF-(kappa)B activity. *Annu. Rev. Immunol.* 18: 621–663.

Kassed CA, Butler TL, Navidomskis MT, Gordon MN, Morgan D, Pennypacker KR. **2003**. Mice expressing human mutant presenilin–1 exhibit decreased activation of NF-kappaB p50 in hippocampal neurons after injury. *Brain Res. Mol. Brain Res.* 110: 152–157.

Kemler I, Fontana A. **1999**. Role of IkappaBalpha and IkappaBbeta in the biphasic nuclear translocation of NF-kappaB in TNFalpha-stimulated astrocytes and in neuroblastoma cells. *Glia* 26: 212–220.

Kolesnick R, Golde DW. **1994**. The sphingomyelin pathway in tumor necrosis factor and interleukin–1 signaling. *Cell* 77: 325–328.

Körner M, Rattner A, Mauxion F, Sen R, Citri Y. **1989**. A brain-specific transcription activator. *Neuron* 3: 563–572.

Kovalenko A, Chable-Bessia C, Cantarella G, Israel A, Wallach D, Courtois G. **2003**. The tumour suppressor CYLD negatively regulates NF-kappaB signalling by deubiquitination. *Nature* 424: 801–805.

Lee SJ, Zhou T, Choi C, Wang Z, Benveniste EN. **2000**. Differential regulation and function of Fas expression on glial cells. *J. Immunol.* 164: 1277–1285.

Lezoualc'h F, Sparapani M, Behl C. **1998**. N-acetyl-serotonin (normelatonin) and melatonin protect neurons against oxidative challenges and suppress the activity of the transcription factor NF-kappaB. *J. Pineal Res.* 24: 168–178.

Lezoualc'h F, Engert S, Berning B, Behl C. **2000**. Corticotropin-releasing hormone-mediated neuroprotection against oxidative stress is associated with the increased release of non-amyloidogenic amyloid beta precursor protein and with the suppression of nuclear factor-kappaB. *Mol. Endocrinol.* 14: 147–159.

Lilienbaum A, Israel A. **2003**. From calcium to NF-kappa B signaling pathways in neurons. *Mol. Cell. Biol.* 23: 2680–2698.

Lin L, DeMartino GN, Greene WC. **1998**. Cotranslational biogenesis of NF-kappaB p50 by the 26S proteasome. *Cell* 92: 819–828.

Lipsky RH, Xu K, Zhu D, Kelly C, Terhakopian A, Novelli A, Marini AM. **2001**. Nuclear factor kappaB is a critical determinant in N-methyl-D-aspartate receptor-mediated neuroprotection. *J. Neurochem.* 78: 254–264.

Malek S, Huxford T, Ghosh G. **1998**. Ikappa Balpha functions through direct contacts with the nuclear localization signals and the DNA binding sequences of NF-kappaB. *J. Biol. Chem.* 273: 25427–25435.

Marchetti L, Klein M, Schlett K, Pfizenmaier K, Eisel UL. **2004**. Tumor necrosis factor (TNF)-mediated neuroprotection against glutamate-induced excitotoxicity is enhanced by N-methyl-D-aspartate receptor activation. Essential role of a TNF receptor 2–mediated phosphatidylinositol 3–kinase-dependent NF-kappa B pathway. *J. Biol. Chem.* 279: 32869–32881.

Massa PT. **1993**. Specific suppression of major histocompatibility complex class I and class II genes in astrocytes by brain-enriched gangliosides. *J. Exp. Med.* 178: 1357–1363.

Mattson MP. **2004**. Pathways towards and away from Alzheimer's disease. *Nature* 430: 631–639.

Mattson MP, Goodman Y, Luo H, Fu W, Furukawa K. **1997**. Activation of NF-kappaB protects hippocampal neurons against oxidative stress-induced apoptosis: evidence for induction of manganese superoxide dismutase and suppression of peroxynitrite production and protein tyrosine nitration. *J. Neurosci. Res.* 49: 681–697.

Meberg PJ, Kinney WR, Valcourt EG, Routtenberg A. **1996**. Gene expression of the transcription factor NF-κB in hippocampus: regulation by synaptic activity. *Mol. Brain Res.* 38: 179–190.

Meffert MK, Baltimore D. **2005**. Physiological functions for brain NF-kappaB. *Trends Neurosci.* 28: 37–43.

Meffert MK, Chang JM, Wiltgen BJ, Fanselow MS, Baltimore D. **2003**. NF-kappaB functions in synaptic signaling and behavior. *Nat. Neurosci.* 6: 1072–8.

Micheau O, Tschopp J. **2003**. Induction of TNF receptor I-mediated apoptosis via two sequential signaling complexes. *Cell* 114: 181–190.

Middleton G, Hamanoue M, Enokido Y, Wyatt S, Pennica D, Jaffray E, Hay RT, Davies AM. **2000**. Cytokine-induced nuclear factor kappa B activation promotes the survival of developing neurons. *J. Cell Biol.* 148: 325–332.

Moerman AM, Mao X, Lucas MM, Barger SW. **1999**. Characterization of a neuronal kappaB-binding factor distinct from NF-kappaB. *Brain Res. Mol. Brain Res.* 67: 303–315.

Nabel G, Baltimore D. 1987. An inducible transcription factor activates expression of human immunodeficiency virus in T cells. *Nature* 326: 711–713.

Nakajima K, Kikuchi Y, Ikoma E, Honda S, Ishikawa M, Liu Y, Kohsaka S. **1998**. Neurotrophins regulate the function of cultured microglia. *Glia* 24: 272–289.

Nickols JC, Valentine W, Kanwal S, Carter BD. **2003**. Activation of the transcription factor NF-kappaB in Schwann cells is required for peripheral myelin formation. *Nat. Neurosci.* 6: 161–167.

Ninan I, Arancio O. **2004**. Presynaptic CaMKII is necessary for synaptic plasticity in cultured hippocampal neurons. *Neuron* 42: 129–141.

Nolan GP, Baltimore D. **1992**. The inhibitory ankyrin and activator Rel proteins. *Curr. Opin. Genet. Dev.* 2: 211–220.

Nonaka M, Chen XH, Pierce JE, Leoni MJ, McIntosh TK, Wolf JA, Smith DH. **1999**. Prolonged activation of NF-kappaB following traumatic brain injury in rats. *J. Neurotrauma* 16: 1023–1034.

O'Neill LAJ, Kaltschmidt C. **1997**. NF-κB: a crucial transcription factor for glial and neuronal cell function. *Trends Neurosci.* 20: 252–258.

Ouyang Y, Kantor D, Harris KM, Schuman EM, Kennedy MB. **1997**. Visualization of the distribution of autophosphorylated calcium/calmodulin-dependent protein kinase II after tetanic stimulation in the CA1 area of the hippocampus. *J. Neurosci.* 17: 5416–5427.

Pistritto G, Franzese O, Pozzoli G, Mancuso C, Tringali G, Preziosi P, Navarra P. **1999**. Bacterial lipopolysaccharide increases prostaglandin production by rat astrocytes via inducible cyclo-oxygenase: evidence for the involvement of nuclear factor kappaB. *Biochem. Biophys. Res. Commun.* 263: 570–574.

Pizzi M, Goffi F, Boroni F, Benarese M, Perkins SE, Liou HC, Spano P. **2002**. Opposing roles for NF-kappa B/Rel factors p65 and c-Rel in the modulation of neuron survival elicited by glutamate and interleukin-1beta. *J. Biol. Chem.* 277: 20717–20723.

Pousset F, Dantzer R, Kelley KW, Parnet P.

2000. Interleukin–1 signaling in mouse astrocytes involves Akt: a study with interleukin–4 and IL–10. *Eur. Cytokine Netw.* 11: 427–434.

Povelones M, Tran K, Thanos D, Ambron RT. 1997. An NF-kappaB-like transcription factor in axoplasm is rapidly inactivated after nerve injury in *Aplysia*. *J. Neurosci.* 17: 4915–4920.

Pyo H, Joe E, Jung S, Lee SH, Jou I. 1999. Gangliosides activate cultured rat brain microglia. *J. Biol. Chem.* 274: 34584–34589.

Rape M, Jentsch S. 2004. Productive RUPture: activation of transcription factors by proteasomal processing. *Biochim. Biophys. Acta* 1695: 209–213.

Rattner A, Korner M, Walker MD, Citri Y. 1993. NF-kappa B activates the HIV promoter in neurons. *EMBO J.* 12: 4261–4267.

Ravati A, Ahlemeyer B, Becker A, Klumpp S, Krieglstein J. 2001. Preconditioning-induced neuroprotection is mediated by reactive oxygen species and activation of the transcription factor nuclear factor-kappaB. *J. Neurochem.* 78: 909–919.

Rossler OG, Bauer I, Chung HY, Thiel G. 2004. Glutamate-induced cell death of immortalized murine hippocampal neurons: neuroprotective activity of heme oxygenase–1, heat shock protein 70, and sodium selenite. *Neurosci. Lett.* 362: 253–257.

Routtenberg A. 2000. It's about time. Memory Consolidation. PE Gold, WT Greenough (Eds), American Psychological Association, 1st edition, Washington, DC.

Saccani S, Marazzi I, Beg AA, Natoli G. 2004. Degradation of promoter-bound p65/RelA is essential for the prompt termination of the nuclear factor kappaB response. *J. Exp. Med.* 200: 107–113.

Savaskan NE, Brauer AU, Kuhbacher M, Eyupoglu IY, Kyriakopoulos A, Ninnemann O, Behne D, Nitsch R. 2003. Selenium deficiency increases susceptibility to glutamate-induced excitotoxicity. *FASEB J.* 17: 112–114.

Schmidt-Ullrich R, Memet S, Lilienbaum A, Feuillard J, Raphael M, Israel A. 1996. NF-κB activity in transgenic mice: developmental regulation and tissue specificity. *Development* 122: 2117–2128.

Schmied R, Ambron RT. 1997. A nuclear localization signal targets proteins to the retrograde transport system, thereby evading uptake into organelles in *Aplysia* axons. *J. Neurobiol.* 33: 151–160.

Schmitz ML, Mattioli I, Buss H, Kracht M. 2004. NF-kappaB: a multifaceted transcription factor regulated at several levels. *Chembiochem* 5: 1348–1358.

Schneider A, Martin-Villalba A, Weih F, Vogel J, Wirth T, Schwaninger M. 1999. NF-kappaB is activated and promotes cell death in focal cerebral ischemia. *Nat. Med.* 5: 554–559.

Scholzke MN, Potrovita I, Subramaniam S, Prinz S, Schwaninger M. 2003. Glutamate activates NF-kappaB through calpain in neurons. *Eur. J. Neurosci.* 18: 3305–3310.

Schwamborn J, Lindecke A, Elvers M, Horejschi V, Kerick M, Rafigh M, Pfeiffer J, Prullage M, Kaltschmidt B, Kaltschmidt C. 2003. Microarray analysis of tumor necrosis factor alpha induced gene expression in U373 human glioblastoma cells. *BMC Genomics* 4: 46.

Schwaninger M, Sallmann S, Petersen N, Schneider A, Prinz S, Libermann TA, Spranger M. 1999. Bradykinin induces interleukin–6 expression in astrocytes through activation of nuclear factor-kappaB. *J. Neurochem.* 73: 1461–1466.

Sen R, Baltimore D. 1986. Inducibility of kappa immunoglobulin enhancer-binding protein Nf-kappa B by a posttranslational mechanism. *Cell* 47: 921–928.

Sharma SK, Carlson EC, Ebadi M. 2003. Neuroprotective actions of selegiline in inhibiting 1–methyl, 4–phenyl, pyridinium ion (MPP+)-induced apoptosis in SK-N-SH neurons. *J. Neurocytol.* 32: 329–343.

Sole C, Dolcet X, Segura MF, Gutierrez H, Diaz-Meco MT, Gozzelino R, Sanchis D, Bayascas JR, Gallego C, Moscat J, Davies AM, Comella JX. 2004. The death receptor antagonist FAIM promotes neurite outgrowth by a mechanism that depends on ERK and NF-kappa B signaling. *J. Cell Biol.* 167: 479–492.

Sparacio SM, Zhang Y, Vilcek J, Benveniste EN. 1992. Cytokine regulation of interleukin–6 gene expression in astrocytes involves activation of an NF-kappa B-like nuclear protein. *J. Neuroimmunol.* 39: 231–242.

Speese SD, Trotta N, Rodesch CK, Aravamudan B, Broadie K. **2003**. The ubiquitin proteasome system acutely regulates presynaptic protein turnover and synaptic efficacy. *Curr. Biol.* 13: 899–910.

Stephenson D, Yin T, Smalstig EB, Hsu MA, Panetta J, Little S, Clemens J. **2000**. Transcription factor nuclear factor-kappa B is activated in neurons after focal cerebral ischemia. *J. Cereb. Blood Flow Metab.* 20: 592–603.

Sulejczak D, Skup M. **2000**. Axoplasmic localisation of the NF kappa B p65 subunit in the rat brain. *Acta Neurobiol. Exp.* 60: 217.

Suzuki T, Mitake S, Okumura-Noji K, Yang JP, Fujii T, Okamoto T. **1997**. Presence of NF-κB-like and IκB-like immunoreactivities in postsynaptic densities. *Neuroreport* 8: 2931–2935.

Tanaka S, Takehashi M, Matoh N, Iida S, Suzuki T, Futaki S, Hamada H, Masliah E, Sugiura Y, Ueda K. **2002**. Generation of reactive oxygen species and activation of NF-kappaB by non-Abeta component of Alzheimer's disease amyloid. *J. Neurochem.* 82: 305–315.

Tegethoff S, Behlke J, Scheidereit C. **2003**. Tetrameric oligomerization of IkappaB kinase gamma (IKKgamma) is obligatory for IKK complex activity and NF-kappaB activation. *Mol. Cell. Biol.* 23: 2029–2041.

Togashi H, Sasaki M, Frohman E, Taira E, Ratan RR, Dawson TM, Dawson VL. **1997**. Neuronal (type I) nitric oxide synthase regulates nuclear factor kappaB activity and immunologic (type II) nitric oxide synthase expression. *Proc. Natl. Acad. Sci. USA* 94: 2676–2680.

Trompouki E, Hatzivassiliou E, Tsichritzis T, Farmer H, Ashworth A, Mosialos G. **2003**. CYLD is a deubiquitinating enzyme that negatively regulates NF-kappaB activation by TNFR family members. *Nature* 424: 793–796.

Vollgraf U, Wegner M, Richter-Landsberg C. **1999**. Activation of AP-1 and nuclear factor-kappaB transcription factors is involved in hydrogen peroxide-induced apoptotic cell death of oligodendrocytes. *J. Neurochem.* 73: 2501–2509.

Wajant H, Henkler F, Scheurich P. **2001**. The TNF-receptor-associated factor family: scaffold molecules for cytokine receptors, kinases and their regulators. *Cell Signal.* 13: 389–400.

Wang MJ, Lin WW, Chen HL, Chang YH, Ou HC, Kuo JS, Hong JS, Jeng KC. **2002**. Silymarin protects dopaminergic neurons against lipopolysaccharide-induced neurotoxicity by inhibiting microglia activation. *Eur. J. Neurosci.* 16: 2103–2112.

Weil R, Schwamborn K, Alcover A, Bessia C, Di Bartolo V, Israel A. **2003**. Induction of the NF-kappaB cascade by recruitment of the scaffold molecule NEMO to the T cell receptor. *Immunity* 18: 13–26.

Wellmann H, Kaltschmidt B, Kaltschmidt C. **2001**. Retrograde transport of transcription factor NF-κB in living neurons. *J. Biol. Chem.* 276: 11821–11829.

Whiteside ST, Israel A. **1997**. I kappa B proteins: structure, function and regulation. *Semin. Cancer Biol.* 8: 75–82.

Wood JN. **1995**. Regulation of NF-kappa B activity in rat dorsal root ganglia and PC12 cells by tumour necrosis factor and nerve growth factor. *Neurosci. Lett.* 192: 41–44.

Yabe T, Wilson D, Schwartz JP. **2001**. NFkappaB activation is required for the neuroprotective effects of pigment epithelium-derived factor (PEDF) on cerebellar granule neurons. *J. Biol. Chem.* 276: 43313–43319.

Yamazaki S, Muta T, Takeshige K. **2001**. A novel IkappaB protein, IkappaB-zeta, induced by proinflammatory stimuli, negatively regulates nuclear factor-kappaB in the nuclei. *J. Biol. Chem.* 276: 27657–27662.

Yang J, Lin Y, Guo Z, Cheng J, Huang J, Deng L, Liao W, Chen Z, Liu Z, Su B. **2001**. The essential role of MEKK3 in TNF-induced NF-kappaB activation. *Nat. Immunol.* 2: 620–624.

Yang L, Lindholm K, Konishi Y, Li R, Shen Y. **2002**. Target depletion of distinct tumor necrosis factor receptor subtypes reveals hippocampal neuron death and survival through different signal transduction pathways. *J. Neurosci.* 22: 3025–3032.

Yaron A, Hatzubai A, Davis M, Lavon I, Amit S, Manning AM, Andersen JS, Mann M, Mercurio F, Ben-Neriah Y. **1998**. Identification of the receptor component of the IkappaBalpha-ubiquitin ligase. *Nature* 396: 590–594.

Yoon SO, Casaccia-Bonnefil P, Carter B, Chao MV. **1998**. Competitive signaling between TrkA and p75 nerve growth factor receptors determines cell survival. *J. Neurosci.* 18: 3273–3281.

Zelenaia O, Schlag BD, Gochenauer GE, Ganel R, Song W, Beesley JS, Grinspan JB, Rothstein JD, Robinson MB. **2000**. Epidermal growth factor receptor agonists increase expression of glutamate transporter GLT–1 in astrocytes through pathways dependent on phosphatidylinositol 3–kinase and transcription factor NF-kappaB. *Mol. Pharmacol.* 57: 667–678.

Zhu Y, Culmsee C, Klumpp S, Krieglstein J. **2004**. Neuroprotection by transforming growth factor-beta1 involves activation of nuclear factor-kappaB through phosphatidylinositol–3–OH kinase/Akt and mitogen-activated protein kinase-extracellular-signal regulated kinase1,2 signaling pathways. *Neuroscience* 123: 897–906.

17
Calcineurin/NFAT Signaling in Development and Function of the Nervous System

Isabella A. Graef, Gerald R. Crabtree, and Fan Wang

Abstract

The four genes that encode the cytoplasmic subunits of NFAT-transcription complexes (NFATc proteins) are both transcription factors and signaling molecules. Calcium stimuli in neurons lead to the activation of calcineurin (CaN) phosphatase activity and the rapid dephosphorylation of NFATc proteins. Once in the nucleus, these proteins assemble on DNA with nuclear partner proteins (NFATns) and form active transcription complexes. This mechanism allows NFAT complexes to function as coincidence detectors and signal integrators. The specificity of Ca^{2+} signaling at a transcriptional level might arise from combinatorial assembly of diverse NFAT complexes. Rapid export from the nucleus following rephosphorylation by GSK3 insulates NFAT transcription from transient Ca^{2+} fluxes and plays a critical role in the decoding of Ca^{2+} signals. Recent studies have indicated that NFAT signaling and transcriptional control play roles in axon outgrowth, synaptogenesis, memory formation and possibly in laying down common tracks for nerves and vessels during development. Genetic studies in mice and humans indicate that NFAT signaling might play a role in schizophrenia and the developmental defects of Down syndrome.

17.1
Biochemistry of NFAT Signaling

17.1.1
Biochemical Basis of Coincidence Detection and Signal Integration by NFAT Transcription Complexes

The NFAT-signaling pathway was defined by a reverse biochemical approach to understand the sequence of events conveying signals from the cell membrane to the nucleus. This approach resulted in the delineation of the pathway shown in Fig. 17.1. The downstream DNA target sequences were first identified, followed by the biochemical purification of the protein complex (NFAT complex) bound to these

Transcription Factors. Edited by Gerald Thiel
Copyright © 2006 WILEY-VCH Verlag GmbH & Co. KGaA, Weinheim
ISBN 3-527-31285-4

target sequences [1–9]. NFATc proteins and their response elements were then used to elucidate many of the remaining steps in the pathway shown in Fig. 17.1. In these studies, the ability of NFAT complexes to activate transcription was found to require the coincidence of a Ca^{2+} signal with a PKC and/or ras signal. The basis of this two-signal requirement was found by biochemical reconstitution experiments to be the assembly of NFAT complexes on the essential regulatory sequences. To detect these NFAT complexes, it is critical to use specialized nuclear extracts that were designed to detect the activity on functional NFAT sites. This procedure is described in detail at the website: Crablab.stanford.edu.index.html. The Ca^{2+} signal is required for the translocation of a Ca^{2+}/CaN and cyclosporine A (CsA)-sensitive subunit (NFATc) into the nucleus, while a ras or PKC signal was necessary to induce or activate the nuclear subunit (NFATn) [3, 5]. The cooperation between these two activities was shown by biochemical reconstitution as well as by experiments with pharmacologic stimulators of the pathway [5, 6]. The structural basis of this coincidence detection is the specialized DNA binding domain of the four genes that encode the cytosolic subunits of these complexes, NFATc1–c4. The NFATc proteins show only weak binding to DNA *in vitro* and thus usually require a partner protein (NFATn) to bind tightly to DNA and form a transcriptionally active complex [10, 11]. Hence, combinatorial assembly of NFAT complexes probably plays a critical role in determining the specificity of transcribed target genes.

The cytoplasmic to nuclear translocation of the four proteins (NFATc1–c4), which make up the cytosolic subunits of the NFAT transcriptional complex, requires a sustained increase of intracellular Ca^{2+} levels. The function of the Ca^{2+} release-activated Ca^{2+} (CRAC) channel, which opens in response to depletion of intracellular stores through the IP_3 receptor [12–15] is essential for prolonged Ca^{2+} signaling in many cell types. Somatic cell mutants defective for the regulation of the CRAC channel generate brief pulses of Ca^{2+} that are not sustained. In these somatic cell mutants, NFATc family members are not maintained in the nucleus and hence are unable to activate NFAT-dependent transcription of target genes [16]. More recent experiments have indicated that other types of Ca^{2+} channels such as L-type voltage-gated Ca^{2+} channels are also capable of activating NFAT-dependent transcription [17–19].

Initial biochemical reconstitution experiments, using cytosolic and nuclear extracts, characterized the nuclear subunits of NFAT transcription complexes as requiring transcription and being constitutively localized in the nucleus [5]. Later studies indicated that NFATc proteins can form complexes with a rather wide variety of different proteins including AP–1, GATA, MEF2 and others that contribute their DNA-binding affinity to the complex [20–23]. To date, partner proteins whose null phenotypes match the predominant NFATc1–4 phenotypes have not been described. The identification of these proteins will likely be critical to gain a thorough understanding of the versatility and full range of gene regulation by this family of transcription factors.

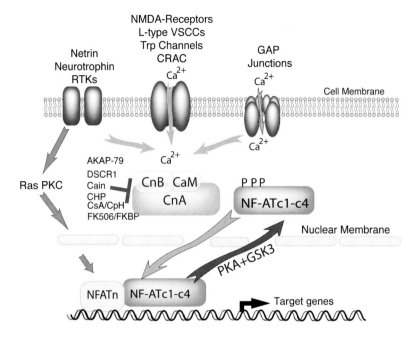

Fig. 17.1 Biochemical aspects of NFAT signaling. NFAT signaling can be activated by receptors that trigger an increase in intracellular Ca^{2+} including neurotrophins, netrins, as well as Ca^{2+} channels such as L-type voltage-sensitive Ca^{2+} channels (VSCC) and NMDA receptors. In addition, genetic studies in mice indicate that gap junctions can activate NFATc translocation and NFAT-dependent transcription. Calcineurin dephosphorylates a number of phosphoserines in the N-terminus of the NFATc1–c4 proteins, exposing nuclear import sequences and leading to nuclear entry. This pathway is inactivated by nuclear kinases such as PKA and GSK3 as well as by a group of proteins that block the activity of calcineurin such as DSCR1. The powerful calcineurin inhibitor, cyclosporine A, binds to cyclophilins to form dominantly acting inhibitory complexes that block calcineurin activity. FK506 binds to FKBP to form dominantly acting inhibitory complexes for calcineurin at nanomolar concentrations. The NFATn proteins are a collection of transcription factors that respond to map kinase signaling as well as to other signaling pathways and bind cooperatively to DNA with NFATc proteins.

17.1.2
The Mechanism of Nuclear Entry of NFATc Proteins.

The defining feature of the NFATc family is the Ca^{2+}/CaN-sensitive translocation domain (Fig. 17.2) that is present in each of the four proteins (Fig. 17.3) [23–25]. This ~300 amino acid domain, located N-terminal to the DNA-binding domain, is encoded by a single exon in all four NFATc proteins [25]. This domain contains two conserved motifs, a serine-rich region (SRR) at its amino-terminus and a repeated motif, the SP repeat, made up of SPXXSPXXSPSSD/ES [26, 27]. The SRRs and the SP repeats are the site of phosphorylation, which make interactions directly with the nuclear localization sequences (NLS) in NFATc proteins, thereby preventing nuclear entry in resting cells. CaN dephosphorylates critical serines in these motifs, thus triggering nuclear entry of NFATc proteins [26,28,29]. Efficient dephosphorylation requires a docking interaction between NFAT and CaN [30–32]. The major docking site for calcineurin is located in the translocation domain and has the consensus sequence PxIxIT (Fig. 17.2). Dephosphorylation of critical phosphoserines in the SRR and SP-repeats by CaN leads to the exposure of the NLS. These studies imply that an allosteric change might alternatively expose and conceal the NLS sequences in NFATc and accompany dephosphorylation of the SRR and the SP repeats by CaN. To date, this suspected allosteric regulatory mechanism has not been documented by structural studies. The N-terminal translocation domain is unstructured in solution, and possibly interacts with the C-terminal domain, indicating that structural studies will be needed on the entire protein to define this switching mechanism [33].

17.1.3
Discrimination of Calcium Signals and the Nuclear Exit of NFATc Proteins

A fine balance between nuclear import and export determines the cytoplasmic-nuclear ratio, and thus the activity of NFATc proteins. Nuclear export appears to depend on: (a) an NES, which is competent for exporting heterologous nuclear proteins to the cytoplasm [34]; and (b) the same group of serines that are essential for nuclear import. The export receptor for phospho-NFATc is most likely Crm1, the target of the small molecule leptomycin B [34–36]. Phosphorylation of these serines in NFATc1/NFATc4 occurs by a two-step mechanism involving sequential phosphorylation by a priming kinase followed by GSK3. When these kinases are active, the cytosolic subunits are retained for only a few minutes in the nucleus and are quickly exported to the cytosol. This mechanism assures that most brief Ca^{2+} signals, such as those that occur during heart muscle contraction or certain rapid neuronal responses, for example, will not activate NFAT-dependent gene transcription. It appears that GSK3 must be inactivated for an effective transcriptional stimulus. Normally, nuclear GSK3 is under the control of AKT and PI–3 kinase signaling [37–41]. In this pathway, which may start with a ras signal, AKT phosphorylates GSK3β on Ser9, thereby inactivating it and allowing accumulation of NFATc family members in the nucleus. Thus, concomitant signals by AKT and Ca^{2+} can lead to continuous nuclear localization of NFATc proteins, and transcription of target genes if an additional NFATn partner protein is present. Perhaps the best evidence for this mechanism of nuclear retention comes from the observation that mice with mutations in NF–1 (a ras-GAP that inactivates ras) show reduced NFATc1 export from the nucleus. This would be consistent with a model in which increased ras activity leads to inhibition of GSK3 (through PI–3 kinase and AKT) and thus to a reduction of NFATc1 export [42].

One issue that has remained unanswered is the nature of the priming kinases for the NFATc family. GSK3 is an unusual kinase in that it requires prior phosphory-

Features of NF-ATc Family Members

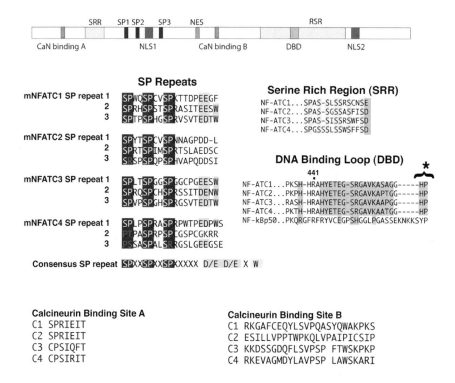

Fig. 17.2 The critical sequences within the NFATc family of proteins that mediate its response to calcineurin and its rapid export from the nucleus. The N-terminal domain of the protein is necessary and sufficient to allow Ca^{2+}/calcineurin-dependent nuclear import and GSK3–dependent export and functions as a potent dominant negative of NFAT function [8]. The SP-repeat and serine-rich regions [24] contain most of the phosphorylation sites for GSK3 and PKA, which are in turn dephosphorylated by calcineurin. The two calcineurin binding sites that probably account for the dominant negative effects of the N-termini are shown as sites A and B. (This figure also appears with the color plates.)

lation of its substrates by another kinase. In the case of NFATc1 and c4 it is relatively clear that these additional phosphorylations provided by GSK3 are essential for nuclear export [17, 43]. Immunodepletion of GSK3 depletes all of the second export kinase activity for NFATc1 and c4 from nuclear extracts of neurons [17, 43]. In contrast, depletion of protein kinase A (PKA) leads only to partial reduction of the priming kinase activity [43, 44]. Clearly, the critical priming kinase must be nuclear, but its identity has not been determined.

In addition to the critical regulatory nuclear kinases, several cytoplasmic maintenance kinases have also been identified, including JNK, Casein-kinase-II, and others

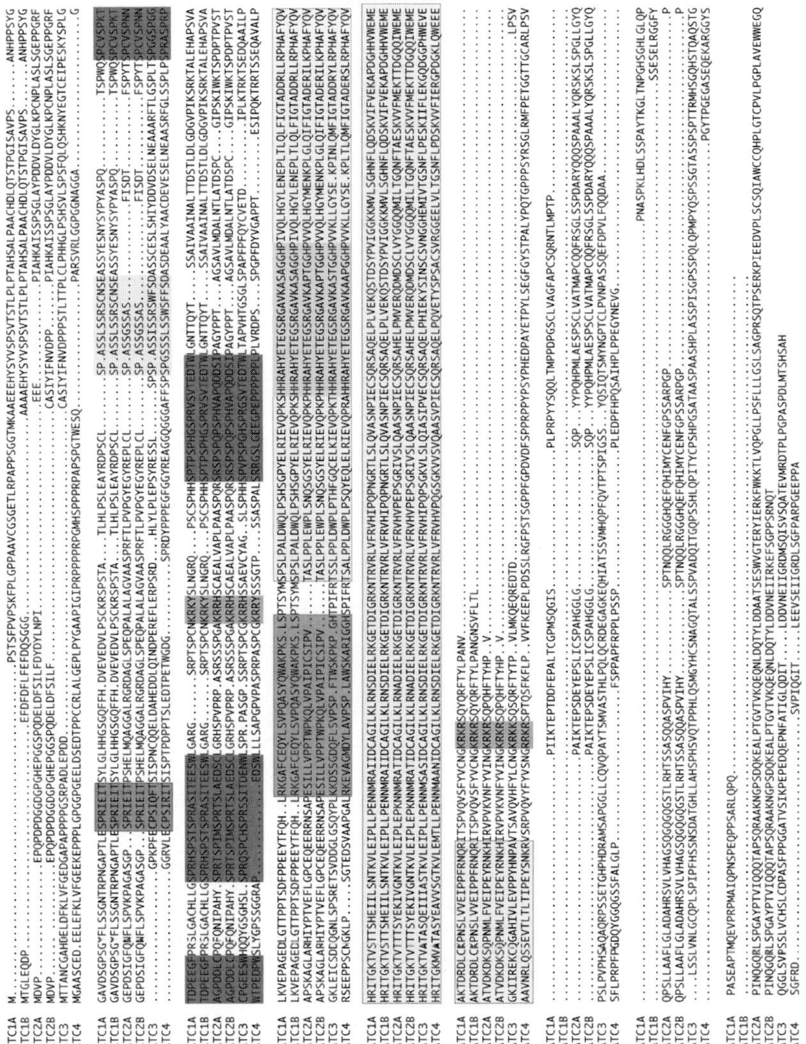

Fig. 17.3 Alignment of the sequences of the four NFATc family members and their splice products. The coloring of the domains in the proteins are taken from Fig. 17.2. (This figure also appears with the color plates.)

[45–48]. These kinases may participate in the discrimination of Ca^{2+} signals by phosphorylating NFATc proteins in the cytoplasm, thus providing constant opposition to brief Ca^{2+} signals that might transiently activate CaN. Consequently, at least three mechanisms appear to be involved in the discrimination of Ca^{2+} signals, thereby insulating transcription of NFAT target genes against Ca^{2+} transients:

- the requirement for a nuclear partner to assemble NFAT transcription complexes;
- the rapid nuclear export of NFATc proteins by GSK3 and a priming kinase; and
- the group of cytoplasmic maintenance kinases that oppose brief Ca^{2+}-dependent activation of CaN.

17.1.4
Combinatorial Assembly of NFAT Transcription Complexes Determines Specificity of Ca^{2+} Responses.

One paradox of Ca^{2+} as a second messenger and signaling molecule is the issue of specificity: how could a molecule so ubiquitous and used for so many purposes play such an important role as the mediator of highly specific transcriptional responses? The answer to this question has been very elusive [49, 50]. In the case of NFAT-dependent transcription, a partial solution to this problem is the very rapid export of these proteins from the nucleus, which insulates transcription from brief Ca^{2+} signals. NFATc proteins have a remarkable ability to sense dynamic changes in intracellular Ca^{2+} and frequencies of Ca^{2+} oscillations in cells [16,51–53]. Another contributor to specificity is probably the requirement for the concomitant activation of a second signaling pathway that inactivates GSK3–dependent export from the nucleus and is needed for the NFATc1–c4 proteins to accumulate to levels sufficient for transcriptional activation. These explanations, however, do not provide an understanding of the more complex question of how very specific patterns of transcription are produced in different cell types.

Much of the transcriptional specificity of Ca^{2+} signaling may be due to the need for different nuclear partners to form NFAT transcription complexes. As mentioned above, NFATc proteins probably do not bind very effectively to DNA *in vivo* by themselves. The structural and biochemical data indicate that NFAT transcriptional complexes function as signal integrators and coincidence detectors. Ca^{2+}/calmodulin-dependent activation of CaN [3] is required in all cell types, while the second signal can be either tissue- or context-specific [22, 23]. One nuclear partner identified at an early stage was AP–1, specifically JunB, sequences of which were obtained from purification using a DNA-binding site [20, 54]. However, recent results have called into question the significance of AP–1 as a general partner. The phenotypes of mice lacking components of AP–1 do not recapitulate the phenotypes of mice lacking NFATc1–c4. These studies and the highly tissue-specific nature of the target genes such as brain-derived neurotrophic factor (BDNF), vascular endothelial growth factor (VEGF) and interleukin (IL)–2 have called attention to the need for a definitive purification of the tissue-specific, cooperating NFATn proteins [3,55,56].

Another layer of specificity of NFAT transcriptional responses to Ca^{2+} signaling is likely imposed by the state of chromatin at the time the signal is received. Much of the genome is inaccessible to signaling pathways. This inaccessibility probably mirrors the inaccessibility of large regions of non-expressed DNA to nucleases in the conventional nuclease sensitivity assays. For many years, the nature of the molecules that determine this inaccessibility have been an enigma. However, recently a neuron-specific ATP-dependent chromatin remodeling complex (nBAF) was discovered that could play a part in at least some of the tissue-specific patterns of accessibility in neurons [57].

17.1.5
Dedication of CaN to NFATc Family Members

One surprising result that has emerged from studies of mice with mutations in the different NFATc proteins is the high degree of similarity to the phenotypes of mice with mutations in the subunits of CaN phosphatase complexes [22,56,58–60]. CaN is a three-subunit enzyme made up of a catalytic subunit encoded by three genes, a regulatory subunit that binds Ca^{2+}, encoded by two genes, and calmodulin. Overexpression studies using a constitutively active form of the catalytic subunit of CaN and studies with the purified enzyme suggested that CaN would dephosphorylate many substrates. Hence, one would expect minor parallels between the phenotypes of CaN-deficient mice and NFATc-deficient mice. However, to date there has been remarkable overlap between these two phenotypes in developmental studies, suggesting that during embryonic development CaN function is in some way dedicated to NFATc1–c4. The basis of this high enzyme-substrate specificity is probably explained by the observation that CaN binds to NFATc proteins remarkably tightly. This was initially observed based on the ability of the N-terminal translocation domain of CaN to function as a dominant negative of calcineurin activity [8]. Two binding sites within the translocation domain have been defined that underlie the interaction between CaN and the NFATc family (see Fig. 17.2) [30,33,61]. These bindings sites are probably sufficiently tight that much of CaN becomes dedicated to NFATc proteins during embryonic neural development when NFATc3 and NFATc4 are highly expressed. In the adult brain, CaN is expressed at very high levels and hence cannot be dedicated to the much less abundant NFATc1–c4. The relative change in abundance of the NFATc protein during neural development probably accounts for a shift in CaN function from transcriptional control in embryos to other aspects of neural function in the adult nervous system.

17.1.6
Evolution of the Genes that Encode the Cytosolic Components, the NFATc Family

We are accustomed to thinking of signaling pathways as ancient and highly conserved. However, NFAT signaling and transcriptional control is perhaps the only membrane-to-nucleus signaling pathway that was created at the dawning of vertebrate life. NFATc proteins with the characteristic Ca^{2+}/CaN-sensitive translocation domain are not present in any invertebrate genome, but are present in the sequenced genomes of all vertebrates. The creation of these genes most likely occurred by the recombination of an exon encoding a Ca^{2+}/CaN-sensitive translocation domain into proximity with exons encoding a rel domain [25]. This exon shuffling apparently occurred about 500 million years ago since all vertebrate genomes contain two to four genes encoding the cytoplasmic NFAT subunits, which contain the vertebrate-specific translocation domain. In contrast, the genomes of all currently sequenced invertebrates including sea urchins, flies, worms, and yeast lack homologues of the NFATc family of transcription factors. Interestingly, the appearance of vertebrates also coincided with the appearance of a group of new receptors and ligands including neurotrophins, neurotrophin receptors, VEGF, the T-cell receptor, the B-cell receptor and others. Signaling by each of these receptors is critically dependent on NFATc family members. These observations suggest that, with the beginning of vertebrate life, calcium signals were redirected to a new transcriptional program. This event might have allowed the new receptors and ligands that also emerged at the origin of vertebrates to initiate organogenesis characteristic of vertebrates. One exception is netrins, which require CaN and NFATc2/c3/c4 to attract commissural axons to the midline [60]. Why should netrins [unc6], which are clearly not a vertebrate invention, need NFAT signaling? The answer may lie in the longer trajectories taken by vertebrate axons. The axons of mice lacking CaN or c2/c3/c4 are much slower in advancing, but do show some measurable rate of advance. Perhaps the NFAT pathway creates a turbo-charged growth cone that is able to transverse the relatively long pathways needed for the uniquely long axonal paths that are features of the vertebrate nervous system?

17.2
Roles of NFAT Signaling in Axonal Outgrowth and Synaptogenesis

The generation of the precise neuronal network laid down during embryonic development requires that axons navigate accurately to their appropriate targets, often traversing long distances before reaching their correct synaptic partner. The proper formation of these connections relies on the neural growth cone to correctly respond to guidance cues and to activate an intricate network of downstream genes. A variety of extracellular cues, including netrins and neurotrophins, stimulate, inhibit, and guide process extension and branching by binding receptors present on the developing neuron [62–64]. The axonal growth cone interprets positive and negative guidance cues and generates signals that determine both the course as well as the rate of axon outgrowth. The growth cone contains a variety of signal transducing receptors that mediate the response to morphogenetic cues and are coupled to second messenger pathways. Several of these second messengers affect the organization of the cytoskeleton, thereby regulating growth cone turning. However, for these guidance cues to be effective in the complex and dynamic field of embryonic morphogenetic signals, the immediate turning events must be coupled with more elaborate responses in the nucleus. Although some of these transcriptional processes may be activated autonomously in neurons simply as a consequence of an early specification event, in other cases their action may be regulated by late environmental signals. Coupling neuronal transcription to extracellular signals allows morphogenetic events such as branching to emerge from the coordination of internal programs of neuronal development with extracellular signals. These considerations raise the question of what sort of mechanisms might coordinate developmental programs with extrinsic signals and how might these signals be conveyed from the cell membrane to the nucleus.

NFATc family members function in the control of neuronal morphogenesis by conveying netrin and neurotrophin signals to the nucleus, leading to the direct activation of genes essential for rapid neurite outgrowth.

Profound defects in sensory axon projections from both cranial ganglia as well as dorsal root ganglia (DRG) were observed in embryos with combined deletions of NFATc2/c3/c4 (Fig. 17.4) [60]. Moreover, these outgrowth defects were phenocopied by pharmacological inhibition of CaN function during early embryonic development *in utero*. While both CaN and NFATc proteins were necessary for neurotrophin-stimulated growth of embryonic sensory neurons, neurotrophin-independent neurite outgrowth was unaffected by lack of CaN/NFAT signaling. The defect of sensory axon outgrowth appeared to be cell-autonomous. Trigeminal ganglia from NFATc triple mutants or wild-type ganglia, when cultured in the presence of CsA and FK506, were severely compromised in their ability to extend axons *in vitro* in response to exogenously added neurotrophins (Fig. 17.4).

Several other axonal projections in the NFAT triple-mutant embryos were also defective. The central branches of spinal sensory neurons from the DRG failed to project longitudinally upon reaching the dorsal spinal cord at the dorsal root entry zone. As a result, the longitudinal tract or dorsal funiculus was absent in triple-mutant embryos.

NFATc triple-mutant embryos also displayed profound disturbances in commissural axon growth [60]. TAG–1–positive axons that emanate from commissural interneurons were much shorter in the NFAT mutant neural tubes, and no TAG–1–positive axons reached the floorplate and crossed the midline in NFAT triple mutants. The extension defects of commissural axons of NFATc triple-mutant mice are similar to defects found in mice mutant in netrin–1 or its receptor, DCC [65, 66]. Inhibition of CaN with FK506 and CsA completely blocked the rapid netrin-induced axon extension from E13 rat dorsal spinal cord explants, while not affecting slow netrin-independent outgrowth. This indicates that the observed inhibitory effect does not represent a general inhibition of outgrowth, but rather inhibition of outgrowth stimulated by netrin/DCC signaling [65, 67]. As had been shown in the case of neurotrophin signaling, netrin is also capable of directly activating NFAT-dependent transcription in cultured cortical neurons [60]. Intracellular Ca^{2+} transients elicited by neurotrophins and netrins have an important role in regulating growth cone motility and axonal growth [68–70], and may provide a link to the activation of CaN/NFAT-dependent transcription. CaN/NFAT signaling thus appears to direct a transcriptional program of axonal outgrowth in response to extracellular cues (Fig. 17.5). Several important questions remain to be addressed including:

- Whether additional axon guidance and outgrowth cues signal through CaN/NFATc?
- How guidance and growth signals from distal neuronal compartments, such as the growth cone, are propagated to the nucleus to activate NFAT-dependent transcription?

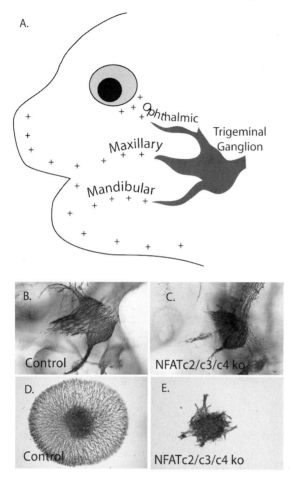

Fig. 17.4 NFAT signaling is essential for axonal outgrowth from the trigeminal ganglia as well as from the dorsal root ganglia (not shown). (A) Schematic representation of the outgrowth of axons from trigeminal ganglia at E10.5. (B,C) Anti-Neurofilament stains of E10.5 control (B) and NFATc2/c3/c4 triple mutant embryos (C). (D,E) Outgrowth of axons from trigeminal explants from (D) control and (E) NFATc2/c3/c4 triple mutant embryos. These studies show that NFAT signaling functions cell-autonomously in neurons, but do not exclude additional roles in the cells surrounding the extending axon (see Fig. 17.6). (Reproduced, with modifications, from [60].)

- What is the identity of the core transcriptional programs that control rapid axon extension in response to these guidance and growth factors?

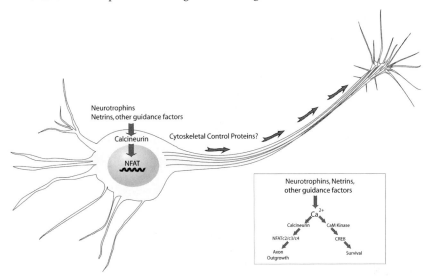

Fig. 17.5 Neurotrophins use calcineurin and NFATc proteins to control the expression of genes that enhance neurite outgrowth in response to guidance signals. Inset: NFAT signaling in sensory neurons appears restricted to a role in neurite and axonal outgrowth and does not appear to be involved in cell survival, which is dependent upon CREB.

The establishment of synapses plays a central role in the creation of functional neuronal circuits. The arrival of an axon at a target cell initiates a series of morphogenic events that lead to the development of a synapse. The establishment of a mature synapse requires the bidirectional exchange of signals between the nerve and its target tissue. Signaling through Ca^{2+}, CaN and NFATc has been shown to be critical in coordinating cell-cell interactions and in shaping the response of the two communicating cells during development [22, 59]. Therefore, it is exciting that recent findings point to a role of CaN/NFATc signaling in presynaptic differentiation [71]. The process by which an axonal growth cone is transformed into a presynaptic terminal is not well understood. Recent studies in zebrafish embryos demonstrated that CaN/NFAT signaling is critical for remodeling the axon tips of olfactory sensory neurons into presynaptic terminals. During synaptogenesis, large, filopodia-rich growth cones are transformed into small presynaptic terminals lacking filopodia. Pharmacologic or genetic inhibition of the CaN/NFAT pathway significantly impaired this remodeling process. Thus, activation of NFAT is probably required for these morphological changes. The challenge for the future will be to understand the integration of CaN/NFAT signaling together with molecules, such as FGF22, Wnt–7a, neuroligin, SynCAM and thrombospondin, which are known to regulate synaptogenesis.

17.3
A Possible Role for NFAT Signaling in Defining Pathways for Both Vessels and Peripheral Nerves

The similarity of the paths followed by peripheral nerves and blood vessels was first noted by Leonardo da Vinci. Later, this observation led to the speculation that vasculogenesis and neurogenesis may use common guidance principles to sculpt the delicate patterning of nerves and vessels to meet the needs of the developing embryo. The arteries and veins of adult mammals follow precise anatomical paths that show relatively little variation between individuals. This stereotyped anatomy suggests that some form of guidance must be provided to the growing vessel. Such guidance cues are likely to involve both positive cues that direct vessels into specific regions, and negative cues that prevent major vessels from developing within structures that require a very specific pattern of vascularization. This is highly reminiscent of how neuronal connections form during embryonic development. Developing axons follow positive and negative guidance cues and extend to their appropriate target regions in a highly stereotyped and directed manner, making very few errors of navigation [64].

Both the nervous and vascular systems are faced with similar tasks during development; they must both navigate precisely within the three-dimensional space of the developing embryo, and both are highly branched networks consisting of separate afferents and efferents. Indeed, they often co-align and follow common anatomic paths that are not simply a matter of mechanical barriers. For example, the phrenic nerve, artery and vein run tightly together over a long path with no obvious physical barriers, suggesting that this coordination may be achieved by crosstalk between the two tissues [72–76].

Recent evidence indicates that the similarities between the two systems might extend also to the molecular level, and that they share not only their anatomic but also their molecular pathways (Fig. 17.6). For example, signaling pathways that previously had only been appreciated for their role in patterning and differentiation of the nervous system have proven critical for the morphogenesis of the vascular system, and vice versa. This list includes the Notch, ephrin-Eph, plexin-semaphorin, Slit-Robo, netrin, neurotrophin and VEGF signaling pathways [77–90]. Most interestingly, several of these receptors utilize Ca^{2+}/CaN/NFATc to communicate their signals to the nucleus [55,60,91,92]. This aspect of CaN/NFATc signaling is even more intriguing in light of the defect seen in patterning of both the vascular and nervous system seen in NFATc-mutant mice.

Mice with null mutations of NFATc3 and NFATc4, as well as mice bearing a mutation in the regulatory subunit of CaN, CnB, die at midgestation due to severe vascular patterning defects. In the mutant embryos, intersomitic vessels form inappropriate branches and invade the somites. Furthermore, excess vessels also penetrate the neural epithelium. These results indicate that signaling through Ca^2/CaN/NFATc3/c4 mediates a negative signal that prevents the aberrant growth of vessels, presumably either by modulating the response of the growing vessel to guidance cues or by controlling the regional expression of vascular guidance cues (Fig. 17.6).

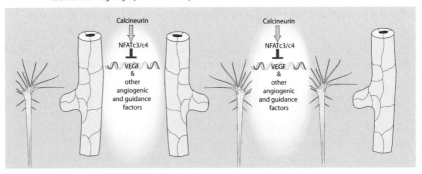

Fig. 17.6 NFAT signaling represses VEGF expression locally to pattern pathways for vessels and nerves. Although NFAT signaling regulates a large number of secreted proteins in the developing embryo, a model is presented with VEGF, which is repressed by calcineurin-NFAT signaling [56, 58] by direct binding of NFAT complexes to essential regulatory regions within this gene. The upstream receptors, which might activate this pathway locally, are not known. An alternative to a localized cytokine or growth factor is the possibility that an NFATn could be a segmental or localized homeodomain protein or other transcription factor. Thus, a localized external signal might not be directly required.

In the mutant embryos, *in-situ* hybridization for VEGF mRNA showed enhanced expression in both somites and the neural tube, which suggested that regionally specific NFAT signaling represses VEGF expression and could thereby modulate the pathway of vessels in certain tissues [58]. On the other hand, in endothelial cells VEGF directly activates the CaN/NFATc pathway [91]. Loss of the normal segmented blood vessel pattern reminiscent of the one seen in the NFATc3/c4– and CnB– mutant mice has been described in mutants of the ephrin-Eph, plexin-semaphorin and netrin pathways [83,84,86, 93–96]. The similarity of the nature of the defect seen in the NFATc mutants and in the mice carrying mutation of the netrin receptor unc5b [86], is particularly exciting when taken together with the observation that the NFATc proteins are downstream effectors of netrin/DCC signaling in the developing nervous system [60]. It is worthy of note, with respect to the role of NFATc signaling in neuronal development, that VEGF has been shown, in addition to its central role in vascular development, to promote neuronal or glial cell survival, proliferation, and axonal outgrowth [96–98].

NFAT signaling is often essential in two communicating cells or tissues, and coordinates the gene expression necessary for, or resulting from, this communication [22]. For example, bidirectional NFAT signaling is essential for the morphogenesis of the heart valves [56]. Therefore, it is stimulating to speculate that the CaN/NFATc pathway could be a critical effector in the bidirectional communication between vessels and nerves, and might contribute to the control of a genetic program which directs the congruent development of vessels and nerves (Fig. 17.6).

17.4
Roles of NFAT Signaling in Later Development: Responses to Spontaneous Activity

Changes in the efficacy of synaptic transmission are important for neuronal function. Activity-dependent modification is thought to be central to both the refinement of neuronal connections in the immature nervous system and memory storage in the mature brain. Many presynaptic and postsynaptic proteins have been implicated in altering synaptic strength, and it has been well established that information processing of signals in the nucleus followed by the activation of gene transcription is required for long-lasting changes of synaptic efficacy.

Although the requirement for CaN in learning and memory is firmly demonstrated, the exact mechanisms underlying its role are not fully understood. CaN can regulate both short- and long-term changes in neuronal plasticity [99–107]. Within neurons, CaN is enriched in dendritic spines and postsynaptic densities in the cell body [108–110]. Because of its high affinity and low dissociation constant for Ca^{2+} and its co-localization with NMDARs, CaN is among the first of the postsynaptic enzymes to be activated after Ca^{2+} influx [111], and is the predominant Ca^{2+}-regulated protein phosphatase in neurons.

NFATc4 is expressed within hippocampal neurons and initiates gene expression in response to synaptic activity. Under basal conditions, NFATc4 is phosphorylated and resides in the cytoplasm. Many Ca^{2+} channels (including L-type, N-type, and P/Q-type Ca^{2+} channels, as well as NMDA-and AMPA-type glutamate receptors) can cause a marked elevation in the bulk intracellular Ca^{2+} levels. However, most of these are ineffective at directly activating NFAT and only Ca^{2+} entry through both *N*-methyl-D-aspartate receptors (NMDARs) and L-type voltage-gated Ca^{2+} channels can trigger translocation of NFATc4 from the cytosol to the nucleus (Fig. 17.7) [17]. The translocation step is dependent upon CaN-mediated dephosphorylation of NFATc4, which causes the unmasking of multiple nuclear localization signals and triggers rapid transport through the nuclear pore complex. Neuronal activity not only triggers nuclear translocation of NFATc4 but it additionally regulates NFATc4–dependent transcription by delaying nuclear export of NFAT through L-type channel-mediated inhibition of GSK3 [17]. The result is prolonged NFATc4 activation and increased NFAT-dependent gene expression. Although L-type channels contribute only a minor component to the synaptic Ca^{2+} currents, they seem to play a major role in coupling synaptic excitation to activation of transcriptional events, that contribute to synaptic plasticity. The privileged roles of the NMDA receptor and the L-type channel in the activation of CaN and NFAT-dependent transcription has not yet been determined, but may be attributable in part to their, subcellular distribution at or near synapses, and the prolonged Ca^{2+} influx that they induce. The latter is required to overcome the rapid export of NFATc proteins from the nucleus.

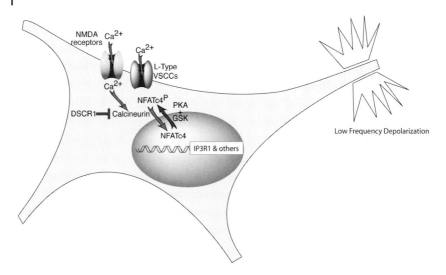

Fig. 17.7 Calcineurin-NFAT signaling in hippocampal neurons responds to L-type channels and NMDA receptor stimulation and regulates genes such as the IP3R1 that could form a positive feedback mechanism. This pathway may be involved in the pathogenesis of schizophrenia or Down syndrome.

Long-lasting, transmission-dependent changes in synaptic strength are thought to form the basis for neuronal changes that contribute to various forms of memory and learning [112]. In the hippocampus, two forms of long-term synaptic plasticity have been demonstrated: long-term potentiation (LTP) and long-term depression (LTD). In the CA1 region of the hippocampus, both LTP and LTD can be induced at the same synapses by different frequencies of stimulation; low-frequency stimulation at 1–5 Hz induces LTD, while high-frequency stimulation at 100 Hz induces LTP. Long-lasting forms of synaptic plasticity and memory require *de novo* gene expression [113, 114]. The Ca^{2+}-activated serine-threonine phosphatase CaN has been shown to be involved in the induction of these two forms of synaptic plasticity in the hippocampus [100–105,107]. Remarkably, NFATc4 translocation in CA1/CA3 hippocampal neurons is more pronounced in response to low-frequency stimuli (5 Hz) as compared to high-frequency stimulation (Fig. 17.7) [17].

The interplay between the regulated nuclear import and export of NFATc4 in hippocampal neurons represents a pathway parallel with others for transcriptional control by Ca^{2+}, including Ca^{2+}-dependent phosphorylation and dephosphorylation of CREB. The participation of multiple pathways may be important for the proper decoding of Ca^{2+} transients that differ in intensity, duration, temporal pattern, or mode of Ca^{2+} entry, allowing these signals to be translated into activation of distinctive sets of genes.

The preferential activation of NFAT-dependent transcription by low-frequency stimulation has also been observed during the communication between motor neurons and myofibers. Signals that are exchanged between a motor neuron and its target muscle initiate the formation and assembly of a highly differentiated presynaptic nerve terminal and the muscle's highly specialized postsynaptic apparatus. The pattern of motor nerve activity determines whether skeletal muscle fibers develop fast- or slow-twitch properties [115]. Recent studies have revealed a key role for CaN signaling in the induction of a program of gene expression associated with the slow-twitch myofibers [116–118]. More specifically, it has been shown that NFAT acts as a nerve activity sensor and controls activity-dependent fiber type specification in skeletal muscle [119]. NFAT activity is induced by electrostimulation with a tonic low-frequency impulse pattern, mimicking the firing pattern of slow motor neurons, but not with a phasic high-frequency pattern typical of fast motor neurons [119].

Neurotrophins (NTs) are important regulators of NFAT-dependent transcription within neurons. NTs were initially described as target-derived trophic factors, and have more recently been found to be regulated by synaptic activity and to, in turn, modulate neuronal excitability, synaptic transmission, and synaptic structure [120–123]. They can thus act as synaptic modulators and link synaptic activity with long-term functional and structural modification of synaptic connections. In particular, BDNF has received much attention for its potential role in activity-dependent synaptic plasticity within the hippocampus. BDNF induces the translocation of NFAT family members from the cytoplasm to the nucleus of cortical neurons. In addition, BDNF is a powerful activator of endogenous NFAT-dependent transcription in both cortical neurons and CA3/CA1 hippocampal neurons [55, 60], and is completely blocked by inhibition of CaN phosphatase. The phosphorylation of Trk-receptors creates docking sites for adaptor proteins that couple these receptors to intracellular signaling cascades, including the ras/MAPK, PI3K/Akt pathway and the PLCγ1 pathways. Examination of the downstream signal transduction pathways revealed that the interaction of Trk receptors with PLCγ1 and SHC was required for activation of NFAT-dependent transcription [60]. A requirement for the Shc-interaction site might reflect the requirement for ras /MAPK or PI3K activation for the induction of the nuclear components of NFAT transcription complexes, which are PKC/ras-dependent [5]. The requirement for the PLCγ1 interaction site of the Trk receptors may relate to PLCγ's ability to stimulate Ca^{2+} release via the generation of IP_3. A large number of different receptors respond to stimulation by generating the second messenger IP_3. IP_3 triggers the release of Ca^{2+} from the endoplasmic reticulum by binding to the IP_3 receptor, thereby causing the rapid influx of extracellular Ca^{2+} via CRAC channels. In addition, Ca^{2+} release from IP_3-sensitive stores can cooperate with Ca^{2+} influx via voltage-gated and receptor-operated calcium channels. Interestingly, CaN/NFAT signaling appears to directly regulate the expression of the neuronal-specific, IP_3 type 1 receptor (IP3R1) [17,55,124], both in response to L-type Ca^{2+} channel- or BDNF-induced NFAT-dependent transcription. Moreover, BDNF activation of NFAT-dependent transcription will also lead to increased BDNF mRNA and protein production, evoking yet another positive feedback loop. Thus, NFAT-controlled gene expression in neurons can provide a positive feedback mechanism capable of altering the amplitude or spatial organization of Ca^{2+} signals. Such positive feedback loops could lead to long-lasting modulation of spatial and temporal patterns of neuronal activity and be critical for the refinement of synaptic connections that occur during neural development.

17.5
The Role of NFAT in Neuronal Survival

During development, neurons are produced in numbers exceeding those seen in the adult brain. Thus, programmed cell death plays an important role in shaping the nervous system. Neuronal cell survival can be modulated by survival factors such as neurotrophins or by spontaneous activity of a neuron itself. The removal of afferents or the blockade of neuronal activity in turn leads to an increase in programmed cell death, suggesting that the depolarization of neurons normally favors survival. Depolarization may increase survival by increasing the sensitivity of neurons to neurotrophic factors [125], or by enhancing the endogenous synthesis of these factors [126]. In accordance with this, culture of several types of neurons in elevated extracellular K^+ concentration leads to enhanced survival. In cerebellar granule neurons, NFATc4 plays a critical role in mediating growth factor- and activity-dependent survival. In the presence of high K^+ in the culture medium, NFATc4 was found in the nucleus of cerebellar granule cells. Upon serum withdrawal and low K^+ conditions, which lead to the subsequent induction neuronal cell death, NFATc4 was exported to the cytoplasm. Genetic knockdown of NFATc4 by RNA interference triggered the apoptosis of granule neurons even under survival conditions of serum growth factors and neuronal activity, while expression of a constitutively active form of NFAT protected granule neurons against cell death. Together, these findings suggest that NFATc4–dependent transcription plays a critical role in promoting the survival of neurons, and also suggests a direct role for NFAT proteins in regulating the expression of pro-survival genes in cerebellar granule cells [127]. The anti-apoptotic kinase Akt might enhance neuronal survival by controlling at least two different transcription factors. On one hand it promotes the nuclear export of the pro-apoptotic transcription factor FOXO3, while on the other hand it increases the amount and/or time of NFATc4 in the nucleus by inactivating the NFAT export kinase GSK3 [128].

In contrast to the role of NFAT-dependent transcription in the survival of cerebellar granule cells, *in-vitro* and *in-vivo* neurotrophin-mediated survival of embryonic sensory neurons does not require signaling via CaN/NFAT [60]. Mice with mutations of NFATc2/c3/c4 did not display any significant increase in neuronal cell death. The lack of effect on cell death is also evident in culture experiments, as blocking CaN/NFAT signaling with FK/CsA provides a reversible block of sensory axon growth from explants in collagen, and does not increase sensory neuron death in low-density dissociated cultures on laminin. Although neurons lacking NFAT signaling are unable to extend axons efficiently in response to neurotrophins, they do not appear to be comprised in the ability of neurotrophins to promote survival. However, mice with mutations in CREB were found to have defects in neurotrophin-induced survival both *in vivo* and *in vitro* [129]. CREB responds to both cAMP and Ca^{2+} signals and activates transcription of genes essential for cell survival. A straightforward synthesis of these results is to propose that neurotrophin signals split into a survival pathway requiring CREB and an outgrowth pathway through CaN and NFATc proteins (see Fig. 17.5). This precise parsing of signals could allow independent control of survival and axonal extension by factors encountered along the paths of axons to their targets. These factors might regulate pathways that intersect with neurotrophin or netrin signaling downstream of their receptors. Such independent control of survival and axonal extension could provide a coincidence detection mechanism, allowing the elimination of neurons whose axons wandered onto an incorrect path. This argument is supported by previous studies in mice with mutations in the pro-apoptotic Bcl–2 family member, Bax. Neuronal cell death in the absence of neurotrophins occurs by apoptosis, and requires the actions of Bax [130]. The targeted deletion of Bax rescues the survival of DRG neurons in NGF- or TrkA-deficient mice. However, the rescued neurons show a dramatic reduction of peripheral sensory connections [131, 132], implying that the signaling mechanisms controlling outgrowth are different from those controlling survival. In future, it will be interesting to determine which other neuronal subtypes utilize the CaN/NFAT pathway to modulate cell survival, which upstream signals regulate CaN/NFAT, which NFAT nuclear partner proteins are involved, and which downstream targets are affected.

17.6
Small Molecule Inhibitors of CaN are Powerful Probes of Neuronal Development

17.6.1
The Mechanism of Action of FK506 and Cyclosporine A

During the 1980s, the drugs FK506 and cyclosporine A (CsA) revolutionized transplant surgery, making possible successful organ transplantation that had been impossible or ineffective before their use [133]. Despite the great importance of these drugs, their mechanism of action was unknown until it was discovered that they inhibited NFAT signaling and the activation of NFAT-dependent genes by blocking the activity of CaN [4,134,135]. These drugs have an almost unique mechanism of action. Their highly lipophilic nature allows them to pass through the cell membrane, the placenta or the blood-brain barrier. Once inside the cell, they bind to the FKBP or cyclophilin class of prolyl isomerases (immunophilins). This interaction is diffusion-limited and of high affinity (about 0.5 nM). The composite surface created by the combination of drug and immunophilin then directly interacts with CaN (see Fig. 17.1) and prevents substrate access [136–140].

Both FK506 and CsA are natural products synthesized by microorganisms, which produce these molecules to block CaN activity in their competitors. This also might explain why both of these chemically dissimilar drugs bind to a prolyl isomerase. Both molecules are very hydrophobic, which allows them to pass through any membrane by lipid permeation. The prolyl isomerase pocket is one of the most hydrophobic pockets in the cell, and hence is ideally suited to the task of binding these drugs. The fact that both compounds completely inhibit CaN activity and NFAT-dependent transcription at concentrations far lower than would be necessary to inhibit prolyl isomerase activity present in cells [4, 135], shows that the inhibition of prolyl isomerase activity has little to do with their mechanism of action at low drug concentrations. At higher concentrations (10 µM and above), they intercalate into membranes, as do most hydrophobic drugs, leading to a wide variety of nonspecific actions. When used at low concentrations, the drugs appear to be two of the most specific inhibitors available. In many ways, these small molecules sparked the origin of the field of chemical biology to probe biologic mechanisms [138, 139]. Indeed, present studies indicate that the phenotypes produced by drug administration to either pregnant or adult mice parallels those produced by deleting the CaN gene [56,58,60]. This test of specificity has been equaled by few (if any) other drugs. These observations strongly indicate that, at low concentrations, FK506 and CsA might have a single target in the mammalian genome.

17.6.2
Use of CsA and FK506 in Studies of Neural Development and Function

The unusual mechanism of action of these drugs underlies some relatively unintuitive aspects of their use. Perhaps the most unintuitive aspect is that the correct concentration to inhibit a CaN-dependent process will vary with the cell type and the developmental window studied. This is because the concentrations of CaN, FKBP and cyclophilin vary widely between different cell types and developmental time points. For example, lymphocytes contain about 5000 molecules of CaN per cell, and about one million molecules of FKBP or cyclophilin. Hence, a small amount of drug that occupies only a few percent of the total FKBP or cyclophilin molecules will be sufficient to completely inhibit CaN in a lymphocyte. In neurons, where CaN expression levels are relatively high (estimated at about 1% of total cellular protein in hippocampal neurons), much higher levels of the drugs are necessary to achieve full inhibition of CaN activity. The high concentration of CaN in the brain is likely to be the reason that few patients treated with these drugs suffer neurological side effects, unless accidentally the drug reaches high levels in the central nervous system. In such cases the patients may develop transient and reversible memory loss, cortical blindness or psychosis as a result [140, 141].

One-way to circumvent the difficulty of dealing with high concentrations of CaN in neurons (often exceeding the concentration of either FKBP or cyclophilin) is to use both FK506 and CsA to fully inhibit CaN activity. This will result in the formation of two types of drug-isomerase complexes, each able to bind to the active site in CaN and preclude substrate access. Indeed, the use of both drugs is essential to block the activity of CaN in embryonic neurons [17, 60].

17.6.3
Assessing CaN Activity

As mentioned above, one problem with using inhibitors of CaN or conditional knockout of its subunits is that it is difficult to know if the enzymatic activity is fully inhibited or ablated, as the produrance time for the mRNA and protein are rather long in neurons (24–48 h). Fortunately, a simple assay was devised to determine if any remaining activity of calcineurin was present in a cell. The NFATc proteins are hyperphosphorylated in the absence of CaN activity. Hence, if CaN activity is completely blocked, a large shift in the mobility of the particular NFATc family members can be observed by Western blotting. This approach has been very effective for checking the activity of CaN in neurons or other tissues of the developing embryo where it is very difficult to inhibit [58].

17.7
NFAT Signaling and Transcriptional Control in Human Disease

17.7.1
Possible Defects in NFAT Signaling in Human Schizophrenia

The hypothesis that CaN-NFAT signaling might play a role in schizophrenia originally came from studies of transplant patients treated with CsA. A small percentage of patients had a neurological reaction that appeared to be similar to schizophrenia [140–144], but was reversible upon discontinuation of the drug. However, these cases were rare and it was not clear that a causal association existed between inhibition of CaN activity and the development of a schizophrenia-like syndrome. More recently, however, human genetic studies have suggested that a well characterized schizophrenia locus on human chromosome 8p21.3 may encompass the gene encoding one of the three catalytic subunits of CaN, CaN-A3 or -Aγ [145]. In addition, one of the other human schizophrenia loci is near the gene for NFATc2 on human chromosome 20q13 [146]. Thus, the possibility exists that a defect in signaling mediated by CaN-A γ and NFATc2 could be associated with an increased risk of schizophrenia. This notion is reinforced by the finding that mice with a knockout of the CaN-Aγ gene show critical aspects of schizophrenia [147]. In a more recent study of 457 cases of schizophrenia in Japan compared to 429 controls, no association could be detected, indicating that perhaps ethnic background might provide modulators of this association [148].

17.7.2
Down Syndrome and NFAT Signaling

The discovery that the Down Syndrome Critical region (DSCR) encodes a gene, *DSCR1* (also called MCIP1 or calcipressin), which encodes a competitive CaN inhibitor with nanomolar binding affinity, led to the hypothesis that the inhibition of

CaN might play a role in Down syndrome. DSCR1 was discovered in a search for proteins that were located in the 3 Mb "critical region" thought to contain the genes, which when overexpressed 1.5–fold rise to the symptoms of Down syndrome [149, 150]. *DSCR1* was the first gene to be found in this region, and was rediscovered in a search for CaN inhibitors in muscle as MCIP [151] and also in yeast [152]. *DSCR1* overexpression blocks NFATc translocation in a number of cell types, and is a powerful inhibitor of NFAT-dependent transcription in several cell types. Remarkably, Olson and colleagues discovered that DSCR1 or MCIP1 was regulated by NFAT transcription complexes through a cluster of 15 tandem NFAT-binding sites within the second intron of the *DSCR1* gene [153]. The transcriptional regulation of an inhibitor of the CaN/NFAT pathway by NFAT itself contributes further to the complex positive and negative feedback controls that modulate this pathway. As positive and negative feedback mechanisms tend to be unstable, these findings raise the possibility that attenuation of NFAT signaling during development might contribute to the pleiotropic developmental defects observed in patients with Down syndrome.

17.8
Conclusion

Understanding the roles of NFAT signaling in the development and function of the nervous system were probably delayed by the unfortunate name given to this pathway, which implied that it was somehow restricted to activated T-lymphocytes. With the discovery of a number of essential roles in the development and function of the nervous system, perhaps a better name would be <u>N</u>eural <u>F</u>actor <u>A</u>ctivated by <u>T</u>rk. Although many further studies need to be done, investigations into the many modulators of the pathway including CaN, GSK3, DSCR1 and others, as well as the many NFAT target genes, should provide fresh insights into aspects of axonal guidance, synaptogenesis, and synaptic plasticity. Finally, the pathway and its modulators will likely be useful and important for understanding the pathogenesis of several human diseases of the nervous system.

Abbreviations

BDNF	brain-derived neurotrophic factor
CaN	calcineurin
CRAC	Ca^{2+} release-activated Ca^{2+}
CsA	cyclosporine A
DRG	dorsal root ganglia
DSCR	Down Syndrome Critical Region
IL	interleukin
LTD	long-term depression
LTP	long-term potentiation

NFAT	nuclear factor of activated T-cells
NLS	nuclear localization sequence
NMDAR	*N*-methyl-D-aspartate receptor
NT	neurotrophin
PKA	protein kinase A
SRR	serine-rich region
VEGF	vascular endothelial growth factor

References

1. D. B. Durand, M. R. Bush, J. G. Morgan, A. Weiss, G. R. Crabtree, *J. Exp. Med.* 165, 395–407 (1987).
2. J.-P. Shaw, et al., *Science* 241, 202–205 (1988).
3. G. R. Crabtree, *Science* 243, 355–361 (1989).
4. E. A. Emmel, et al., *Science* 246, 1617–1620 (1989).
5. W. M. Flanagan, B. Corthesy, R. J. Bram, G. R. Crabtree, *Nature* 352, 803–807 (1991).
6. W. M. Flanagan, G. R. Crabtree, *J. Biol. Chem.* 267, 399–406 (1992).
7. J. P. Northrop, K. S. Ullman, G. R. Crabtree, *J. Biol. Chem.* 268, 2917–2923 (1993).
8. J. P. Northrop, et al., *Nature* 369, 497–502 (1994).
9. P. G. McCaffrey, et al., *Science* 262, 750–754 (1993).
10. S. A. Wolfe, et al., *Nature* 385, 172–176 (1997).
11. L. Chen, J. N. Glover, P. G. Hogan, A. Rao, S. C. Harrison, *Nature* 392, 42–48. (1998).
12. D. E. Clapham, *Cell* 80, 259–268 (1995).
13. M. M. Winslow, J. R. Neilson, G. R. Crabtree, *Curr. Opin. Immunol.* 15, 299–307 (2003).
14. A. B. Parekh, J. W. Putney, Jr., *Physiol. Rev.* 85, 757–810 (2005).
15. R. S. Lewis, *Biochem. Soc. Trans.* 31, 925–929 (2003).
16. A. T. Serafini, et al., *Immunity* 3, 239–250 (1995).
17. I. A. Graef, et al., *Nature* 401, 703–708 (1999).
18. M. Kumai, et al., *Development* 127, 3501–3512 (2000).
19. A. Badou, et al., *Science* 307, 117–121 (2005).
20. J. Jain, P. G. McCaffrey, V. E. Valge-Archer, A. Rao, *Nature* 356, 801–804 (1992).
21. J. D. Molkentin, et al., *Cell* 93, 215–228 (1998).
22. G. R. Crabtree, E. N. Olson, *Cell* 109 Suppl, S67–S79 (2002).
23. P. G. Hogan, L. Chen, J. Nardone, A. Rao, *Genes Dev.* 17, 2205–2232 (2003).
24. S. N. Ho, et al., *J. Biol. Chem.* 270, 19898–19907 (1995).
25. I. A. Graef, J. M. Gastier, U. Francke, G. R. Crabtree, *Proc. Natl. Acad. Sci. USA* 98, 5740–5745 (2001).
26. C. R. Beals, N. A. Clipstone, S. N. Ho, G. R. Crabtree, *Genes Dev.* 11, 824–834 (1997).
27. H. Okamura, et al., *Mol Cell* 6, 539–550 (2000).
28. K. T.-Y. Shaw, et al., *Proc. Natl. Acad. Sci. USA* 92, 11205–11209 (1995).
29. C. M. Porter, M. A. Havens, N. A. Clipstone, *J. Biol. Chem.* 275, 3543–3551 (2000).
30. J. Aramburu, et al., *Molecular Cell* 1, 627–637 (1998).
31. J. Aramburu, et al., *Science* 285, 2129–2133 (1999).
32. J. Liu, K. Arai, N. Arai, *J. Immunol.* 167, 2677–2687 (2001).
33. S. Park, M. Uesugi, G. L. Verdine, *Proc. Natl. Acad. Sci. USA* 97, 7130–7135 (2000).
34. J. D. Klemm, C. R. Beals, G. R. Crabtree, *Curr. Biol.* 7, 638–644 (1997).
35. R. H. Kehlenbach, A. Dickmanns, L.

Gerace, *J. Cell Biol.* 141, 863–874 (1998).
36. H. Okamura, et al., *Molecular Cell* 6, 539–550 (2000).
37. D. A. Cross, D. R. Alessi, P. Cohen, M. Andjelkovich, B. A. Hemmings, *Nature* 378, 785–789 (1995).
38. V. Stambolic, J. R. Ruel, J. R. Woodgett, *Curr. Biol.* 6, 1664–1668 (1997).
39. P. C. van Weeren, K. M. de Bruyn, A. M. de Vries-Smits, J. van Lint, B. M. Burgering, *J. Biol. Chem.* 273, 13150–13156 (1998).
40. B. W. Doble, J. R. Woodgett, *J. Cell. Sci.* 116, 1175–1186 (2003).
41. A. Ali, K. P. Hoeflich, J. R. Woodgett, *Chem. Rev.* 101, 2527–2540 (2001).
42. A. D. Gitler, et al., *Nat. Genet.* 33, 75–79 (2003).
43. C. R. Beals, C. M. Sheridan, C. W. Turck, P. Gardner, G. R. Crabtree, *Science* 275, 1930–1934 (1997).
44. C. M. Sheridan, E. K. Heist, C. R. Beals, G. R. Crabtree, P. Gardner, *J. Biol. Chem.* 277, 48664–48676 (2002).
45. J. Zhu, et al., *Cell* 93, 851–861 (1998).
46. C. W. Chow, C. Dong, R. A. Flavell, R. J. Davis, *Mol. Cell. Biol.* 20, 5227–5234 (2000).
47. C. M. Porter, M. A. Havens, N. A. Clipstone, *J. Biol. Chem.* 275, 3543–3551 (2000).
48. H. Okamura, et al., *Mol. Cell. Biol.* 24, 4184–4195 (2004).
49. M. J. Berridge, *Neuron* 21, 13–26 (1998).
50. M. J. Berridge, M. D. Bootman, P. Lipp, *Nature* 395, 645–648 (1998).
51. R. E. Dolmetsch, K. Xu, R. S. Lewis, *Nature* 392, 933–936 (1998).
52. R. E. Dolmetsch, R. S. Lewis, C. C. Goodnow, J. I. Healy, *Nature* 386, 855–858 (1997).
53. W. Li, J. Llopis, M. Whitney, G. Zlokarnik, R. Y. Tsien, *Nature* 392, 936–941 (1998).
54. K. S. Ullman, et al., *Transplant. Proc.* 23, 2845 (1991).
55. R. D. Groth, P. G. Mermelstein, *J. Neurosci.* 23, 8125–8134 (2003).
56. C. P. Chang, et al., *Cell* 118, 649–663 (2004).
57. I. Olave, W. Wang, Y. Xue, A. Kuo, G. R. Crabtree, *Genes Dev.* 16, 2509–2517 (2002).
58. I. A. Graef, F. Chen, L. Chen, A. Kuo, G. R. Crabtree, *Cell* 105, 863–875 (2001).
59. I. A. Graef, F. Chen, G. R. Crabtree, *Curr. Opin. Genet. Dev.* 11, 505–512 (2001).
60. I. A. Graef, et al., *Cell* 113, 657–670 (2003).
61. C. Loh, et al., *J. Biol. Chem.* 271, 10884–10891 (1996).
62. R. J. Giger, A. L. Kolodkin, *Cell* 105, 1–4 (2001).
63. E. J. Huang, L. F. Reichardt, *Annu. Rev. Neurosci.* 24, 677–736 (2001).
64. M. Tessier-Lavigne, C. S. Goodman, *Science* 274, 1123–1133 (1996).
65. A. Fazeli, et al., *Nature* 386, 796–804 (1997).
66. T. Serafini, et al., *Cell* 87, 1001–1014 (1996).
67. K. Keino-Masu, et al., *Cell* 87, 175–185 (1996).
68. K. L. Lankford, P. C. Letourneau, *J. Cell Biol.* 109, 1229–1243 (1989).
69. K. Hong, M. Nishiyama, J. Henley, M. Tessier-Lavigne, M. Poo, *Nature* 403, 93–98 (2000).
70. G. L. Ming, et al., *Nature* 417, 411–418 (2002).
71. T. Yoshida, M. Mishina, *J. Neurosci.* 25, 3067–3079 (2005).
72. Y. Mukouyama, D. Shin, S. Britsch, M. Taniguchi, D. J. Anderson, *Cell* 109, 693–705 (2002).
73. Y. Honma, et al., *Neuron* 35, 267–282 (2002).
74. H. Gerhardt, et al., *J. Cell Biol.* 161, 1163–1177 (2003).
75. Y. S. Mukouyama, H. P. Gerber, N. Ferrara, C. Gu, D. J. Anderson, *Development* 132, 941–952 (2005).
76. D. Bates, et al., *Dev. Biol.* 255, 77–98 (2003).
77. P. Kermani, et al., *J. Clin. Invest.* 115, 653–663 (2005).
78. H. Kim, Q. Li, B. L. Hempstead, J. A. Madri, *J. Biol. Chem.* 279, 33538–33546 (2004).
79. Y. H. Youn, J. Feng, L. Tessarollo, K. Ito, M. Sieber-Blum, *Mol. Cell. Neurosci.* 24, 160–170 (2003).

80 D. von Schack, et al., *Nat. Neurosci.* 4, 977–978 (2001).
81 M. J. Donovan, et al., *Development* 127, 4531–4540 (2000).
82 O. Behar, J. A. Golden, H. Mashimo, F. J. Schoen, M. C. Fishman, *Nature* 383, 525–528 (1996).
83 A. D. Gitler, M. M. Lu, J. A. Epstein, *Dev. Cell* 7, 107–116 (2004).
84 C. Gu, et al., *Science* 307, 265–268 (2005).
85 J. A. Alva, M. L. Iruela-Arispe, *Curr. Opin. Hematol.* 11, 278–283 (2004).
86 X. Lu, et al., *Nature* 432, 179–186 (2004).
87 N. D. Lawson, et al., *Development* 128, 3675–3683 (2001).
88 J. Torres-Vazquez, et al., *Dev. Cell* 7, 117–123 (2004).
89 C. Gu, et al., *Dev. Cell* 5, 45–57 (2003).
90 K. W. Park, et al., *Dev. Biol.* 261, 251–267 (2003).
91 A. L. Armesilla, et al., *Mol. Cell. Biol.* 19, 2032–2043 (1999).
92 G. L. V. Hernandez, O.V. Iniguez, M.A. Lorenzo, E. Martinez-Martinez, S. Grau, R. Fresno M, Redondo, J. M., *J. Exp. Med.* 193, 607–620 (2001).
93 H. U. Wang, Z. F. Chen, D. J. Anderson, *Cell* 93, 741–753 (1998).
94 R. H. Adams, et al., *Genes Dev.* 13, 295–306 (1999).
95 C. Ruhrberg, et al., *Genes Dev.* 16, 2684–2698 (2002).
96 P. M. Helbling, D. M. Saulnier, A. W. Brandli, *Development* 127, 269–278 (2000).
97 M. Sondell, G. Lundborg, M. Kanje, *J. Neurosci.* 19, 5731–5740 (1999).
98 A. Eichmann, F. Le Noble, M. Autiero, P. Carmeliet, *Curr. Opin. Neurobiol.* 15, 108–115 (2005).
99 I. M. Mansuy, M. Mayford, B. Jacob, E. R. Kandel, M. E. Bach, *Cell* 92, 39–49 (1998).
100 D. G. Winder, I. M. Mansuy, M. Osman, T. M. Moallem, E. R. Kandel, *Cell* 92, 25–37 (1998).
101 H. Zeng, et al., *Cell* 107, 617–629 (2001).
102 G. Malleret, et al., *Cell* 104, 675–686 (2001).
103 Y. M. Lu, I. M. Mansuy, E. R. Kandel, J. Roder, *Neuron* 26, 197–205 (2000).
104 M. Zhuo, et al., *Proc. Natl. Acad. Sci. USA* 96, 4650–4655 (1999).
105 I. M. Mansuy, *Biochem. Biophys. Res. Commun.* 311, 1195–1208 (2003).
106 J. Wang, et al., *J. Neurosci.* 23, 826–836 (2003).
107 R. M. Mulkey, S. Endo, S. Shenolikar, R. C. Malenka, *Nature* 369, 486–488 (1994).
108 M. Morioka, S. Nagahiro, K. Fukunaga, E. Miyamoto, Y. Ushio, *Neuroscience* 78, 673–684 (1997).
109 S. Halpain, A. Hipolito, L. Saffer, *J. Neurosci.* 18, 9835–9844 (1998).
110 A. Sik, N. Hajos, A. Gulacsi, I. Mody, T. F. Freund, *Proc. Natl. Acad. Sci. USA* 95, 3245–3250 (1998).
111 J. H. Wang, R. Desai, *Biochem. Biophys. Res. Commun.* 72, 926–932 (1976).
112 R. C. Malenka, M. F. Bear, *Neuron* 44, 5–21 (2004).
113 P. V. Nguyen, T. Abel, E. R. Kandel, *Science* 265, 1104–1107 (1994).
114 C. H. Bailey, D. Bartsch, E. R. Kandel, *Proc. Natl. Acad. Sci. USA* 93, 13445–13452 (1996).
115 E. N. Olson, R. S. Williams, *Cell* 101, 689–692 (2000).
116 E. R. Chin, et al., *Genes Dev.* 12, 2499–2509 (1998).
117 F. J. Naya, et al., *J. Biol. Chem.* 275, 4545–4548 (2000).
118 A. L. Serrano, et al., *Proc. Natl. Acad. Sci. USA* 98, 13108–13113 (2001).
119 K. J. McCullagh, et al., *Proc. Natl. Acad. Sci. USA* 101, 10590–10595 (2004).
120 B. Rudy, B. Kirschenbaum, A. Rukenstein, L. A. Greene, *J. Neurosci.* 7, 1613–1625 (1987).
121 S. S. Lesser, N. T. Sherwood, D. C. Lo, *Mol. Cell. Neurosci.* 10, 173–183 (1997).
122 V. Lessmann, K. Gottmann, R. Heumann, *Neuroreport* 6, 21–25 (1994).
123 A. M. Lohof, N. Y. Ip, M. M. Poo, *Nature* 363, 350–353 (1993).
124 A. A. Genazzani, E. Carafoli, D. Guerini, *Proc. Natl. Acad. Sci. USA* 96, 5797–5801 (1999).
125 A. Meyer-Franke, M. R. Kaplan, F. W.

Pfrieger, B. A. Barres, *Neuron* 15, 805–819 (1995).

126 A. Ghosh, J. Carnahan, M. E. Greenberg, *Science* 263, 1618–1623 (1994).

127 A. B. Benedito, et al., *J. Biol. Chem.* 280, 2818–2825 (2005).

128 S. R. Datta, A. Brunet, M. E. Greenberg, *Genes Dev.* 13, 2905–2927 (1999).

129 B. E. Lonze, A. Riccio, S. Cohen, D. D. Ginty, *Neuron* 34, 371–385 (2002).

130 T. L. Deckwerth, et al., *Neuron* 17, 401–411 (1996).

131 A. M. Davies, *Curr. Biol.* 10, R374–R376 (2000).

132 T. D. Patel, A. Jackman, F. L. Rice, J. Kucera, W. D. Snider, *Neuron* 25, 345–357 (2000).

133 D. M. Canafax, N. L. Ascher, *Clin. Pharm.* 2, 515–524 (1983).

134 J. Liu, et al., *Cell* 66, 807–815 (1991).

135 N. A. Clipstone, G. R. Crabtree, *Nature* 357, 695–697 (1992).

136 G. D. Van Duyne, R. F. Standaert, P. A. Karplus, S. L. Schreiber, J. Clardy, *Science* 252, 839–842 (1991).

137 J. P. Griffith, et al., *Cell* 82, 507–522 (1995).

138 S. L. Schreiber, *Bioorg. Med. Chem.* 6, 1127–1152 (1998).

139 R. L. Strausberg, S. L. Schreiber, *Science* 300, 294–295 (2003).

140 W. O. Bechstein, *Transplant. Int.* 13, 313–326 (2000).

141 J. M. Gijtenbeek, M. J. van den Bent, C. J. Vecht, *J. Neurol.* 246, 339–346 (1999).

142 E. F. Wijdicks, *Liver Transpl.* 7, 937–942 (2001).

143 E. F. Wijdicks, R. H. Wiesner, R. A. Krom, *Neurology* 45, 1962–1964 (1995).

144 P. Neuhaus, et al., *Transplant. Int.* 7 Suppl 1, S27–S31 (1994).

145 D. J. Gerber, et al., *Proc. Natl. Acad. Sci. USA* 100, 8993–8998 (2003).

146 R. Freedman, et al., *Am. J. Med. Genet.* 105, 794–800 (2001).

147 T. Miyakawa, et al., *Proc. Natl. Acad. Sci. USA* 100, 8987–8992 (2003).

148 Y. Kinoshita, et al., *J. Neural Transm.* 112, 1255–62 (2005).

149 J. J. Fuentes, et al., *Hum. Mol. Genet.* 9, 1681–1690 (2000).

150 J. J. Fuentes, et al., *Hum. Mol. Genet.* 4, 1935–1944 (1995).

151 B. Rothermel, et al., *J. Biol. Chem.* 275, 8719–8725 (2000).

152 J. Gorlach, et al., *EMBO J.* 19, 3618–3629 (2000).

153 J. Yang, et al., *Circ. Res.* 87, E61–E68 (2000).

18
Stimulus-Transcription Coupling in the Nervous System: The Zinc Finger Protein Egr–1

Oliver G. Rössler, Luisa Stefano, Inge Bauer, and Gerald Thiel

Abstract

The biosynthesis of the zinc finger transcription factor Egr–1 is stimulated by many extracellular signaling molecules including hormones, neurotransmitters, and growth and differentiation factors, indicating that the *Egr–1* gene is a convergence point for many intracellular signaling cascades. Moreover, synaptic activity induces the expression of Egr–1, suggesting that Egr–1 orchestrates activity-regulated neuronal transcriptional programs. The Egr–1 protein links cellular signaling cascades with changes in the gene expression pattern of Egr–1–responsive target genes. Many biological functions have been attributed to Egr–1. In the nervous system, Egr–1 may control neuronal plasticity, neuronal cell death, and the proliferation of astrocytes.

18.1
Introduction

The zinc finger transcription factor Egr–1 (Sukhatme et al., 1988), also known as zif268 (Christy et al., 1988), NGFI-A (Milbrandt, 1987), Krox24 (Lemaire et al., 1988) and TIS8 (Lim et al., 1987), has been independently discovered by several investigators searching for genes essential for growth, proliferation, or differentiation. The *Egr–1* gene belongs to a group of early response genes, as stimulation with many environmental signals including growth factors, hormones, and neurotransmitters strongly and rapidly, induces *Egr–1* gene expression. The *Egr–1* gene therefore functions as a convergence point for many signaling cascades, and regulation of the *Egr–1* gene has been used as a model system to study stimulus-transcription coupling. Egr–1 couples extracellular signals to long-term responses by altering gene expression of Egr–1 target genes. The gene products of those target genes are then responsible for the physiological alterations that result from cellular stimulation and Egr–1 biosynthesis. In theory, the research on Egr–1 merges two broad fields of molecular research, described as "signal transduction" and "control of gene expression".

18.2
Modular Structure of Egr–1

Transcription factors have a modular structure; that is, the different functions such as DNA-binding, activation or repression of transcription can be attributed to distinct regions within the molecule. The modular structure of Egr–1 is depicted in Fig. 18.1. The DNA-binding domain of Egr–1 contains three zinc finger motifs. The structure of a complex formed between these three zinc fingers and its cognate DNA-binding site has been solved (Pavletich and Pabo, 1991) and subsequently used as a framework for understanding how zinc fingers recognize DNA (Jamieson et al., 1996; Elrod-Erickson and Pabo, 1999). Each zinc finger domain consists of an antiparallel β-sheet and an α-helix held together by a zinc ion and hydrophobic residues. The structure of the Egr–1 zinc finger domain provided the framework for design of novel DNA-binding proteins (Greisman and Pabo, 1997). Egr–1 preferentially binds to the GC-rich sequence 5'-GCGGGGGCG–3' (Christy and Nathans, 1989b; Cao et al., 1993). The transcription factors Sp1 and Sp3 also bind with their zinc finger DNA-binding domains to GC-rich sequences. A comparison of Sp1/Sp3 and Egr–1 binding specificities to DNA reveals that the DNA-binding sites of Egr–1 (GCG GGG GCG = A B A) and Sp1 (GGG GCG GGG = B A B) are similar and appear as a rearrangement of one another (A B A versus B A B) (Kriwacki et al., 1992). The similarity of the DNA-binding sites of Sp1, Sp3, and Egr–1 has triggered the hypothesis that they compete for the same DNA-binding site. A detailed analysis of Sp1/Sp3 and Egr–1–responsive genes has revealed that there are genuine Sp1/Sp3 or Egr–1 controlled genes showing no cross-regulation by Sp1/Sp3 and Egr–1 through the same DNA-binding site (Al-Sarraj et al., 2005). However, this does not exclude that some genes contain overlapping Sp1/Egr–1 or Sp3/Egr–1 binding sites where competition for a DNA-binding occurs. This has been shown, for example, for the human angiotensin-converting enzyme gene (Day et al., 2004; Al Sarraj et al., 2005).

The transcriptional activation domain of Egr–1 is, in contrast to its DNA-binding domain, not as well characterized. Gashler et al. (1993) mapped an extensive N-terminal activation domain, and similar results have been reported by the present authors (Thiel et al., 2000). An inhibitory domain between the activation domain and the DNA-binding domain was identified that functions as a binding site for two transcriptional cofactors termed NGFI-A binding proteins 1 and 2 (NAB1, NAB2) (Russo et al., 1995; Svaren et al., 1996). Both NAB1 and NAB2 block the biological activity of Egr–1 (Russo et al., 1995; Svaren et al., 1996; Thiel et al., 2000). A fusion protein consisting of NAB1 and a heterologous DNA-binding domain was shown to function very well as a transcriptional repressor (Thiel et al., 2000), indicating that NAB1 only needs to be recruited to the transcription unit, either by protein-protein or by DNA-protein interaction, to repress transcription. The discovery of the co-repressor proteins NAB1 and NAB2 produced a further level of complexity for the understanding of the function of Egr–1 because induction of transcription of the *Egr–1* gene may have no biological effect when the transcriptional activator function of Egr–1 is neutralized by NAB1 or NAB2. The concentrations of both co-repressors

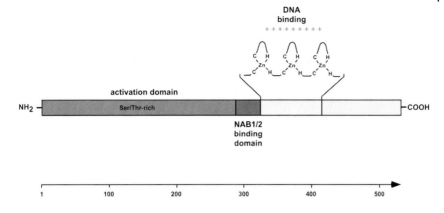

Fig. 18.1 Modular structure of the zinc finger transcription factor Egr–1. The Egr–1 protein contains an extended transcriptional activation domain on the N-terminus and a DNA-binding domain, consisting of three zinc finger motifs. Additionally, an inhibitory domain has been mapped between the activation and DNA-binding domains that functions as a binding site for the transcriptional co-repressor proteins NAB1 and NAB2. (This figure also appears with the color plates.)

in a particular cell is thus of extreme importance for Egr–1 function. In PC12 cells, for instance, overexpression of NAB2 inhibits nerve growth factor-induced differentiation (Qu et al., 1998). The expression of the *NAB2* gene is controlled by Egr–1 (Ehrengruber et al., 2000), indicating that Egr–1 controls its biological activity in a negative feedback loop via the synthesis of NAB2.

18.3
Intracellular Signaling Cascades Converging at the *Egr–1* Gene

Originally, induction of Egr–1 biosynthesis was observed following the stimulation of receptor tyrosine kinases triggering the activation of mitogenic signaling cascades. Later, it was thoroughly demonstrated that Egr–1 is synthesized following stimulation of epidermal growth factor (EGF) receptor, platelet-derived growth factor receptor, fibroblast growth factor (FGF) receptor, or Trk neurotrophin receptors (Milbrandt, 1987; Mundschau et al., 1994; Kaufmann and Thiel, 2001; Rössler and Thiel, 2004). Integral parts of this signaling cascades are receptor tyrosine kinases, adapter proteins such as Grb2 (growth factor receptor-bound–2) or Shc (Src-homology–2 containing), the nucleotide exchange factor Sos (son-of-sevenless), the G-protein Ras and a cascade of protein kinases in the order Raf → MEK (mitogen-activated protein kinase (MAPK) kinase and extracellular signal regulated protein kinase kinase) → ERK (extracellular signal-regulated protein kinase) (Fig. 18.2). This pathway is known as the ERK/MAPK cascade.

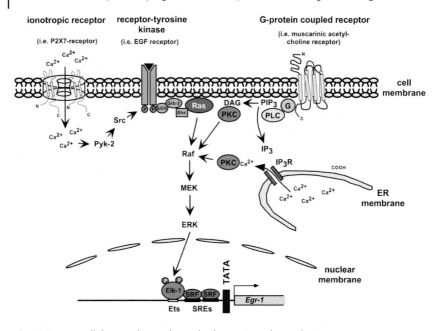

Fig. 18.2 Intracellular signaling pathways leading to Egr–1 biosynthesis. Ligand binding to receptor tyrosine kinases leads to receptor dimerization and intracellular *trans*-autophosphorylation of key tyrosine residues. The phosphotyrosyl residues function as docking sites for SH2–containing adapter proteins such as Grb2 (growth factor-receptor-bound 2). The nucleotide exchange factor Sos (son-of-sevenless) is recruited and activates the G-protein Ras. GTP-bound Ras in turn activates the protein kinase Raf via recruitment to the plasma membrane, leading to the sequential phosphorylation and activation of the protein kinases MEK and ERK. Ligands that bind to G-protein-coupled receptors (GPCR) stimulate ERK activation via activation of protein kinase C or transactivation of the EGF receptor. Protein kinase C can directly or indirectly stimulate the activity of Raf via phosphorylation. Transactivation of the EGF receptor may be accomplished by cytosolic tyrosine kinases of the src-family or via the activation of membrane-bound metalloproteinases. Likewise, an increase in the intracellular Ca^{2+}-concentration as a result of $P2X_7$-receptor stimulation triggers Egr–1 biosynthesis via transactivation of the EGF receptor and activation of ERK. (This figure also appears with the color plates.)

Ligands that bind and activate G-protein-coupled receptors (GPCRs) are also potent inducers of the Egr–1 biosynthesis. Distinct intracellular signaling cascades connect GPCR-stimulation with enhanced Egr–1 transcription. These pathways share a final activation of the protein kinase ERK. Stimulation of G_q-coupled GPCR can activate ERK via a protein kinase C-dependent pathway involving direct or indirect activation of Raf (Kolch et al., 1993; Schonwasser et al., 1998; Corbit et al., 2003) (see Fig. 18.2). The signal transduction of muscarinic acetylcholine receptors in neuroblastoma cells functions via this scheme (Grimes and Jope, 1999; O. Rössler and G. Thiel, unpublished observations). Stimulation of GPCR can also induce a "transactivation" of receptor tyrosine kinases, in particular of the EGF receptor (Daub et al., 1997;

Wetzker and Böhmer, 2003). As a result, the ERK/MAPK signaling cascade is activated, leading to the biosynthesis of Egr–1. In human glioma cells, stimulation of the G-protein-coupled neurokinin receptor–1 by substance P induces the biosynthesis of Egr–1. The up-regulation of Egr–1 synthesis is completely blocked by AG1487, an EGF receptor-specific tyrosine kinase inhibitor (Al-Sarraj and Thiel, 2002), indicating that transactivation of the EGF receptor is essential for substance P/neurokinin 1 receptor-induced activation of Egr–1 biosynthesis. Several mechanisms have been proposed to explain the cross-talk between GPCR-stimulation and subsequent receptor tyrosine kinase activation (Leserer et al., 2000; Wetzker and Böhmer, 2003). An increase in the intracellular Ca^{2+} concentration, for example, can activate the Ca^{2+}-dependent cytosolic tyrosine kinase Pyk2 that may trigger transactivation of the EGF receptor. Alternatively, GPCR-stimulation may activate membrane-bound metalloproteinases, leading to the liberation of EGF receptor ligands. Thus, several routes are available in the cell for the signal transfer from GPCR to ERK/MAPK, and cell type-specific differences are obvious.

Finally, we recently discovered that stimulation of the ionotropic receptor $P2X_7$ by ATP or ATP analogues induces the biosynthesis of Egr–1. The signaling cascade following P2X7 receptor stimulation involves an influx of Ca^{2+}, transactivation of the EGF receptor, and the subsequent activation of the ERK/MAPK cascade. In addition, Ca^{2+}/calmodulin-dependent protein kinases may also be involved in the signal transfer from the $P2X_7$ receptor to the Egr–1 gene (L. Stefano, O.G. Rössler and G. Thiel, manuscript in preparation). In summary, the fact that ligands that stimulate either receptor tyrosine kinases, GPCR, or ionotropic receptors also stimulate Egr–1 biosynthesis indicates that the *Egr–1* gene functions as a point of convergence of intracellular signaling cascades involved in stimulus-dependent regulation of gene transcription.

18.4
The Egr–1 Promoter

Transcription factors bound to the genetic elements within the 5'-flanking region of the *Egr–1* gene are targets for the signaling cascades that regulate stimulus-induced Egr–1 transcription. The landmark transcriptional regulatory elements within the human Egr–1 promoter are depicted in Fig. 18.3. Most importantly, the human Egr–1 promoter contains five serum response elements (SREs) encompassing the consensus sequence $CC(A/T)_6GG$, also termed CArG box (Christy and Nathans, 1988a; Sakamoto et al., 1991; Bauer et al., 2005). Two kinds of transcription factors are required for SRE-mediated activity – the serum response factor (SRF), and the ternary complex factor. The ternary complex factors Elk–1 (Ets-like protein–1), SAP–1 and SAP–2/Net/ERP contact DNA and also bind to the SRF. Elk–1 is a major nuclear target of mitogen-activated protein kinases, and Elk–1 phosphorylation leads to enhanced DNA-binding activity, ternary complex formation and SRE-mediated transcription (Shaw and Saxton, 2003). Thus, Elk–1 connects the activation of mitogen-activated protein kinases with enhanced transcription of genes containing

SREs in their regulatory region. Accordingly, expression of a constitutively active mitogen-activated protein kinase kinase, the kinase responsible for the phosphorylation and activation of extracellular signal-regulated protein kinase (ERK), strongly stimulates Egr–1 promoter activity (Kaufmann et al., 2001). Likewise, synthesis of Egr–1 mRNA induced by serum, platelet-derived growth factor, or phorbol 12–myristate 13–acetate was shown to be almost completely dependent on the activation of the Ras-Raf-MEK-ERK signaling pathway (Gineitis and Treisman, 2001). Transcriptional activation of Egr–1 is often preceded by an activation of Elk–1. In stimulated glutamatergic corticostriatal neurons, for example, a strict spatiotemporal connection between Elk–1 activation and Egr–1 mRNA synthesis has been demonstrated (Sgambato et al. 1998). Likewise, nitric oxide (NO) donor-induced activation of Elk–1 preceded transcription of the *Egr–1* gene in human neuroblastoma cells (Cibelli et al., 2002).

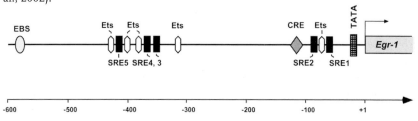

Fig. 18.3 Control elements of the human *Egr–1* gene. The transcriptional start is indicated by the arrow. The 5'-flanking region contains a TATA box, five serum-response elements (SRE), encompassing the sequence CC(A/T)$_6$GG and binding sites for ternary complex factors (Ets) belonging to the Ets family of transcription factors. The Ets binding sites are characterized by a conserved GGAA core. A cyclic AMP response element (CRE) is present that contains one mismatch in comparison to the consensus sequence. Egr–1 also binds to its own promoter via the EBS site. The sequence is available from the public database (accession # AJ245926.1 and X12617.1).

The SREs occur in two clusters in the Egr–1 promoter – a distal 5' cluster of three SREs and a proximal 3' cluster of two SREs. In addition, multiple binding sites for ternary complex factors (Ets) are adjacent to the CArG boxes having the Ets consensus core sequence GGAA/T (Fig. 18.3). *In-vitro* protein-DNA-binding experiments revealed that Elk–1 and SAP–1 are able to form ternary complexes with the SRF on both the proximal and distal SRE clusters of the Egr–1 promoter (Watson et al., 1997). The formation of a quaternary complex occurred on the distal SREs due to the presence of two ternary complex binding sites flanking two CArG boxes (SREs # 3 and 4). The proximal SRE cluster encompassing a single Ets binding site was able to recruit a ternary, but not a quaternary, complex *in vitro*. These differences shed light on the fact that the proximal and distal SRE clusters have a different impact on stimulus-induced gene transcription of the *Egr–1* gene, despite the fact that both the distal as well as the proximal SRE clusters couple enhanced ERK activity with transcriptional up-regulation. The biological diversity of the five SREs of the Egr–1 promoter has been noticed by us and others in recent years. In human neuroblastoma cells, for instance, NO donors induced *Egr–1* gene transcription only when the distal SREs were present in the transcription unit (Cibelli et al., 2002). The upstream SREs also represent a convergence point for shear stress, phorbol ester or urea administration, leading to an up-regulation of *Egr–1* gene transcription (Cohen et al., 1996; Schwachtgen et al., 1998; Bauer et al., 2005). The proximal SREs, in contrast, are required for induction of *Egr–1* gene transcription by epidermal growth factor or thrombin (Tsai et al., 2001; Wu et al., 2002; Bauer et al., 2005).

The Egr–1 promoter contains a cAMP response element (CRE) that encompasses the sequence 5'-TCACGTCA-3'. This motif shows a one base difference in comparison to the canonical CRE sequence 5'-TGACGTCA-3'. The functionality of this element has been a matter of controversy. The fact that forskolin, an activator of adenylate cyclase, did not stimulate Egr–1 promoter activity in human U87 glioma cells (Meyer et al., 2002) was used as an argument that the CRE in the context of the Egr–1 promoter does not function as a cAMP-inducible enhancer element. Rather, the CRE was suggested to control Egr–1 gene transcription via activation of the p38/stress-activated protein kinase–2–mediated signaling cascade (Rolli et al., 1999). However, activation of adenylate cyclase by forskolin up-regulated Egr–1 biosynthesis in PC12 cells and pancreatic β-cells (Josefsen et al., 1999; Tsai et al., 2001), indicating that the *Egr–1* gene is responsive to elevated cAMP concentrations. These examples shed light on the fact that a genetic element such as the CRE within the Egr–1 promoter is not *per se* active or inactive, but rather reflects cell type-specific differences in the concentrations of signaling molecules or transcriptional activators or repressors. Using a constitutively active mutant of CREB, it was shown unequivocally that the Egr–1 promoter is a target for CREB (Bauer et al., 2005). In contrast, a constitutively active ATF2 mutant only weakly elevated Egr–1 promoter activity (J. Al Sarraj et al., 2005), indicating that CREB, but not ATF2, belongs to the major regulators of *Egr–1* gene transcription.

Interestingly, the Egr–1 protein can bind to its own gene, via the EBS sequence 5'-CGCCCCCGC-3'. As a result, Egr–1 down-regulates the transcription of its own gene (Cao et al., 1993). Although the molecular mechanism of this repressive activity is unknown, this is a further negative feedback loop, in addition to the induction of NAB2 synthesis by Egr–1, that allows only a transient – but not a sustained – synthesis of Egr–1.

18.5
Lessons from *Egr–1*–Deficient Mice

Transgenic mice containing an inactivated *Egr–1* gene have been generated (Lee et al., 1996; Topilko et al., 1997; Mataga et al., 2001). Egr–1$^{-/-}$-homozygous mice are viable, but have a reduced body size. The homozygous mice are infertile due to defects in hormone regulation. The pituitary gland showed a reduced size, in particular affecting the anterior lobe. Homozygous Egr–1$^{-/-}$-mice had a clear reduction in the number of growth hormone-positive cells, and luteinizing hormone (LH)β-expression was completely blocked (Lee et al., 1996; Topilko et al., 1997). A histological examination of hippocampal anatomy using cellular, neuronal, and presynaptic markers revealed that the basic neuronal architecture of the hippocampus is not affected by inactivation of the *Egr–1* gene (Jones et al., 2001). However, behavioral tasks as well as plasticity-related tests revealed that loss of Egr–1 has a tremendous impact on the consolidation of memories and the reconsolidation of recognition memory and long-lasting synaptic enhancement in the hippocampus (Wei et al., 2000; Jones et al., 2001; Bozon et al., 2003; Lee et al., 2004; see also below). The relatively mild phenotype of Egr–1$^{-/-}$-mice may be the result of functional redundancy. The Egr–1–related transcription factors Egr–2, Egr–3, and Egr–4 (Joseph et al., 1988; Patwardhan et al., 1991; Crosby et al., 1992) that bind with their zinc finger domains to Egr–1 target genes may compensate for the loss of Egr–1 function. Alternatively, Egr–1 may be part of a cellular stress response program, as shown by the fact that cellular stressors such as radiation, oxygen deprivation, or reactive oxygen intermediates are powerful inducers of Egr–1 biosynthesis. Thus, the loss-of-function phenotype may only be obvious in stressful situations where Egr–1 governs a coordinated program of gene transcription that allows the stressed cells to acquire an adaptive phenotype.

18.6
Egr–1 Regulates Synaptic Plasticity in the Nervous System

The expression of Egr–1 in the nervous system is regulated by synaptic activity. Accordingly, high-frequency stimulation of excitatory synapses of the hippocampus activates Egr–1 biosynthesis (Cole et al., 1989). Egr–1 functions as a signaling molecule downstream from the NMDA receptor, and contributes to long-lasting synaptic enhancement in the hippocampus (Wei et al., 2000). Clear evidence has been provided that Egr–1 controls synaptic plasticity. The *Egr–1* gene is rapidly transcribed in hippocampal neurons following the induction of long-term potentiation. Experiments performed with mutant mice with a targeted disruption of the *Egr–1* gene revealed that Egr–1 is essential for the maintenance of late long-term potentiation and expression of long-term memory (Jones et al., 2001), indicating that Egr–1 plays a key role in the transition from short- to long-term synaptic plasticity. Loss of Egr–1 expression, however, does not affect short-term memory processes, as shown by the ability of homozygous Egr–1$^{-/-}$-mice to perform olfactory discrimination in social transmission of food preference, and visual discrimination in an object recognition task (Jones et al., 2001). In addition to its role in the consolidation of new memory, Egr–1 has recently been shown to be involved in reconsolidation of recognition memory following reactivation (Bozon et al., 2003). Experiments involving the infusion of antisense oligonucleotides into the hippocampus of rats confirmed the role of Egr–1 in the reconsolidation of long-term contextual fear memory, but questioned the involvement of Egr–1 in the consolidation process (Lee et al., 2004). Infusion of Egr–1 antisense oligonucleotides into the amygdala was shown to interfere with learning and memory processes of fear (Malkani et al., 2004), indicating that Egr–1 plays in the amygdala an important role in long-term learning and memory of fear.

Hippocampal neurons respond to different sensory stimuli including noxious or painful somatosensory stimulation, indicating that the hippocampus is involved in the formation of spatial memory associated with potentially dangerous sensory experience. An analysis of activity-dependent expression of Egr–1 in hippocampal neurons revealed that Egr–1 is required for tissue-injury-induced plastic changes in the hippocampus (Wei et al., 2000). These data indicate that Egr–1 may serve as a regulator of nociception or pain-related plasticity within the hippocampus.

18.7
Correlation Between Proliferation of Astrocytes and Egr–1 Biosynthesis

Since the discovery of the *Egr–1* gene as an "early growth response gene" (Sukhatme et al., 1988), research has been directed towards the function of Egr–1 in growth and proliferation. Egr–1 biosynthesis is strongly stimulated by signals that activate the mitogen-activated/extracellular signal-regulated protein kinase ERK (Kaufmann et al., 2001), the protein kinase that is activated following mitogenic stimulation. In the nervous system, a key feature of astrocyte reactivity is their proliferative response. Mitogens such as EGF and basic-FGF (bFGF) stimulate DNA synthesis in glioma cells and primary astrocytes via activation of the ERK/MAPK signaling cascade (Kaufmann and Thiel, 2001; Riboni et al., 2001; O. Rössler, S. Mayer, G. Thiel, unpublished observations). Amplification and/or mutations of the EGF receptor have been frequently reported in human malignant gliomas, the most common primary tumor of the adult central nervous system. EGF stimulation of astrocytes and glioma cells strongly enhances Egr–1 biosynthesis (Kaufmann and Thiel, 2001; O.G. Rössler, S. Mayer, G. Thiel, unpublished observations). Hence, Egr–1 is most likely an important "late" component of the EGF-initiated signaling cascades in glioma cells, and functions as a "third messenger" connecting growth factor stimulation with changes in gene transcription. Moreover, the expression of the EGF receptor is controlled by Egr–1 via the sequence 5'-GCGGGGGCC–3', encompassing nucleotides –433 to –425 of the EGFR gene promoter (Nishi et al., 2002).

Glial cells synthesize high levels of bFGF that in turn stimulates autocrine growth (Riboni et al., 2001). Interestingly, the human bFGF gene is transactivated by Egr–1 via two Egr–1 binding sites in the proximal promoter of the human bFGF gene (Biesiada et al., 1996; Wang et al., 1997). These sites, which encompass the sequences 5'-CTCCCCCGC–3' and 5'-GCGGGGGTG–3', deviate by one nucleotide from the consensus sequence. The stimulation of bFGF synthesis by Egr–1 indicates the existence of a growth stimulatory autocrine loop, since stimulation of the bFGF receptor by its cognate ligand in turn enhances Egr–1 synthesis via ERK (Wang et al., 1997) (Fig. 18.4).

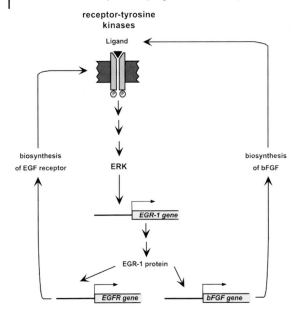

Fig. 18.4 An autocrine loop involving receptor tyrosine kinase-triggered Egr–1 biosynthesis may regulate cell growth. Stimulation of the receptor tyrosine kinases such as the EGF or bFGF receptors triggers the biosynthesis of Egr–1. Egr–1, in turn, binds to the regulatory region of the EGF receptor gene, leading to an up-regulation of EGF receptor expression. Egr–1 also transactivates the gene encoding bFGF, leading to an enhanced biosynthesis of the growth factor bFGF.

The fact that mitogenic stimulation by EGF or bFGF includes an activation of ERK (Tournier et al., 1994; Kaufmann and Thiel, 2001), the kinase that strongly stimulates Egr–1 gene transcription, implies that the induction of Egr–1 biosynthesis is an integral part of the mitogenic pathway and continues the mitogenic signaling cascade via stimulation of growth factor or growth factor receptor synthesis. However, it is worth stating that there are reports to the contrary, attributing to Egr–1 a growth inhibitory function. Egr–1 has been characterized as a tumor suppressor for human glioma cells, based on the fact that Egr–1 overexpression negatively regulates tumor growth (Huang et al., 1995). Likewise, the expression levels of Egr–1 are significantly down-regulated in astrocytomas and glioblastomas (Calogero et al., 2001). Additionally, adenoviral-mediated expression of Egr–1 almost completely abolished the growth of tumor cells *in vitro* (Calogero et al., 2004). Mechanistically, it has been proposed that Egr–1 transactivates the transforming growth factor (TGF)–β1 gene. Newly synthesized TGF-β1, in turn, stimulates the synthesis of the cyclin-dependent protein kinase inhibitor p21$^{WAF1/Cip1}$, leading to cell cycle arrest (Calogero et al., 2001). Further studies are required to solve the discrepancy between the proposed growth-promoting role of Egr–1 for astrocytes and a growth-suppressing role for brain tumor cells.

18.8
Egr–1: A "Pro-apoptotic Protein" for Neurons?

Despite the discovery of Egr–1 as a growth-promoting protein, several reports have been published in recent years describing Egr–1 as a "pro-apoptotic protein" (Liu et al., 1998). In particular, the failure of attempts to generate PC12 cells overexpressing Egr–1 (Qu et al., 1998) suggests that Egr–1 either represses growth or promotes apoptosis.

Which role, if any, Egr–1 plays within apoptotic signaling cascades in neurons is currently not known. According to one scenario, direct transactivation of the *PTEN* gene by Egr–1 should explain the apoptotic activity of Egr–1. PTEN (phosphatase and tensin homologue deleted on chromosome 10) is a lipid phosphatase that cleaves the 3'-phosphate from phosphatidylinositol 3,4,5–trisphosphate (PtdIns(3,4,5)P$_3$) to generate phosphatidylinositol 4,5–bisphosphate (PtdIns(4,5)P$_2$) (Fig. 18.5). PTEN opposes the action of phosphoinositide 3–kinase (PI3–kinase) that phosphorylates the 3'-OH group of the inositol ring in inositol phospholipids. PI3–kinase regulates cell survival via activation of protein kinase B/Akt, and constitutive activation of PI3–kinase induces cellular transformation. Thus, PTEN functions as a tumor suppressor protein by antagonizing the PI3–kinase/Akt signaling cascade. Egr–1 was proposed to directly regulate expression of the *PTEN* gene via a functional 5'-GCGGCGGCG-3' Egr–1 binding site within the 5'-untranslated region (Virolle et al., 2001). A correlation was observed between the biosynthesis of Egr–1 induced by UV or γ-radiation and an up-regulation of PTEN mRNA and protein. In mouse embryonic fibroblasts, derived from Egr–1 knockout-mice, PTEN was not synthesized following UV-radiation, and UV-induced apoptosis was blocked (Virolle et al., 2001). This study, which defines Egr–1 as a positive activator of the *PTEN* gene, places Egr–1 in the regulatory circuit for growth inhibition and apoptosis. The relationship between Egr–1 biosynthesis and PTEN expression in neurons has not been analyzed. Likewise, other proposed scenarios to explain an apoptotic activity of Egr–1, involving a cross-talk between Egr–1 and p53/p73 or c-Jun (Levkovitz and Baraban, 2001; Pignatelli et al., 2003), require further experimental proof. The activity of Egr–1 may also depend on the cell type and the nature of the cytotoxic stimulus. Thus, results obtained with non-neuronal cells may not be directly transferable to the environment of the nervous system.

Injection of the excitotoxin kainic acid is used as an *in-vivo* experimental model of temporal lobe epilepsy. Epileptic seizures are manifest in specific brain regions, including the piriform and entorhinal cortex, amygdaloid complex and hippocampus. Injection of kainic acid is accompanied by a strong and rapid up-regulation of Egr–1 in the hippocampus, cortex, and amygdala (Honkaniemi and Sharp, 1999; Mataga et al., 2001). The expression of Egr–1 lasts for 24 hours in the CA1 and CA3 subfield of the hippocampus (Honkaniemi and Sharp, 1999), the brain area that is massively damaged by kainate. The kainate-induced up-regulation of Egr–1 may represent a stress response, where the expression of Egr–1 target genes is intended to protect the stressed cells from the neurotoxic challenge. Alternatively, the expression of Egr–1 may be an integral part of an apoptotic signaling cascade in neurons

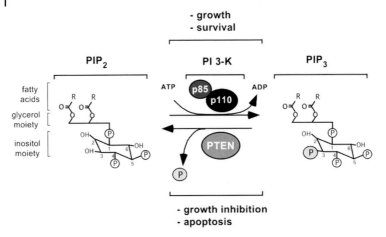

Fig. 18.5 Biological activity of the lipid phosphatase PTEN. Activation of growth factor receptor tyrosine kinases triggers the activation of phosphoinositide 3–kinase (PI3–K). Phosphatidylinositol phosphates are composed of a glycerol moiety and a membrane-associated phosphatidic acid, linked to a cytosolic phosphorylated inositol group. PI3–K phosphorylates phosphatidylinositol 4,5–bisphosphate at the D3 position, generating phosphatidylinositol 3,4,5–trisphosphate (PtdIns(3,4,5)P_3). This metabolite functions as an activator of protein kinase B/Akt, leading to enhanced cell survival and/or cell growth. The lipid phosphatase PTEN removes the D3 phosphate from PtdIns(3,4,5)P_3 and therefore opposes the action of PI3–kinase.

that coordinates transcription of the genes encoding pro-apoptotic proteins. Experiments performed with transgenic mice harboring an inactivated *Egr–1* gene revealed that the presence or absence of Egr–1 has no effect on the rate of kindling, an experimental model of epileptogenesis, and associated mossy fiber sprouting in the hippocampus (Zheng et al., 1998).

18.9
Conclusions and Future Prospects

Considerable progress has been made in the identification of intracellular signaling pathways leading to Egr–1 expression. This progress is the first step to understanding how stimulation of the cells of the nervous system influences the gene expression pattern. To gain better insight into the function of Egr–1 in the nervous system, it will be necessary to identify the target genes for Egr–1 in neurons, astrocytes, and microglia cells. The identification of those target genes should provide clues about the functions of Egr–1 in orchestrating synaptic activity-induced neuronal transcriptional programs. Moreover, we expect to learn more about the role of Egr–1 in the regulation of genes encoding growth-promoting or apoptosis- inducing proteins. The availability of *Egr–1* knockout mice will be very helpful in investigating changes in physiological parameters in the absence of Egr–1. However, additional questions are sometimes raised by these mouse models, mainly because it is difficult to exclude that other transcription factors, related to Egr–1, may compensate for the loss of Egr–1. Therefore, the analysis of Egr–1–lacking cells derived from Egr–1$^{-/-}$-mice must be supplemented by a dominant-negative approach to exclude compensation by related transcription factors.

Acknowledgments

The authors thank Libby Guethlein for critical reading of the manuscript. They also acknowledge a fellowship from the European Graduate School of Neuroscience (EURON) to Luisa Stefano. Research related to the function of Egr–1 in the nervous system is supported by the Deutsche Forschungsgemeinschaft (grants # Th 377/10–1 and SFB 530, C14).

Abbreviations

CRE	cAMP response element
EGF	epidermal growth factor
ERK	extracellular signal-regulated protein kinase
FGF	fibroblast growth factor
GPCR	G-protein-coupled receptor
Grb2	growth factor receptor-bound–2
LH	luteinizing hormone
MAPK	mitogen-activated protein kinase
PDGF	platelet-derived growth factor
Shc	Src-homology–2 containing
Sos	son-of-sevenless
SRE	serum response element
SRF	serum response factor
TGF	transforming growth factor

References

Al-Sarraj A, Thiel G. 2002. Substance P induced biosynthesis of the zinc finger transcription factor Egr–1 in human glioma cells requires activation of the EGF receptor and of extracellular signal-regulated protein kinase. *Neurosci Lett* 332: 111–114.

Al-Sarraj A, Day RM, Thiel G. 2005. Specificity of transcriptional regulation by the zinc finger transcription factors Sp1, Sp3 and Egr–1. *J Cell Biochem* 94:153–167.

Al Sarraj J., Vinson C., Han J. and Thiel G. 2005 Regulation of GTP cyclohydrolase I gene transcription by basic region leucine zipper transcription factors. *J. Cell. Biochem*, in press.

Bauer I, Hohl M, Al-Sarraj A, Vinson C, Thiel G. 2005. Transcriptional activation of the Egr–1 gene mediated by tetradecanoylphorbol acetate and extracellular signal-regulated protein kinase. *Arch Biochem Biophys,* 438: 36–52.

Biesiada E, Razandi M, Levin ER. **1996**. Egr–1 activates basic fibroblast growth factor transcription. Mechanistic implications for astrocyte proliferation. *J Biol Chem* 271: 18576–18581.

Bozon B, Davis S, Laroche S. **2003**. A requirement for the immediate early gene zif268 in reconsolidation of recognition after retrieval. *Neuron* 40: 695–701.

Calogero A, Arcella A, De Gregorio G, Porcellini A, Mercola D, Liu C, Lombari V, Zani M, Giannini G, Gagliardi FM, Caruso R, Gulino A, Frati L, Ragona G. **2001**. The early growth response gene *EGR–1* behaves as a suppressor gene that is down-regulated independent of ARF/Mdm2 but not p53 alterations in fresh human gliomas. *Clin Cancer Res* 7: 2788–2796.

Calogero A, Lombari V, De Gregorio G, Porcellini A, Ucci S, Arcella A, Caruso R, Gagliardi FM, Gulino A, Lanzellta G, Frati L, Mercola D, Ragona G. **2004**. Inhibition of cell growth by EGR–1 in human primary cultures from malignant glioma. *Cancer Cell Int* 4: 1.

Cao X, Mahendran R, Guy GR, Tan YH. **1993**. Detection and characterization of cellular Egr–1 binding to ist recognition site. *J Biol Chem* 268: 16949–16957.

Christy B, Nathans D. **1989a**. Functional serum response elements upstream of the growth factor-inducible gene *zif268*. *Mol Cell Biol* 9: 4889–4895.

Christy B, Nathans D. **1989b**. DNA-binding site of the growth factor-inducible protein Zif268. *Proc Natl Acad Sci USA* 86: 8737–8741.

Christy B, Lau LF, Nathans D. **1988**. A gene activated in mouse 3T3 cells by serum growth factors encodes a protein with "zinc finger" sequences. *Proc Natl Acad Sci USA* 85: 7857–7861.

Cibelli G, Policastro V, Rössler OG, Thiel G. **2002**. Nitric oxide-induced programmed cell death in human neuroblastoma cells is accompanied by the synthesis of Egr–1, a zinc finger transcription factor. *J Neurosci Res* 67: 450–460.

Cohen DM, Gullans SR, Chin WW. **1996**. Urea inducibility of *egr–1* in murine inner medullary collecting duct cells is mediated by the serum response element and adjacent Ets motifs. *J Biol Chem* 271: 12903–12908.

Cole AJ, Saffen DW, Baraban JM, Worley PF. **1989**. Rapid increase of the immediate-early gene messenger RNA in hippocampal neurons by synaptic NMDA receptor activation. *Nature* 340: 474–476.

Corbit KC, Trakul N, Eves EM, Diaz B, Marshall M, Rosner MR. **2003**. Activation of Raf-1 signaling by protein kinase C through a mechanism involving Raf kinase inhibitory protein. *J Biol Chem* 278: 13061–13068.

Crosby SD, Veile RA, Donis-Keller H, Baraban JM, Bhat RV, Simburger KS, Milbrandt J. **1992**. Neural-specific expression, genomic structure, and chromosomal localization of the gene encoding the zinc-finger transcription factor NGFI-C. *Proc Natl Acad Sci USA* 89: 4739–4743.

Daub H, Wallasch C, Lankenau A, Herrlich A, Ullrich A. **1997**. Signal characteristics of G protein-transactivated EGF receptor. *EMBO J* 16: 7032–7044

Day RM, Thiel G, Lum J, Chévere RD, Yang Y, Stevens J, Silbert L, Fanburg BL. **2004**. Hepatocyte growth factor regulates angiotensin converting enzyme expression. *J Biol Chem* 279: 8792–8801.

Ehrengruber MU, Muhlebach SG, Söhrman S, Leutenegger CM, Lester HA, Davidson N. **2000**. Modulation of early growth response (EGR) transcription factor-dependent gene expression by using recombinant adenovirus. *Gene* 258: 63–69.

Elrod-Erickson M, Pabo CO. **1999**. Binding studies with mutants of Zif268. Contribution of individual side chains to binding affinity and specificity in the Zif268 zinc finger DNA complex. *J Biol Chem* 274: 19281–19285.

Gashler AL, Swaminathan S, Sukhatme VP. **1993**. A novel repressor module, an extensive activation domain, and a bipartite nuclear localization signal defined in the immediate-early transcription factor egr–1. *Mol Cell Biol* 13: 4556–4571.

Gineitis D, Treisman R. **2001**. Differential usage of signal transduction pathways defines two types of serum response factor target gene. *J Biol Chem* 276: 24531–24539.

Greisman HA, Pabo CO. **1997**. A general strategy for selecting high-affinity zinc finger proteins for diverse DNA target sites. *Science* 275: 657–661.

Grimes CA, Jope RS. **1999**. Cholinergic stimulation of early growth response–1 DNA binding activity requires protein kinase C and mitogen-activated protein kinase kinase activation and is inhibited by sodium valproate in SH-SY5Y cells. *J Neurochem* 73: 1384–1392.

Honkaniemi, J. and Sharp, FR. (**1999**). Prolonged expression of zinc finger immediate-early gene mRNAs and decreased protein synthesis following kainic acid induced seizures. *Eur J Neurosci* 11: 10–17.

Huang R-P, Liu C, Fan Y, Mercola D, Adamson ED. **1995**. Egr–1 negatively regulates human tumor cell growth via the DNA-binding domain. *Cancer Res* 55:5054–5062.

Jamieson AC, Wang H, Kim S-H. **1996**. A zinc finger directory for high-affinity DNA recognition. *Proc Natl Acad Sci USA* 93: 12834–12839.

Jones MW, Errington ML, French PJ, Fine PJ, Bliss TVP, Garel S, Charney P, Bozon B, Laroche S, Davis S. **2001**. A requirement for the immediate early gene *Zif268* in the expression of late LTP and long-term memories. *Nature Neurosci* 4: 289–296.

Josefsen K, Sorensen LR, Buschard K, Birkenbach M. **1999**. Glucose induces early growth response gene (Egr–1) expression in pancreatic beta cells. *Diabetologia* 42: 195–203.

Joseph LJ, Le Beau MM, Jamieson GA, Acharya S, Shows T, Rowley JD, Sukhatme VP. **1988**. Molecular cloning, sequencing, and mapping of EGR2, a human early growth response gene encoding a protein with "zinc-binding finger" structure. *Proc Natl Acad Sci USA* 85: 7164–7168.

Kaufmann K, Thiel G. **2001**. Epidermal growth factor and platelet-derived growth factor induce expression of Egr–1, a zinc finger transcription factor, in human malignant glioma cells. *J Neurol Sci* 189: 83–91.

Kaufmann K, Bach K, Thiel G. **2001**. Extracellular signal-regulated protein kinases Erk1/Erk2 stimulate expression and biological activity of the transcriptional regulator Egr–1. *Biol Chem* 382: 1077–1081.

Kolch W, Heidecker G, Kochs G, Hummel R, Vahidi H, Mischak H, Finkenzeller G, Marmé D, Rapp UR. **1993**. Protein kinase C alpha activates RAF-1 by direct phosphorylation. *Nature* 364: 249–252.

Kriwacki RW, Schultz SC, Steitz TA, Cardonna JP. **1992**. Sequence-specific recognition of DNA by zinc-finger peptides derived from the transcription factor Sp1. *Proc Natl Acad Sci USA* 89: 9759–9763.

Lee JL, Everitt BJ, Thomas KL. **2004**. Independent cellular processes for hippocampal memory consolidation and reconsolidation. *Science* 304: 839–843.

Lee S, Sadovsky Y, Swirnoff AH, Polish JA, Goda P, Gavrilina G, Milbrandt J. **1996**. Luteinizing hormone deficiency and female infertility in mice lacking the transcription factor NGFI-A (egr–1). *Science* 273: 1219–1221.

Lemaire P, Revelant O, Bravo R, Charnay P. **1988**. Two mouse genes encoding potential transcription factors with identical DNA-binding domains are activated by growth factors in cultured cells. *Proc Natl Acad Sci USA* 85: 4691–4695.

Leserer M, Gschwind A, Ullrich A. **2000**. Epidermal growth factor receptor signal transactivation. *IUBMB Life* 49: 405–409.

Levkovitz Y. and Baraban J.M. **2001**. A dominant negative inhibitor of the Egr–1 family of transcription regulatory factors suppresses cerebellar granule cell apoptosis by blocking c-Jun activation. *J Neurosci* 21: 5893–5901.

Lim RW, Varnum BC, Herschman HR. **1987**. Cloning of tetradecanoyl phorbol ester-induced ‚primary response' sequences and their expression in density-arrested Swiss 3T3 cells and a TPA non-proliferative variant. *Oncogene* 1: 263–270.

Liu C, Rangnekar VM, Adamson E, Mercola D. **1998**. Suppression of growth and transformation and induction of apoptosis by Egr–1. *Cancer Gene Ther* 5: 3–28.

Malkani S, Wallace KJ, Donley MP, Rosen JB. **2004**. An egr–1 (zif268) antisense oligonucleotide infused into the amygdala disrupts fear conditioning. *Learn Mem* 11: 617–624.

Mataga N, Fujishima S, Condie BG, Hensch TK. 2001. Experience-dependent plasticity of mouse visual cortex in the absence of the neuronal activity-dependent marker *egr–1/zif268*. *J Neurosci* 15: 9724–9732.

Meyer RG, Küpper J-H, Kandolf R, Rodemann HP. 2002. Early growth response–1 gene (*Egr–1*) promoter induction by ionizing radiation in U87 malignant glioma cells *in vitro*. *Eur J Biochem* 269: 337–346.

Milbrandt J. 1987. A nerve growth factor-induced gene encodes a possible transcriptional regulatory factor. *Science* 238: 797–799.

Mundschau LJ, Forman LW, Weng H, Faller DV. 1994. Platelet-derived growth factor (PDGF) induction of egr–1 is independent of PDGF receptor autophosphorylation on tyrosine. *J Biol Chem* 269:16137–16142.

Nishi H, Nishi KH, Johnson AC. 2002. Early growth response–1 gene mediates up-regulation of epidermal growth factor receptor expression during hypoxia. *Cancer Res* 62: 827–834.

Pavletich NP, Pabo CO. 1991. Zinc finger-DNA recognition: crystal structure of a zif268–DNA complex at 2.1Å. *Science* 252: 809–817.

Patwardhan S, Gashler A, Siegel MG, Chang LC, Joseph LJ, Shows TB, Le Beau MM, Sukhatme VP. 1991. EGR3, a novel member of the Egr family of genes encoding immediate-early transcription factors. *Oncogene* 6: 917–928.

Pignatelli M, Luna-Medina R., Pérez-Rendón A, Santos A, Perez-Castillo A. 2003. The transcription factor early growth response factor–1 (EGR–1) promotes apoptosis in neuroblastoma cells. *Biochem J* 373: 739–746.

Qu Z, Wolfraim LA, Svaren J, Ehrengruber MU, Davidson N, Milbrandt J. 1998. The transcriptional corepressor NAB2 inhibits NGF-induced differentiation of PC12 cells. *J Cell Biol* 142: 1075–1082.

Riboni L, Viani P, Bassi R, Giussani P, Tettamanti G. 2001. Basic fibroblast growth factor-induced proliferation of primary astrocytes. *J Biol Chem* 276: 12797–12804.

Rössler OG, Thiel G. 2004. Brain-derived neurotrophic factor, epidermal growth factor, or A-Raf induced growth of HaCaT keratinocytes requires extracellular signal-regulated kinase. *Am J Physiol – Cell Physiol* 286: C1118–C1129.

Rolli M, Kotlyarov A, Sakamoto KM, Gaestel M, Neininger A. 1999. Stress-induced stimulation of early growth response gene–1 by p38/stress-activated protein kinase 2 is mediated by a cAMP-responsive promoter element in a MAPKAP kinase–2 independent manner. *J Biol Chem* 274: 19559–19564.

Russo MW, Sevetson BR, Milbrandt J. 1995. Identification of NAB1, a repressor of NGFI-A and Krox20–mediated transcription. *Proc Natl Acad Sci USA* 92:6873–6877.

Sakamoto KM, Bardeleben C, Yates KE, Golde MA, Gasson JC. 1991. 5' upstream sequence and genomic structure of the human primary response gene, Egr–1/TIS8. *Oncogene* 6: 867–871.

Schonwasser DC, Marais RM, Marshall CJ, Parker P. 1998. Activation of the mitogen-activated protein kinase/extracellular signal-regulated kinase pathway by conventional, novel, and atypical protein kinase C isotypes. *Mol Cell Biol* 18: 790–798.

Schwachtgen J-L, Houston P, Campbell C., Sukhatme V., Braddock M. 1998. Fluid shear stress activation of *egr–1* transcription in cultured human endothelial and epithelial cells is mediated via the extracellular signal-regulated kinase 1/2 mitogen-activated protein kinase pathway. *J Clin Invest* 101: 2540–2549.

Sgambato V, Vanhoutte P, Pagès C, Rogard M, Hipskind R, Besson M-J, Caboche J. 1998. In vivo expression and regulation of Elk–1, a target of the extracellular-regulated kinase signaling pathway, in the adult rat brain. *J Neurosci* 18: 214–226.

Shaw PE, Saxton J. 2003. Ternary complex factors: prime nuclear targets for mitogen-activated protein kinases. *Int J Biochem Cell Biol* 35: 1210–1226.

Sukhatme VP, Cao X, Chang LC, Tsai-Morris C-H, Stamenkovich D, Ferreira PCP, Cohen DR, Edwards SA, Shows TB, Curran T, Le Beau MM, Adamson ED. 1988. A zinc finger-encoding gene coregulated with c-fos during growth and differentiation, and after cellular depolarization. *Cell* 53: 37–43.

Svaren J, Sevetson BR, Apel ED, Zimonjic DB, Popescu NC, Milbrandt J. **1996**. NAB2, a corepressor of NGFI-A (egr–1) and Krox20, is induced by proliferative and differentiative stimuli. *Mol Cell Biol* 16: 3545–3553.

Thiel G, Kaufmann K, Magin A, Lietz M, Bach K, Cramer M. **2000**. The human transcriptional repressor protein NAB1: expression and biological activity. *Biochim Biophys Acta* 1493: 289–301.

Topilko P, Schneider-Maunoury S, Levi G, Trembleau A, Gourji D, Driancourt M-A, Rao CV, Charnay P. **1997**. Multiple pituitary and ovarian defects in *Krox–24* (*NGFI-A, Egr–1*)-targeted mice. *Mol Endocrinol* 12: 107–122.

Tournier C, Pomerance M, Gavaret JM, Pierre M. **1994**. MAP kinase cascade in astrocytes. *Glia* 10: 81–88.

Tsai JC, Liu L, Zhang J, Spokes KC, Topper JN, Aird WC. **2001**. Epidermal growth factor induces Egr–1 promoter activity in hepatocytes in vitro and in vivo. *Am J Physiol – Gastrointest Liver Physiol* 281: G1271–G1278.

Virolle T, Adamson ED, Baron D, Mercola D, Mustelin T, de Belle I. **2001**. The Egr–1 transcription factor directly activates PTEN during irradiation-induced signalling. *Nature Cell Biol* 3: 1124–1128.

Wang D, Mayo MW, Baldwin AS, Jr. **1997**. Basic fibroblast growth factor transcriptional autoregulation requires Egr–1. *Oncogene* 14: 2291–2299.

Watson DK, Robinson L, Hodge DR, Kola I, Papas TS, Seth A. **1997**. FLI1 and EWS-FLI1 function as ternary complex factors and ELK1 and SAP1a function as ternary and quaternary complex factors on the Egr1 promoter serum response elements. *Oncogene* 14: 213–221.

Wei F, Xu ZC, Qu Z, Milbrandt J, Zhuo M. **2000**. Role of EGR1 in hippocampal synaptic enhancement induced by tetanic stimulation and amputation. *J Cell Biol* 149: 1325–1333.

Wetzker R, Böhmer FD. **2003**. Transactivation joins multiple tracks to the ERK/MAPK cascade. *Nature Rev Mol Cell Biol* 4: 651–657.

Wu S-Q, Minami T, Donovan DJ, Aird WC. **2002**. The proximal serum response element in the Egr–1 promoter mediates response to thrombin in primary human endothelial cells. *Blood* 100: 4454–4461.

Zheng D, Butler LS, McNamara JO. **1998**. Kindling and associated mossy fibre sprouting are not affected in mice deficient of *NGFI-A/NGFI-B* genes. *Neuroscience* 83: 251–258.

Part III
Transcription Factors in Neuronal Diseases

19
The Presenilin/γ-Secretase Complex Regulates Production of Transcriptional Factors: Effects of FAD Mutations

Nikolaos K. Robakis and Philippe Marambaud

Abstract

First described for its β-amyloid-producing function, the presenilin (PS)-dependent γ-secretase complex has since been shown to catalyze the ε-cleavage of type I transmembrane cell surface proteins producing soluble peptides containing the intracellular domains (ICDs) of the cleaved proteins. These peptides act as regulators of gene expression, suggesting that the PS/γ-secretase processing of cell-surface proteins is a key factor in surface-to-nucleus signal transduction and communication. Signal-induced gene expression mediates neuronal responses to environmental changes, and is a key event in neuronal survival and synaptic function. Many of the cell-surface proteins cleaved by the PS/γ-secretase system are cell-surface receptors involved in diverse functions ranging from development and differentiation to cell-cell adhesion and communication. Thus, the cleavage controlled by the PS/γ-secretase system may be used to transmit information to the interior of the cell, including the nucleus, about specific changes taking place in the extracellular environment. Familial Alzheimer's disease (FAD) mutations may interfere with this flow of information by inhibiting the ε-cleavage of cell-surface proteins and thus down-regulating production of biologically active ICDs. Here, we discuss how the study of the genetic and molecular aspects of Alzheimer's disease (AD) reveals a dual role for the PS/γ-secretase complex in transcriptional regulation and in AD pathogenesis.

19.1
Introduction

Alzheimer's disease (AD) is a progressive neurodegenerative disorder of the central nervous system (CNS) leading to the most common form of dementia. The neuropathology of AD is characterized by large numbers of brain neuritic plaques (NPs), neurofibrillary tangles (NFTs), and neuronal loss in the striatum and cortex of the CNS. NPs are complex extracellular structures containing at their core amyloid depositions of fibrillar A β-protein (Aβ) surrounded by reactive astrocytes, microglia,

Transcription Factors. Edited by Gerald Thiel
Copyright © 2006 WILEY-VCH Verlag GmbH & Co. KGaA, Weinheim
ISBN 3-527-31285-4

and dystrophic neurites. NFTs accumulate intracellularly and consist mainly of paired helical filaments of over-phosphorylated tau protein (Selkoe, 2001). The pathogenesis of the disease is complex, and it may be driven by both environmental and genetic factors. Most AD cases are sporadic and occur after the ages of 65 or 70 years. Sporadic late-onset AD has no clear genetic or environmental etiology, although inheritance of the E4 allele of apolipoprotein E represents a genetic risk factor for late-onset dementia (Pericak-Vance et al., 1991). A small percentage of all AD cases (ca. 5%) have a clear genetic etiology and are classified as familial AD (FAD); this usually occurs at an earlier age and follows a more rapid, progressive course than the sporadic form of AD. Three genetic loci have been linked to early-onset FAD, including the genes encoding the amyloid precursor protein (APP), the precursor of the Aβ peptides (Goldgaber et al., 1987; Kang et al., 1987; Robakis et al., 1987; Tanzi et al., 1987), PS1 (Sherrington et al., 1995), and PS2 (Levy-Lahad et al., 1995; Rogaev et al., 1995). The brain histopathology and clinical course of the disease, however, seem similar in all AD cases, suggesting the involvement of common cellular mechanisms in all forms of AD.

Our understanding of the molecular basis of AD has benefited from observations that many mutations linked to FAD increase production of the Aβ-peptides that aggregate to form the amyloid depositions. These observations supported the amyloid plaque theories of AD, including the cerebrovascular (Glenner, 1988) and the neuritic plaque (Hardy and Higgins, 1992) theories that form the basis of most strategies aimed at treating this disorder. Nevertheless, the amyloid theories remain controversial because of the weak correlation between brain amyloid load and several parameters of neurological dysfunction, including degree of dementia, loss of synapse, loss of neurons, and the distribution of neuronal and cytoskeletal abnormalities in the brain (Terry et al., 1981; Braak and Braak, 1991; Dickson et al., 1995; Neve and Robakis, 1998). Moreover, amyloid depositions at levels similar to those seen in AD are often detected in normal individuals who have no evidence of cognitive impairment (Davies, 2000; Neve and Robakis, 1998; Roses, 1994). These discrepancies prompted the development of the soluble oligomer Aβ theories of AD, which proposed that soluble oligomers of extracellular (Klein et al., 2001) or intracellular (Wilson et al., 1999) Aβ42 not detected by the classical amyloid stains, represent the neurotoxic forms of Aβ. Indeed, recent reports have suggested that soluble oligomeric Aβ may interfere with synaptic plasticity and memory formation (Dodart et al., 2002; Walsh et al., 2002; Cleary et al., 2005). Although the soluble Aβ theories are also consistent with the data that FAD-linked mutations increase the production of Aβ peptides, no correlation has been found between Aβ increase and age of onset associated with specific FAD mutations. The identification of neurotoxic soluble Aβ oligomers specific to AD may provide strong support to the soluble Aβ theories of AD (for reviews, see Hardy and Selkoe, 2002; and Robakis, 2003).

19.2
Processing of APP and FAD

Most cellular APP is processed through the non-amyloidogenic secretory pathway that cleaves APP within the Aβ sequence (Anderson et al., 1991). This cleavage is catalyzed by several constitutive and inducible proteases, collectively called α-secretases, including TACE (TNFα converting enzyme or ADAM–17) and ADAM–10, both members of the disintegrin and metalloprotease family of proteases (Buxbaum et al., 1998; Lammich et al., 1999; Lopez-Perez et al., 2001). In the amyloidogenic pathway, sequential processing of APP by the endoproteolytic activities of β-secretase and γ-secretase leads to the production of Aβ peptides. β-Secretase is a membrane-bound aspartyl protease, named BACE1, that is homologous to the pepsin family (Citron, 2004). This protease acts at the N-terminus of the Aβ sequence of APP, liberating an amyloid precursor of 99 residues, termed C99. This APP fragment contains the entire Aβ sequence plus the transmembrane and cytoplasmic domains of APP (Fig. 19.1). In the amyloidogenic processing, C99 is further processed by the PS/γ-secretase complex at several sites located close to the middle of the transmembrane sequence of APP, producing various Aβ peptides each containing 39 to 43 amino-acid residues. The major cleavages however, take place either after Val40, producing the 40–amino acids Aβ40, the most abundant Aβ peptide, or after Ala42, producing the highly fibrilogenic peptide Aβ42 that represents about 10% of all Aβ (Fig. 19.1). The PS/ γ-secretase system has an unusual aspartyl protease activity formed by a multicomponent high molecular-weight complex that includes presenilins, nicastrin, APH–1, and PEN–2 (for a review, see Haass, 2004).

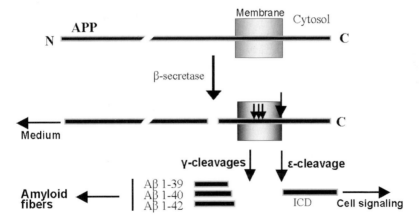

Fig. 19.1 Secretase processing of amyloid precursor protein (APP), illustrating the sites of β- and γ—secretase cleavages important for the production of Aβ peptides, and ε-cleavage that is the final step in the production of biologically active intracellular domains (ICD).

Recent evidence shows that, in addition to the amyloidogenic γ-cleavages of APP, the PS/γ-secretase system promotes the ε-cleavage of several type I transmembrane proteins, including APP, Notch1 receptor, E-cadherin, N-cadherin and CD44 (Schroeter et al., 1998; De Strooper et al., 1999; Gu et al., 2001; Sastre et al., 2001; Yu et al., 2001; Lammich et al., 2002; Marambaud et al., 2002; Weidemann et al., 2002). The ε-cleavage results in the release of soluble cytosolic peptides containing the intracellular domains (ICDs) of the cleaved substrate proteins. Although this cleavage is also sensitive to γ-secretase inhibitors, it takes place further downstream from the amyloidogenic γ-cleavages at a site closer to the membrane/cytoplasm interface than the γ-cleavages (see Fig. 19.1). In certain cases, like E- and N-cadherin, the ε-cleavage is greatly stimulated by calcium influx (Marambaud et al., 2002, 2003). To date, at least 16 cell-surface type I transmembrane proteins and one multipass transmembrane receptor have been shown to be processed by the PS/γ-secretase system, producing soluble peptides containing the ICD sequence of the cleaved proteins. Research conducted during the past few years has shown that these peptides may migrate to the nucleus where they act as regulators of gene expression, while others remain in the cytoplasm where they may regulate the metabolism of transcription factors.

Several groups have reported evidence that many PS1 FAD mutations inhibit the processing of type I transmembrane proteins and receptors, including N-cadherin, APP and Notch–1, at the ε-cleavage site, thus reducing production of the ICDs (Song et al., 1999; Nakajima et al., 2000; Chen et al., 2002; Moehlmann et al., 2002; Marambaud et al., 2003; Walker et al., 2005). These data support an alternative hypothesis of AD based on the observation that ICD peptides may have important biological functions, and that inhibition of their production in neuronal cells may be detrimental to neuronal function and survival (Fortini, 2003; Robakis, 2003). Thus, recent findings on the transcriptional and signal transduction properties of the ICD peptides are important not only because they shed light on the biological function of the PS/γ-secretase proteolytic system but also because they implicate transcriptional and signaling dysregulations in the mechanisms of AD neurodegeneration. Below, we discuss several paradigms illuminating the biological function of these peptides and their potential involvement in AD.

19.3
The Presenilins

Mutations in the genes encoding PS1, and its close homologue PS2, are responsible for most genetic forms of AD. To date, 135 mutations have been found in the *PS1* gene and 10 in the *PS2* gene (see http://molgen-www.uia.ac.be/ADMutations). As a result, the biological functions of these proteins and their potential involvement in neurodegeneration have been the subject of intense investigation. The genes encoding PS1 and PS2 are located on chromosomes 14 and 1, respectively (Levy-Lahad et al., 1995; Rogaev et al., 1995; Sherrington et al., 1995). PS1 is ubiquitously expressed, and in the brain it is found at high levels in neuronal cells (Sherrington et al., 1995; Elder et al., 1996). Both proteins are multipass transmembrane peptides containing nine hydrophobic regions, and although their exact membrane topology is still under investigation, a commonly accepted model features eight transmembrane domains where the N- and C-termini are both located in the cytoplasm (Doan et al., 1996; Li and Greenwald, 1996; De Strooper et al., 1997). PS1 holoprotein contains 467 residues, and in the cell it undergoes constitutive endoproteolysis between the transmembrane domains VI and VII, by an unknown catalytic activity termed preseninilase (Thinakaran et al., 1996; Campbell et al., 2003). This endoproteolytic event may occur in the endoplasmic reticulum (ER), and leads to the formation of a stable PS heterodimer (Podlisny et al., 1997; Ratovitski et al., 1997; Capell et al., 1998; Zhang et al., 1998). Although early studies indicated that presenilins are mostly localized in the ER and Golgi (Kovacs et al., 1996; Walter et al., 1996), more recent studies showed that in several cell systems and in tissues PS1 is also found at the cell surface (Georgakopoulos et al., 1999; Ray et al., 1999a). Indeed, in cells that form cadherin-dependent cell-cell contacts – such as confluent epithelial cells and neuronal synapses – PS1 is found at the plasma membrane where it co-localizes and interacts directly with the adhesion receptor E-cadherin (Georgakopoulos et al., 1999; Baki et al., 2001). These data are important because they demonstrate that PS1 is found in a subcellular compartment where γ-secretase-mediated cleavages of cell-surface receptors occur (Chyung et al., 2004). PS1 fragments, which may constitute the biologically active form of PS, are found in complexes with many cellular proteins including APP (Weidemann et al., 1997; Xia et al., 1997), Notch–1 (Ray et al., 1999b), β-catenin (Zhou et al., 1997; Tesco et al., 1998; Yu et al., 1998; Georgakopoulos et al., 1999), cadherins (Georgakopoulos et al., 1999; Baki et al., 2001), and telencephalin (Annaert et al., 2001). Many of the proteins that form complexes with PSs have been shown to be cleaved by the γ-secretase activity at the γ- or ε-sites, or both. In addition to their γ-secretase-related functions, presenilins have been shown to have γ-secretase-independent functions, including stabilization of the adherens junctions (Baki et al., 2001), destabilization of β-catenin (Kang et al., 2002), and the stimulation of the PI3K/Akt cell survival pathway (Baki et al., 2004).

19.4
The Notch1 ICD (NICD) Mediates Transcriptional and Developmental Functions Associated with Notch1 Receptor

Upon ligand binding, Notch1 protein is cleaved by the PS/γ-secretase system to produce the biologically active peptide NICD, which has been shown to migrate to the nucleus where it acts as a transcriptional cofactor of the CSL protein family (Iso et al., 2003). In Notch signaling, CSL (also known as RBP-Jx) promotes transcription of target genes such as the *HES* (Hairy/Enhancer of Split) gene family of basic helix-loop-helix transcriptional regulators. HES, a downstream effector of Notch signal transduction, regulates cell fate decision by controlling cell cycle and differentiation. Genetic screening revealed that sel–12, the PS homologue of *Caenorhabditis elegans* (Levitan and Greenwald, 1995), is a facilitator lin–12, the worm homologue of Notch1 (Levitan et al., 1996; Baumeister et al., 1997). Loss of sel–12 function results in an egg-laying defect phenotype due to a deficient lin–12 signaling. PS1 expression rescues this defect (Levitan et al., 1996; Baumeister et al., 1997). PS-null mutants of *Drosophila* display developmental defects resembling those observed in flies lacking Notch (Struhl and Greenwald, 1999; Ye et al., 1999). These mutants produce no NICD, which suggests that the phenotype caused by the PS1 mutations may be due to the inhibition of NICD (De Strooper et al., 1999; Song et al., 1999; Struhl and Greenwald, 1999; Ye et al., 1999). In mice, PS1 deficiency leads to perinatal lethality (Shen et al., 1997; Wong et al., 1997). The PS1 knockout mice display developmental abnormalities consistent with loss of Notch function (Shen et al., 1997; Wong et al., 1997). To examine the effects of a PS1 loss-of-function in late embryogenesis, a brain-specific PS1 transgene was used to rescue the embryonic lethality at the perinatal stage (Wang et al., 2003). The authors reported that lack of PS1 expression in the kidneys of these mice results in nephrogenesis defect and in reduced expression of several Notch target genes, such as Hesr1, Delta like–1 (Dll1), and Jag1 (Wang et al., 2003). On the other hand, kidney defects could also result from dysfunction of cadherin-based adhesion due to the absence of PS1 (see below).

Immunopurification of PS1–containing complexes led to the discovery of nicastrin as a cofactor of the PS/γ-secretase system (Yu et al., 2000). Nicastrin suppression in *Drosophila* and *C. elegans* leads to defects resembling the phenotypes induced by presenilin deficiency (Goutte et al., 2000; Yu et al., 2000; Chung and Struhl, 2001; Hu et al., 2002; Lopez-Schier and St Johnston, 2002). Two other cofactors of the PS/γ-secretase system, APH–1 (anterior pharynx-defective phenotype) and PEN–2 (presenilin enhancer), were identified by screening for enhancers of a presenilin-dependent Notch-deficient phenotype (Francis et al., 2002; Goutte et al., 2002). Suppression of APH–1 or PEN–2 in *C. elegans* promotes an egg-laying defect similar to the phenotype caused by sel–12 loss-of-function mutations (Levitan and Greenwald, 1995; Goutte et al., 2002). Together, these studies indicate that PS1 plays a fundamental role *in vivo* in Notch-mediated transcription and signaling during embryonic development. Although these data suggest the involvement of NICD in these Notch-related functions, rescue experiments to test whether NICD is able to rectify these defects have not been reported.

19.5
Transcriptional Function of the APP ICD (AICD)

Similar to Notch–1, APP seems to regulate transcription via the release of its PS/γ-secretase-derived intracellular domain, AICD. Fe65 is a multidomain nuclear adaptor protein that regulates transcription by binding to nuclear histone acetyltransferase Tip60. Fe65 binds the cytoplasmic domain of intact APP in the cytosol, thus preventing its nuclear translocation and transcriptional activity (Minopoli et al., 2001). Upon ε-cleavage of APP and AICD release, Fe65 dissociates from APP and translocates to the nucleus where it binds Tip60 and the nucleosome assembly factor SET to promote transcription (Cao and Sudhof, 2001; Cupers et al., 2001; Kimberly et al., 2001; Telese et al., 2005). It is not clear, however, whether AICD itself translocates to the nucleus to form a transcriptionally active complex with Fe65 and Tip60 (Baek et al., 2002) or is degraded in the cytoplasm (Cao and Sudhof, 2004).

Several candidate gene targets of the AICD-dependent transcription complex have been proposed, such as the IL–1β-regulated gene *KAI1* (Baek et al., 2002), *ace–1* and *ace–2* (Bimonte et al., 2004). The *ace–1* and *ace–2* genes code for the two major acetylcholinesterase activities in the worm, and mutations in the Fe65 homologue in *C. elegans* FEH–1 were found to reduce expression of *ace–1* and *ace–2*. FEH–1 binds the APP homologue APL–1, and in the worm down-regulation of either FEH–1 or APL–1 leads to identical phenotypes of dysregulation of pharyngeal activity (Zambrano et al., 2002). AICD may also act as a transcriptional repressor in Notch signaling by binding the cytosolic adaptor proteins Numb and Numb-like (Roncarati et al., 2002). Furthermore, cytoplasmic fragments of APP bind to and promote the proteasomal degradation of the nuclear factor PAT1 (Gao and Pimplikar, 2001). This latest observation resembles the mechanism of transcriptional regulation by N-cadherin, where the γ-secretase-derived intracellular domain of N-cadherin binds to the transcriptional co-activator CBP to promote its proteasomal degradation, ultimately leading to repression of CBP-dependent gene expression (see below). Together, the available data suggest that AICD plays a central role in the transcriptional regulation of specific genes.

19.6
PS1 and β-Catenin-Mediated Transcription

The use of the yeast two-hybrid system and of co-immunoprecipitation techniques has led to the identification of catenins as PS1 interacting proteins (Zhou et al., 1997; Tesco et al., 1998; Yu et al., 1998; Georgakopoulos et al., 1999). The PS1 sequence that mediates the complex with β-catenin has been mapped within PS1 residues 330 to 360 (Saura et al., 2000; Soriano et al., 2001). β-Catenin plays a dual role in cell-cell adhesion and in transactivation. Indeed, β-catenin is a component of the cadherin-based adherens junctions, where it functions to link cell-surface cadherins to the actin cytoskeleton (see also below). Furthermore, β-catenin functions as a downstream effector for the Wnt/wingless signaling cascade, where it interacts with transcription factors of the lymphoid enhancer binding factor (LEF)/T-cell-specific factor (TCF) family, regulating gene expression and cell-fate decisions during embryonic development (Ben-Ze'ev and Geiger, 1998). Mutations in various components of the β-catenin signaling pathway leading to increased activation of β-catenin transcriptional activity are frequently observed during tumor progression (Roose and Clevers, 1999). Several lines of evidence indicate that PS1 represses β-catenin signaling. PS1 deficiency in *Drosophila* results in the cytosolic accumulation of the β-catenin homologue, *armadillo* (Noll et al., 2000), and genetic screening identified *Drosophila* PS1 as a Wnt/wingless signaling repressor (Cox et al., 2000). Furthermore, loss of PS1 expression in embryonic fibroblasts leads to β-catenin accumulation and to an increase in LEF-dependent cyclin D1 transcription (Soriano et al., 2001). Strikingly, reduced expression of PS1 in an animal model leads to enhanced β-catenin signaling and skin tumorigenesis (Xia et al., 2001). PS1 was proposed to promote β-catenin phosphorylation and degradation in the ubiquitin/proteasome system via a mechanism that bypasses the canonical Wnt/Axin/APC pathway (Kang et al., 2002). It is still not clear, however, what is the mechanism of association of PS1 and β-catenin. That both proteins bind directly to cadherins (Baki et al., 2001) suggests that cadherins may mediate the physical and functional interactions between PS1 and β-catenin (Serban et al., 2005).

19.7
PS1 is a Critical Regulator of Cadherin-Dependent Cell-Cell Adhesion and Signal Transduction

By using immunocytochemistry and immunoprecipitation techniques, PS1 was found in adherens junctions, where it interacts with several members of the cadherin/catenin cell-cell adhesion complex (Georgakopoulos et al., 1999; Baki et al., 2001). Classic cadherins, including epithelial (E)– and neural (N)-cadherins, belong to the cadherin superfamily of cell adhesion molecules. Through calcium-dependent homophilic interactions between their extracellular domains, these cell-surface proteins mediate cell-cell adhesion and control morphogenesis and structural integrity of almost all solid tissues (Gumbiner, 1996). Cadherin-based adherens junctions (AJ) are specialized forms of cellular adhesive contacts at which cadherins form complexes with cytosolic catenins which in turn bind the actin cytoskeleton. PS1 binds directly the juxtamembrane domain of E-cadherin and stabilizes the E-cadherin complexes with β- and γ-catenins (Baki et al., 2001). Moreover, PS1 stabilizes the cytoskeletal association of the AJ complexes and increases calcium-dependent cell-cell aggregation (Baki et al., 2001). In contrast, during embryonic development or under conditions of cell-cell dissociation, PS1 promotes a γ-secretase-dependent proteolysis of E-cadherin leading the disassembly of AJs (Marambaud et al., 2002). As previously reported for APP and Notch, E-cadherin processing is severely affected by knockout of the *PS1* gene in mouse embryos and by the γ-secretase inhibitor, L–685,458 and at least some of the phenotypes of PS1–knockout mice may be due to cadherin malfunction. Unlike APP and Notch, which are cleaved within the transmembrane domain, the γ-secretase activity cleaves E-cadherin at the predicted membrane/cytosol interface (Marambaud et al., 2002).

The remodeling of cell-cell interactions is a central determinant for several functions, including morphogenesis, tissue repair, cell migration or cell death, and implies disassembly of the cadherin-based AJ (Nelson and Nusse, 2004). The major catalytic activities involved in the degradation of cell-surface cadherins are the matrix metalloproteinases (MMPs). Under conditions of cell-cell dissociation, E-cadherin is cleaved by an unknown MMP to produce a membrane-bound C-terminal fragment termed E-Cad/CTF1. E-Cad/CTF1, which is anchored to the actin cytoskeleton through its association with β- and α-catenins, is then cleaved by γ-secretase to produce a soluble C-terminal ICD fragment, termed E-Cad/CTF2. This process dissociates β- and α-catenins from the AJ complexes, thereby increasing their cytosolic pools (Marambaud et al., 2002). Thus, the PS1–dependent γ-secretase activity plays a critical role in AJ disassembly and in the regulation of the soluble levels of the transcriptional factor β-catenin. PS1 has therefore two distinct functions in cell-cell adhesion and dissociation: (a) it binds cadherins, thus stabilizing the cadherin/catenin complex and stimulating cell-cell adhesion; and (b) under conditions of cell-cell dissociation, PS1 promotes the γ-secretase cleavage of E-cadherin leading to AJ disassembly.

In the CNS and during the early postnatal period, cadherins promote brain circuitry through their involvement in axon guidance, target recognition, and synaptogenesis (Tepass et al., 2000). Cadherins are also expressed in mature synapses where they constitute a main structural component and may regulate synaptic plasticity (Murase and Schuman, 1999; Huntley et al., 2002). Indeed, N-cadherin was found to control synaptic strength by affecting long-term potentiation (LTP), the most studied form of synaptic plasticity of the hippocampus (Tang et al., 1998). Changes in synaptic strength that coordinate synaptic plasticity implies a remodeling of the synaptic architecture and most likely disassembly of cadherin-based AJ at the synapse (Murase and Schuman, 1999). PS1 was found in complexes with brain E- and N-cadherins, and has been localized at synaptic contacts (Georgakopoulos et al., 1999). Recent results show that stimulation of the NMDA receptor (NMDAR) promotes the γ-secretase processing of N-cadherin in primary neurons, suggesting a physiological relevance of this pathway during synaptic activation (Marambaud et al., 2003).

NMDAR is a glutamate- and voltage-gated channel which is permeable to calcium ions. This receptor is localized on the postsynaptic membrane at excitatory synapses and plays a central role in synaptic transmission (Platenik et al., 2000). Activation of NMDAR promotes signaling cascades leading to phosphorylation and activation of CREB transcription factor (Greengard, 2001), a mechanism that controls synaptic plasticity and may underlie learning and memory (Kandel, 2001). CREB transcriptional activity depends on its ability to recruit the transcriptional cofactor CBP (CREB binding protein) (Impey and Goodman, 2001). Recent studies conducted by our group revealed that N-Cad/CTF2, the N-cadherin ICD derived through the γ-secretase-dependent ε-cleavage of N-cadherin, promotes proteasomal degradation of CBP, thus inhibiting CREB transcriptional function. Indeed, N-Cad/CTF2 binds endogenous CBP and promotes its translocation to the cytoplasm where CBP is ubiquitinated and degraded in the proteasome (Marambaud et al., 2003). Consistent with its role in CBP metabolism, N-Cad/CTF2 suppresses CRE-mediated transactivation and c-*fos* expression, two indicators of CREB-mediated transcription. In contrast, genetic and pharmacological approaches aimed at inhibiting PS1–dependent γ-secretase activity decreased N-Cad/CTF2 production and stimulated CREB-mediated transcription (Marambaud et al., 2003). Together, these data show that the PS1/γ-secretase system regulates the activity of the CBP/CREB transcription pathway, and reveal novel mechanisms that involve processing of cell-surface transmembrane proteins in the regulation of gene expression (for a review, see Rao and Finkbeiner, 2003).

Numerous FAD-linked PS1 mutations were found to severely inhibit the γ-secretase cleavage of N-cadherin, resulting in an up-regulation of CBP/CREB-mediated transcriptional activity (Marambaud et al., 2003). These novel findings, together with the observation that FAD-linked PS1 mutations overstimulate LTP (Parent et al., 1999; Schneider et al., 2001), raise the possibility that by inhibiting cleavage of synaptic N-cadherin PS, FAD mutants may impair neuronal signaling and transcription thus affecting synaptic plasticity and neuronal survival. Recently, conditional knockout mice lacking both presenilins in the postnatal forebrain were generated in order to study presenilin function in the adult brain. The authors reported that presenilin deficiency at 2 months of age, leads to a defect in hippocampal learning and memory by affecting LTP and NMDAR-dependent synaptic activity. These defects were accompanied by a selective decrease in synaptic levels of NMDAR subunits (NR1 and NR2A) and of αCaMKII (Saura et al., 2004). Furthermore, CBP and c-Fos expression was reduced in these mice, which suggests that presenilins act in the adult brain as positive regulators of CRE-dependent transcription. These results challenge the conclusion that CRE-dependent gene expression is increased in $PS1^{-/-}$ embryonic brains (Marambaud et al., 2003). Although these two studies are difficult to compare (adult versus embryonic brains; and $PS1\&2^{-/-}$ versus $PS1^{-/-}$ genotypes), the possibility cannot be excluded that depending on the developmental or degenerative states, presenilins promote different signaling events leading either to a transcriptional stimulation by facilitating CBP expression (Saura et al., 2004), or to a transcriptional repression by stimulating CBP proteasomal degradation (Marambaud et al., 2003).

Dysregulation of CRE-dependent transcription is believed to contribute to the neuronal dysfunction in polyglutamine-related neurodegenerative disorders such as Huntington disease (HD) (Ross, 2002). Recent studies have demonstrated that HD transgenic mice, which develop massive neurodegeneration, exhibit an overstimulation of CRE-dependent transcription in neurons (Obrietan and Hoyt, 2004). An abnormal increase in CRE-dependent transcription may therefore contribute to the neurodegeneration of AD and hence may follow a pathogenic mechanism similar to those observed in other neurodegenerative disorders.

19.8
Conclusions

Emerging evidence indicates a central role for presenilins in diverse signaling pathways leading to transcriptional regulation. PS1 destabilizes the transcription factor β-catenin, stimulates PI3K/Akt signaling, and controls the release of transcriptional cofactors derived from the γ-secretase-dependent cleavages of cell-surface proteins including APP, cadherins, and Notch receptor. These observations illuminate the important roles that PSs play in intercellular communication, signal transduction, and gene expression. Moreover, findings that FAD-linked mutations inhibit the production of factors involved in cell-surface-to-nucleus communication and in gene expression provide support for the hypothesis that transcriptional dysregulation contributes to the neurodegeneration and memory defects characteristic of AD.

Abbreviations

AD	Alzheimer's disease
AJ	adherens junctions
APP	amyloid precursor protein
CBP	CREB binding protein
CNS	central nervous system
CRE	cAMP-responsive-element
ER	endoplasmic reticulum
FAD	Familial Alzheimer's disease
HD	Huntington disease
ICD	intracellular domain
LEF	lymphoid enhancer binding factor
LTP	long-term potentiation
MMP	matrix metalloproteinase
NFT	neurofibrillary tangle
NMDA	*N*-methyl-D-aspartate
NMDAR	*N*-methyl-D-aspartate receptor
NP	brain neuritic plaque
TCF	T-cell-specific factor

References

Anderson, J.P., Esch, F.S., Keim, P.S., Sambamurti, K., Lieberburg, I., Robakis, N.K. (**1991**) Exact cleavage site of Alzheimer amyloid precursor in neuronal PC–12 cells. *Neurosci Lett*, 128, 126–128.

Annaert, W.G., Esselens, C., Baert, V., Boeve, C., Snellings, G., Cupers, P., Craessaerts, K., De Strooper, B. (**2001**) Interaction with telencephalin and the amyloid precursor protein predicts a ring structure for presenilins. *Neuron*, 32, 579–589.

Baek, S.H., Ohgi, K.A., Rose, D.W., Koo, E.H., Glass, C.K., Rosenfeld, M.G. (**2002**) Exchange of N-CoR corepressor and Tip60 coactivator complexes links gene expression by NF-kappaB and beta-amyloid precursor protein. *Cell*, 110, 55–67.

Baki, L., Marambaud, P., Efthimiopoulos, S., Georgakopoulos, A., Wen, P., Cui, W., Shioi, J., Koo, E., Ozawa, M., Friedrich, V.L., Jr., Robakis, N.K. (**2001**) Presenilin–1 binds cytoplasmic epithelial cadherin, inhibits cadherin/p120 association, and regulates stability and function of the cadherin/catenin adhesion complex. *Proc Natl Acad Sci USA*, 98, 2381–2386.

Baki, L., Shioi, J., Wen, P., Shao, Z., Schwarzman, A., Gama-Sosa, M., Neve, R., Robakis, N.K. (**2004**) PS1 activates PI3K thus inhibiting GSK–3 activity and tau overphosphorylation: effects of FAD mutations. *EMBO J*, 23, 2586–2596.

Baumeister, R., Leimer, U., Zweckbronner, I., Jakubek, C., Grunberg, J., Haass, C. (**1997**) Human presenilin–1, but not familial Alzheimer's disease (FAD) mutants, facilitate *Caenorhabditis elegans* Notch signalling independently of proteolytic processing. *Genes Funct*, 1, 149–159.

Ben-Ze'ev, A., Geiger, B. (**1998**) Differential molecular interactions of beta-catenin and plakoglobin in adhesion, signaling and cancer. *Curr Opin Cell Biol*, 10, 629–639.

Bimonte, M., Gianni, D., Allegra, D., Russo, T., Zambrano, N. (**2004**) Mutation of the feh–1 gene, the *Caenorhabditis elegans* orthologue of mammalian Fe65, decreases the expression of two acetylcholinesterase genes. Eur J Neurosci, 20, 1483–1488.

Braak, H., Braak, E. (**1991**) Neuropathological stageing of Alzheimer-related changes. *Acta Neuropathol (Berl)*, 82, 239–259.

Buxbaum, J.D., Liu, K.N., Luo, Y., Slack, J.L., Stocking, K.L., Peschon, J.J., Johnson, R.S., Castner, B.J., Cerretti, D.P., Black, R.A. (**1998**) Evidence that tumor necrosis factor alpha converting enzyme is involved in regulated alpha-secretase cleavage of the Alzheimer amyloid protein precursor. *J Biol Chem*, 273, 27765–27767.

Campbell, W.A., Reed, M.L., Strahle, J., Wolfe, M.S., Xia, W. (**2003**) Presenilin endoproteolysis mediated by an aspartyl protease activity pharmacologically distinct from gamma-secretase. *J Neurochem*, 85, 1563–1574.

Cao, X., Sudhof, T.C. (**2001**) A transcriptionally [correction of transcriptively] active complex of APP with Fe65 and histone acetyltransferase Tip60. *Science*, 293, 115–120.

Cao, X., Sudhof, T.C. (**2004**) Dissection of amyloid-beta precursor protein-dependent transcriptional transactivation. *J Biol Chem*, 279, 24601–24611.

Capell, A., Grunberg, J., Pesold, B., Diehlmann, A., Citron, M., Nixon, R., Beyreuther, K., Selkoe, D.J., Haass, C. (**1998**) The proteolytic fragments of the Alzheimer's disease-associated presenilin-1 form heterodimers and occur as a 100–150–kDa molecular mass complex. *J Biol Chem*, 273, 3205–3211.

Chen, F., Gu, Y., Hasegawa, H., Ruan, X., Arawaka, S., Fraser, P., Westaway, D., Mount, H., St George-Hyslop, P. (**2002**) Presenilin 1 mutations activate gamma 42–secretase but reciprocally inhibit epsilon-secretase cleavage of amyloid precursor protein (APP) and S3–cleavage of notch. *J Biol Chem*, 277, 36521–36526.

Chung, H.M., Struhl, G. (**2001**) Nicastrin is required for Presenilin-mediated transmembrane cleavage in *Drosophila*. *Nat Cell Biol*, 3, 1129–1132.

Chyung, J.H., Raper, D.M., Selkoe, D.J. (**2005**) Gamma-secretase exists on the plasma membrane as an intact complex that accepts substrates and effects intramembrane cleavage. *J Biol Chem*, 280, 4383–4392.

Citron, M. (**2004**) Beta-secretase inhibition for the treatment of Alzheimer's disease – promise and challenge. *Trends Pharmacol Sci*, 25, 92–97.

Cleary, J.P., Walsh, D.M., Hofmeister, J.J., Shankar, G.M., Kuskowski, M.A., Selkoe, D.J., Ashe, K.H. (**2005**) Natural oligomers of the amyloid-beta protein specifically disrupt cognitive function. *Nat Neurosci*, 8, 79–84.

Cox, R.T., McEwen, D.G., Myster, D.L., Duronio, R.J., Loureiro, J., Peifer, M. (**2000**) A screen for mutations that suppress the phenotype of *Drosophila* armadillo, the beta-catenin homolog. *Genetics*, 155, 1725–1740.

Cupers, P., Orlans, I., Craessaerts, K., Annaert, W., De Strooper, B. (**2001**) The amyloid precursor protein (APP)-cytoplasmic fragment generated by gamma-secretase is rapidly degraded but distributes partially in a nuclear fraction of neurones in culture. *J Neurochem*, 78, 1168–1178.

Davies, P. (**2000**) A very incomplete comprehensive theory of Alzheimer's disease. *Ann N Y Acad Sci*, 924, 8–16.

De Strooper, B., Annaert, W., Cupers, P., Saftig, P., Craessaerts, K., Mumm, J.S., Schroeter, E.H., Schrijvers, V., Wolfe, M.S., Ray, W.J., Goate, A., Kopan, R. (**1999**) A presenilin–1-dependent gamma-secretase-like protease mediates release of Notch intracellular domain. *Nature*, 398, 518–522.

De Strooper, B., Beullens, M., Contreras, B., Levesque, L., Craessaerts, K., Cordell, B., Moechars, D., Bollen, M., Fraser, P., George-Hyslop, P.S., Van Leuven, F. (**1997**) Phosphorylation, subcellular localization, and membrane orientation of the Alzheimer's disease-associated presenilins. *J Biol Chem*, 272, 3590–3598.

Dickson, D.W., Crystal, H.A., Bevona, C., Honer, W., Vincent, I., Davies, P. (**1995**) Correlations of synaptic and pathological markers with cognition of the elderly. *Neurobiol Aging*, 16, 285–298; discussion 298–304.

Doan, A., Thinakaran, G., Borchelt, D.R., Slunt, H.H., Ratovitsky, T., Podlisny, M., Selkoe, D.J., Seeger, M., Gandy, S.E., Price, D.L., Sisodia, S.S. (**1996**) Protein topology of presenilin 1. *Neuron*, 17, 1023–1030.

Dodart, J.C., Bales, K.R., Gannon, K.S., Greene, S.J., DeMattos, R.B., Mathis, C., DeLong, C.A., Wu, S., Wu, X., Holtzman, D.M., Paul, S.M. (**2002**) Immunization reverses memory deficits without reducing brain Abeta burden in Alzheimer's disease model. *Nat Neurosci*, 5, 452–457.

Elder, G.A., Tezapsidis, N., Carter, J., Shioi, J., Bouras, C., Li, H.C., Johnston, J.M., Efthimiopoulos, S., Friedrich, V.L., Jr., Robakis, N.K. (**1996**) Identification and neuron specific expression of the S182/presenilin I protein in human and rodent brains. *J Neurosci Res*, 45, 308–320.

Fortini, M.E. (**2003**) Neurobiology: double trouble for neurons. *Nature*, 425, 565–566.

Francis, R., McGrath, G., Zhang, J., Ruddy, D.A., Sym, M., Apfeld, J., Nicoll, M., Maxwell, M., Hai, B., Ellis, M.C., Parks, A.L., Xu, W., Li, J., Gurney, M., Myers, R.L., Himes, C.S., Hiebsch, R., Ruble, C., Nye, J.S., Curtis, D. (**2002**) aph-1 and pen-2 are required for Notch pathway signaling, gamma-secretase cleavage of betaAPP, and presenilin protein accumulation. *Dev Cell*, 3, 85–97.

Gao, Y., Pimplikar, S.W. (**2001**) The gamma-secretase-cleaved C-terminal fragment of amyloid precursor protein mediates signaling to the nucleus. *Proc Natl Acad Sci USA*, 98, 14979–14984.

Georgakopoulos, A., Marambaud, P., Efthimiopoulos, S., Shioi, J., Cui, W., Li, H.C., Schutte, M., Gordon, R., Holstein, G.R., Martinelli, G., Mehta, P., Friedrich, V.L., Jr., Robakis, N.K. (**1999**) Presenilin-1 forms complexes with the cadherin/catenin cell-cell adhesion system and is recruited to intercellular and synaptic contacts. *Mol Cell*, 4, 893–902.

Glenner, G.G. (**1988**) Alzheimer's disease: its proteins and genes. *Cell*, 52, 307–308.

Goldgaber, D., Lerman, M.I., McBride, O.W., Saffiotti, U., Gajdusek, D.C. (**1987**) Characterization and chromosomal localization of a cDNA encoding brain amyloid of Alzheimer's disease. *Science*, 235, 877–880.

Goutte, C., Hepler, W., Mickey, K.M., Priess, J.R. (**2000**) aph-2 encodes a novel extracellular protein required for GLP-1–mediated signaling. *Development*, 127, 2481–2492.

Goutte, C., Tsunozaki, M., Hale, V.A., Priess, J.R. (**2002**) APH-1 is a multipass membrane protein essential for the Notch signaling pathway in *Caenorhabditis elegans* embryos. *Proc Natl Acad Sci USA*, 99, 775–779.

Greengard, P. (**2001**) The neurobiology of slow synaptic transmission. *Science*, 294, 1024–1030.

Gu, Y., Misonou, H., Sato, T., Dohmae, N., Takio, K., Ihara, Y. (**2001**) Distinct intramembrane cleavage of the beta-amyloid precursor protein family resembling gamma-secretase-like cleavage of Notch. *J Biol Chem*, 276, 35235–35238.

Gumbiner, B.M. (**1996**) Cell adhesion: the molecular basis of tissue architecture and morphogenesis. *Cell*, 84, 345–357.

Haass, C. (**2004**) Take five-BACE and the gamma-secretase quartet conduct Alzheimer's amyloid beta-peptide generation. *EMBO J*, 23, 483–488.

Hardy, J., Selkoe, D.J. (**2002**) The amyloid hypothesis of Alzheimer's disease: progress and problems on the road to therapeutics. *Science*, 297, 353–356.

Hardy, J.A., Higgins, G.A. (**1992**) Alzheimer's disease: the amyloid cascade hypothesis. Science, 256, 184–185.

Hu, Y., Ye, Y., Fortini, M.E. (**2002**) Nicastrin is required for gamma-secretase cleavage of the *Drosophila* Notch receptor. *Dev Cell*, 2, 69–78.

Huntley, G.W., Benson, D.L., Colman, D.R. (**2002**) Structural remodeling of the synapse in response to physiological activity. *Cell*, 108, 1–4.

Impey, S., Goodman, R.H. (**2001**) CREB signaling – timing is everything. *Sci STKE*, 2001, PE1.

Iso, T., Kedes, L., Hamamori, Y. (**2003**) HES and HERP families: multiple effectors of the Notch signaling pathway. *J Cell Physiol*, 194, 237–255.

Kandel, E.R. (**2001**) The molecular biology of memory storage: a dialogue between genes and synapses. *Science*, 294, 1030–1038.

Kang, D.E., Soriano, S., Xia, X., Eberhart, C.G., De Strooper, B., Zheng, H., Koo, E.H. (**2002**) Presenilin couples the paired phosphorylation of beta-catenin independent of axin: implications for beta-catenin activation in tumorigenesis. *Cell*, 110, 751–762.

Kang, J., Lemaire, H.G., Unterbeck, A., Salbaum, J.M., Masters, C.L., Grzeschik, K.H., Multhaup, G., Beyreuther, K., Muller-Hill, B. (1987) The precursor of Alzheimer's disease amyloid A4 protein resembles a cell-surface receptor. *Nature*, 325, 733–736.

Kimberly, W.T., Zheng, J.B., Guenette, S.Y., Selkoe, D.J. (2001) The intracellular domain of the beta-amyloid precursor protein is stabilized by Fe65 and translocates to the nucleus in a notch-like manner. *J Biol Chem*, 276, 40288–40292.

Klein, W.L., Krafft, G.A., Finch, C.E. (2001) Targeting small Abeta oligomers: the solution to an Alzheimer's disease conundrum? *Trends Neurosci*, 24, 219–224.

Kovacs, D.M., Fausett, H.J., Page, K.J., Kim, T.W., Moir, R.D., Merriam, D.E., Hollister, R.D., Hallmark, O.G., Mancini, R., Felsenstein, K.M., Hyman, B.T., Tanzi, R.E., Wasco, W. (1996) Alzheimer-associated presenilins 1 and 2: neuronal expression in brain and localization to intracellular membranes in mammalian cells. *Nat Med*, 2, 224–229.

Lammich, S., Kojro, E., Postina, R., Gilbert, S., Pfeiffer, R., Jasionowski, M., Haass, C., Fahrenholz, F. (1999) Constitutive and regulated alpha-secretase cleavage of Alzheimer's amyloid precursor protein by a disintegrin metalloprotease. *Proc Natl Acad Sci USA*, 96, 3922–3927.

Lammich, S., Okochi, M., Takeda, M., Kaether, C., Capell, A., Zimmer, A.K., Edbauer, D., Walter, J., Steiner, H., Haass, C. (2002) Presenilin-dependent intramembrane proteolysis of CD44 leads to the liberation of its intracellular domain and the secretion of an Abeta-like peptide. *J Biol Chem*, 277, 44754–44759.

Levitan, D., Doyle, T.G., Brousseau, D., Lee, M.K., Thinakaran, G., Slunt, H.H., Sisodia, S.S., Greenwald, I. (1996) Assessment of normal and mutant human presenilin function in *Caenorhabditis elegans*. *Proc Natl Acad Sci USA*, 93, 14940–14944.

Levitan, D., Greenwald, I. (1995) Facilitation of lin-12-mediated signalling by sel-12, a *Caenorhabditis elegans* S182 Alzheimer's disease gene. *Nature*, 377, 351–354.

Levy-Lahad, E., Wasco, W., Poorkaj, P., Romano, D.M., Oshima, J., Pettingell, W.H., Yu, C.E., Jondro, P.D., Schmidt, S.D., Wang, K., et al. (1995) Candidate gene for the chromosome 1 familial Alzheimer's disease locus. *Science*, 269, 973–977.

Li, X., Greenwald, I. (1996) Membrane topology of the *C. elegans* SEL–12 presenilin. *Neuron*, 17, 1015–1021.

Lopez-Perez, E., Zhang, Y., Frank, S.J., Creemers, J., Seidah, N., Checler, F. (2001) Constitutive alpha-secretase cleavage of the beta-amyloid precursor protein in the furin-deficient LoVo cell line: involvement of the pro-hormone convertase 7 and the disintegrin metalloprotease ADAM10. *J Neurochem*, 76, 1532–1539.

Lopez-Schier, H., St Johnston, D. (2002) *Drosophila* nicastrin is essential for the intramembranous cleavage of notch. *Dev Cell*, 2, 79–89.

Marambaud, P., Shioi, J., Serban, G., Georgakopoulos, A., Sarner, S., Nagy, V., Baki, L., Wen, P., Efthimiopoulos, S., Shao, Z., Wisniewski, T., Robakis, N.K. (2002) A presenilin–1/gamma-secretase cleavage releases the E-cadherin intracellular domain and regulates disassembly of adherens junctions. *EMBO J*, 21, 1948–1956.

Marambaud, P., Wen, P.H., Dutt, A., Shioi, J., Takashima, A., Siman, R., Robakis, N.K. (2003) A CBP binding transcriptional repressor produced by the PS1/epsilon-cleavage of N-cadherin is inhibited by PS1 FAD mutations. *Cell*, 114, 635–645.

Minopoli, G., de Candia, P., Bonetti, A., Faraonio, R., Zambrano, N., Russo, T. (2001) The beta-amyloid precursor protein functions as a cytosolic anchoring site that prevents Fe65 nuclear translocation. *J Biol Chem*, 276, 6545–6550.

Moehlmann, T., Winkler, E., Xia, X., Edbauer, D., Murrell, J., Capell, A., Kaether, C., Zheng, H., Ghetti, B., Haass, C., Steiner, H. (2002) Presenilin–1 mutations of leucine 166 equally affect the generation of the Notch and APP intracellular domains independent of their effect on Abeta 42 production. *Proc Natl Acad Sci USA*, 99, 8025–8030.

Murase, S., Schuman, E.M. (1999) The role of cell adhesion molecules in synaptic plasticity and memory. *Curr Opin Cell Biol*, 11, 549–553.

Nakajima, M., Shimizu, T., Shirasawa, T.

(2000) Notch–1 activation by familial Alzheimer's disease (FAD)-linked mutant forms of presenilin–1. *J Neurosci Res*, 62, 311–317.

Nelson, W.J., Nusse, R. (2004) Convergence of Wnt, beta-catenin, and cadherin pathways. *Science*, 303, 1483–1487.

Neve, R.L., Robakis, N.K. (1998) Alzheimer's disease: a re-examination of the amyloid hypothesis. *Trends Neurosci*, 21, 15–19.

Noll, E., Medina, M., Hartley, D., Zhou, J., Perrimon, N., Kosik, K.S. (2000) Presenilin affects arm/beta-catenin localization and function in *Drosophila*. *Dev Biol*, 227, 450–464.

Obrietan, K., Hoyt, K.R. (2004) CRE-mediated transcription is increased in Huntington's disease transgenic mice. *J Neurosci*, 24, 791–796.

Parent, A., Linden, D.J., Sisodia, S.S., Borchelt, D.R. (1999) Synaptic transmission and hippocampal long-term potentiation in transgenic mice expressing FAD-linked presenilin 1. *Neurobiol Dis*, 6, 56–62.

Pericak-Vance, M.A., Bebout, J.L., Gaskell, P.C., Jr., Yamaoka, L.H., Hung, W.Y., Alberts, M.J., Walker, A.P., Bartlett, R.J., Haynes, C.A., Welsh, K.A., et al. (1991) Linkage studies in familial Alzheimer disease: evidence for chromosome 19 linkage. *Am J Hum Genet*, 48, 1034–1050.

Platenik, J., Kuramoto, N., Yoneda, Y. (2000) Molecular mechanisms associated with long-term consolidation of the NMDA signals. *Life Sci*, 67, 335–364.

Podlisny, M.B., Citron, M., Amarante, P., Sherrington, R., Xia, W., Zhang, J., Diehl, T., Levesque, G., Fraser, P., Haass, C., Koo, E.H., Seubert, P., St George-Hyslop, P., Teplow, D.B., Selkoe, D.J. (1997) Presenilin proteins undergo heterogeneous endoproteolysis between Thr291 and Ala299 and occur as stable N- and C-terminal fragments in normal and Alzheimer brain tissue. *Neurobiol Dis*, 3, 325–337.

Rao, V.R., Finkbeiner, S. (2003) Secrets of a secretase: N-cadherin proteolysis regulates CBP function. *Cell*, 114, 533–535.

Ratovitski, T., Slunt, H.H., Thinakaran, G., Price, D.L., Sisodia, S.S., Borchelt, D.R. (1997) Endoproteolytic processing and stabilization of wild-type and mutant presenilin. *J Biol Chem*, 272, 24536–24541.

Ray, W.J., Yao, M., Mumm, J., Schroeter, E.H., Saftig, P., Wolfe, M., Selkoe, D.J., Kopan, R., Goate, A.M. (1999a) Cell surface presenilin–1 participates in the gamma-secretase-like proteolysis of Notch. *J Biol Chem*, 274, 36801–36807.

Ray, W.J., Yao, M., Nowotny, P., Mumm, J., Zhang, W., Wu, J.Y., Kopan, R., Goate, A.M. (1999b) Evidence for a physical interaction between presenilin and Notch. *Proc Natl Acad Sci USA*, 96, 3263–3268.

Robakis, N.K. (2003) An Alzheimer's disease hypothesis based on transcriptional dysregulation. *Amyloid*, 10, 80–85.

Robakis, N.K., Ramakrishna, N., Wolfe, G., Wisniewski, H.M. (1987) Molecular cloning and characterization of a cDNA encoding the cerebrovascular and the neuritic plaque amyloid peptides. *Proc Natl Acad Sci USA*, 84, 4190–4194.

Rogaev, E.I., Sherrington, R., Rogaeva, E.A., Levesque, G., Ikeda, M., Liang, Y., Chi, H., Lin, C., Holman, K., Tsuda, T., et al. (1995) Familial Alzheimer's disease in kindreds with missense mutations in a gene on chromosome 1 related to the Alzheimer's disease type 3 gene. *Nature*, 376, 775–778.

Roncarati, R., Sestan, N., Scheinfeld, M.H., Berechid, B.E., Lopez, P.A., Meucci, O., McGlade, J.C., Rakic, P., D'Adamio, L. (2002) The gamma-secretase-generated intracellular domain of beta-amyloid precursor protein binds Numb and inhibits Notch signaling. *Proc Natl Acad Sci USA*, 99, 7102–7107.

Roose, J., Clevers, H. (1999) TCF transcription factors: molecular switches in carcinogenesis. *Biochim Biophys Acta*, 1424, M23–M37.

Roses, A.D. (1994) Apolipoprotein E affects the rate of Alzheimer disease expression: beta-amyloid burden is a secondary consequence dependent on APOE genotype and duration of disease. *J Neuropathol Exp Neurol*, 53, 429–437.

Ross, C.A. (2002) Polyglutamine pathogenesis: emergence of unifying mechanisms for Huntington's disease and related disorders. *Neuron*, 35, 819–822.

Sastre, M., Steiner, H., Fuchs, K., Capell, A., Multhaup, G., Condron, M.M., Teplow, D.B., Haass, C. (2001) Presenilin-dependent gamma-secretase processing of beta-amyloid precursor protein at a site corresponding to the S3 cleavage of Notch. *EMBO Rep*, 2, 835–841.

Saura, C.A., Choi, S.Y., Beglopoulos, V., Malkani, S., Zhang, D., Rao, B.S., Chattarji, S., Kelleher, R.J., 3rd, Kandel, E.R., Duff, K., Kirkwood, A., Shen, J. (**2004**) Loss of presenilin function causes impairments of memory and synaptic plasticity followed by age-dependent neurodegeneration. *Neuron*, 42, 23–36.

Saura, C.A., Tomita, T., Soriano, S., Takahashi, M., Leem, J.Y., Honda, T., Koo, E.H., Iwatsubo, T., Thinakaran, G. (**2000**) The nonconserved hydrophilic loop domain of presenilin (PS) is not required for PS endoproteolysis or enhanced abeta 42 production mediated by familial early onset Alzheimer's disease-linked PS variants. *J Biol Chem*, 275, 17136–17142.

Schneider, I., Reverse, D., Dewachter, I., Ris, L., Caluwaerts, N., Kuiperi, C., Gilis, M., Geerts, H., Kretzschmar, H., Godaux, E., Moechars, D., Van Leuven, F., Herms, J. (**2001**) Mutant presenilins disturb neuronal calcium homeostasis in the brain of transgenic mice, decreasing the threshold for excitotoxicity and facilitating long-term potentiation. *J Biol Chem*, 276, 11539–11544.

Schroeter, E.H., Kisslinger, J.A., Kopan, R. (**1998**) Notch–1 signalling requires ligand-induced proteolytic release of intracellular domain. *Nature*, 393, 382–386.

Selkoe, D.J. (**2001**) Alzheimer's disease: genes, proteins, and therapy. *Physiol Rev*, 81, 741–766.

Serban, G., Kouchi, Z., Baki, L., Georgakopoulos, A., Litterst, C.M., Shioi, J., Robakis, N.K. (**2005**) Cadherins mediate both the association between PS1 and beta-catenin and the effects of PS1 on beta-catenin stability. *J Biol Chem*, August 26 [Epub ahead of print].

Shen, J., Bronson, R.T., Chen, D.F., Xia, W., Selkoe, D.J., Tonegawa, S. (**1997**) Skeletal and CNS defects in Presenilin–1-deficient mice. *Cell*, 89, 629–639.

Sherrington, R., Rogaev, E.I., Liang, Y., Rogaeva, E.A., Levesque, G., Ikeda, M., Chi, H., Lin, C., Li, G., Holman, K., et al. (**1995**) Cloning of a gene bearing missense mutations in early-onset familial Alzheimer's disease. *Nature*, 375, 754–760.

Song, W., Nadeau, P., Yuan, M., Yang, X., Shen, J., Yankner, B.A. (**1999**) Proteolytic release and nuclear translocation of Notch–1 are induced by presenilin–1 and impaired by pathogenic presenilin–1 mutations. *Proc Natl Acad Sci USA*, 96, 6959–6963.

Soriano, S., Kang, D.E., Fu, M., Pestell, R., Chevallier, N., Zheng, H., Koo, E.H. (**2001**) Presenilin 1 negatively regulates beta-catenin/T cell factor/lymphoid enhancer factor–1 signaling independently of beta-amyloid precursor protein and notch processing. *J Cell Biol*, 152, 785–794.

Struhl, G., Greenwald, I. (**1999**) Presenilin is required for activity and nuclear access of Notch in *Drosophila*. *Nature*, 398, 522–525.

Tang, L., Hung, C.P., Schuman, E.M. (**1998**) A role for the cadherin family of cell adhesion molecules in hippocampal long-term potentiation. *Neuron*, 20, 1165–1175.

Tanzi, R.E., Gusella, J.F., Watkins, P.C., Bruns, G.A., St George-Hyslop, P., Van Keuren, M.L., Patterson, D., Pagan, S., Kurnit, D.M., Neve, R.L. (**1987**) Amyloid beta protein gene: cDNA, mRNA distribution, and genetic linkage near the Alzheimer locus. *Science*, 235, 880–884.

Telese, F., Bruni, P., Donizetti, A., Gianni, D., D'Ambrosio, C., Scaloni, A., Zambrano, N., Rosenfeld, M.G., Russo, T. (**2005**) Transcription regulation by the adaptor protein Fe65 and the nucleosome assembly factor SET. *EMBO Rep*, 6, 77–82.

Tepass, U., Truong, K., Godt, D., Ikura, M., Peifer, M. (**2000**) Cadherins in embryonic and neural morphogenesis. *Nat Rev Mol Cell Biol*, 1, 91–100.

Terry, R.D., Peck, A., DeTeresa, R., Schechter, R., Horoupian, D.S. (**1981**) Some morphometric aspects of the brain in senile dementia of the Alzheimer type. *Ann Neurol*, 10, 184–192.

Tesco, G., Kim, T.W., Diehlmann, A., Beyreuther, K., Tanzi, R.E. (**1998**) Abrogation of the presenilin 1/beta-catenin interaction and preservation of the heterodimeric presenilin 1 complex following caspase activation. *J Biol Chem*, 273, 33909–33914.

Thinakaran, G., Borchelt, D.R., Lee, M.K., Slunt, H.H., Spitzer, L., Kim, G., Ratovitsky, T., Davenport, F., Nordstedt, C., Seeger, M., Hardy, J., Levey, A.I., Gandy, S.E., Jenkins, N.A., Copeland, N.G., Price, D.L., Sisodia, S.S. (1996) Endoproteolysis of presenilin 1 and accumulation of processed derivatives in vivo. *Neuron*, 17, 181–190.

Walker, E.S., Martinez, M., Brunkan, A.L., Goate, A. (2005) Presenilin 2 familial Alzheimer's disease mutations result in partial loss of function and dramatic changes in Abeta 42/40 ratios. *J Neurochem*, 92, 294–301.

Walsh, D.M., Klyubin, I., Fadeeva, J.V., Cullen, W.K., Anwyl, R., Wolfe, M.S., Rowan, M.J., Selkoe, D.J. (2002) Naturally secreted oligomers of amyloid beta protein potently inhibit hippocampal long-term potentiation in vivo. *Nature*, 416, 535–539.

Walter, J., Capell, A., Grunberg, J., Pesold, B., Schindzielorz, A., Prior, R., Podlisny, M.B., Fraser, P., Hyslop, P.S., Selkoe, D.J., Haass, C. (1996) The Alzheimer's disease-associated presenilins are differentially phosphorylated proteins located predominantly within the endoplasmic reticulum. *Mol Med*, 2, 673–691.

Wang, P., Pereira, F.A., Beasley, D., Zheng, H. (2003) Presenilins are required for the formation of comma- and S-shaped bodies during nephrogenesis. *Development*, 130, 5019–5029.

Weidemann, A., Eggert, S., Reinhard, F.B., Vogel, M., Paliga, K., Baier, G., Masters, C.L., Beyreuther, K., Evin, G. (2002) A novel epsilon-cleavage within the transmembrane domain of the Alzheimer amyloid precursor protein demonstrates homology with Notch processing. *Biochemistry*, 41, 2825–2835.

Weidemann, A., Paliga, K., Durrwang, U., Czech, C., Evin, G., Masters, C.L., Beyreuther, K. (1997) Formation of stable complexes between two Alzheimer's disease gene products: presenilin-2 and beta-amyloid precursor protein. *Nat Med*, 3, 328–332.

Wilson, C.A., Doms, R.W., Lee, V.M. (1999) Intracellular APP processing and A beta production in Alzheimer disease. *J Neuropathol Exp Neurol*, 58, 787–794.

Wong, P.C., Zheng, H., Chen, H., Becher, M.W., Sirinathsinghji, D.J., Trumbauer, M.E., Chen, H.Y., Price, D.L., Van der Ploeg, L.H., Sisodia, S.S. (1997) Presenilin 1 is required for Notch1 and DII1 expression in the paraxial mesoderm. *Nature*, 387, 288–292.

Xia, W., Zhang, J., Kholodenko, D., Citron, M., Podlisny, M.B., Teplow, D.B., Haass, C., Seubert, P., Koo, E.H., Selkoe, D.J. (1997) Enhanced production and oligomerization of the 42–residue amyloid beta-protein by Chinese hamster ovary cells stably expressing mutant presenilins. *J Biol Chem*, 272, 7977–7982.

Xia, X., Qian, S., Soriano, S., Wu, Y., Fletcher, A.M., Wang, X.J., Koo, E.H., Wu, X., Zheng, H. (2001) Loss of presenilin 1 is associated with enhanced beta-catenin signaling and skin tumorigenesis. *Proc Natl Acad Sci USA*, 98, 10863–10868.

Ye, Y., Lukinova, N., Fortini, M.E. (1999) Neurogenic phenotypes and altered Notch processing in *Drosophila* Presenilin mutants. *Nature*, 398, 525–529.

Yu, C., Kim, S.H., Ikeuchi, T., Xu, H., Gasparini, L., Wang, R., Sisodia, S.S. (2001) Characterization of a presenilin-mediated amyloid precursor protein carboxyl-terminal fragment gamma. Evidence for distinct mechanisms involved in gamma -secretase processing of the APP and Notch1 transmembrane domains. *J Biol Chem*, 276, 43756–43760.

Yu, G., Chen, F., Levesque, G., Nishimura, M., Zhang, D.M., Levesque, L., Rogaeva, E., Xu, D., Liang, Y., Duthie, M., St George-Hyslop, P.H., Fraser, P.E. (1998) The presenilin 1 protein is a component of a high molecular weight intracellular complex that contains beta-catenin. *J Biol Chem*, 273, 16470–16475.

Yu, G., Nishimura, M., Arawaka, S., Levitan, D., Zhang, L., Tandon, A., Song, Y.Q., Rogaeva, E., Chen, F., Kawarai, T., Supala, A., Levesque, L., Yu, H., Yang, D.S., Holmes, E., Milman, P., Liang, Y., Zhang, D.M., Xu, D.H., Sato, C., Rogaev, E., Smith, M., Janus, C., Zhang, Y., Aebersold, R., Farrer, L.S., Sorbi, S., Bruni, A., Fraser, P., St George-Hyslop, P. (2000) Nicastrin modulates presenilin-mediated notch/glp-1 signal transduction and betaAPP processing. *Nature*, 407, 48–54.

Zambrano, N., Bimonte, M., Arbucci, S., Gianni, D., Russo, T., Bazzicalupo, P. (**2002**) feh–1 and apl–1, the *Caenorhabditis elegans* orthologues of mammalian Fe65 and beta-amyloid precursor protein genes, are involved in the same pathway that controls nematode pharyngeal pumping. *J Cell Sci*, 115, 1411–1422.

Zhang, J., Kang, D.E., Xia, W., Okochi, M., Mori, H., Selkoe, D.J., Koo, E.H. (**1998**) Subcellular distribution and turnover of presenilins in transfected cells. *J Biol Chem*, 273, 12436–12442.

Zhou, J., Liyanage, U., Medina, M., Ho, C., Simmons, A.D., Lovett, M., Kosik, K.S. (**1997**) Presenilin 1 interaction in the brain with a novel member of the Armadillo family. *Neuroreport*, 8, 2085–2090.

20
Transcriptional Abnormalities in Huntington's Disease

Dimitri Krainc

Abstract

Huntington's disease (HD) is an inherited neurodegenerative disease caused by a glutamine repeat expansion in huntingtin protein. Transcriptional dysregulation has emerged as a potentially important pathogenic mechanism in HD. Regulation of transcription in eukaryotic cells involves an orchestrated interplay of chromatin-packed genes and protein complexes that control chromatin dynamics, transcriptional initiation, and transcription elongation. In HD, mutant huntingtin may interfere with several of these processes to cause transcriptional dysregulation. Mutant huntingtin-directed transcriptional repression involves gene-specific activator proteins such as Sp1 and CREB and selective components of the core transcription apparatus, including TFIID and TFIIF. In addition, the interference of chromatin modification by mutant huntingtin has been demonstrated through the use of histone deacetylase (HDAC) inhibitors. Deregulation of gene transcription by mutant huntingtin leads to repression of target genes involved in pathogenesis of HD such as dopamine D_2 receptor. In addition, mutant huntingtin causes disruption of mitochondrial function by inhibiting gene expression of PGC–1α, a transcriptional co-activator that regulates several metabolic processes, including mitochondrial biogenesis and respiration. Mutant huntingtin represses PGC–1α expression by interfering with the CREB/TAF4 transcriptional pathway critical for the regulation of PGC–1α promoter. Inhibition of PGC–1α expression by mutant huntingtin leads to defects in energy metabolism and dysfunction of neurons that are most vulnerable to metabolic stress, such as striatum. Such disruption in energy homeostasis in HD may lead to early abnormalities in multiple cellular functions, and ultimately results in neurodegeneration. These studies suggest that transcriptional deregulation may occur before the development of disease symptoms in HD.

20.1
Introduction

Huntington's disease (HD) is a progressive and fatal neurological disorder that is characterized phenotypically by involuntary movements and psychiatric disturbances (Vonsattel and DiFiglia, 1998). The gene *huntingtin*, which is mutated in HD patients, contains an expanded polyglutamine repeat within exon 1 of the gene. The number of polyglutamine diseases continues to grow, and they share several common features, including neurodegeneration, a dominant pattern of inheritance and genetic anticipation (Zoghbi and Orr, 2000). Despite the widespread tissue distribution of the protein for each of these disease genes, the affected region is primarily the brain, and the regions of neuronal loss are somewhat selective and specific for each given disease (Zoghbi and Orr, 2000). In HD, mutant huntingtin is expressed ubiquitously but selective cell loss is observed in the brain, particularly in the caudate and putamen of the striatum (Vonsattel and DiFiglia, 1998)

The functional significance of the expanded polyglutamine tract is not well understood. Most other proteins containing polyglutamine-rich regions function as transcription factors (Alba and Guigo, 2004). Interactions of mutant huntingtin with several transcription factors have been demonstrated, suggesting that mutant huntingtin may be directly involved in regulation of gene transcription (Sugars and Rubinsztein, 2003).

20.2
Mutant Huntingtin Interferes with Specific Components of General Transcriptional Machinery

It has been well established, first in the case of Sp1 (Courey and Tjian, 1988) and subsequently with other transcription factors, that activation domains are often composed of glutamine-rich protein interfaces (Gerber et al., 1994). Thus, transcription factor interactions with other cellular factors may be disrupted by mutant huntingtin bearing polyQ expansions. Indeed, mutant huntingtin has been shown to interact directly with a number of nuclear transcription factors (Okazawa, 2003). Recent DNA microarray studies detected changes in gene expression profiles in HD transgenic mice at early stages, suggesting that transcription of select genes had already been altered even when mice showed only minimal abnormalities (Luthi-Carter et al., 2002a, 2002b). Analysis of the affected regulatory sequences revealed that select Sp1–dependent transcription pathways were disrupted. This hypothesis is consistent with recent *in-vivo* observations that mutant huntingtin may target Sp1 and its co-activator TAF4 (Tanese et al., 1991; Chen et al., 1994) through direct protein-protein interactions to disrupt transcription (Dunah et al., 2002; Li et al., 2002). Remarkably, in primary striatal neurons, co-expression of Sp1 and TAF4 resulted in a significant rescue of mutant huntingtin-induced inhibition of transcription (Dunah et al., 2002). These studies suggested that a soluble form of mutant huntingtin may interfere with specific components of activated and general transcriptional machinery in the early stages of HD.

20.2 Mutant Huntingtin Interferes with Specific Components of General Transcriptional Machinery

Human TAF4 is one of at least 12 TATA-binding protein (TBP)-associated factors (TAF$_{II}$s) in TFIID (Albright and Tjian, 2000). The transcription initiation factor TFIID is recruited to the core promoter through its interaction with specific activators such as Sp1, and binds to the TATA element at core promoters. A series of transcription factor interactions, involving TFIIA, TFIIB, TFIID, TFIIE, TFIIF, TFIIH, CRSP (also called mediator), and RNA polymerase II (Pol II) subsequently leads to the formation of the preinitiation complex (PIC) and transcriptional activation (Ryu et al., 1999). Human TFIIF consists of two subunits, RAP30 and RAP74, that bind RNA Pol II directly and help recruit the enzyme to a preformed TFIID/TFIIB complex (Conaway et al., 2000). TFIIF also stimulates the rate of RNA Polymerase II elongation.

A chromatin-based *in-vitro* transcription system was adopted (Naar et al., 1999; Lemon et al., 2001) to dissect potential mechanisms by which mutant huntingtin might repress transcription (Zhai et al., 2005). As expected, high levels of transcription from the assembled chromatin template were obtained when purified recombinant Sp1 was added to a well-defined set of basal transcription factors which included purified recombinant TFIIA, TFIIB, TFIIE, TFIIF, and affinity-purified components TFIID, TFIIH, RNA Pol II, and CRSP. In order to assess the effect of mutant huntingtin on Sp1–dependent transcription, the N-terminal portion of either wild-type or mutant huntingtin was expressed in *Escherichia coli*, and purified. Adding purified normal huntingtin did not significantly alter the levels of transcription. In contrast, adding mutant huntingtin resulted in a dramatic decrease in transcription. Thus, purified recombinant mutant huntingtin fragment is capable of inhibiting Sp1–dependent transcription in a well-defined *in-vitro* reaction, and this repression appears to depend on the presence of an expanded polyQ tract in huntingtin.

To better understand the mechanisms of transcriptional repression by mutant huntingtin, it was systematically tested which components present in the defined *in-vitro* transcription system could rescue the inhibition caused by mutant huntingtin (Zhai et al., 2005). It has been reported that mutant huntingtin can disrupt the interaction between Sp1 and its co-activator TAF4, and that the repression of the dopamine D$_2$ receptor promoter activity by mutant huntingtin can be reversed by co-expression of Sp1 and TAF4 in cultured striatal cells (Dunah et al., 2002). Based on these *in-vivo* results, it was anticipated that the addition of Sp1 and/or TAF4 (in the form of TFIID complex) should rescue the transcriptional repression induced by mutant huntingtin. To rescue this level of repression, an additional two- to four-fold of each individual basal factor was added to transcription reactions containing mutant huntingtin. As expected, the addition of purified Sp1 or the TAF4–containing TFIID complex was able to efficiently rescue the repression caused by mutant huntingtin. These findings indicated that the *in-vitro* transcription system largely recapitulates the transcriptional repression observed for mutant huntingtin *in vivo* (Dunah et al., 2002). In addition, these *in-vitro* assays using purified transcription components provide evidence that most likely both Sp1 and TFIID are directly targeted by mutant huntingtin for repression. A recent study demonstrated that mutant huntingtin could structurally destabilize TBP through a polyQ-mediated interaction (Schaffar et al., 2004). Because TFIID contains TBP and multiple TAFs with distinct essential transcriptional functions, mutant huntingtin may target multiple components of TFIID.

In order to screen for potential novel targets of mutant huntingtin, an attempt was made to rescue the *in-vitro* repression by adding back an excess of other components of the core transcription machinery. Surprisingly, the addition of excess TFIIF was found to reverse the repression, suggesting that TFIIF may also be targeted by mutant huntingtin as part of the repression mechanism. Addition of the other basal factors and cofactors, including TFIIA, TFIIB, TFIIE, TFIIH, RNA Pol II, and CRSP, had little or no effect on the repression by mutant huntingtin (Zhai et al., 2005).

A key observation made in HD patients is that longer polyQ expansions in huntingtin are associated with more severe symptoms and earlier age of onset (Rubinsztein et al., 1993). Therefore, it is important to determine whether mutant huntingtin with varying polyQ length would also differ in their ability to repress transcription *in vitro*. To test this potential correlation, the N-terminal fragment of huntingtin carrying 25, 46, or 97 polyQs were expressed and purified. When these purified huntingtin polypeptides were tested in transcription assays, their ability to repress transcription differed substantially. The addition of normal huntingtin fragment had no measurable effect on Sp1–dependent transcription, while mutant huntingtin inhibited transcription. When the number of polyQ in huntingtin was increased to 97, the efficiency of mutant huntingtin to repress transcription increased dramatically. The addition of huntingtin with 97Q resulted in a 90% decrease in the levels of transcription. Even lower amounts of huntingtin with 120QP were sufficient to repress transcription efficiently (Zhai et al., 2005).

Next, it was determined whether transcriptional repression by mutant huntingtin carrying different numbers of polyQ targeted the same components of the transcription apparatus. Transcriptional repression by mutant huntingtin with 97 polyQ can be rescued by adding more TFIIF, TFIID, and Sp1. The addition of other basal transcription factors was not able to alleviate the repressive effect of this mutant huntingtin. Likewise, repression by the mutant huntingtin with 46 polyQ can also be rescued by the addition of more TFIIF, TFIID and Sp1, but not the other basal transcription factors. Thus, the same set of factors (i.e., Sp1, TFIID, and TFIIF) was capable of reversing the transcriptional repression mediated by mutant huntingtin carrying different numbers of polyQ. These results suggest that mutant huntingtin with different numbers of polyQ likely utilizes the same molecular targets and mechanisms to repress transcription (Zhai et al., 2005).

These data strongly suggest that mutant huntingtin can inhibit Sp1–dependent transcription by interfering with the functions of Sp1, TFIID as well as TFIIF. The ability of Sp1 and TFIID to rescue the transcriptional repression was expected, since mutant huntingtin can interact with Sp1 and TAF4 to disrupt this activator-co-activator pair *in vivo* (Dunah et al., 2002).

These *in-vitro* studies identified TFIIF as a novel direct target in mutant huntingtin-mediated transcriptional repression (Zhai et al., 2005). Although there have been several reports linking TFIIF to the function of transcription activators and repressors (Frejtag et al., 2001), this study provides the first direct connection between

TFIIF and transcriptional repression induced by a polyQ expansion protein. The RAP30 subunit of TFIIF, which interacts with huntingtin, consists of three functional domains. The N-terminal domain of RAP30 is thought to bind RAP74 subunit of TFIIF (Tan et al., 1995), the central region binds RNA Pol II (McCracken and Greenblatt, 1991), and the C-terminal domain binds DNA (Garrett et al., 1992). It was found that RAP74 and mutant huntingtin compete for binding to RAP30, and that the regions of RAP30 required for binding to RAP74 and huntingtin overlap with each other. Because RAP30 lacks a glutamine-rich domain, its interaction with mutant huntingtin is likely mediated through an alternative mechanism. The crystal structure of the N-terminal fragments of RAP30 and RAP74 had been shown to adopt a triple-barrel structure with multiple beta-sheets (Gaiser et al., 2000). Since mutant huntingtin favors the formation of an intramolecular beta-sheet structure (Perutz et al., 2002), it is possible that the RAP30–mutant huntingtin interaction involves contact between beta-sheet structures. Such a structure-based interference mechanism is consistent with our finding that expansion of glutamines in mutant huntingtin enhanced its affinity for RAP30. This can also explain why beta-sheet reactive agents such as Congo red have protective effects in polyQ disease models (Sanchez et al., 2003), presumably by blocking the toxic interactions between mutant huntingtin and proteins with certain beta-sheet configurations. Thus, mutant huntingtin may target not only polyQ-containing proteins, but also non-polyQ proteins with specific beta-sheet structures.

An important finding of the *in-vitro* study is that mutant huntingtin has a higher affinity for RAP30 than normal huntingtin, and may compete with RAP74 for interaction with RAP30. Because an intact TFIIF complex is required for efficient initiation and elongation of transcription at least for some promoters, it was hypothesized that TFIIF dissociation could contribute to the transcriptional inhibition by mutant huntingtin. Inside a cell, there is likely a balance of RAP30, RAP74, and huntingtin. In normal cells, normal huntingtin is mostly cytoplasmic and its affinity for RAP30 is low. Such conditions would favor the formation of active TFIIF complexes. In HD cells, however, mutant huntingtin accumulates in both the cytoplasm and nucleus, and has a much higher affinity for RAP30. Thus, the equilibrium shifts toward favoring RAP30–mutant huntingtin interaction, resulting in less TFIIF formed in the cytoplasm and more TFIIF disrupted in the nucleus. Such a shift would likely result in a general decrease of transcription in HD cells, as has been observed. In several DNA microarray studies, the level of RNA Pol II large subunits has been shown to increase in mutant HD brain (Luthi-Carter et al., 2002a). Since the role of TFIIF in transcription is dependent on its interaction with RNA Pol II, it is speculated that elevated levels of RNA Pol II subunits in HD cells may arise as a compensatory mechanism triggered by decreased levels of TFIIF (Zhai et al., 2005).

In addition, it was shown that overexpression of RAP30 is able to abrogate transcriptional repression and rescue the cellular toxicity induced by mutant huntingtin in primary striatal neurons. There are two potential ways in which RAP30 overexpression could neutralize the toxic properties of mutant huntingtin. The first way is for RAP30 to interact with mutant huntingtin and compete it away from other huntingtin interacting partners. This is certainly possible, since the interaction between mutant huntingtin and RAP30 is polyQ-dependent. The second way is for RAP30 to drive the formation of more TFIIF complexes, thereby potentiating transcription of important genes involved in neuronal survival. The finding that overexpression of RAP30 can at least partially rescue the mutant huntingtin-mediated transcriptional inhibition supports this mechanism. In addition, overexpression of RAP74 alone could induce significant cellular toxicity in striatal neurons, suggesting that the chronic release of free RAP74 from TFIIF may contribute to the progressive nature of HD pathogenesis. The finding that overexpression of RAP30 can rescue the mutant huntingtin and RAP74–induced cellular toxicity also favors a mechanism in which RAP30 can protect the neurons by promoting TFIIF complex formation. To better understand how much the TFIIF-mediated mechanism contributes to HD pathogenesis, it will be important to identify those genes whose transcription is particularly sensitive to both mutant huntingtin and RAP74.

Taking these observations together with previous *in-vivo* studies, it was suggested that in normal cells Sp1 is recruited to GC-box containing promoters through its DNA-binding domain. Once bound to DNA, Sp1 utilizes its multiple glutamine-rich activation domains to target components of the basal transcription machinery, one of which is TAF4, a subunit of TFIID. In a stepwise recruiting process involving TFIIA, TFIID, TFIIB, TFIIE, TFIIF, TFIIH, RNA Pol II, and CRSP, the PIC is then formed on activated promoters to potentiate transcription. In HD cells, soluble nuclear mutant huntingtin fragment is free to bind Sp1 through direct protein interactions, thus sequestering it from binding to GC-boxes. Furthermore, mutant huntingtin can also prevent Sp1–mediated recruitment of TFIID through its interaction with TAF4. In the case where there is already an Sp1–TFIID complex formed at the promoter, mutant huntingtin could subsequently disrupt the stepwise PIC assembly by targeting TFIIF, an essential transcription factor important for initiation, promoter escape, and elongation at certain promoters. It is anticipated that for different potential target genes, mutant huntingtin will have differential effects because these multiple transcription factor targets may be differentially required for critical functions and rate-limiting transactions at specific gene promoters. Therefore, this model will undergo further refinements as more gene regulatory targets for mutant huntingtin are identified and their molecular consequences determined.

20.3

Mutant Huntingtin Disrupts Sp1–TAF4 Transcriptional Pathway

In order to better understand the relevance of transcriptional repression in HD it is important to analyze target genes that show decreased expression in HD as a result of such repression. One such target is the dopamine D_2 receptor that is strongly regulated by Sp1–TAF4–mediated transcriptional pathway (Dunah et., 2005). Using the yeast two-hybrid system and co-immunoprecipitation experiments, it was found that both Sp1 and TAF4 interact with full-length huntingtin. The interactions between Sp1 and huntingtin are stronger in the presence of expanded polyglutamine repeat compared to the non-expanded repeat length, whereas the interactions between TAF4 and huntingtin are not influenced by polyglutamine tract length. Although the glutamine-rich regions of Sp1 and TAF4 are sufficient for their interaction with huntingtin, presence of the C-terminal DNA-binding domain of Sp1 or the conserved C-terminal domain of TAF4 results in stronger interaction. These results indicated that polyglutamine expansion enhances the interaction of Sp1, but not TAF4, with huntingtin.

To establish whether huntingtin interacts with Sp1 and TAF4 in human brain, co-immunoprecipitation studies were performed using extracts from the caudate nucleus of Grade 1 HD brain with anti-Sp1, anti-TAF4 or anti-huntingtin antibodies. Both anti-Sp1 and anti-TAF4 antibodies precipitated huntingtin protein. In addition, the anti-huntingtin antibody co-immunoprecipitated substantial amounts of Sp1 and TAF4 proteins. These results indicated that both endogenous Sp1 and TAF4 interact strongly with huntingtin in the early stages of pathology in post-mortem HD brain.

To determine the effects of huntingtin on the binding of Sp1 to DNA, an electrophoretic mobility shift assay (EMSA) with purified Sp1 and huntingtin proteins was performed. Using a consensus Sp1 binding site as probe, a 70% decrease in Sp1 binding to DNA in the presence of mutant and 20% with the wild-type huntingtin was found. To examine whether mutant huntingtin affects Sp1 binding to the D_2 promoter in striatal cells, EMSA was performed using nuclear extracts from primary striatal neurons transfected with wild-type or mutant huntingtin. Using a region of the D_2 dopamine receptor promoter as a probe, decreased Sp1 binding was found in extracts expressing mutant huntingtin compared to wild-type huntingtin. Downregulation of D_2 receptor expression has also been reported in the striata of presymptomatic HD patients. Therefore, EMSA was performed using nuclear extracts isolated from the caudate and hippocampus of Grade 1 and Grade 4 HD brains. In Grades 1 and 2 of HD there is mild to moderate neuronal loss in the caudate nucleus, whereas the hippocampus remains relatively unaffected until later in the course of disease. In Grade 4, the striatum, as well as other brain regions, is severely atrophic and is depleted of 95% or more of its neurons. Using Sp1 binding sites in the D_2 promoter as a labeled probe, a significant decrease in the DNA-binding activity of Sp1 was found in the caudate nucleus of Grade 1 HD brain compared to control brain. Similar decreases in Sp1 binding were found in extracts from Grade 4 HD brain, suggesting early and persistent inhibition of Sp1 function. Interestingly, when EMSA was performed with nuclear extracts from the hippocampus of Grade 1 HD brain, a statistically insignificant increase, rather than a decrease, in Sp1 binding was observed. This caudate-specific inhibition of Sp1 function may be, in part, due to the preferential accumulation of mutant huntingtin in the striatum in early stages of HD.

To further establish the role of huntingtin in Sp1–mediated transcription, primary striatal neurons were transfected with D_2 promoter-reporter gene constructs along with mutant or normal huntingtin. While huntingtin with normal glutamine repeats had no significant effect on promoter activity, mutant huntingtin showed significant inhibition. To determine whether the inhibition of Sp1–mediated transcription was dependent on increased levels of huntingtin relative to endogenous

Sp1 and TAF4, Sp1 and/or TAF4 were overexpressed together with huntingtin. Overexpression of Sp1 or TAF4 alone did not significantly alter the inhibitory effects of mutant huntingtin, whereas co-expression of TAF4 and Sp1 resulted in complete reversal of huntingtin-induced inhibition of D_2 promoter activity. These effects of Sp1 and TAF4 were dependent on Sp1 binding to the D_2 promoter, since no effect was seen when the Sp1 functional site was deleted, or when an Sp1 expression vector lacking the DNA-binding domain was used. Similarly, no effect on promoter activity was seen with the N-terminus of TAF4, which does not affect Sp1 activity or does not interact with huntingtin. Taken together, these observations suggest that mutant huntingtin specifically represses D_2 promoter activity in a Sp1/TAF4–dependent manner.

Huntingtin has previously been suggested to interfere with gene transcription by depleting transcription factors from their normal location and sequestering them into nuclear aggregates. To examine whether the function of Sp1 is compromised through sequestration into nuclear inclusions, immunocytochemistry on transgenic HD mice and human post-mortem HD brains was performed. Nuclear and cytoplasmic inclusions were strongly labeled with an antibody to the N-terminus of huntingtin that specifically labels huntingtin aggregates, whereas Sp1 or TAF4 staining was not detected in these inclusions; this suggested that the soluble rather than the aggregated form of huntingtin interacts with Sp1. This finding was confirmed by the Western blot analysis, which showed a robust increase of Sp1 protein in the soluble fraction of caudate tissue from post-mortem HD brain. Sp1 protein levels were also increased in the cerebral cortex, but decreased in the hippocampus. Together, the DNA-binding and protein-expression data suggest that the decreased function of Sp1 in HD is not due to sequestration of Sp1 into aggregates but rather due to inhibition of Sp1 by soluble mutant huntingtin.

20.4
Deregulation of CRE-Dependent Transcription in HD

These findings and the recent report that atrophin (which causes another polyglutamine disease, DRPLA) also binds to the C-terminal domain of TAF4 (Shimohata et al., 2000), suggest that by competing with the critical protein interaction surface of TAF4, polyglutamine stretches may interfere with the coupling of activator-mediated signals to the basal transcriptional machinery. These contributions of TAF4 as well as other TAFs to gene transcription are likely to be promoter- and cell type-specific.

Interestingly, TAF4 also directly interacts with CREB to regulate CRE-dependent transcription (Saluja et al., 1998). Several studies have indicated that CREB/CBP transcriptional pathway is deregulated in HD (Sugars and Rubinsztein, 2003), suggesting that both Sp1– and CRE-mediated transcriptional pathways may be disrupted by mutant huntingtin (Fig. 20.1).

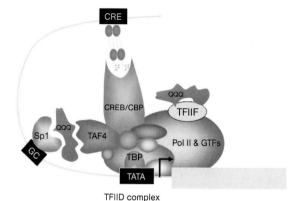

Fig. 20.1 Model of potential mechanisms used by mutant huntingtin (QQQ) to disrupt Sp1– and CRE-mediated transcription. In normal cells, transcription factors Sp1 and CREB bind to GC-box and CRE sequences, respectively. Sp1 and CREB/CBP target TAF4 and recruit TFIID and other components of the general transcription machinery to form a productive preinitiation complex. In HD cells, mutant huntingtin may target multiple components of the general transcription machinery for repression. First, mutant huntingtin can sequester Sp1 and prevent it from binding to GC-box sequences in the promoter. Second, mutant huntingtin can target TAF4 in the TFIID complex, and therefore impair the recruitment of TFIID by Sp1 and CREB. Third, mutant huntingtin disrupts the TFIIF complex formation and thus interferes with transcription initiation, promoter escape, and elongation. (This figure also appears with the color plates.)

In order to examine whether CRE-dependent transcriptional pathways are involved in HD pathogenesis, it is important to analyze specific CRE-regulated target genes that show altered expression in HD. One such target of mutant huntingtin is PGC–1α (Peroxisome proliferator-activated receptor Gamma Coactivator–1 alpha) (L. Cui et al., unpublished results). In these studies, it was shown that mutant huntingtin interferes with energy metabolism by transcriptional repression of PGC–1α, and that PGC–1α plays a crucial role in mediating survival of striatal neurons in cell culture and animal models of HD.

PGC–1α is a transcriptional co-activator that powerfully regulates several metabolic processes, including mitochondrial biogenesis and oxidative phosphorylation (Puigserver et al., 1998; Wu et al., 1999; Puigserver and Spiegelman, 2003). Ectopic expression of PGC–1α induces both nuclear and mitochondrial mRNAs encoding respiratory subunits, and also activates the program of mitochondrial biogenesis (Wu et al., 1999; Lin et al., 2002b). In addition to inducing respiratory subunit mRNAs, PGC–1α increases COXIV and cytochrome c protein levels as well as the steady-state level of mitochondrial DNA (mtDNA). Increased oxygen uptake in differentiated myotubes and a visible increase in mitochondrial number accompany these changes. Overproduction of PGC–1α in cardiac myocytes had similar effects on the induction of respiratory subunits, oxidative enzymes, oxygen uptake and mitochondrial biogenesis as observed for other cell types (Lehman et al., 2000).

These studies were extended to the mouse, where expression of PGC–1α from a cardiac-specific promoter resulted in massive proliferation of enlarged mitochondria in the heart (Lehman and Kelly, 2002; Russell et al., 2004). The results lend further support to the idea that PGC–1α serves to integrate transcription factors found in a variety of cells and tissues into a program of mitochondrial biogenesis. Recent studies have indicated that the metabolic role of PGC–1α extends beyond the regulation of mitochondrial biogenesis and thermogenesis. For example, PGC–1α has been implicated in the regulation of hepatic gluconeogenesis and skeletal myofiber type specification (Yoon et al., 2001; Lin et al. 2002b).

Altered energy metabolism has long been implicated in HD pathogenesis. Striatal hypometabolism was detected in asymptomatic HD subjects (Grafton et al., 1992; Kuwert et al., 1993; Antonini et al., 1996), and activities of complexes of the electron transport chain were selectively reduced in caudate and putamen of advanced-grade HD patients (Gu et al., 1996; Browne et al., 1997). Moreover, there is evidence of a direct interaction of huntingtin protein with mitochondria, by demonstrating the localization of huntingtin both within degenerating mitochondria, and on their surfaces (Yu et al., 2003; Browne and Beal, 2004). Functional changes in mitochondria caused by mutant huntingtin have also recently been shown by the demonstration that polyglutamine-containing polypeptides can influence mitochondrial calcium handling (Panov et al., 2002, 2003; Brustovetsky et al., 2003). Additional evidence that energetic deficits may contribute to neurodegeneration in HD came from studies showing that agents which enhance energy production in the brain are neuroprotective (Browne and Beal, 2004). Importantly, the specific mechanisms by which the polyglutamine-containing proteins may inhibit mitochondrial function and oxidative metabolism are not clear.

The predominant pathological feature of HD is progressive degeneration of the caudate and putamen nuclei. By end stage, when most of the caudate and putamen neurons are lost and the striatum is severely atrophic and gliotic, degeneration is also evident in several other brain regions, including the white matter. On the other hand, early neuropathological changes in HD predominantly affect the caudate nucleus (Vonsattel and DiFiglia, 1998). Compelling evidence for a potential striatal dysfunction comes from findings that striatal hypometabolism precedes the bulk of tissue loss in asymptomatic HD subjects (Browne and Beal, 2004).

To determine whether PGC–1α expression may be altered in a region-specific manner, PGC–1α mRNA was examined in post-mortem brain samples isolated from presymptomatic HD patients. A 30% decrease in PGC–1α mRNA was found in the caudate nucleus, whereas no significant changes in PGC–1α expression were detected in the hippocampus or cerebellum in these brain samples. These results indicated that PGC–1α expression is specifically decreased in the caudate, the first region affected in HD. Next, it was tested whether these changes in PGC–1α levels were accompanied by changes in the expression of several mitochondrial genes that were decreased in brain samples of PGC–1α knockout mice. A 25–30% decrease in expression of mRNAs was found for the ubiquinol-cytochrome c reductase complex (involved in electron transport), cox7a1 (a subunit of the cytochrome c oxidase complex), and ATP5I (a component of the F0F1 ATP synthetase complex) in HD brain as compared to controls. No significant changes in mitochondrial genes were observed in the cerebellum and hippocampus from the same brain tissues. These findings suggest that decreased expression of PGC–1α and its mitochondrial targets may represent an early dysfunction in HD pathogenesis.

In order to further examine the function of PGC–1α in the context of HD, PGC–1α levels were analyzed in a well-established model of striatal neuronal cell lines that were generated from knock-in HD mice (Trettel et al., 2000; Wheeler et al., 2000, 2002). In these animals, a mutant huntingtin containing 111 glutamines is expressed under the control of an endogenous promoter. The cognate mutant striatal neuronal cell lines (STHdh^{Q111}) express endogenous mutant huntingtin and display dominant mutant phenotypes, including decreased cAMP synthesis and reduced ATP levels consistent with altered energy metabolism and aberrant PKA-CREB signaling (Gines et al., 2003). The precise mechanism underlying these energy deficits in STHdh^{Q111} cells has not been determined. To begin addressing this question, expression of PGC–1α and its mitochondrial targets were compared in wild-type STHdhQ7 and mutant STHdh^{Q111} cells. Expression of both PGC–1α and mitochondrial genes such as cytochrome c and COXIV were significantly decreased in the STHdh^{Q111} cells. The PGC–1α mRNA level is reduced approximately 10–fold in mutant cells compared to the wild-type control. These results suggest that decreased levels of PGC–1α might play a role in mediating energy defects in mutant STHdh^{Q111} striatal cells.

In order to elucidate the mechanisms of PGC–1α down-regulation observed in HD brain and STHdh^{Q111} cells, it was investigated whether mutant huntingtin might repress the PGC–1α promoter. The PGC–1α promoter-reporter, encompassing the regions between +78 and –2533 of the mouse PGC–1α gene, was transfected into striatal cell lines from the wild-type and HD mutant mice. A dramatic inhibition of the reporter was observed in the mutant cells (STHdh^{Q111}) compared to the wild-type cells (STHdh^{Q7}). In order to examine whether mutant huntingtin directly inhibits the reporter, wild-type cells (STHdh^{Q7}) were transfected with increasing concentrations of normal and mutant huntingtin along with the PGC–1α reporter. Transfected mutant huntingtin significantly repressed the PGC–1α reporter in a dose-dependent manner, whereas increasing concentrations of normal huntingtin activated the reporter. These results suggested that mutant huntingtin modulates the expression of the PGC–1α gene at the level of transcription.

Since STHdh^{Q111} cells exhibit decreased cAMP levels (Gines et al., 2003), it was next investigated whether treatment of these cells with cAMP could affect the reporter activity driven by the PGC–1α promoter. Treatment of STHdh^{Q111} mutant cells with 8–bromo-cAMP was able significantly to up-regulate the expression of the PGC–1α reporter. Importantly, these effects were dependent on the CRE binding site in the PGC–1α proximal promoter, which is known to be critically important in its regulation in several tissues (Herzig et al., 2001; Handschin et al., 2003).

The reduced expression of PGC–1α mRNA in STHdh^{Q111} cells, together with the fact that the PGC–1α reporter was up-regulated by cAMP, prompted us to examine whether endogenous PGC–1α mRNA was modulated by cAMP in STHdh^{Q111} cells. Treatment of these cells with cAMP significantly up-regulated PGC–1α mRNA.

Taken together, these results suggest that PGC–1α gene repression by mutant huntingtin involves CRE-dependent transcription.

To begin a direct examination of the mechanism of regulation of the PGC–1α promoter by mutant huntingtin, primary striatal neurons were transfected with PGC–1α promoter-reporter gene constructs along with mutant or normal huntingtin. Although huntingtin with normal glutamine repeats had no significant effect on the promoter activity, mutant huntingtin produced significant inhibition by more than 75% compared to the wild-type huntingtin. To determine whether the inhibition of CRE-mediated transcription was dependent on increased levels of huntingtin relative to those of endogenous CREB, the latter was overexpressed together with huntingtin. It was found that overexpression of CREB that is constitutively active (Du et al., 2000) significantly reduced, but did not rescue, the inhibitory effects of mutant huntingtin on PGC–1α promoter-reporter.

It had been shown previously that huntingtin interacts with TAF4, at least in part, in the context of TFIID complex and that TAF4 plays a critical role in mediating mutant huntingtin effects on Sp1–driven gene transcription (Dunah et al., 2002; Zhai et al., 2005). In addition to its function in Sp1 transcription, TAF4 also mediates coupling of CREB to TFIID and the basal transcriptional machinery (Saluja et al., 1998; Shimohata et al., 2000). Therefore, we examined whether CREB/TAF4 is required for maximal expression of the PGC–1α promoter. Co-expression of TAF4 and CREB resulted in complete reversal of huntingtin-induced inhibition of PGC–1α promoter activity, indicating that mutant huntingtin interferes with the CREB/TAF4 transcriptional pathway, while TAF4 interacts with other TAFs primarily via its C-terminal domain; this suggests the importance of this domain for the interactions of TAF4 with the TFIID complex. To examine whether the effects of CREB and TAF4 on PGC–1α were dependent on intact C-terminus domain of TAF4, the C-terminal mutant of TAF4 were co-expressed in primary striatal neurons. No significant effect of TAF4 on PGC–1α promoter activity was observed when the C-terminus of TAF4 was deleted, suggesting that the C-terminus domain of TAF4 plays a critical role in the repression of PGC–1α promoter by mutant huntingtin. Importantly, the C-terminus domain of TAF4 appears to be targeted by mutant huntingtin (Dunah et al., 2002) and another polyglutamine disease protein, atrophin (Shimohata et al., 2000); this suggests that, by competing for a common interaction domain in TAF4, polyglutamine proteins may disrupt CREB-dependent transcription.

Histone deacetylases (HDACs) work in concert with histone acetyl transferases (HATs) to modify chromatin and regulate transcription (Marks et al., 2001). Recent studies in cell culture, yeast, and *Drosophila* models of polyglutamine disease have indicated that HDAC inhibitors might provide a useful class of agents to ameliorate the transcriptional changes in HD (Hughes et al., 2001; McCampbell et al., 2001; Steffan et al., 2001; Ferrante et al., 2003; Hockly et al., 2003). On the other hand, specific targets of HDACs that mediate these protective effects have not been identified. Since our results suggest that mutant huntingtin down-regulates PGC–1α expression by transcriptional mechanisms, we tested whether HDAC inhibitors may alter PGC–1α expression in the context of HD. As a first step, we analyzed the activity of the PGC–1α promoter in the wild-type and mutant striatal cell lines treated with HDAC inhibitors. It was found that treatment of the mutant cells with 4–phenylbutyrate (4PBA) or trichostatin A (TSA) up-regulated the activity of the reporter by more than 20– and 10–fold, respectively. To examine whether endogenous PGC–1α mRNA levels was also affected by the treatment with HDAC inhibitors, PGC–1α mRNA was measured in the mutant cells treated with HDAC inhibitors. In the presence of 4PBA or TSA, mutant striatal cell lines exhibited an approximate two-fold induction of the PGC–1α mRNA as compared to the untreated cells.

Treatments with HDAC inhibitor sodium butyrate extended survival and delayed neuropathological abnormalities in transgenic HD mice (R6/2). Sodium butyrate also protected against 3–nitropropionic acid (3–NP) neurotoxicity in this mouse model of HD (Ferrante et al., 2003). Since 3–NP acts as a mitochondrial toxin, these results in HD mice suggested that the benefits of treatments by HDAC inhibitors might involve an improvement in mitochondrial function. Therefore, we examined whether treatments of R6/2 mice with sodium butyrate may affect the expression of PGC–1α mRNA. Daily treatments of R6/2 mice with sodium butyrate were found to lead to an approximately 80% increase in PGC–1α mRNA expression in the striatum of the HD mice as compared to untreated mice. In order to investigate whether these changes in PGC–1α mRNA expression may be due to changes in PGC–1α transcription *in vivo*, RNA polymerase II chromatin immuno precipitations (RNAPol-ChIP) were performed. This procedure uses the presence of RNA polymerase II within the coding region of the target gene as a way of measuring active transcription (Sandoval et al., 2004). Using this approach, a decreased expression of PGC–1α gene in R6/2 brain was found compared to normal littermates. Treatment of R6/2 mice with 4PBA for 2 days resulted in up-regulation of PGC–1α transcription to almost normal levels. When genes that are not altered in HD mice (e.g., GAPDH and albumin) were analyzed by RNAPol-ChIP, no change was observed in polymerase II binding between the normal and HD mice.

Taken together, these results suggest that changes in PGC–1α mRNA observed in HD tissues and cells may be due to alterations in PGC–1α gene transcription. Moreover, PGC–1α may represent at least one biologically significant target of HDAC inhibitors that could contribute to the benefit observed with these treatments in HD mice.

Extensive evidence suggests a broad role of PGC–1α as a master regulator of mitochondrial function in several tissues and cell types (Puigserver and Spiegelman, 2003). On the other hand, the function of PGC–1α in neurons has not been investigated. To further establish the role of PGC–1α in neuronal function and in huntingtin-induced cell death, we analyzed striatal tissues and cell cultures isolated from the PGC–1α knockout mice (PGC–1α$^{-/-}$). These mice exhibit abnormal movements and specific neuropathological lesions that are predominantly found in the striata, strongly suggesting that the neurological phenotype is caused by the striatal lesions (Lin et al., 2004). To test whether deficiency of PGC–1α leads to altered expression of genes involved in the energy metabolism of the brain, a gene array analysis was performed using brain samples isolated from PGC–1α$^{-/-}$ mice. The gene list that covers over 300 mitochondrial genes was generated through Gene Ontology (GO) annotations for genes involved in mitochondrial function (Lin et al., 2004). When the wild-type mice were compared with PGC–1α$^{-/-}$ mice, expression of a large number of mitochondrial genes were down-regulated in PGC–1α$^{-/-}$ mouse brain, including genes involved in electron transport, the tricarboxylic acid (TCA) cycle and oxidative phosphorylation. In order to validate the gene array results, a real-time PCR analysis of a subset of genes from each category was performed. We found that succinyl-CoA synthetase, ubiquinol-cytochrome c reductase complex, cox7a1 and ATP5I were significantly decreased in the knockout mice as compared to the wild-type mice, underscoring the importance of PGC–1α in expression of genes involved in mitochondrial function.

We next examined the mitochondrial staining of primary striatal neurons isolated from PGC–1α$^{-/-}$ mice. Consistent with the decreased mitochondrial gene expression in these mice, decreased mitochondrial staining in striatal neurons was observed in isolated cells from PGC–1α$^{-/-}$ mice as compared to their wild-type littermates. In addition, primary neurons isolated from PGC–1α$^{-/-}$ mice showed decreased neurites as compared to their wild-type littermates (Lin et al., 2004). These findings are consistent with the primarily axonal pathology observed in the striatum of PGC–1α$^{-/-}$ mice (Lin et al., 2004).

Taken together, these results indicate that PGC–1α plays a very important role in the regulation of genes involved in energy metabolism of the brain. In addition, these data indicate that a compromise in mitochondrial function may be responsible for the specific lesions seen in the striata of PGC–1α$^{-/-}$ mice (Lin et al., 2004).

We next examined whether the loss of PGC–1α expression through the mutant huntingtin protein rendered cells more susceptible to the neurotoxin 3–nitropropionic acid (3–NP), which is known to induce HD-like symptoms. 3–NP is a complex II inhibitor that selectively induces striatal lesions when administered systemically in humans, primates and rodents. The distribution and pathology of the 3–NP-induced striatal lesions closely resemble those seen in HD (Ludolph et al., 1991; Brouillet and Hantraye, 1995; Browne and Beal, 2004). Primary striatal cultures from PGC–1α$^{-/-}$ mice were treated with 3–NP. As shown previously (Galas et al., 2004), treatment of wild-type striatal neurons with 3–NP led to an approximately two-fold increase in neuronal death as compared to untreated neurons. When striatal neurons from PGC–1α$^{-/-}$ mice were treated with 3–NP under the same experimental conditions, approximately 30% greater toxicity was observed in these neurons compared to those from wild-type littermates. These results demonstrated that PGC–1α deficiency renders striatal neurons more susceptible to additional metabolic insults.

Based on these findings, it was hypothesized that PGC–1α might protect normal neurons subjected to metabolic stressors. In order to test this hypothesis, PGC–1α mRNA levels were measured in primary striatal neurons exposed to 3–NP. Interestingly, significant elevation of PGC–1α mRNA was noted within 3 hours and peaked at 6 hours after exposure to 3–NP, which suggested that metabolic stress leads to rapid up-regulation of PGC–1α transcription. Since these data suggest that mutant huntingtin inhibits PGC–1α transcription, the question was asked as to whether

20.4 Deregulation of CRE-Dependent Transcription in HD

normal up-regulation of PGC–1α might be inhibited in the presence of mutant huntingtin. Cultured primary striatal neurons were isolated from normal and mutant knock-in HD mice (140 CAG) (Menalled et al., 2002) and treated with 3–NP. Levels of PGC–1α mRNA determined at 6 hours after the treatment were significantly increased in normal striatal cells, but no increase was detected in mutant knock-in cells. These results further demonstrated that inhibition of PGC–1α transcription by mutant huntingtin may negatively affect the ability of neurons to respond to metabolic demands in HD.

Since PGC–1α regulates mitochondrial respiration (Puigserver and Spiegelman, 2003), the decreased levels of PGC–1α may lead to mitochondrial dysfunction that is characteristic of striatal HD cells (Gines et al., 2003). We therefore directly assessed mitochondrial energy metabolism in wild-type and homozygous mutant STHdh^{Q111} cells using the MTT assay to assess mitochondrial function. This method is based on the reduction of the soluble tetrazolium salt to insoluble formazan product by mitochondrial succinic dehydrogenase. The results revealed that formazan production was significantly reduced (by approximately 35%) in mutant cells compared to wild-type cells. To determine whether PGC–1α was capable of correcting this defect in mitochondrial function, STHdh^{Q111} cells were infected with a recombinant adenovirus expressing PGC–1α (Lin et al., 2002a). Adenoviral-mediated expression of PGC–1α significantly reversed the mitochondrial defect as determined by MTT assay, suggesting that up-regulation of PGC–1α can rescue the effect of mutant huntingtin on mitochondrial function.

We then analyzed primary striatal cultures that recapitulate features of the neurodegenerative process that occurs in HD. Transfection of full-length mutant huntingtin induced neuronal cell death, whereas the wild-type huntingtin did not show any toxic effects, as reported previously (Saudou et al., 1998; Dunah et al., 2002). However, when mutant huntingtin is co-expressed in striatal neurons together with PGC–1α, the toxicity of mutant huntingtin is significantly abrogated. These data suggest that expression of PGC–1α in striatal neurons substantially rescues the deleterious effects of mutant huntingtin on mitochondrial function and cellular toxicity. Alternatively, it was possible that a lack of PGC–1α might increase the toxic effects of mutant huntingtin. To investigate this possibility, primary striatal neurons were isolated from the PGC–1α null mice and transfected with normal or mutant huntingtin. While transfection of normal huntingtin did not lead to increased toxicity, transfection of mutant huntingtin resulted in significantly increased toxicity in PGC–1α knockout cells compared to normal cells. Together, these results indicate that PGC–1α protects striatal neurons from mutant huntingtin-induced toxicity.

Since these data showed that CREB and TAF4 were able to completely reverse the repressive effects of mutant huntingtin on PGC–1α promoter activity, we tested whether striatal neurons could be protected against mutant huntingtin by transfecting constitutively active CREB along with TAF4. It was found that co-expression of CREB and TAF4 resulted in a statistically significant protection of striatal cells from the effects of mutant huntingtin. Co-expression of CREB with TAF4 lacking its C-terminus domain failed to block huntingtin-induced cell death, suggesting that intact interactions of TAF4 with TFIID and other nuclear factors are required for its protective function in HD. Our previous results had demonstrated that co-expression of TAF4 and Sp1 also protected striatal cells from mutant huntingtin-mediated cell death (Dunah et al., 2002), which suggested that TAF4 might play a central role in mediating huntingtin-induced toxicity. Together, these data demonstrate that CREB/TAF4–dependent regulation of PGC–1α gene, at least in part, mediates the protective effects of PGC–1α in striatal neurons.

In order to further test the hypothesis that PGC–1α might be protective in HD, lentiviral vector expressing PGC–1α was administered directly into the striatum of R6/2 transgenic HD mice. The R6/2 model was chosen because it has a well-characterized progressive phenotype that recapitulates many of the neuropathological features observed in HD patients, such as striatal atrophy, cellular atrophy and huntingtin-positive aggregates. It has been shown previously that compounds which affect energetic pathways significantly improved the behavioral and neuropathological phenotype in R6/2 mice. For example, creatine treatment extended survival and improved neuronal atrophy in R6/2 mice, suggesting that metabolic dysfunction plays a pathogenic role in HD mice (Dedeoglu et al., 2003; Browne and Beal, 2004).

Since these results suggested that the steady-state levels and active transcription of PGC–1α were decreased in R6/2, we tested whether complementing PGC–1α might be neuroprotective. The striata of R6/2 mice aged 5.5 weeks were injected with lentivirus vector expressing PGC–1α. Using unbiased stereological methods, neuronal volumes were examined at 3.5 weeks after the injection. When analyzing a series of sections spanning the injected region of striatum, a statistically significant increase (by 27.8%) in mean neuronal volume was observed in striata injected with PGC–1α as compared to contralateral striata that were not injected. No significant change in cell volume was observed in neurons injected with control lentiviral vector expressing green fluorescent protein (GFP). Previous studies have shown that R6/2 mice exhibit progressive atrophy of striatal neurons from about 3 to 12 weeks of age, with an almost 40% overall decrease in area measurements. In addition, at the age of 9 weeks (when this analysis was performed), striatal neuronal areas in R6/2 mice decrease by about 30% compared to littermate transgene-negative mice (Dedeoglu et al., 2003). Here, we show that the administration of PGC–1α to R6/2 mice at least partially prevents neuronal degeneration. Moreover, the findings in HD mice were in agreement with other studies in cell culture, showing that overexpression of PGC–1α protects striatal neurons from toxicity by mutant huntingtin.

Taken together, these data strongly suggest that decreased levels of PGC–1α mediate neuronal toxicity in HD. In addition, the fact that PGC–1α is specifically decreased in the striatum of presymptomatic HD brain provides a possible explanation for differential vulnerability of the striatum in the early stages of HD.

This conclusion is underscored by the fact that the PGC–1α knockout mice show most severe lesions in the striatum despite the uniform absence in PGC–1α expression in all brain regions (Lin et al., 2004). PGC–1α has been shown to play an important role in the regulation of oxidative metabolism and mitochondrial biology in a number of tissues, but its role in brain function has not been investigated (Puigserver and Spiegelman, 2003). Interestingly, PGC–1α knockout mice show a

20.4 Deregulation of CRE-Dependent Transcription in HD

neurological phenotype characterized by abnormal movements accompanied by specific lesions in the striatum of the knockout mice (Lin et al., 2004). We found decreased expression of a number of mitochondrial genes in brain samples of the PGC–1α knockout mice, which suggested that alterations in mitochondrial function may be responsible for the observed neuropathology in striata of these mice. Although the neuropathologic changes observed in the knockout mice appear severe compared to those found in HD, it is striking that these mice show such prominent abnormalities in the striatum, the region most affected in HD. This finding suggests an intriguing possibility that the complete absence of PGC–1α seen in the knockout mice results in more dramatic striatal pathology compared to the lesions caused by only partial down-regulation of PGC–1α in HD.

We show that a possible mechanism for the observed down-regulation of PGC–1α in HD may involve transcriptional repression of PGC–1α promoter by mutant huntingtin. The mouse PGC–1α promoter contains a putative TATA box and a full palindromic consensus CRE site positioned at –130. The CRE on PGC–1α promoter has been shown to be a direct target for cAMP and CREB action *in vivo* (Herzig et al., 2001). Mutant huntingtin represses CRE-mediated transcription of PGC–1α by interfering with CREB/TAF4 transcriptional pathway in striatal neurons. Reduced CRE-dependent transcription has been previously observed as an early abnormality in HD pathogenesis, but the target genes of the CRE-dependent transcription in HD have not been determined (Gines et al., 2003; Sugars and Rubinsztein, 2003; Sugars et al., 2004). Moreover, ablation of CREB and CREM in the postnatal CNS produces a progressive neurodegeneration in striatum that is reminiscent of HD, suggesting that CREB might play an important role in HD pathogenesis (Mantamadiotis et al., 2002). Our data show that overexpression of CREB and TAF4 rescues the toxicity by mutant huntingtin in striatal neurons, underscoring the importance of CREB and TAF4 in neuronal survival in HD. These findings do not exclude a possibility that other target genes, in addition to PGC–1α, mediate the survival effects of CREB and TAF4 in HD. Mutant huntingtin has been shown previously to directly repress transcription of Sp1–driven genes by uncoupling the interactions of activator (Sp1) with the co-activator TAF4 (Dunah et al., 2002; Zhai et al., 2005). Since both Sp1 and CREB interact with TAF4, it is hypothesized that Sp1– or CREB-mediated interactions with TAF4 are directly targeted by mutant huntingtin depending on the promoter context. Huntingtin was found to interact with both the glutamine-rich central domain and the C-terminal domain of TAF4, and that the C-terminal domain is required for protection against huntingtin-induced transcriptional dysregulation and neuronal cell death (Dunah et al., 2002). The conserved C-terminal domain of TAF4 participates in a number of protein-protein interactions, including several TAFs in the TFIID complex (Gangloff et al., 2000; Furukawa and Tanese, 2000; Asahara et al., 2001). These findings, and the recent report that atrophin also binds to the C-terminal domain of TAF4 (Shimohata et al., 2000), suggests that TAF4, probably in the context of TFIID, represents a key target of polyglutamine proteins such as huntingtin.

These findings show that transcriptional down-regulation of PGC–1α leads to decreased mitochondrial function and increased susceptibility of striatal cells to

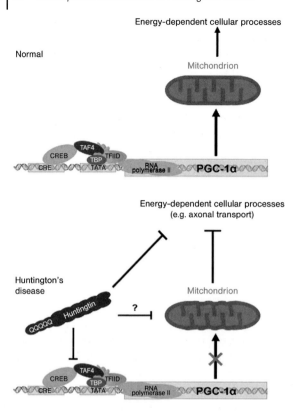

Fig. 20.2 Model for regulation of PGC–1α in Huntington's disease (HD). Upper panel: In a normal state, PGC–1α regulates metabolic programs and maintains energy homeostasis in the CNS. Lower panel: In HD, mutant huntingtin interferes with CREB and TAF4 regulation of PGC–1α transcription that leads to inhibited expression of PGC–1α. Inhibition of PGC–1α expression limits the ability of the vulnerable neurons to adequately respond to energy demands in HD. Direct interactions of mutant huntingtin with mitochondria may also contribute to defects in energy metabolism in HD. (This figure also appears with the color plates.)

mutant huntingtin-induced toxicity. Recent evidence suggests that mutant huntingtin may also interact directly with mitochondria. Degenerated mitochondria were detected in transgenic HD mouse models before other marked pathological changes were detected within neurons (Yu et al., 2003). Moreover, the evidence of a direct interaction of huntingtin protein with mitochondria was presented, by showing the localization of huntingtin within and on the surface of degenerating mitochondria (Panov et al., 2002; Yu et al., 2003). In addition, a recent study has demonstrated that polyglutamines can influence mitochondrial calcium handling and that expanded polyglutamines alter mitochondrial membrane depolarization, making cells more vulnerable to metabolic stress (Panov et al., 2002; Brustovetsky et al., 2003; Choo et al., 2004).

Here, a model is proposed where mutant huntingtin affects mitochondrial function by interfering with PGC–1α expression (Fig. 20.2). It is hypothesized that, in the normal state, PGC–1α regulates metabolic programs and maintains energy homeostasis in the CNS, whereas inhibition of PGC–1α by mutant huntingtin leads to defects in energy metabolism and dysfunction of neurons that are most vulnerable to metabolic stress, such as striatum. Our model does not exclude a possibility that mutant huntingtin causes neurodegeneration by multiple mechanisms, as suggested by numerous other studies. Instead, our results indicate that inhibition of PGC–1α function by mutant huntingtin limits the ability of striatal neurons to respond adequately to metabolic demands in HD. Such disruption in energy homeostasis in HD may lead to subtle disruptions of multiple cellular functions, and ultimately results in neurodegeneration. These data also suggest that stimulation of the pathways of energy metabolism controlled by PGC–1α could provide potential clinical benefit at an early stage of HD.

20.5 Summary

In summary, these results demonstrate that transcriptional changes occur in HD. On the other hand, much remains to be elucidated to understand fully the consequences of transcriptional deregulation in the pathogenesis of HD. CRE and Sp1 pathways are clearly impaired, but it is not entirely clear whether these are primary or secondary responses of the huntingtin mutation. The recent data described above suggest that there is a direct effect of mutant huntingtin on the transcriptional machinery, and that this could actually represent a key primary event in the onset of HD (Dunah et al., 2002; Zhai et al., 2005). In addition, rescuing some of these transcriptional deficits could be of therapeutic benefit. For example, histone deacetylase inhibitors reduce polyQ toxicity in *Drosophila* models (Steffan et al., 2001), and recent data showing beneficial effects of this class of compound in a HD mouse model (Ferrante et al., 2003; Hockly et al., 2003) suggest that these compounds deserve serious consideration for further study, as they could potentially be used in patients. Finally, recent data showed that transcriptional deregulation may be directly linked to dysfunction of energy metabolism in HD (L. Cui et al., unpublished results), further suggesting that disruption of transcription might occur early in the disease pathogenesis. It will be important to study individual transcriptional pathways and to identify new targets, which may be amenable to treatments in HD. If new neuroprotective treatments may be developed that would directly affect transcriptional pathways, they could represent a proof that deregulation of transcription plays a key role in HD.

Abbreviations

3–NP	3–nitropropionic acid
4PBA	4–phenylbutyrate
EMSA	electrophoretic mobility shift assay
GFP	green fluorescent protein
GO	Gene Ontology
HAT	histone acetyl transferase
HD	Huntington's disease
HDAC	histone deacetylase
mtDNA	mitochondrial DNA
PGC–1α	Peroxisome proliferator-activated receptor Gamma Coactivator–1 alpha
PIC	preinitiation complex
RNAPol-ChIP	RNA polymerase II chromatin immunoprecipitations
TAF	TBP-associated factor
TBP	TATA-binding protein
TSA	trichostatin A

Acknowledgments

I particularly thank members of my laboratory, Hunkyung Jeong, Libin Cui, Anthone Dunah, April Griffin, Fran Borovecki and Katrin Lindenberg. I also thank our collaborators Naoko Tanese, Weigou Zhai, Anne B. Young, Robert Tjian, Jiandie Li, Bruce Spiegelman, Maral Mouradian for their help with experiments, constructive suggestions and comments on the manuscript.

References

Alba, M.M., Guigo, R. (2004). Comparative analysis of amino acid repeats in rodents and humans. *Genome Res* 14, 549–554.

Albright, S.R., Tjian, R. (2000). TAFs revisited: more data reveal new twists and confirm old ideas. *Gene* 242, 1–13.

Antonini, A., Leenders, K.L., Spiegel, R., Meier, D., Vontobel, P., Weigell-Weber, M., Sanchez-Pernaute, R., de Yebenez, J.G., Boesiger, P., Weindl, A., Maguire, R.P. (1996). Striatal glucose metabolism and dopamine D2 receptor binding in asymptomatic gene carriers and patients with Huntington's disease. *Brain* 11, 2085–2095.

Asahara, H., Santoso, B., Guzman, E., Du, K., Cole, P.A., Davidson, I., Montminy, M. (2001). Chromatin-dependent cooperativity between constitutive and inducible activation domains in CREB. *Mol Cell Biol* 21, 7892–7900.

Brouillet, E., Hantraye, P. (1995). Effects of chronic MPTP and 3–nitropropionic acid in nonhuman primates. *Curr Opin Neurol* 8, 469–473.

Browne, S.E., Beal, M.F. (2004). The energetics of Huntington's disease. *Neurochem Res* 29, 531–546.

Browne, S.E., Bowling, A.C., MacGarvey, U., Baik, M.J., Berger, S.C., Muqit, M.M., Bird, E.D., Beal, M.F. (1997). Oxidative damage and metabolic dysfunction in Huntington's disease: selective vulnerability of the basal ganglia. *Ann Neurol* 41, 646–653.

Brustovetsky, N., Brustovetsky, T., Purl, K.J.,

Capano, M., Crompton, M., Dubinsky, J.M. (**2003**). Increased susceptibility of striatal mitochondria to calcium-induced permeability transition. *J Neurosci* 23, 4858–4867.

Choo, Y.S., Johnson, G.V., MacDonald, M., Detloff, P.J., Lesort, M. (**2004**). Mutant huntingtin directly increases susceptibility of mitochondria to the calcium-induced permeability transition and cytochrome c release. *Hum Mol Genet* 13, 1407–1420.

Chen, J.L., Attardi, L.D., Verrijzer, C.P., Yokomori, K., Tjian, R. (**1994**). Assembly of recombinant TFIID reveals differential coactivator requirements for distinct transcriptional activators. *Cell* 79, 93–105.

Conaway, J.W., Shilatifard, A., Dvir, A., Conaway, R.C. (**2000**). Control of elongation by RNA polymerase II. *Trends Biochem Sci* 25, 375–380.

Courey, A.J., Tjian, R. (**1988**). Analysis of Sp1 in vivo reveals multiple transcriptional domains, including a novel glutamine-rich activation motif. *Cell* 55, 887–898.

Dedeoglu, A., Kubilus, J.K., Yang, L., Ferrante, K.L., Hersch, S.M., Beal, M.F., Ferrante, R.J. (**2003**). Creatine therapy provides neuroprotection after onset of clinical symptoms in Huntington's disease transgenic mice. *J Neurochem.* 85, 1359–1367.

Du, K., Asahara, H., Jhala, U.S., Wagner, B.L., Montminy, M. (**2000**). Characterization of a CREB gain-of-function mutant with constitutive transcriptional activity in vivo. *Mol Cell Biol* 20, 4320–4327.

Dunah, A.W., Jeong, H., Griffin, A., Kim, Y.M., Standaert, D.G., Hersch, S.M., Mouradian, M.M., Young, A.B., Tanese, N., Krainc, D. (**2002**). Sp1 and TAFII130 transcriptional activity disrupted in early Huntington's disease. *Science* 296, 2238–2243.

Ferrante, R.J., Kubilus, J.K., Lee, J., Ryu, H., Beesen, A., Zucker, B., Smith, K., Kowall, N.W., Ratan, R.R., Luthi-Carter, R., Hersch, S.M. (**2003**). Histone deacetylase inhibition by sodium butyrate chemotherapy ameliorates the neurodegenerative phenotype in Huntington's disease mice. *J Neurosci* 23, 9418–9427.

Frejtag, W., Zhang, Y., Dai, R., Anderson, M.G., Mivechi, N.F. (**2001**). Heat shock factor-4 (HSF-4a) represses basal transcription through interaction with TFIIF. *J Biol Chem* 276, 14685–14694.

Furukawa, T., Tanese, N. (**2000**). Assembly of partial TFIID complexes in mammalian cells reveals distinct activities associated with individual TATA box-binding protein-associated factors. *J Biol Chem* 275, 29847–29856.

Galas, M.C., Bizat, N., Cuvelier, L., Bantubungi, K., Brouillet, E., Schiffmann, S.N., Blum, D. (**2004**). Death of cortical and striatal neurons induced by mitochondrial defect involves differential molecular mechanisms. *Neurobiol Dis* 15, 152–159.

Gangloff, Y.G., Werten, S., Romier, C., Carre, L., Poch, O., Moras, D., Davidson, I. (**2000**). The human TFIID components TAF(II)135 and TAF(II)20 and the yeast SAGA components ADA1 and TAF(II)68 heterodimerize to form histone-like pairs. *Mol Cell Biol* 20, 340–351.

Gaiser, F., Tan, S., Richmond, T.J. (**2000**). Novel dimerization fold of RAP30/RAP74 in human TFIIF at 1.7 A resolution. *J Mol Biol* 302, 1119–1127.

Garrett, K.P., Serizawa, H., Hanley, J.P., Bradsher, J.N., Tsuboi, A., Arai, N., Yokota, T., Arai, K., Conaway, R.C., Conaway, J.W. (**1992**). The carboxyl terminus of RAP30 is similar in sequence to region 4 of bacterial sigma factors and is required for function. *J Biol Chem* 267, 23942–23949.

Gerber, H.P., Seipel, K., Georgiev, O., Hofferer, M., Hug, M., Rusconi, S., Schaffner, W. (**1994**). Transcriptional activation modulated by homopolymeric glutamine and proline stretches. *Science* 263, 808–811.

Gines, S., Seong, I.S., Fossale, E., Ivanova, E., Trettel, F., Gusella, J.F., Wheeler, V.C., Persichetti, F., MacDonald, M.E. (**2003**). Specific progressive cAMP reduction implicates energy deficit in presymptomatic Huntington's disease knock-in mice. *Hum Mol Genet* 12, 497–508.

Grafton, S.T., Mazziotta, J.C., Pahl, J.J., St George-Hyslop, P., Haines, J.L., Gusella, J., Hoffman, J.M., Baxter, L.R., Phelps, M.E. (**1992**). Serial changes of cerebral glucose metabolism and caudate size in persons at risk for Huntington's disease. *Arch Neurol* 49, 1161–1167.

Gu, M., Gash, M.T., Mann, V.M., Javoy-Agid, F., Cooper, J.M., Schapira, A.H. (**1996**). Mitochondrial defect in Huntington's disease caudate nucleus. *Ann Neurol* 39, 385–389.

Handschin, C., Rhee, J., Lin, J., Tarr, P.T., Spiegelman, B.M. (**2003**). An autoregulatory loop controls peroxisome proliferator-activated receptor gamma coactivator 1alpha expression in muscle. *Proc Natl Acad Sci USA* 100, 7111–7116.

Herzig, S., Long, F., Jhala, U.S., Hedrick, S., Quinn, R., Bauer, A., Rudolph, D., Schutz, G., Yoon, C., Puigserver, P., et al. (**2001**). CREB regulates hepatic gluconeogenesis through the coactivator PGC–1. *Nature* 413, 179–183.

Hockly, E., Richon, V.M., Woodman, B., Smith, D.L., Zhou, X., Rosa, E., Sathasivam, K., Ghazi-Noori, S., Mahal, A., Lowden, P.A., et al. (**2003**). Suberoylanilide hydroxamic acid, a histone deacetylase inhibitor, ameliorates motor deficits in a mouse model of Huntington's disease. *Proc Natl Acad Sci USA* 100, 2041–2046.

Hughes, R.E., Lo, R.S., Davis, C., Strand, A.D., Neal, C.L., Olson, J.M., Fields, S. (**2001**). Altered transcription in yeast expressing expanded polyglutamine. *Proc Natl Acad Sci USA* 98, 13201–13206.

Kuwert, T., Lange, H.W., Boecker, H., Titz, H., Herzog, H., Aulich, A., Wang, B.C., Nayak, U., Feinendegen, L.E. (**1993**). Striatal glucose consumption in chorea-free subjects at risk of Huntington's disease. *J Neurol* 241, 31–36.

Lehman, J.J., Kelly, D.P. (**2002**). Transcriptional activation of energy metabolic switches in the developing and hypertrophied heart. *Clin Exp Pharmacol Physiol* 29, 339–345.

Lehman, J.J., Barger, P.M., Kovacs, A., Saffitz, J.E., Medeiros, D.M., Kelly, D.P. (**2000**). Peroxisome proliferator-activated receptor gamma coactivator–1 promotes cardiac mitochondrial biogenesis. *J Clin Invest* 106, 847–856.

Lemon, B., Inouye, C., King, D.S., Tjian, R. (**2001**). Selectivity of chromatin-remodelling cofactors for ligand-activated transcription. *Nature* 414, 924–928.

Li, S.H., Cheng, A.L., Zhou, H., Lam, S., Rao, M., Li, H., Li, X.J. (**2002**). Interaction of Huntington disease protein with transcriptional activator Sp1. *Mol Cell Biol* 22, 1277–1287.

Lin, J., Puigserver, P., Donovan, J., Tarr, P., Spiegelman, B.M. (**2002a**). Peroxisome proliferator-activated receptor gamma coactivator 1beta (PGC–1beta), a novel PGC-1–related transcription coactivator associated with host cell factor. *J Biol Chem* 277, 1645–1648.

Lin, J., Wu, H., Tarr, P.T., Zhang, C.Y., Wu, Z., Boss, O., Michael, L.F., Puigserver, P., Isotani, E., Olson, E.N., et al. (**2002b**). Transcriptional co-activator PGC–1 alpha drives the formation of slow-twitch muscle fibres. *Nature* 418, 797–801.

Lin, J., Wu, P., Tarr, P., St-Pierre, J., Zhang, J., Mootha, V.K., Jager, S., Vianna, C., Reznick, R., Manieri, M., et al. (**2004**). Defects in adaptive energy metabolism with hyperactivity in PGC–1alpha mutant mice. *Cell* 119, 121–135.

Ludolph, A.C., He, F., Spencer, P.S., Hammerstad, J., Sabri, M. (**1991**). 3–Nitropropionic acid-exogenous animal neurotoxin and possible human striatal toxin. *Can J Neurol Sci* 18, 492–498.

Luthi-Carter, R., Hanson, S.A., Strand, A.D., Bergstrom, D.A., Chun, W., Peters, N.L., Woods, A.M., Chan, E.Y., Kooperberg, C., Krainc, D., et al. (**2002a**). Dysregulation of gene expression in the R6/2 model of polyglutamine disease: parallel changes in muscle and brain. *Hum Mol Genet* 11, 1911–1926.

Luthi-Carter, R., Strand, A.D., Hanson, S.A., Kooperberg, C., Schilling, G., La Spada, A.R., Merry, D.E., Young, A.B., Ross, C.A., Borchelt, D.R., Olson, J.M. (**2002b**). Polyglutamine and transcription: gene expression changes shared by DRPLA and Huntington's disease mouse models reveal context-independent effects. *Hum Mol Genet* 11, 1927–1937.

Mantamadiotis, T., Lemberger, T., Bleckmann, S.C., Kern, H., Kretz, O., Martin Villalba, A., Tronche, F., Kellendonk, C., Gau, D., Kapfhammer, J., et al. (**2002**). Disruption of CREB function in brain leads to neurodegeneration. *Nat Genet* 31, 47–54.

Marks, P., Rifkind, R.A., Richon, V.M., Breslow, R., Miller, T., Kelly, W.K. (**2001**). Histone deacetylases and cancer: causes and therapies. *Nat Rev Cancer* 1, 194–202.

McCampbell, A., Taye, A.A., Whitty, L., Penney, E., Steffan, J.S., Fischbeck, K.H. (**2001**). Histone deacetylase inhibitors reduce polyglutamine toxicity. *Proc Natl Acad Sci USA* 98, 15179–15184.

McCracken, S., Greenblatt, J. (**1991**). Related RNA polymerase-binding regions in human RAP30/74 and *Escherichia coli* sigma 70. *Science* 253, 900–902.

Menalled, L.B., Sison, J.D., Wu, Y., Olivieri, M., Li, X.J., Li, H., Zeitlin, S., Chesselet, M.F. (**2002**). Early motor dysfunction and striosomal distribution of huntingtin microaggregates in Huntington's disease knock-in mice. *J Neurosci* 22, 8266–8276.

Naar, A.M., Beaurang, P.A., Zhou, S., Abraham, S., Solomon, W., Tjian, R. (**1999**). Composite co-activator ARC mediates chromatin-directed transcriptional activation. *Nature* 398, 828–832.

Okazawa, H. (**2003**). Polyglutamine diseases: a transcription disorder? *Cell Mol Life Sci* 60, 1427–1439.

Panov, A.V., Burke, J.R., Strittmatter, W.J., Greenamyre, J.T. (**2003**). In vitro effects of polyglutamine tracts on Ca2+-dependent depolarization of rat and human mitochondria: relevance to Huntington's disease. *Arch Biochem Biophys* 410, 1–6.

Panov, A.V., Gutekunst, C.A., Leavitt, B.R., Hayden, M.R., Burke, J.R., Strittmatter, W.J., Greenamyre, J.T. (**2002**). Early mitochondrial calcium defects in Huntington's disease are a direct effect of polyglutamines. *Nat Neurosci* 5, 731–736.

Perutz, M.F., Finch, J.T., Berriman, J., Lesk, A. (**2002**). Amyloid fibers are water-filled nanotubes. *Proc Natl Acad Sci USA* 99, 5591–5595.

Puigserver, P., Spiegelman, B.M. (**2003**). Peroxisome proliferator-activated receptor-gamma coactivator 1 alpha (PGC–1 alpha): transcriptional coactivator and metabolic regulator. *Endocr Rev* 24, 78–90.

Puigserver, P., Wu, Z., Park, C.W., Graves, R., Wright, M., Spiegelman, B.M. (**1998**). A cold-inducible coactivator of nuclear receptors linked to adaptive thermogenesis. *Cell* 92, 829–839.

Rubinsztein, D.C., Barton, D.E., Davison, B.C., Ferguson-Smith, M.A. (**1993**). Analysis of the huntingtin gene reveals a trinucleotide-length polymorphism in the region of the gene that contains two CCG-rich stretches and a correlation between decreased age of onset of Huntington's disease and CAG repeat number. *Hum Mol Genet* 2, 1713–1715.

Russell, L.K., Mansfield, C.M., Lehman, J.J., Kovacs, A., Courtois, M., Saffitz, J.E., Medeiros, D.M., Valencik, M.L., McDonald, J.A., Kelly, D.P. (**2004**). Cardiac-specific induction of the transcriptional coactivator peroxisome proliferator-activated receptor gamma coactivator–1alpha promotes mitochondrial biogenesis and reversible cardiomyopathy in a developmental stage-dependent manner. *Circ Res* 94, 525–533.

Ryu, S., Zhou, S., Ladurner, A.G., Tjian, R. (**1999**). The transcriptional cofactor complex CRSP is required for activity of the enhancer-binding protein Sp1. *Nature* 397, 446–450.

Saluja, D., Vassallo, M.F., Tanese, N. (**1998**). Distinct subdomains of human TAFII130 are required for interactions with glutamine-rich transcriptional activators. *Mol Cell Biol* 18, 5734–5743.

Sanchez, I., Mahlke, C., Yuan, J. (**2003**). Pivotal role of oligomerization in expanded polyglutamine neurodegenerative disorders. *Nature* 421, 373–379.

Sandoval, J., Rodriguez, J.L., Tur, G., Serviddio, G., Pereda, J., Boukaba, A., Sastre, J., Torres, L., Franco, L., Lopez-Rodas, G. (**2004**). RNAPol-ChIP: a novel application of chromatin immunoprecipitation to the analysis of real-time gene transcription. *Nucleic Acids Res* 32, 1–8.

Saudou, F., Finkbeiner, S., Devys, D., Greenberg, M.E. (**1998**). Huntingtin acts in the nucleus to induce apoptosis but death does not correlate with the formation of intranuclear inclusions. *Cell* 95, 55–66.

Schaffar, G., Breuer, P., Boteva, R., Behrends, C., Tzvetkov, N., Strippel, N., Sakahira, H., Siegers, K., Hayer-Hartl, M., Hartl, F.U. (**2004**). Cellular toxicity of polyglutamine expansion proteins: mechanism of transcription factor deactivation. *Mol Cell* 15, 95–105.

Schmittgen, T.D. (**2001**). Real-time quantitative PCR. *Methods* 25, 383–385.

Shimohata, T., Nakajima, T., Yamada, M., Uchida, C., Onodera, O., Naruse, S., Kimura, T., Koide, R., Nozaki, K., Sano, Y., Ishiguro, H., Sakoe, K., Ooshima, T., Sato, A., Ikeuchi, T., Oyake, M., Sato, T., Aoyagi, Y., Hozumi, I., Nagatsu, T., Takiyama, Y., Nishizawa, M., Goto, J., Kanazawa, I., Davidson, I., Tanese, N., Takahashi, H., Tsuji, S. (**2000**) Expanded polyglutamine stretches interact with TAFII130, interfering with CREB-dependent transcription. *Nat Genet* 26, 29–36.

Steffan, J.S., Bodai, L., Pallos, J., Poelman, M., McCampbell, A., Apostol, B.L., Kazantsev, A., Schmidt, E., Zhu, Y.Z., Greenwald, M., et al. (**2001**). Histone deacetylase inhibitors arrest polyglutamine-dependent neurodegeneration in *Drosophila*. *Nature* 413, 739–743.

Sugars, K.L., Rubinsztein, D.C. (**2003**). Transcriptional abnormalities in Huntington disease. *Trends Genet* 19, 233–238.

Sugars, K.L., Brown, R., Cook, L.J., Swartz, J., Rubinsztein, D.C. (**2004**). Decreased cAMP response element-mediated transcription: an early event in exon 1 and full-length cell models of Huntington's disease that contributes to polyglutamine pathogenesis. *J Biol Chem* 279, 4988–4999.

Tan, S., Conaway, R.C., Conaway, J.W. (**1995**). Dissection of transcription factor TFIIF functional domains required for initiation and elongation. *Proc Natl Acad Sci USA* 92, 6042–6046.

Tanese, N., Pugh, B.F., Tjian, R. (**1991**). Coactivators for a proline-rich activator purified from the multisubunit human TFIID complex. *Genes Dev* 5, 2212–2224.

Trettel, F., Rigamonti, D., Hilditch-Maguire, P., Wheeler, V.C., Sharp, A.H., Persichetti, F., Cattaneo, E., MacDonald, M.E. (**2000**). Dominant phenotypes produced by the HD mutation in STHdh(Q111) striatal cells. *Hum Mol Genet* 9, 2799–2809

Vonsattel, J.P., DiFiglia, M. (**1998**). Huntington disease. *J Neuropathol Exp Neurol* 57, 369–384.

Wheeler, V.C., Gutekunst, C.A., Vrbanac, V., Lebel, L.A., Schilling, G., Hersch, S., Friedlander, R.M., Gusella, J.F., Vonsattel, J.P., Borchelt, D.R., MacDonald, M.E. (**2002**). Early phenotypes that presage late-onset neurodegenerative disease allow testing of modifiers in Hdh CAG knock-in mice. *Hum Mol Genet* 11, 633–640.

Wheeler, V.C., White, J.K., Gutekunst, C.A., Vrbanac, V., Weaver, M., Li, X. J., Li, S.H., Yi, H., Vonsattel, J.P., Gusella, J.F., et al. (**2000**). Long glutamine tracts cause nuclear localization of a novel form of huntingtin in medium spiny striatal neurons in HdhQ92 and HdhQ111 knock-in mice. *Hum Mol Genet* 9, 503–513.

Wu, Z., Puigserver, P., Andersson, U., Zhang, C., Adelmant, G., Mootha, V., Troy, A., Cinti, S., Lowell, B., Scarpulla, R.C., Spiegelman, B.M. (**1999**). Mechanisms controlling mitochondrial biogenesis and respiration through the thermogenic coactivator PGC–1. *Cell* 98, 115–124.

Yoon, J.C., Puigserver, P., Chen, G., Donovan, J., Wu, Z., Rhee, J., Adelmant, G., Stafford, J., Kahn, C.R., Granner, D.K., et al. (**2001**). Control of hepatic gluconeogenesis through the transcriptional coactivator PGC–1. *Nature* 413, 131–138.

Yu, Z.X., Li, S.H., Evans, J., Pillarisetti, A., Li, H., Li, X. J. (**2003**). Mutant huntingtin causes context-dependent neurodegeneration in mice with Huntington's disease. *J Neurosci* 23, 2193–2202.

Zuccato, C., Tartari, M., Crotti, A., Goffredo, D., Valenza, M., Conti, L., Cataudella, T., Leavitt, B.R., Hayden, M.R., Timmusk, T., Rigamonti, D., Cattaneo, E. (**2003**). Huntingtin interacts with REST/NRSF to modulate the transcription of NRSE-controlled neuronal genes. *Nat Genet* 35, 76–83.

Zhai, W., Jeong, H., Krainc, D., Tjian, R. (**2005**). In vitro analysis of huntingtin-mediated transcriptional repression reveals novel target. *Cell* (in press).

Zoghbi, H.Y., Orr, H.T. (**2000**). Glutamine repeats and neurodegeneration. *Annu Rev Neurosci* 23, 217–247.

Index

a

addiction 233, 260, 266
ALS, see amyotrophic lateral sclerosis
Alzheimer's disease 106, 235, 331, 342, 399, 400
amyloid precursor protein (APP) 400–406
amyotrophic lateral sclerosis (ALS) 271
Aniridia 31, 32
ankyrin repeat 329–330
AP–1 208, 354, 260, 263–264, 267, 272, 359
apoptosis 222, 248–249, 261–264, 268–269, 272, 291–295, 331–333, 342
APP, see amyloid precursor protein 400–406
aspirin 338
ataxin–1 321
ATF1 207–209, 222, 227
ATF2 208, 263, 385
ATF3 208, 218
ATF4 208, 253
atrophin 424

b

BACE1 401
basic helix-loop-helix protein 3–18, 148, 164
bending 182, 186
Bergmann glia 191
β-catenin 311–312, 320
bHLH domain 5–6
BMP, see bone morphogenetic protein
bone morphogenetic protein (BMP) 9, 56, 61, 64
Brn–2 194
bromodomain 212

c

cabin1 287, 301
Ca^{2+}/calmodulin-dependent protein kinase II (CamKII) 341, 408
Ca^{2+}/calmodulin-dependent protein kinase IV (CamKIV) 216, 248, 254, 300–301, 313, 321
calcineurin 249, 296, 301, 353–362, 364–374
calcipressin 373
CamKII, see Ca^{2+}/calmodulin-dependent protein kinase II
CamKIV, see Ca^{2+}/calmodulin-dependent protein kinase IV
cAMP 9, 55, 207, 214–215, 217, 220, 224–226, 230, 235–236, 243, 249–250, 253–254, 370, 427, 433
cAMP-dependent protein kinase, see protein kinase A (PKA)
cAMP-response element (CRE) 55, 207–208, 210–211, 217, 219, 222–226, 228, 230–231, 233, 236, 251, 253–254, 384–385, 408, 424–425, 427–428, 433, 435
CArG box 95, 98–99, 383–384
casein kinase II 297, 357
CBP, see CREB-binding protein
CCHS, see congenital central hypoventilation syndrome
C/EBP 208, 218, 224–225, 243–255
c-Fos 217, 225, 232, 243–244, 259–260, 264–265, 267–269, 408
c-Jun 243, 259–273, 293–294, 389
c-Jun N-terminal protein kinase (JNK) 261–264, 266–273, 357
chromatin 113, 117, 120, 123, 212, 229, 359, 360, 428
chronic pain disorder 130, 138
ciliary neurotrophic factor (CNTF) 247, 336
circadian rhythm 231, 316–317
CLS, see Coffin-Lowry syndrome
CNTF, see ciliary neurotrophic factor
Coffin-Lowry syndrome (CLS) 229
congenital central hypoventilation syndrome (CCHS) 60
CoREST 119–120
CRAC channel 354–355, 369
CRE, see cAMP-response element
CREB 207–212, 214–236, 243, 252–254, 364–265, 300, 331, 342, 368, 370–371, 407–408, 417, 424–425, 427–428, 431–435

CREB–2 224, 225
CREB-binding protein (CBP) 15, 55, 194, 207–208, 210–216, 220, 228–229, 234–235, 405, 407–408
c-rel 328–329
CREM 207–209, 211, 216, 222, 226–227, 235
CsA, *see* cyclosporine A
CtBP 300
cyclosporine A (CsA) 354–355, 372–373

d

Dlx 35, 139
Dbx2 168, 169
Delta 8, 11
DNA methylation 120
Down syndrome 353, 373
dopamine-β-hydroxylase 53–58, 61–63, 66–67
dopaminergic neuron 16, 35, 38–39, 82, 151, 154
doublecortin 38
DSCR1 368, 373–374

e

E47 7
E-box 5, 6
EGF receptor 387
Egr–1 98–100, 104, 217–218, 220, 225–226, 264–265, 267–269, 379–388
Elk–1 96–97, 101, 103–106, 263, 383–384
embryonic stem cell 183
engrailed1 28
enteric nervous system 189
enteric neuron 62
ependymal cells 4, 14
ERK, *see* extracellular signal-regulated protein kinase
ETRX, *see* X-linked alpha-thalassemia
ETS 153
excitotoxicity 267, 298, 331, 342
extracellular signal-regulated protein kinase (ERK) 28–29, 96–97, 104, 216, 224–225, 247, 254, 261, 269, 381–384, 387, 388
eye development 29–30

f

Fgf, *see* fibroblast growth factor
fibroblast growth factor 2 (Fgf2) 247, 271
fibroblast growth factor 7 (Fgf7) 35
fibroblast growth factor 8 (Fgf8) 4, 16
FK506 355, 362, 371–372
FKBP 355, 371–372
FOXO3 370

g

G9a 120
GABAergic neuron 35, 40, 129–132, 134–139, 193
Gata2 61, 64, 67, 146–150, 154, 156–157
Gata3 61, 64, 67, 146–150, 154, 156–157
gliogenesis 15, 173, 189–190, 247
GluR2 135, 137, 267
glutamate decarboxylase 130, 137
glutamatergic neuron 35, 40, 129–132, 134–139
granule neuron 292, 296–298, 300, 310, 314–316, 318, 320–321, 339–340, 342, 370
groucho 7, 15, 24, 164
GSK3 356–358, 367, 370, 374
gustucin 54, 68

h

Hand2 55, 61, 62, 64, 67
HAT, *see* histone acetyltransferase
HDAC, *see* histone deacetylase
helix-turn-helix motif 24
Her5 16
HerII/Him 16
Hes protein 3–21, 403
Hes1 5, 8–18, 185
Hes3 5, 8–13, 15–18
Hes5 5, 8, 10–15, 17–18, 65, 148
Hes6 15
Hes7 10
Hesr1 13, 16
Hesr2 3, 16
heterochromatin 120–121
heterochromatin protein 1 120–121, 124
HGM, *see* high-mobility-group (domain)
HIF1α 313
high-mobility-group (HGM) domain 181
hindbrain development 4
hippocampal neuron 225, 226, 229, 262, 340, 372
histone acetylation 117–118
histone acetyltransferase (HAT) 212–213, 229, 233–235
histone deacetylase (HDAC) 7, 113, 117–121, 229, 235, 288, 297, 300–301, 417, 428–429
histone methylation 120, 121
histone methyltransferase 113, 120
HMG group protein 181
homeobox 24, 53–54, 129, 131, 133
homeodomain 75–76, 146, 150, 154, 163–165, 168
Huntington Disease 124, 235, 417–435

i
ICER   211, 216, 218, 231, 232
Id   7
Id2   65
IκB   328–335, 338–340, 342
IκB kinase   331–333, 342
integrins   40
interleukin–1   249, 261, 334–335
IP$_3$ receptor   354, 368, 369
Irx3   172
ischemia   268, 269
islet–1   65, 76–77, 80, 83, 87–88
islet–2   77, 80, 82–83, 87
isthmic organizer   4, 10–11, 16, 84, 87

j
Jagged   8, 404
JIP, *see* JNK interacting protein
JNK, *see* c-Jun N-terminal protein kinase
JNK interacting protein (JIP)   261–262, 264, 267, 269–270
JunB   260, 264, 267–269, 359
JunD   260

k
kainic acid   124, 267, 335, 389

l
L-type calcium channel   248, 354–355, 340–341, 367–369
L1   40
leucine zipper   208–210, 243, 246
Lhx1   77, 80, 82–83, 88
Lhx2   77, 80, 83–85, 87–88
Lhx3   77, 80–81, 83, 87–88
Lhx4   77, 80–81, 83, 87
Lhx5   77, 82–83, 85
Lhx6   77, 82, 85
Lhx8   77, 82
LIM domain   76
LIM-domain-binding-protein Ldb1   85, 86
LIM-homeodomain protein   75–84, 86–88
lipopolysaccharide (LPS)   249, 333, 336
Lmx1a   84
Lmx1b   59, 64, 82–84, 87, 143, 146–157
locus coeruleus   16, 57, 64, 66, 233
long-term depression (LTD)   105, 342, 368
long-term potentiation (LTP)   104–105, 224, 226–229, 234, 253, 260, 266, 342, 368, 386, 407–408
LPS, *see* lipopolysaccharide
LTD, *see* long-term depression
LTP, *see* long-term potentiation

m
MADS-box protein   95
MADS domain   286–288, 299–300
MafB   263
Mash1   7, 12, 15, 35, 54, 61–65, 67, 131–132, 136–137, 143, 146–149, 156–157, 188, 291, 293, 300
Math3   12, 15
MeCP2   120
medulloblastoma   192
MEF2   285–301, 354
MEKK3   333
memory   229–231, 234, 250–254, 260, 266, 327, 340, 342, 353, 386, 408
midbrain development   4
Mitf   31
motor neuron   83, 87, 88, 145–148, 163, 167, 169–174, 177, 190, 264, 271–273, 368
Müller glia   15–16, 18
Myc   244
myelin basic protein   164, 174–176, 191
myelination   163, 177, 191

n
N-CAM   40, 53–55
NAB1   380
NAB2   380, 381
N-box   5, 6
N-CoR   312
nerve growth factor (NGF)   250, 264, 271, 333, 336, 371
NGF1–A   379
Nestin   185, 194
Net1   96–97
Netrin   361–362, 365–366
neural crest stem cell   187, 193
neural stem cell   247
neural stem cell
 –role of Hes bHLH proteins   3–5, 8, 10, 11–14, 17–18
 –role of pax6   33
 –role of REST   123
 –role of sox   181, 185, 192
neuroepithelial cell   3, 10, 12, 13, 163, 167
neuroepithelial stem cell   184, 187–188, 190, 194
neurogenesis   131, 137–138, 148, 223–224, 247, 291–292
Neurogenin 1   15, 131–132, 136–138, 173, 177
Neurogenin 2   15, 35, 40, 63, 131–132, 136–138, 148, 173, 177
neuron-restrictive silencer factor (NRSF)   114

neuronal activity   129, 139
neuronal migration   98, 101, 172, 174
neuroprotection   221, 229, 234, 267, 270, 295–296, 327, 331, 342, 370
neurosphere   12
neurotransmitter   129, 137, 138, 250–251, 254
NFAT   301, 353–374
NF-κB   9, 244, 267, 327–342
NGF, see nerve growth factor
Nicastrin   404
Nkx2.1   82
Nkx2.1   164–168
Nkx2.2   37, 59, 65, 83, 143, 146–148, 151, 156, 157, 163–172, 174–175, 177
Nkx2.9   59, 147, 163–167, 172
Nkx6.1   65, 163–174, 177
Nkx6.2   65, 163–167, 169–173, 175–177
Nkx6.3   165–167, 172
NMDA receptor   214, 222, 224–226, 248, 254, 269, 293, 300, 341, 343, 355, 367, 368, 407–408
noradrenergic differentiation   55, 58, 61–64, 66–67
noradrenergic neuron   57, 59, 61–64, 67
noradrenergic-specific cis-acting element   55
Notch   8, 10–11, 13–14, 185, 402–406, 408
NRSF, see neuron-restrictive silencer factor
nTS   64
nucleosome   117

o

oculomotor nuclei   16
olfactory bulb   37, 39
olfactory neuron   83
Olig2   35, 65, 164, 171–175, 177
oligodendrocyte   163, 173–177, 189–191, 194

p

p38 protein kinase   28–29, 96, 261, 264, 269, 294–295, 298, 301
p53   212, 248, 342–343
paired-box   23–24, 26
paired homeodomain   24, 26–27, 36, 53, 66
parasympathetic neuron   61–62
pax1   24–25
pax2   24–25, 29, 31, 131–132, 135, 137, 139
pax3   24–25, 28, 169, 194
pax4   24–25
pax5   24–25
pax6   23–41, 65, 136, 168–169, 194
pax6 (5a)   25, 27–29, 36, 39

pax7   24–25, 169
pax8   24–25
pax9   24–25
PEA3   153
Pet1   83, 143, 146–150, 152, 153–157
phosphatidylinositol 3–kinase (PI3K)   248, 254, 356, 369, 403, 408
Phox2   53–68, 134, 147–149, 172, 188
PI3K, see phosphatidylinositol 3–kinase
Pitx3   154
PKA, see protein kinase A
presenilin   399, 401–408
proprioceptive neuron   153
proteasome   8, 10, 86, 255, 312, 328, 332, 335, 339, 341, 406–408
protein kinase A (PKA, cAMP-dependent protein kinase)   8, 55, 104, 210, 213–216, 224–226, 235, 251, 254, 297, 355, 357
protein kinase C   8, 216, 254, 354, 369, 382
protein phosphatase 1   214, 215
protein phosphatase 2A   214
PTEN   389, 390
ptx3   82
Purkinje neuron   8, 307, 312–318, 320–321, 339
Purkinje cell protein 2   310–312

r

radial glia   3, 4, 10–14, 33, 35, 191
RBP-J   8, 403
receptor tyrosine kinase   382, 383
reelin   35, 38
RelA   329
RelB   328–329
rel-homology domain   328, 330
repressor   211, 216
REST   113–117, 119–120, 122–125
ret   58, 60, 62–63, 172
reticulon–1   35
reticulospinal neuron   83
retina   15–16, 27–31, 264
retinoblastoma protein   28
retinoic acid receptor   307
Rett's syndrome (RT)   229
RGS4   54, 68
ribosomal S6 kinase (RSK)   214–215, 234, 248, 254
RORα   307–311, 313–322
Rnx   64
RSK, see ribosomal S6 kinase
RT, see Rett's syndrome
RTS, see Rubinstein-Taybi syndrome
Rubinstein-Taybi syndrome (RTS)   229, 233

s

Sap–1 96–97
SCAR1, *see* spinocerebellar ataxia type 1
schizophrenia 373
Schwann cell 184, 189–191
seizure 260, 267
serotonergic neuron 65, 83,143–157, 174
serotonin transporter 154
serum response element 383–385
serum response factor 95–106, 286–288, 383–384
Shh, *see* sonic hedgehog
Sin3A 117
sonic hedgehog (Shh) 9, 146, 167–168, 173, 316, 318, 320–321
somite segmentation 10
Sox 28, 61, 173, 175, 181–197
Sp1 380, 417–420, 422–425, 428, 432–433, 436
Sp3 380
spinal cord 130, 138, 153, 164, 168, 173–175, 185, 190
spinocerebellar ataxia type 1 (SCAR1) 321
Sry protein 181
STAT3 14
stroke 223–224, 260
substance P 383
sumoylation 296
suprachiasmatic nucleus 231, 316–317
sympathetic neurons 61, 63, 66, 264, 271
synapsin I 116, 119, 122
synaptophysin 116, 119, 122–123, 338

t

TAF4 417–418, 422–425, 428. 431, 433–434
tau 400
Tlx1 129, 133–139
Tlx2 133–134
Tlx3 129, 131–139
TATA-box binding protein (TBP) 28, 87, 212
TBP, *see* TATA-box binding protein
Tbx5 29
Tenascin-c 40
ternary complex factor 95–97, 383–384
TLE/Grg 7, 15
TNFα, *see* tumor necrosis factor α
TRAIL 331
transcriptional repression 7–8, 24, 113–114, 116–120, 154, 164, 166, 211, 216, 300, 408
transforming growth factor α 35
transforming growth factor β 299
Trp channel 355
tryptophan hydroxylase 154
tumor necrosis factor α (TNFα) 249, 260–261, 331–335, 340, 342
tyrosine hydroxylase 56, 61–63, 82, 154

v

vascular endothelial growth factor (VEGF) 366
VEGF, *see* vascular endothelial growth factor
vesicular glutamate transporter 129–131, 135, 137
vesicular inhibitory amino acid transporter (Viaat) 130, 135
Viaat, *see* vesicular inhibitory amino acid transporter
visceral sensory neuron 63

w

Wnt1 4, 9, 16
WRPW domain 5–7

x

X-linked alpha-thalassemia (ETRX) 229

z

Zic2 83
zif268 379
zinc finger 149, 308, 380